Springer Proceedings in Physics

Volume 204

The series Springer Proceedings in Physics, founded in 1984, is devoted to timely reports of state-of-the-art developments in physics and related sciences. Typically based on material presented at conferences, workshops and similar scientific meetings, volumes published in this series will constitute a comprehensive up-to-date source of reference on a field or subfield of relevance in contemporary physics. Proposals must include the following:

- name, place and date of the scientific meeting
- a link to the committees (local organization, international advisors etc.)
- scientific description of the meeting
- list of invited/plenary speakers
- an estimate of the planned proceedings book parameters (number of pages/ articles, requested number of bulk copies, submission deadline).

More information about this series at http://www.springer.com/series/361

Didier Sébilleau · Keisuke Hatada
Hubert Ebert
Editors

Multiple Scattering Theory for Spectroscopies

A Guide to Multiple Scattering Computer Codes—Dedicated to C. R. Natoli on the Occasion of his 75th Birthday

Springer

Editors
Didier Sébilleau
Institut de Physique de Rennes (IPR)
Univ Rennes, CNRS, UMR 6251
Rennes
France

Hubert Ebert
Department Chemie
Ludwig-Maximilians-Universität München
Munich, Bavaria
Germany

Keisuke Hatada
Department Chemie
Ludwig-Maximilians-Universität München
Munich, Bavaria
Germany

ISSN 0930-8989 ISSN 1867-4941 (electronic)
Springer Proceedings in Physics
ISBN 978-3-030-08860-6 ISBN 978-3-319-73811-6 (eBook)
https://doi.org/10.1007/978-3-319-73811-6

Printed on acid-free paper

This Springer imprint is published by the registered company Springer Nature Switzerland AG
The registered company address is: Gewerbestrasse 11, 6330 Cham, Switzerland

This book is dedicated to Calogero R. Natoli, more widely known as Rino Natoli, on the occasion of his 75th birthday. A conference in his honour was organized in Rennes (France) on July the 1st–2nd 2016, as a recognition of his outstanding lifetime achievements in the multiple scattering modeling of spectroscopies and his tutoring work in the training of two generations of multiple scattering physicists. Part 2 of this book gathers the contributions to this conference directly related to multiple scattering.

Foreword

It is indeed a pleasure and an honor to write the foreword to this book on multiple scattering theory, both for its deep and useful content, but also to honor Rino Natoli, an outstanding scientist who has contributed so much to this field over a long and continuing career.

As an experimentalist involved in various aspects of photoelectron spectroscopy who has made use of multiple scattering methods through my entire career, I cannot underestimate the importance of the theoretical developments, methods, and computer codes discussed in this volume. For me personally, this spans from the simulation and interpretation of photoelectron diffraction and holography data for bulk- and surface-atomic structure studies, using methods developed in collaboration with John Rehr, a close collaborator of Rino Natoli over many years, to the much more complicated challenge of simulating and interpreting angle-resolved photoemission results aimed at the most detailed determinations of electronic structure, in collaboration with Hubert Ebert, Jürgen Braun, and Ján Minár, who are authors in this volume. Being able to accurately describe the final states and intensities of electrons photoemitted from either core or valence states of atoms, molecules, solids, surfaces, or interfaces is absolutely crucial to all applications of photoelectron spectroscopy, including more recently greatly expanded studies of both spin and time resolution. Recent developments in synchrotron radiation and free-electron laser light sources, as well as high-harmonic generation sources, and time-of-flight spectrometers and spin detectors have opened up enormously exciting new areas for photoelectron spectroscopy in which we are really dealing with "big data", and it is essential that we have "big theory" to describe all this. This volume addresses that, and the fact that, the word "photoelectron" appears over 170 times in it and the word "photoemission" over 150 times attests to the many past, present, and future contributions of multiple scattering in the interpretation of such data.

But beyond this area of spectroscopy also come structure studies with low-energy electron diffraction (for which many early developments were made) and the X-ray absorption and X-ray emission spectroscopies, including X-ray absorption extended fine structure (EXAFS, XAFS mentioned over 80 times), X-ray magnetic circular dichroism (XMCD-over 70 times) and linear dichroism, and resonant inelastic X-ray scattering, forefront synchrotron radiation and free-electron laser techniques whose understanding also critically depends on multiple scattering and is discussed elsewhere in this book. From a fundamental point of view, the multiple scattering methods discussed here are also very closely related to, and based in many ways upon, the Korringa–Kohn–Rostoker method (KKR mentioned over 180 times) that is frequently used in calculating electronic structure.

The book is also enhanced by articles by experimentalists describing their uses of these techniques and illustrating directly the importance of accurate theory for the interpretation of data.

So this volume, with its intent to put in one place an introduction to the various multiple scattering methods and available programs for using them in various spectroscopies, is most timely and useful, and I am grateful to Didier Sébilleau, Keisuke Hatada, and Hubert Ebert for editing it, and to all of the authors for their authoritative and useful chapters. This book will be a highly useful reference to a very broad scientific community.

Davis, USA Charles S. Fadley
March 2017

Preface

The purpose of this book is to provide the users of multiple scattering (MS) codes modeling spectroscopies with all the necessary background in order to understand the theoretical framework behind these codes. It is intended for researchers, as well as students who want to run a multiple scattering code and want to go further than just *button pressing*. It is designed as more or less self-contained as no preliminary knowledge of scattering theory is requested.

This book is composed of two parts which reflect the two events it grew out of. Part I provides basic knowledge of MS theory and of its use in the description of standard spectroscopies used in materials and surface science, either at advanced facilities (synchrotron radiation centers and free-electron lasers essentially) or in academic/industrial research laboratories. It contains the lectures that were given at the EUSpec training school organized in Rennes (France) from the June 27 to 30, 2016. These lectures contained all the background requested to understand the physics behind the codes FPMS, GNXAS, MCMS, MSSPEC, and MXAN, on which the attendees had hands-on during the Rennes school, and the code SPR-KKR which was studied at the EUSpec Plzeň (Czech Republic) Training School in February 2015. Part II deals with more focused aspects of the use of MS to model spectroscopies. It is based on some of the talks that were given at the joint MSNano-EUSpec conference that followed the Rennes Training school. It addresses specific issues related to MS and is intended for readers who want to go further in the understanding of MS theory for spectroscopies.

We hope this book will be helpful to the spectroscopy community to better understand the *dark side of the MS computer codes*.

Finally, we want to thank all the students/researchers who attended the school and the conference, and hope these two events will have given them a better grasp on the codes they use or want to use.

Rennes, France — Didier Sébilleau
Rennes, France — Keisuke Hatada
Munich, Germany — Hubert Ebert
March 2017

The original version of the bookfrontmatter was revised: Revised acknowledgements content has been incorporated. The erratum to this bookfrontmatter is available at https://doi.org/10.1007/978-3-319-73811-6_30

Acknowledgements

This book is based upon work from COST Action MP1306 EUSpec. COST (European Cooperation in Science and Technology) is a funding agency for research and innovation networks. Our Actions help connect research initiatives across Europe and enable scientists to grow their ideas by sharing them with their peers. This boosts their research, career, and innovation.

The editors would also like to acknowledge funding by European FP7MSNano network under Grant Agreement No. PIRSES-GA-2012-317554 and by European FP7 MS-BEEM under Grant Agreement No. PIEF-GA-2013-625388.

In addition, the organizers of the training school and of the conference would like to thank also Institut de Physique de Rennes, Université de Rennes-1, Rennes Métropole, and Université Bretagne Loire for their participation in the funding of the two events. They also want to thank all the people who participated in the organization. www.cost.eu

Funded by the Horizon 2020 Framework Programme of the European Union

Contents

Part II Extended knowledge

Contributors

Maurizio Benfatto LNF INFN, Frascati, Italy

Nadejda Bouldi UPMC Univ Paris 06, CNRS, UMR 7590, IRD, MNHN, Institut de Minéralogie, de Physique des Matériaux et de Cosmochimie (IMPMC), Paris, France

Jürgen Braun Department Chemie, Ludwig-Maximilians-Universität München, Munich, Bavaria, Germany

Christian Brouder UPMC Univ Paris 06, CNRS, UMR 7590, IRD, MNHN, Institut de Minéralogie, de Physique des Matériaux et de Cosmochimie (IMPMC), Paris, France

Giovanni Chillemi CINECA, SuperComputing Applications and Innovation Department, Roma, Italy

Rakesh Choubisa Department of Physics, Birla Institute of Technology and Science, Pilani, Rajasthan, India

Hiroshi Daimon Nara Institute of Science and Technology (NAIST), Takayama, Ikoma, Nara, Japan

Andrea Di Cicco Physics Division, School of Science and Technology, University of Camerino, Camerino (MC), Italy

Igor Di Marco Division of Materials Theory, Department of Physics and Astronomy, Uppsala University, Uppsala, Sweden

Hubert Ebert Department Chemie, Ludwig-Maximilians-Universität München, Munich, Bavaria, Germany

Takashi Fujikawa Graduate School of Advanced Integration Science, Chiba University, Inage, Chiba, Japan

Keisuke Hatada Department Chemie, Ludwig-Maximilians-Universität München, Munich, Bavaria, Germany; Faculty of Science, University of Toyama, Toyama, Japan

Fabio Iesari Physics Division, School of Science and Technology, University of Camerino, 62032 Camerino (MC), Italy; Faculty of Science, University of Toyama, Gofuku 3190, Toyama 930-8555, Japan

Thomas Jaouen Département de Physique and Fribourg Center for Nanomaterials, Université de Fribourg, Fribourg, Switzerland

Yves Joly Inst NEEL, CNRS & Univ. Grenoble Alpes, Grenoble, France

Joshua J. Kas University of Washington, Seattle, WA, USA

Vladislav Kochetov Faculty of Physics, Southern Federal University, Zorge Str. 5, 344090 Rostov-on-Don, Russia; Université de Rennes-1, 35042 Rennes Cedex, France

Jindřich Kolorenč Institute of Physics, Czech Academy of Sciences, Praha 8, Czech Republic

Naoki Komiya Graduate School of Advanced Integration Science, Chiba University, Chiba, Japan

Peter Krüger Graduate School of Engineering, Chiba University, Chiba, Japan

Sergiy Mankovsky Department Chemie, Ludwig-Maximilians-Universität München, Munich, Bavaria, Germany

Marco Minicucci Physics Division, School of Science and Technology, University of Camerino, Camerino (MC), Italy

Ján Minár New Technologies-Research Center, University of West Bohemia, Plzeň, Czech Republic

Calogero R. Natoli Theory Group, INFN-Laboratori Nazionali di Frascati, Frascati, Italy

Kaori Niki Graduate School of Advanced Integration Science, Chiba University, Inage, Chiba, Japan

Alexander Novakovich Institute of Physics, Southern Federal University, Rostov-on-Don, Russia

Fukiko Ota Graduate School of Advanced Integration Science, Chiba University, Chiba, Japan

Elisabetta Pace LNF INFN, Frascati, Italy

Cristiano Padrin CINECA, SuperComputing Applications and Innovation Department, Roma, Italy

Krisztián Palotás Department of Theoretical Physics, Budapest University of Technology and Economics, Budafoki út 8., Budapest 1111, Hungary; Slovak Academy of Sciences, Institute of Physics, Dúbravská cesta 9., 84511 Bratislava, Slovakia

John J. Rehr University of Washington, Seattle, WA, USA

Joaquín García Ruiz Instituto de Ciencia de Materiales de Aragòn CSIC-Univ, Zaragoza, Spain

Nico Sanna CINECA, SuperComputing Applications and Innovation Department, Roma, Italy

Giovanni Stefani Dipartimento di Scienze and Unitá CNISM Universitá Roma Tre, Roma, Italy

László Szunyogh Department of Theoretical Physics and MTA-BME Condensed Matter Research Group, Budapest University of Technology and Economics, Budapest, Hungary

Didier Sébilleau Institut de Physique de Rennes (IPR), Univ Rennes, CNRS, UMR 6251, Rennes, France

Anna Taranukhina Faculty of Physics, Southern Federal University, Rostov-on-Don, Russia

Angela Trapananti Physics Division, School of Science and Technology, University of Camerino, 62032 Camerino (MC), Italy; Istituto Officina dei Materiali (IOM), Consiglio Nazionale delle Ricerche (CNR), Perugia, Italy

Junqing Xu Department für Geo- und Umweltwissenschaften, Ludwig Maximilians Universität München, München; Germany

Rudolf Zeller Institute for Advanced Simulation, Forschungszentrum Jülich GmbH and JARA, Jülich, Germany

Ondřej Šipr Institute of Physics, Czech Academy of Sciences, Praha, Czech Republic; Academy of Science of Czech Republic, Prague 6, Czech Republic

Acronyms

Spectroscopies

AED — Auger electron diffraction: spectroscopy based on electron diffraction where the Auger electrons are monitored

APECS — Auger photoelectron coincidence spectroscopy: spectroscopy based on photoemission where the photoelectron and the Auger electron ejected via the corresponding deexcitation channel are measured in coincidence

ARBIS — Angular-resolved Brehmsstrahlung isochromat spectroscopy (BIS): spectroscopy often improperly termed inverse photoemission. It is the angle-resolved form of BIS where a beam of electrons is focused onto a sample. Within the sample, the electrons can decelerate emitting a photon. Photons of fixed energy are then detected

ARPES — Angle-resolved photoemission spectroscopy: spectroscopy where an incident photon excites an electron on a given atom. The ejected electron is measured by an angle-resolving detector

BEEM — Ballistic electron emission microscopy: microscopy derived from STM where only electrons that travel ballistically (i.e., those elastically scattered) are considered

EELS — Electron energy loss spectroscopy: spectroscopy where the electrons of an incident beam directed at the sample are used to excite electrons on specific atoms. The scattered electrons, which have lost energy in the process, are then probed

EXAFS — Extended x-ray absorption fine structure: spectroscopy using the higher energy part of the x-ray absorption spectrum (usually $\approx$50 eV above the absorbing edge)

HAXPES — Hard x-ray photoelectron spectroscopy: photoemission using *hard* x-rays with energies of several keVs

LEED Low-energy electron diffraction: a technique for the determination of the surface structure of single-crystalline materials by bombardment with a collimated beam of low-energy electrons (20–200 eV) and observation of diffracted electrons as spots on a fluorescent screen

NMR Nuclear magnetic resonance: spectroscopy based on the absorption and reemission of an electromagnetic wave by nuclei in a magnetic field

PED Photoelectron diffraction: spectroscopy based on photoemission where a feature of the photoemission spectrum (core level peak, Auger peak, valence peak, plasmon peak, etc.) is monitored as a function of either the energy or the emission angles

REXS Resonant elastic x-ray scattering: x-ray scattering spectroscopy where the incident photon excites an electron to a virtual state and is reemitted at the same energy in the deexcitation process

RXS Resonant x-ray scattering: generic term covering both resonant elastic and inelastic x-ray scattering

SPLEED Spin-polarized version of LEED

STM Scanning tunneling microscopy: used to image surfaces at the atomic level. It is based on the quantum tunneling from the tip to the substrate

UPS Ultraviolet photoemission spectroscopy: photoemission using incoming photons in the range of a few tens of eV

XANES X-ray absorption near-edge structure: spectroscopy using the lower energy part of the x-ray absorption spectrum, including the rising edge

XAS X-ray absorption spectroscopy: generic term covering both the XANES and the EXAFS regimes

XLD X-ray linear dichroism: spectroscopy making use of the difference in the absorption of x-rays between two different beams with orthogonal polarization

XMCD X-ray magnetic circular dichroism: spectroscopy making use of the difference in the absorption of x-rays between two (right and left) circularly polarized beams, when the sample is in a magnetic field

XPS X-ray photoemission spectroscopy: photoemission using incoming photons in the range of keVs

XRD X-ray diffraction: a photon-in, photon-out spectroscopy with conservation of the photon energy away from an absorbing edge of the sample

Methods

CPA Coherent potential approximation: method for finding the Green's function of an effective medium, when the potential varies randomly from site to site in the sample. It is used to describe disordered materials

DFT	Density functional theory: computational quantum mechanical method, whereby the ground-state properties of a many-body system are assumed to be functionals of only the electron density
DMFT	Dynamical mean field theory: method used to describe the electronic structure of strongly correlated systems
EMB	Embedded cluster method: a method to describe extended defects in an otherwise perfect solid
GGA	Generalized gradient approximation: improvement of the LDA where the exchange and correlation functional depends both on the electron density and its gradient
GW	Approximation of the many-body problem where the self-energy is calculated at the lowest diagrammatic order in terms of the single-particle Green's function G and the screened Coulomb interaction W
IPA	Independent particle approximation: approximation where electron–electron interactions are neglected or treated in an average way
KKR	Korringa–Kohn–Rostoker method: the original multiple scattering method developed by these authors
LDA	Local density approximation: approximation of the DFT, where the exchange and correlation functional depends solely upon the value of the electron density at each point in space
LFM	Ligand field multiplet: many-electron computational method for the electronic structure and spectra of atoms or ions in a field created by the surrounding (ligand) atoms
LMTO	Linear muffin-tin orbital method: *ab initio* electronic structure method using muffin-tin orbitals as the basis set
LSDA	Local spin density approximation: approximation where the exchange and correlation functional depends only on the local electron spin density (see LDA)
MCMS-GF	Multichannel multiple scattering theory: computational method taking into account electron correlations by combining the MS-GF and close-coupling formalism
MCMS	Multichannel multiple scattering: multiple scattering theory taking into account correlation effects
MD	Molecular dynamics: an N-body computer simulation method to mimic the movements of interacting atoms and molecules
MS-GF	Multiple scattering theory: computational method for solving the Dyson equation
MST	Multiple scattering theory: computational method for solving the Schrödinger equation, covering both bound and continuum extended states
PAW	Projector-augmented wave: electronic structure method, generalization of the pseudopotential method, transforming the rapidly oscillating valence wave functions into smooth wave functions and providing a way to calculate all-electron properties from them

PCI	Post-collision interaction: interaction between two or more outgoing electrons after the excitation process has taken place
PWIA	Plane wave impulse approximation: approximation identifying all-electron beams within the sample as plane waves and assuming that the target does not intervene in the electron–electron interaction
RPA	Random-phase approximation: it is one of the oldest non-perturbative methods for computing the ground-state correlation energy of many-electron systems and screening of the long-range Coulomb interaction
SKKR	Screened version of the Korringa-Kohn-Rostoker method for a faster computational algorithm for layered systems

Codes

CONTINUUM	A multiple scattering code, based on the Muffin-Tin approximation, to model x-ray absorption and resonant elastic x-ray scattering
ES2MS	Interface package for passing self-consistent charge densities and potentials from electronic structure codes to multiple scattering codes
FDMNES	Finite difference method near-edge spectroscopy: finite difference and multiple scattering program to model XAS, RXS and other spectroscopies
FEFF	Automated program for *ab initio* multiple scattering calculations of XAS, EELS and non-resonant inelastic x-ray scattering
FPMS	Full potential multiple scattering: a full potential code to model x-ray absorption and other spectroscopies
GAUSSIAN	Computational chemistry code using Gaussian orbitals as the basis set
GNXAS	Extended suite of programs for advanced x-ray absorption data analysis: EXAFS modeling and analysis program
GPAW	DFT code based on the projector-augmented wave method and the atomic simulation environment
KKRnano	Nanostructures focused KKR electronic structure code
MCMS	Multichannel multiple scattering: a computer program to compute XANES spectra for correlated systems
MC-MSGF-SP	A multichannel version of the MSGF-SP package to compute XANES taking into account electron correlations
MSGF-SP	Multiple scattering: a computer package to model XANES, REXS, RXS, and other spectroscopies
MsSpec	Multiple scattering spectroscopies: a computer package modeling several spectroscopies
MXAN	Minuit XANES: code calculating XANES spectra and fitting them to the experiment using the Minuit minimization algorithm

SKKR	Screened KKR: Fully relativistic KKR code including the screening method for the computation of the electronic structure, magnetic and electronic transport properties of layered and low-dimensional materials
SPEA-MEM	Scattering pattern extraction algorithm using maximum entropy method: computer program allowing to reconstruct the atomic image from a photoelectron hologram
SPR-KKR	Spin-polarized relativistic KKR: a computer code to compute the electronic structure of materials and the cross section of several spectroscopies
SPR-TB-KKR	Tight-binding version of the SPR-KKR code
VASP	Vienna *ab initio* simulation package: electronic structure package based on pseudopotentials

A list of computer packages allowing to model spectroscopies is available at http://www.euspec.eu/code-database at the EUSpec website.

Part I
Basic knowledge

This part corresponds to the lectures given during the first day of the EUSpec Training School on multiple scattering codes organized at University of Rennes 1 from June 27 to 30, 2016. The three following days were devoted to hands-on training on five codes: MXAN, GNXAS, MSSPEC, FPMS, and MCMS. These lectures have been written to be self-contained, and therefore no prior knowledge of scattering theory is required to read them. For the sake of conveniency, two of the original lectures have been merged into a single chapter.

Group photograph of the participants and lecturers at the EUSpec training school on multiple scattering codes

Chapter 1
Introduction to (Multiple) Scattering Theory

Didier Sébilleau

Abstract In this introductory chapter, we develop the scattering theory necessary to understand the theoretical models underlying the multiple scattering codes. First, the elementary theory is presented and it is then formalized to introduce the different operators whose matrix elements are computed in the codes. Then, we extend the theory to the case of a collection of potentials, i.e. multiple scattering. Finally we outline the way cross-sections can be derived from the multiple scattering framework and give some practical examples.

1.1 Introduction

In this chapter, we develop all the theoretical tools necessary to understand the scattering framework involved in multiple scattering codes. This framework will be defined for *local*, *short-ranged* potentials, or more specifically for potentials satisfying the so-called *Ikebe conditions* [1] which ensure both that the potential decreases sufficiently fast so that beyond a certain distance it can be considered as zero, and that it is not too strongly singular at the origin (in other words that the spectrum of the Hamiltonian H of the problem has a lower bound) [2]. Note that the first condition excludes Coulomb potentials. In practice, this will not be a problem as in materials, the Coulomb potentials are sufficiently screened so that they can be safely considered as short-ranged. In addition, we will suppose the potential considered to be *Hermitian* so that the eigenvalues of H are *real*.

Furthermore, we will only consider *elastic scattering* so that scattering by potential V does not change the energy of the electron. Finally, we will restrict ourselves in this chapter to *spherically symmetric potentials*.

D. Sébilleau (✉)
Institut de Physique de Rennes (IPR), Univ Rennes, CNRS, UMR 6251,
F-35000 Rennes, France
e-mail: didier.sebilleau@univ-rennes1.fr

D. Sébilleau et al. (eds.), *Multiple Scattering Theory for Spectroscopies*,
Springer Proceedings in Physics 204,
https://doi.org/10.1007/978-3-319-73811-6_1

1.2 Elementary Scattering Theory

1.2.1 The Asymptotical Behaviour of the Wave Function

We consider a potential V satisfying the conditions listed in Sect. 1.1. We choose the center of this potential as the origin of space. Being short-ranged and spherically symmetric, we denote a its range so that $V(r) = 0$ if $r \geq a$. The interaction of an incoming electron with V can be described by the reduced Schrödinger equation (i.e. where we have multiplied the equation by $2m/\hbar^2$. Alternatively, we can use Rydberg atomic units for which $\hbar^2 = 2m = 1$)

$$\left[\Delta + k^2 - V(r)\right] \psi_k(\boldsymbol{r}) = 0 , \tag{1.1}$$

which we want to solve.

As soon as the particle has interacted with the potential and is *outside the potential range a*, it is again a free particle. Therefore, it is possible to deduce its asymptotical form which is given by [3]

$$\psi_k(\boldsymbol{r}) \underset{r \to +\infty}{\longrightarrow} \alpha \left[\mathrm{e}^{\mathrm{i}\boldsymbol{k}\cdot\boldsymbol{r}} + f_k(\theta) \frac{\mathrm{e}^{\mathrm{i}kr}}{r} \right] , \tag{1.2}$$

where θ is the scattering angle, as explicited in Fig. 1.1.

The amplitude modulating the outgoing spherical wave is called the *scattering amplitude* or the *scattering factor*. For a non-spherical potential, it will depend also on the azimuthal angle φ, in addition to θ and k. This expression is very general and does not depend on the structure of the potential, only the expression of the scattering amplitude is potential-dependent.

Here, we have supposed an incoming plane wave. In practice, we never really have a single particle but rather a beam of incoming particles. In this context, the asymptotic equation becomes clear: far away from the potential, the effect of this potential is to suppress some particles from the incoming beam and to redistribute them as spherical waves [4].

1.2.2 The Radial Equation for the Spherically Symmetric Problem

As the potential is spherically symmetric, we can decompose the differential operator Δ into spherical coordinates

$$\Delta = \frac{1}{r^2}\left[\frac{\partial}{\partial r}\left(r^2\frac{\partial}{\partial r}\right) + \frac{1}{\sin\theta}\frac{\partial}{\partial\theta}\left(\sin\theta\frac{\partial}{\partial\theta}\right) + \frac{1}{\sin^2\theta}\frac{\partial^2}{\partial\varphi^2}\right] , \tag{1.3}$$

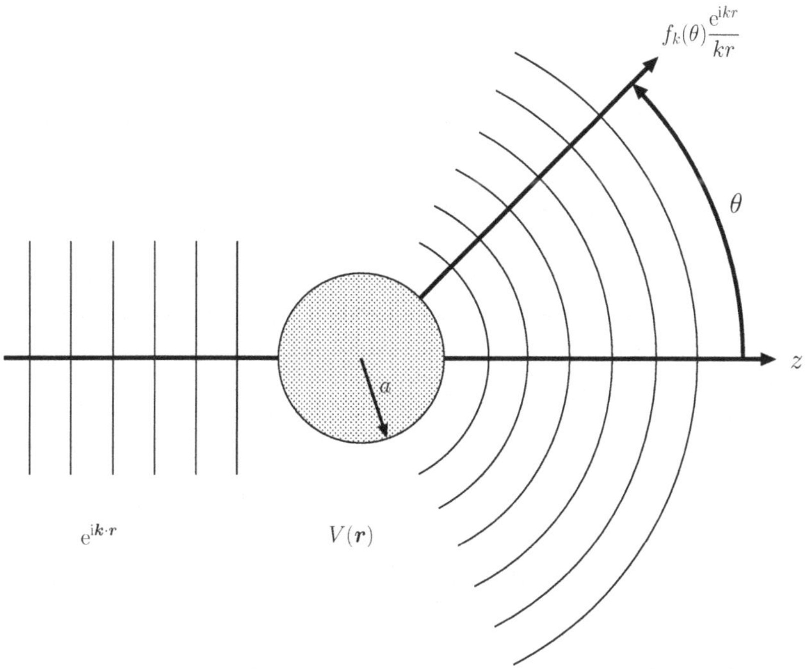

Fig. 1.1 Geometry of the scattering of an electron by a potential $V(\boldsymbol{r})$ of range a

which we rewrite as

$$\Delta = \boldsymbol{\nabla}^2 = \boldsymbol{\nabla}_r^2 + \frac{1}{r^2}\,\boldsymbol{\nabla}_{\theta,\varphi}^2 \,. \tag{1.4}$$

But the angular momentum operator $\boldsymbol{\ell} = \boldsymbol{r} \times \boldsymbol{p}$ is such that it satisfies

$$\boldsymbol{\nabla}_{\theta,\varphi}^2 = -\frac{1}{\hbar^2}\,\boldsymbol{\ell}^2 \,. \tag{1.5}$$

Therefore, the eigenfunctions of $\boldsymbol{\nabla}_{\theta,\varphi}^2$ are the *complex spherical harmonics* $Y_\ell^m(\theta,\varphi)$ [5]. More precisely, we have

$$\boldsymbol{\nabla}_{\theta,\varphi}^2\, Y_\ell^m(\theta,\varphi) = -\ell(\ell+1)\, Y_\ell^m(\theta,\varphi) \,. \tag{1.6}$$

From now on, in order to simplify the notation, we introduce the short-hand notations $L \equiv (\ell, m)$ and $\hat{\boldsymbol{r}} = (\theta, \varphi)$ so that we rewrite the spherical harmonics as $Y_L(\hat{\boldsymbol{r}})$.

Making use of the eigenvalues of $\boldsymbol{\nabla}_{\theta,\varphi}^2$ in (1.1) and simplifying by $Y_L(\hat{\boldsymbol{r}})$, we are left with the sole radial equation to solve

$$\left(\nabla_r^2 - \frac{\ell(\ell+1)}{r^2} + k^2 - V(r)\right) R_\ell(r) = 0 \,. \tag{1.7}$$

This equation can be viewed as a one-dimensional Schrödinger equation with the constraint that $r \geq 0$, and whose potential is $V(r) + \ell(\ell+1)/r^2$. The last term is called the *centrifugal barrier*.

Let us consider first the free electron case. The radial equation (1.7) reduces to the spherical Bessel equation [6] which has two linearly independent solutions to be chosen among $\left(j_\ell(kr), n_\ell(kr), h_\ell^{(1)}(kr), h_\ell^{(2)}(kr)\right)$. $j_\ell(kr)$ is *regular* at origin, which means that it does not diverge there. By contrast, $n_\ell(kr)$ is *irregular* at the origin. However, for a free particle, the wave function must be finite everywhere so that only the general solution $C_\ell \; j_\ell(kr)$ is physically meaningful.

Then, we consider the case of the electron after its scattering by V and *outside of the range of the potential.* We have still a zero potential, but now the general solution of the radial equation for $r > a$ must be of the form $R_\ell(r) = C_\ell^{(1)} \; j_\ell(kr) + C_\ell^{(2)} h_\ell^{(1)}(kr)$, or expressed as the linear combination of any other couple of independent spherical Bessel functions. Then, it can be demonstrated that asymptotically, we must have [7]

$$\begin{cases} \text{without } V: \lim\limits_{r \to +\infty} R_\ell(r) = C_\ell \dfrac{\sin\left(kr - \ell\dfrac{\pi}{2}\right)}{kr} \\[2em] \text{with } V \quad : \lim\limits_{r \to +\infty} R_\ell(r) = \dfrac{C_\ell^{(1)}}{\cos\delta_\ell} \dfrac{\sin\left(kr - \ell\dfrac{\pi}{2} + \delta_\ell\right)}{kr} \,. \end{cases} \tag{1.8}$$

δ_ℓ is called the *phase shift* of order ℓ of potential V. The behaviour of these two radial wave functions is represented in Fig. 1.2.

So, δ_ℓ represents the shift in the asymptotical radial wave due to the presence of the potential when the incoming wave is of angular momentum ℓ.

Solving now the radial Schrödinger equation for $r > a$ for the different choices of the linearly independent spherical Bessel wave functions gives

$$R_\ell(r) = A_\ell \left[j_\ell(kr) + \mathrm{i} \sin\delta_\ell \, \mathrm{e}^{\mathrm{i}\delta_\ell} \, h_\ell^{(1)}(kr) \right] \tag{1.9a}$$

$$= \frac{A_\ell}{2} \left[h_\ell^{(2)}(kr) + \mathrm{e}^{2\mathrm{i}\delta_\ell} \, h_\ell^{(1)}(kr) \right] \tag{1.9b}$$

$$= C_\ell^{(1)} \left[j_\ell(kr) - \tan\delta_\ell \, n_\ell(kr) \right] \,. \tag{1.9c}$$

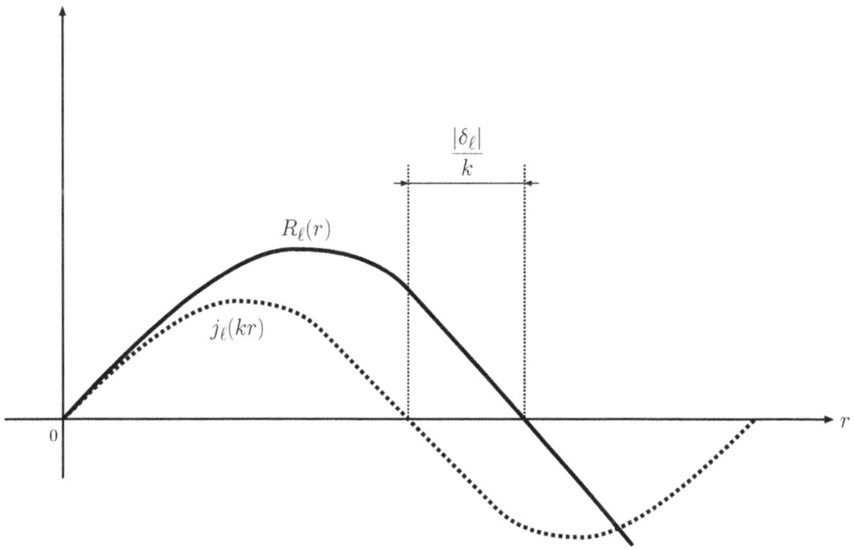

Fig. 1.2 Comparison of the asymptotical form radial wave function with or without the presence of potential V

1.2.3 Partial Wave Expansions

Expansions of physical quantities in terms of L are called *partial wave expansions*. The quantities $\mathrm{i}^\ell j_\ell(kr)Y_L(\hat{\boldsymbol{r}})$ will arise frequently within our theoretical framework: they are called *free spherical waves* as they are the partial wave solutions of the free electron Hamiltonian H_0. Therefore, when dealing with H_0, we will be able to use two kind of basis functions which satisfy the corresponding Schrödinger equation, the *plane waves* $\{|\boldsymbol{k}\rangle\}$ and the *spherical waves* $\{|kL\rangle\}$. They are related through the partial wave expansion known as *Bauer's formula* [8]

$$\mathrm{e}^{\mathrm{i}\boldsymbol{k}\cdot\boldsymbol{r}} = 4\pi \sum_L \mathrm{i}^\ell\, j_\ell(kr)\, Y_L(\hat{\boldsymbol{r}})\, Y_L^*(\hat{\boldsymbol{k}}) \;, \tag{1.10}$$

and the reverse formula

$$\mathrm{i}^\ell\, j_\ell(kr)\, Y_L(\hat{\boldsymbol{r}}) = \frac{1}{4\pi} \int \mathrm{e}^{\mathrm{i}\boldsymbol{k}\cdot\boldsymbol{r}}\, Y_L(\hat{\boldsymbol{k}})\, \mathrm{d}\hat{\boldsymbol{k}} \;. \tag{1.11}$$

These equations allow us to make the difference between the two bases $\{|\boldsymbol{k}\rangle\}$ and $\{|kL\rangle\}$ solution of the free electron Schrödinger equation:

- $|\boldsymbol{k}\rangle$ represents a free particle of well defined energy k^2 and momentum $\hbar\boldsymbol{k}$. However, it does not have a precise angular momentum

- $|kL\rangle$ represents the same particle with the same energy but with a well defined angular momentum. But its momentum is not known as the corresponding wave function is an integral over all possible directions of the momentum

The scattering amplitude can also be expressed as a partial wave expansion

$$f_k(\theta_{\hat{r}}^{\hat{k}}) = \frac{4\pi}{k} \sum_L \sin\delta_\ell \, e^{i\delta_\ell} \, Y_L(\hat{\boldsymbol{r}}) \, Y_L^*(\hat{\boldsymbol{k}}) \; . \tag{1.12}$$

The quantity $\sin\delta_\ell \, e^{i\delta_\ell}$ is often denoted t_ℓ for reasons that will be explicited in Sect. 1.3.3.1.

All these expansions involve a summation over ℓ that must be truncated at a certain point ℓ_{max} for practical reasons. As simple way to estimate this truncation value can be obtained by looking at the radial Schrödinger equation (1.7). Indeed, we see that for the particle to be scattered by the potential V, it must first overcome the centrifugal barrier $\ell(\ell+1)/r^2$. Therefore, its kinetic energy k^2 must satisfy $k^2 > \ell(\ell+1)/r^2$. This implies directly that partial waves with ℓ so that $k^2 < \ell(\ell+1)/a^2$ will not contribute to the scattering process. This gives for ℓ_{max} the estimation

$$ka \approx \sqrt{\ell_{max}(\ell_{max}+1)} \; . \tag{1.13}$$

1.2.4 The Scattering Amplitude

The scattering amplitude $f_k(\theta)$ is an important *asymptotic* quantity that gives us a good insight into the scattering process. As explicited in (1.2) and Fig. 1.1, it is the amplitude in the direction θ of the spherical wave resulting from the scattering of an initial plane wave. Although in practice at atomic level, the initial waves are not plane waves, it does still provides us with a useful information on the outcome of the scattering process. It is easily calculated by means of equation (1.12).

Figure 1.3 gives polar plots of $|f_k(\theta)|$ for Ni and O at kinetic energies ranging from 60–1000 eV [9]. Without doing any further calculations, it already tells us that while at lower energies, the scattering is important in several directions, at higher kinetic energies, it will be strongly dominated by the forward direction. This means than in practice, for energies above 500 eV, we expect the electrons to be scattered essentially along the rows of atoms.

1.2.5 Calculation of the Phase Shifts

We see from (1.9) and (1.12) that the only unknown in order to solve the scattering problem are the phase shifts δ_ℓ. Equation (1.9) gives us the solution of the scattering problem in terms of the δ_ℓ in the asymptotic zone. Let us call $\mathcal{R}_\ell(r)$ the radial solution

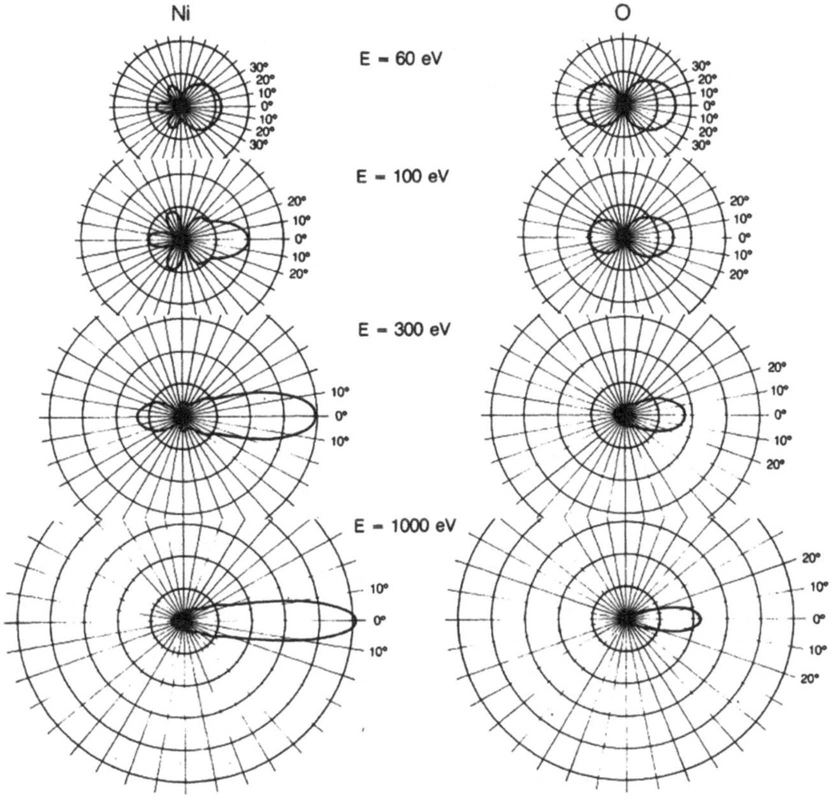

Fig. 1.3 Examples of modulus of the scattering factor $|f_k(\theta)|$ for Ni (left) and O (right) at several kinetic energies [9]

of the Schrödinger equation *inside* the potential sphere, while the solution *outside* is given by (1.9). The phase shifts can be obtained by matching $\mathcal{R}_\ell(r)$ and $R_\ell(r)$ at the surface of the potential sphere of radius a. This is sketched in Fig. 1.4.

If we introduce the *Wronskian* W of two functions f and g by

$$W[f, g] = f\frac{\mathrm{d}g}{\mathrm{d}r} - g\frac{\mathrm{d}f}{\mathrm{d}r}, \tag{1.14}$$

we find the coefficients of (1.9) given by

$$t_\ell = \mathrm{i}\left.\frac{W[j_\ell, \mathcal{R}_\ell]}{W[h_\ell^{(1)}, \mathcal{R}_\ell]}\right|_{r=a} \tag{1.15a}$$

$$\mathrm{e}^{\mathrm{i}\delta_\ell} = -\left.\frac{W[h_\ell^{(2)}, \mathcal{R}_\ell]}{W[h_\ell^{(1)}, \mathcal{R}_\ell]}\right|_{r=a} \tag{1.15b}$$

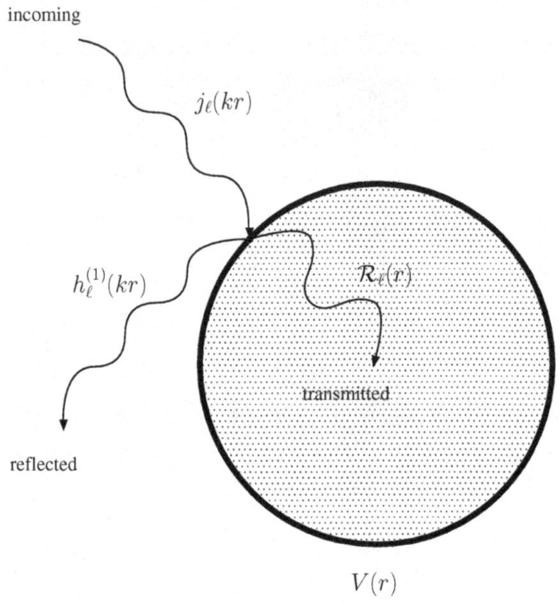

Fig. 1.4 Physical interpretation of the matching of the wave functions at the surface of the potential

$$\tan \delta_\ell = \left. \frac{W[j_\ell, \mathcal{R}_\ell]}{W[n_\ell, \mathcal{R}_\ell]} \right|_{r=a} . \tag{1.15c}$$

Alternatively, t_ℓ can also be obtained from [10]

$$t_\ell = -k \int_0^a j_\ell(kr) V(r) \mathcal{R}_\ell(r)\, r^2 \mathrm{d}r . \tag{1.16}$$

1.3 Formal Scattering Theory

We start by recasting the Schrödinger equation in the form

$$\left(k^2 - H\right) \psi_k(\boldsymbol{r}) = 0 . \tag{1.17}$$

This allows us to introduce the *Green's operator* in the complex plane by

$$G(z) = (z - H)^{-1} . \tag{1.18}$$

Instead of finding ways to solve the Schrödinger equation (1.17), scattering theory deals with the solution of Green's operator. In the mathematical physics literature, this operator is referred to as the *resolvent*. In the case of spectroscopies, where we are essentially interested into the continuum states, we call it the *propagator* for reasons that will become clear later.

From (1.18), by making it act on eigenfunctions of H, we can deduce directly two important properties of $G(z)$:

- $G(z)$ and H have the *same eigenfunctions*
- the eigenvalues of H are the *poles* of $G(z)$.

This demonstrates formally that H and G contain the same information and that therefore solving $G(z)$ will give us the solutions of the Schrödinger equation (1.17). It however leads to a problem as this also means that $G(z)$ is not defined for the eigenvalues of H, which are the quantities we are ultimately interested in. To overcome the division by zero induced by this definition, we redefine the Green's operator by

$$G^{\pm}(z) = \lim_{\varepsilon \to 0^+} (z - H \pm \mathrm{i}\varepsilon)^{-1} , \tag{1.19}$$

so that we approach the eigenvalues of H either from above (+ solution) or from below (− solution).

This definition is explicited in Fig. 1.5.

We represent schematically the fact that H and G contain the same information by

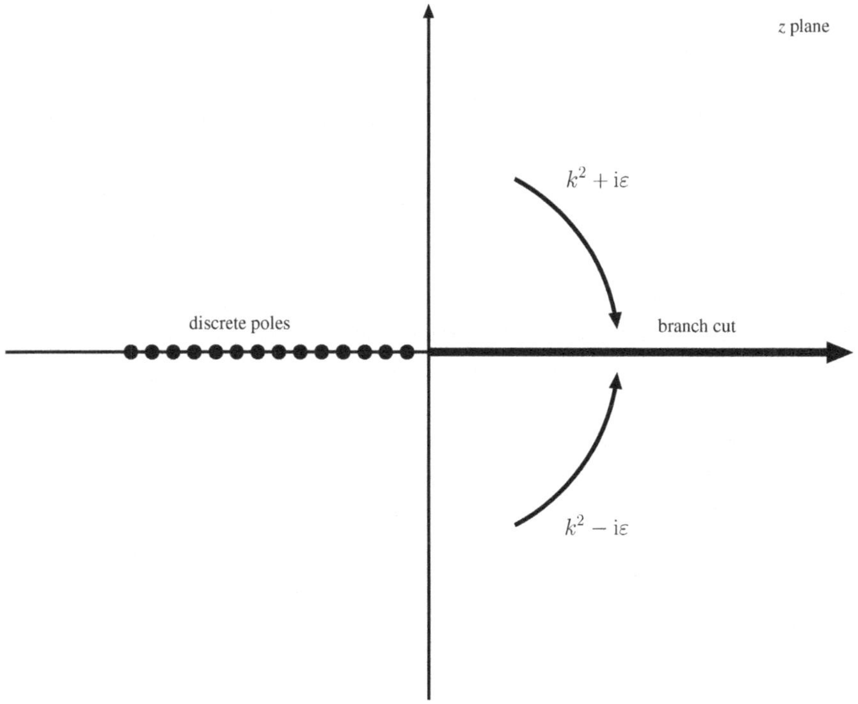

Fig. 1.5 Poles of the resolvent $G(z)$ and definition of $G^{\pm}(k^2)$

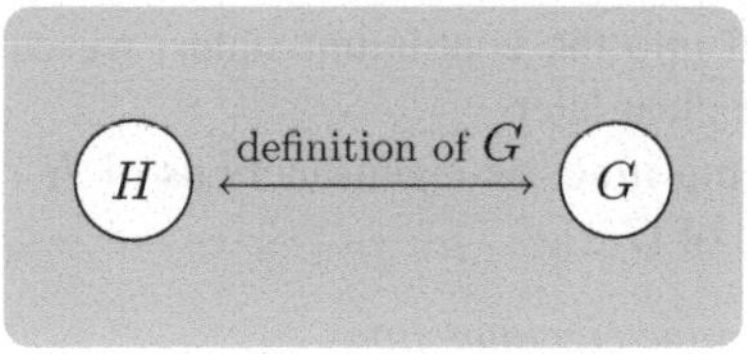

1.3.1 The Free Electron Propagator

We consider first the Green's operator $G_0^{\pm}(k^2)$ in the absence of a potential, i.e. for free electrons. It can be shown that the matrix elements of this operator in the space coordinates, also called the *Green's function*, are given by [11]

$$\langle \boldsymbol{r}'|G_0^{\pm}(k^2)|\boldsymbol{r}\rangle = G_0^{\pm}(\boldsymbol{r}',\boldsymbol{r};k^2) = -\frac{1}{4\pi}\frac{e^{\pm ik|\boldsymbol{r}-\boldsymbol{r}'|}}{|\boldsymbol{r}-\boldsymbol{r}'|} . \tag{1.20}$$

We see that this equation represents the amplitude at $\boldsymbol{r}'$ of a spherical wave issued from $\boldsymbol{r}$. Therefore, $G_0^{+}(k^2)$ represents the propagation of an *outgoing wave* and $G_0^{-}(k^2)$ that of an *incoming wave*. Hence the name free electron *propagator* usually given to the Green's function and by extension to the operator G_0.

This Green's function can be expanded into spherical waves (the partial waves we have defined in Sect. 1.2) as [12]

$$G_0^{+}(\boldsymbol{r}',\boldsymbol{r};k^2) = -ik\sum_L j_\ell(kr_<)h_\ell^{(1)}(kr_>)Y_L(\hat{\boldsymbol{r}}')Y_L^*(\hat{\boldsymbol{r}}) . \tag{1.21}$$

Here, $\boldsymbol{r}_<$ and $\boldsymbol{r}_>$ represent respectively the smaller and the larger of $\boldsymbol{r}$ and $\boldsymbol{r}'$. L is a short-hand notation for the angular momentum indices (ℓ, m).

Because we are dealing with short-ranged potentials, H_0 is valid outside the range of the potential. Therefore, the solutions $|\varphi_k\rangle$ of H_0 are called *asymptotical states* as they represent the state of the system when we are far way from the interaction zone.

1.3.2 The Full Propagator

In the presence of an external potential V, the full propagator G can be rewritten as

$$G(z) = (z-H)^{-1} = (z-H_0-V)^{-1} = (G_0^{-1}(z)-V)^{-1} , \tag{1.22}$$

where for the sake of simplicity, we have dropped the $^{\pm}$. Some simple manipulations of the previous equations give

$$G(z) = G_0(z) + G_0(z)VG(z) \tag{1.23a}$$

$$= G_0(z) + G(z)VG_0(z)\ . \tag{1.23b}$$

This equation that relates the full propagator $G(z)$ to the free electron propagator $G_0(z)$ and to the potential V is called *Dyson equation*.

A related equation, the *Lippmann–Schwinger equation*, relates the eigenfunctions of H and H_0. Indeed, noticing that $G_0V|\psi_k\rangle$ is a particular solution of H, and denoting $|\varphi_k\rangle$ the general solution of H_0 at energy k^2, we can use the well-known theorem on differential equations [13] to write

$$\underbrace{|\psi_k\rangle}_{\text{general solution of } H} = \underbrace{|\varphi_k\rangle}_{\text{general solution of } H_0} + \underbrace{G_0V|\psi_k\rangle}_{\text{particular solution of } H}\ . \tag{1.24}$$

G_0V is called the *kernel* of the Lippmann–Schwinger equation.

For both equations (Dyson and Lippmann–Schwinger), we note that we have the unknown quantity ($G(z)$ and $|\psi_k\rangle$ respectively) on both sides. This gives us two straightforward ways to solve these equations. In the case of the Dyson equation, this gives

1. Expand:

$$G(z) = G_0(z) + G_0(z)VG_0(z) + G_0(z)VG_0(z)VG_0(z) + \cdots \tag{1.25}$$

which is a Born-type expansion

2. Factorize:

$$G(z) = [I - G_0(z)V]^{-1}\ G_0(z) = G_0(z)\ [I - VG_0(z)]^{-1}\ , \tag{1.26}$$

where I represents the identity.

In the first case, we have a perturbative expansion for which convergence has to be properly considered. In the second case, we need to invert a matrix which, depending of its size, can be time-consuming.

An important point to note is that the bound states $|\phi_b\rangle$ of H and the continuum scattering states $|\psi_k\rangle$ form a basis of the Hilbert space $\mathcal{H}$ associated to the Hamiltonian H. Applying their closure relation to definition (1.22) of the full propagator gives the so-called *Lehmann spectral representation* [14]

$$G(z) = \sum_b \frac{|\phi_b\rangle\langle\phi_b|}{z - k_b^2} + \int \frac{|\psi_k\rangle\langle\psi_k|}{z - k^2}\, \mathrm{d}k^2\ . \tag{1.27}$$

Using the well-known formula [15]

$$\lim_{\varepsilon\to 0^+} \frac{1}{x - x_0 \pm \mathrm{i}\varepsilon} = \mathcal{P}\left(\frac{1}{x - x_0}\right) \mp \mathrm{i}\pi\delta(x - x_0)\ , \tag{1.28}$$

where $\mathcal{P}$ stands for Cauchy principal part which can be defined as [15]

$$\mathcal{P}\left(\frac{1}{x - x_0}\right) = \frac{1}{2} \lim_{\varepsilon \to 0^+} \left[\frac{1}{x - x_0 + \mathrm{i}\varepsilon} + \frac{1}{x - x_0 - \mathrm{i}\varepsilon}\right], \tag{1.29}$$

we obtain the relation between the density of states $n(k^2)$ and the Green's operator

$$n(k^2) = \mp\frac{1}{\pi} \operatorname{Tr}\left(\Im\left[G^{\pm}(k^2)\right]\right) . \tag{1.30}$$

Here, Tr represents the trace and $\Im$ the imaginary part.
This expression gives a straightforward way to compute the density of states.

1.3.3 The Transition Operator

In many instances, scattering theory relies on a third approach to solve the Dyson and the Lippmann–Schwinger equations. Indeed, we can introduce a new, unknown operator T, called the *transition operator*, by

$$VG(z) = T(z)G_0(z) \tag{1.31a}$$
$$G(z)V = G_0(z)T(z) . \tag{1.31b}$$

At this stage, it is just a mathematical trick to remove the unknown quantity $G(z)$ on the right-hand side and replace it by the equally unknown quantity $T(z)$ we intend to compute. From the definition (1.31), it can be demonstrated that we have

$$T(z) = V + VG(z)V . \tag{1.32}$$

This representation is important because it shows us that $T(z)$ and $G(z)$ have a similar structure and that $T(z)$ contains all the information encoded into $G(z)$.
In terms of the transition operator, our two fundamental equations now write

- Dyson equation:

$$G^{\pm}(k^2) = G_0^{\pm}(k^2) + G_0^{\pm}(k^2)T^{\pm}(k^2)G_0^{\pm}(k^2) \tag{1.33}$$

- Lippmann–Schwinger equation:

$$|\psi_k^{\pm}\rangle = |\varphi_k\rangle + G_0^{\pm}(k^2)T^{\pm}(k^2)|\varphi_k\rangle \tag{1.34}$$

The $|\psi_k^{\pm}\rangle$ are the *scattering states* solution of H. Their eigenvalues belong to the *continuum* (branch cut in Fig. 1.5). Note that these scattering states are *orthonormal* [16]

$$\langle \psi_k^+ | \psi_{k'}^+ \rangle = \langle \varphi_k | \varphi_{k'} \rangle \,, \tag{1.35}$$

and likewise for the incoming states $|\psi_k^-\rangle$.

In our approach to scattering theory, we will replace the evaluation of $G(z)$ by that of $T(z)$ so that we can now sketch this approach as

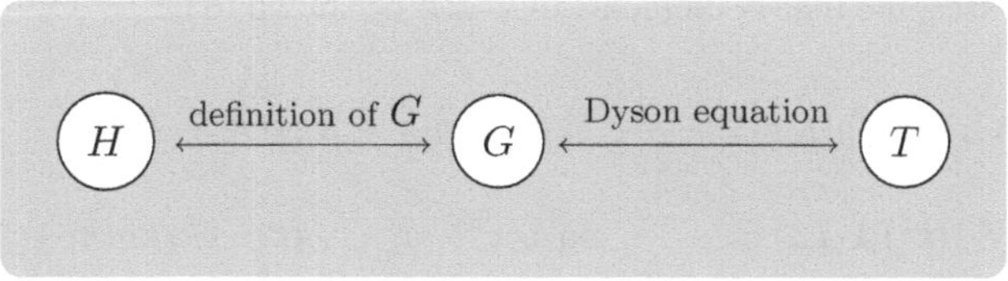

In the scattering process, the *transition probability* $W_{i \to f}$ for the system to evolve from the asymptotical state $|\varphi_i\rangle$ to the asymptotical state $|\varphi_f\rangle$ under the effect of potential V is given by [17]

$$W_{i \to f} = \frac{2\pi}{\hbar} \left| \langle \varphi_f | T^+ | \varphi_i \rangle \right|^2 \delta(E_f - E_i) \,. \tag{1.36}$$

This explains why T is called the transition operator.

We note also that the Hermitian conjugates are given by

$$G^\dagger(z) = G(z^*) \tag{1.37}$$

$$T^\dagger(z) = T(z^*) \,, \tag{1.38}$$

so that, for instance,

$$\left[G_0^-(k^2) \right]^\dagger = G_0^+(k^2) \quad \text{if } k^2 \text{ is real.} \tag{1.39}$$

The difference between T^+ and T^- is called the *generalized optical theorem*. It writes as [18]

$$T^+ - T^- = T^+ \left[G_0^+ - G_0^- \right] T^- \tag{1.40a}$$

$$= V \left[G^+ - G^- \right] V \,. \tag{1.40b}$$

It is an important theorem that ensures the conservation of the number of particles. Moreover, it is the theorem which allows to relate photoemission to x-ray absorption.

1.3.3.1 Matrix Elements in Standard Bases

Computing the matrix elements of the transition operator T will allow us to make the connection with the partial wave theory we have developed in Sect. 1.2. Let us consider the *orthonormalized* plane wave and spherical wave bases [19]

$$\langle \boldsymbol{r}|\boldsymbol{k}\rangle = \frac{1}{(2\pi)^{\frac{3}{2}}}\, \mathrm{e}^{\mathrm{i}\boldsymbol{k}\cdot\boldsymbol{r}} \tag{1.41}$$

$$\langle \boldsymbol{r}|kL\rangle = k\sqrt{\frac{2}{\pi}}\, \mathrm{i}^{\ell}\, j_{\ell}(kr) Y_L(\hat{\boldsymbol{r}})\ . \tag{1.42}$$

In these bases, the matrix elements of T can be shown to be

$$\langle \boldsymbol{K}'|T^{+}(k^2)|\boldsymbol{K}\rangle = -\frac{1}{2\pi^2}\, f_k(\theta_{\hat{\boldsymbol{K}}}^{\hat{\boldsymbol{K}}'})\, \delta(k-K)\delta(K-K') \tag{1.43}$$

$$\langle k_2 L_2|T^{+}(k^2)|k_1 L_1\rangle = -\frac{2k}{\pi}\, \sin\delta_{\ell_1}\mathrm{e}^{\mathrm{i}\delta_{\ell_1}}\, \delta_{L_1 L_2}\, \delta(k_1-k_2)\, \delta(k_1-k)\ . \tag{1.44}$$

We recognize in the first equation the scattering amplitude defined in Sect. 1.2. The second equation involves the quantity $t_\ell = \sin\delta_{\ell_1}\mathrm{e}^{\mathrm{i}\delta_{\ell_1}}$ defined in the same section.

1.3.4 The Møller Wave Operator

We can factorize (1.33) and (1.34) to make an operator $\left[I + G_0^{\pm}(k^2)T^{\pm}(k^2)\right]$ appear. This is the so-called *Møller wave operator* $\Omega^{(\pm)}(k^2)$. It can be written under the different forms

$$\Omega^{(\pm)}(k^2) = \left[I + G_0^{\pm}(k^2)T^{\pm}(k^2)\right] = \left[I + G^{\pm}(k^2)V\right] = \left[I - G_0^{\pm}(k^2)V\right]^{-1}\ . \tag{1.45}$$

We can now recast our different equations in terms of $\Omega^{(\pm)}(k^2)$. This gives

- Dyson equation:

$$G^{\pm}(k^2) = \Omega^{(\pm)}(k^2)\, G_0^{\pm}(k^2) = G_0^{\pm}(k^2)\left[\Omega^{(\mp)}(k^2)\right]^{\dagger} \tag{1.46}$$

- Lippmann–Schwinger equation:

$$|\psi_k^{\pm}\rangle = \Omega^{(\pm)}(k^2)|\varphi_k\rangle \tag{1.47}$$

- Definition of T:

$$T^{\pm}(k^2) = V\Omega^{(\pm)}(k^2) = \left[\Omega^{(\mp)}(k^2)\right]^{\dagger} V \tag{1.48}$$

This representation allows us to make a few important remarks:

- $\Omega^{(\pm)}$ maps the free electron propagator $G_0^{\pm}$ onto the full electron propagator $G^{\pm}$
- $\Omega^{(\pm)}$ maps the asymptotical states $|\varphi_k\rangle$ onto the scattering states

- $T^{\pm}|\varphi_k\rangle = V|\psi_k^{\pm}\rangle$ so that T acts on the asymptotical states. Moreover, its action on these states gives the same result as the action of V on the scattering states.

We note that the Møller wave operators satisfy

$$\Omega^{(\pm)\dagger}\,\Omega^{(\pm)} = I \ . \tag{1.49}$$

However, they are generally not invertible as we have [20]

$$\Omega^{(\pm)}\,\Omega^{(\pm)\dagger} = I - \Lambda \ . \tag{1.50}$$

Here $\Lambda = \sum_b |\phi_b\rangle\,\langle\phi_b|$ is called the *unitary deficiency*. It is the projector on the subspace generated by the bound states $\{|\phi_b\rangle\}$.

1.3.5 Use of Outgoing and Ingoing States

We have seen in Sect. 1.3.1 that the $^+$ and $^-$ solutions were corresponding respectively to outgoing and incoming waves. We address now the important question of which solution to use in practical cases. The Møller wave operator $\Omega^{(\pm)}(k^2)$ gives us a useful tool to answer this question, as it allows us to construct the solution from the known asymptotical state. Therefore, depending which asymptotical state is known to us (the one in the past or the one in the future, considered with respect to the time of the interaction), we will use one solution or the other [21]. This is explicited in Fig. 1.6. Here, we have taken the time at which the interaction takes place as the origin of time.

From this interpretation, we see that if we know the state $|\varphi_k^{\text{in}}\rangle$ of the electron in the past (for instance because we use an electron gun to send a beam of particles onto the sample), we will be able to construct $|\psi_k^{+}\rangle$ states. By contrast, if we know the state $|\varphi_k^{\text{out}}\rangle$ of the electron in the future (because we measure it with a detector, as in the case of photoemission for instance), we will only be able to construct the $|\psi_k^{-}\rangle$ solution. This solution is often termed as the *time-reversed LEED* solution as it is the Hermitian conjugate of the LEED (Low-Energy Electron Diffraction) state constructed from the knowledge of the asymptotical state of the electron emitted by the electron gun.

Note that some theoretical frameworks, especially in nuclear physics, use the *scattering operator* S rather than the transition operator T. This scattering operator is defined by [21]

$$|\varphi_k^{\text{out}}\rangle = S\,|\varphi_k^{\text{in}}\rangle \ . \tag{1.51}$$

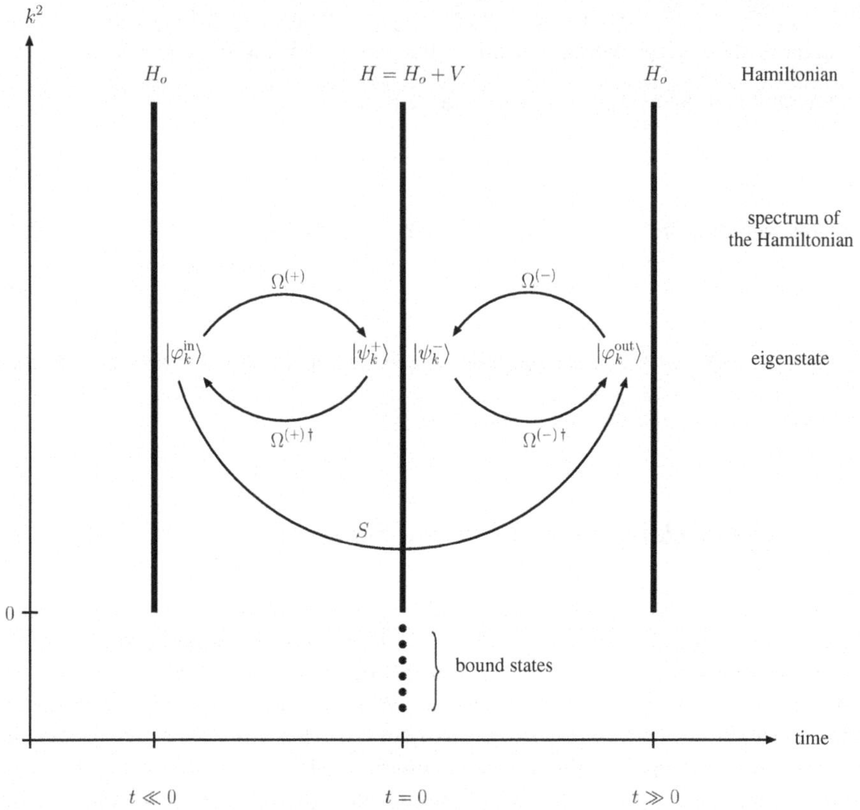

Fig. 1.6 Physical interpretation of outgoing and ingoing states for the scattering by a short-ranged potential V

Note also that the matrix elements of S in the spherical wave basis (1.42) $S_{LL'}$ is given by $S_\ell \, \delta_{LL'}$, with $S_\ell = \exp[2\mathrm{i}\delta_\ell]$. This is the same coefficient as the one arising in (1.9)b.

1.3.6 Two-Potential Formula

Before generalizing this scattering theoretic approach to the case of a superposition of N potentials, let us look first at what happens when we have just two potentials. For this, we partition the Hamiltonian H as

$$H \;=\; H_0 + V_1 + V_2 \;=\; H_1 + V_2 \;, \tag{1.52}$$

where we suppose that we know how to compute the eigensolutions $|\chi_k^\pm\rangle$ of $H_1 = H_0 + V_1$. In order to simplify the notation, we index by $_i$ and $_f$ respectively the *initial* and *final* states.

Then, it can be demonstrated that [22]

$$\langle\varphi_f|T^+|\varphi_i\rangle = \langle\varphi_f|T_1^+|\varphi_i\rangle + \langle\chi_f^-|V_2|\psi_i^+\rangle \tag{1.53a}$$

$$= \langle\chi_f^-|V_1|\varphi_i\rangle + \langle\chi_f^-|V_2|\psi_i^+\rangle \tag{1.53b}$$

$$= \langle\varphi_f|T_1^+|\varphi_i\rangle + \langle\psi_f^-|V_2|\chi_i^+\rangle \tag{1.53c}$$

$$= \langle\varphi_f|V_1|\chi_i^+\rangle + \langle\psi_f^-|V_2|\chi_i^+\rangle\ , \tag{1.53d}$$

or equivalently, in terms of operators

$$T^+ \;=\; T_1^+ + \omega_1^{(-)\,\dagger}\, V_2\, \Omega^{(+)} \;=\; T_1^+ + \Omega^{(-)\,\dagger}\, V_2\, \omega_1^{(+)}\ , \tag{1.54}$$

where we have noted $\omega_1^{(\pm)}$ the Møller wave operator for V_1 alone and $\Omega^{(\pm)}$ the one for $V = V_1 + V_2$.

This formula is of particular interest when the two potentials are of a very different nature, one of them leading to *elastic scattering* and the other to *inelastic* scattering. This is the case for instance when V_1 represents the potential of a cluster of atoms and V_2 a Coulomb interaction leading to the ejection of an electron on a given atom. In this case, the transition from $|\varphi_i\rangle$ to $|\varphi_f\rangle$ cannot arise under V_1 alone as V_1 cannot account for the ejection of the second electron. Consequently, the transition matrix element $\langle\varphi_f|T_1^+|\varphi_i\rangle$ must be zero and we are left with only the second term in (1.53) and (1.54).

Most of the time, however, the calculation of $|\psi_i^+\rangle$ and $|\psi_f^-\rangle$ remains beyond our possibilities. In this case, we can write these states in terms of the Lippmann–Schwinger equation

$$|\psi^\pm\rangle \;=\; |\chi^\pm\rangle + G_1^\pm V_2|\psi^\pm\rangle\ , \tag{1.55}$$

where $G_1^\pm$ is the propagator in the presence on potential V_1 alone, and perform a *Born expansion* of (1.55) to obtain

$$|\psi^\pm\rangle \;=\; |\chi^\pm\rangle + G_1^\pm V_2|\chi^\pm\rangle + \cdots\ , \tag{1.56}$$

If we retain only the first term in (1.56), we have the so-called *distorted wave Born approximation* which is generally used to compute cross-sections in the two-potential case. In terms of the transition operator, this gives [38]

$$T^+ \;=\; T_1^+ + \omega_1^{(-)\,\dagger}\, V_2\, \omega_1^{(+)}\ , \tag{1.57}$$

The term *distorded wave* comes from the fact that $|\chi^\pm\rangle$ corresponds to the asymptotic wave $|\varphi\rangle$ distorded by potential V_1.

1.4 Multiple Scattering Theory

So far, we have considered the scattering of an electron by a single potential. The center of this potential was assumed to be at the origin of space. If we want to generalize the previous approach to the case of a collection of N potentials distributed all over the space, clearly their respective positions have to be taken into account into the theoretical model. One way to do this is to make use of the *translation operator*.

1.4.1 The Translation Operator

We introduce the translation operator $\mathcal{T}(\boldsymbol{a})$ corresponding to a translation by a vector $\boldsymbol{a}$ by its action on the space states $|\boldsymbol{r}\rangle$:

$$\mathcal{T}(\boldsymbol{a})|\boldsymbol{r}\rangle = |\boldsymbol{r} + \boldsymbol{a}\rangle \tag{1.58a}$$

$$\langle \boldsymbol{r}|\mathcal{T}(\boldsymbol{a}) = \langle \boldsymbol{r} - \boldsymbol{a}| \; . \tag{1.58b}$$

Its action on plane wave states $|\boldsymbol{K}\rangle$ is given by

$$\mathcal{T}(\boldsymbol{a})|\boldsymbol{K}\rangle = \mathrm{e}^{-\mathrm{i}\boldsymbol{K}\cdot\boldsymbol{a}} \; |\boldsymbol{K}\rangle \tag{1.59a}$$

$$\langle \boldsymbol{K}|\mathcal{T}(\boldsymbol{a}) = \mathrm{e}^{-\mathrm{i}\boldsymbol{K}\cdot\boldsymbol{a}} \; \langle \boldsymbol{K}| \; . \tag{1.59b}$$

We can now take into account the position of each individual potential V_i located at $\boldsymbol{R}_i$ from the origin by writing [19]

$$V = \sum_i \bar{V}_i \; , \tag{1.60}$$

with

$$\bar{V}_i = \mathcal{T}(\boldsymbol{R}_i) V_i \mathcal{T}(-\boldsymbol{R}_i) \; . \tag{1.61}$$

Equation (1.61) is the operator's equivalent of the so-called *addition theorems* for wave functions [23]. Its geometry is given in Fig. 1.7. Its interpretation is simple. When making $\bar{V}_i$ act on a state, first we translate the potential to the origin of space with $\mathcal{T}(-\boldsymbol{R}_i)$ so that we can apply the scattering theory developed in Sect. 1.3. Then, V_i which is the potential referred to the origin can act on the state. Finally, we move back the potential to its original position with $\mathcal{T}(\boldsymbol{R}_i)$. Therefore, $\bar{V}_i$ is the *potential at site i centered at* $\boldsymbol{R}_i$.

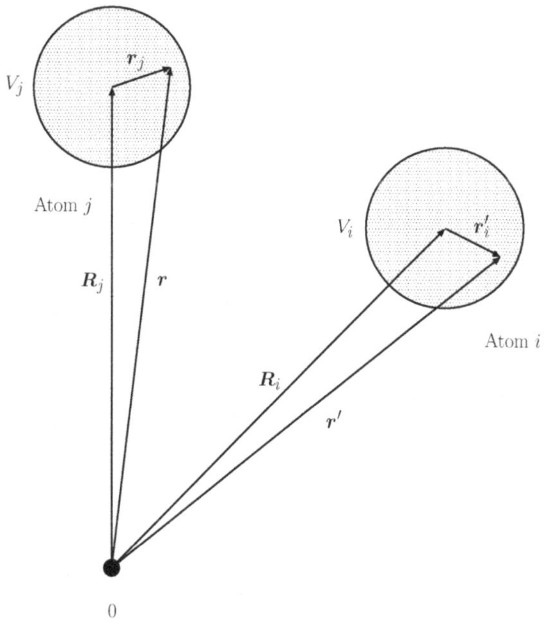

Fig. 1.7 Geometry of potentials V_i and V_j with respect to the origin

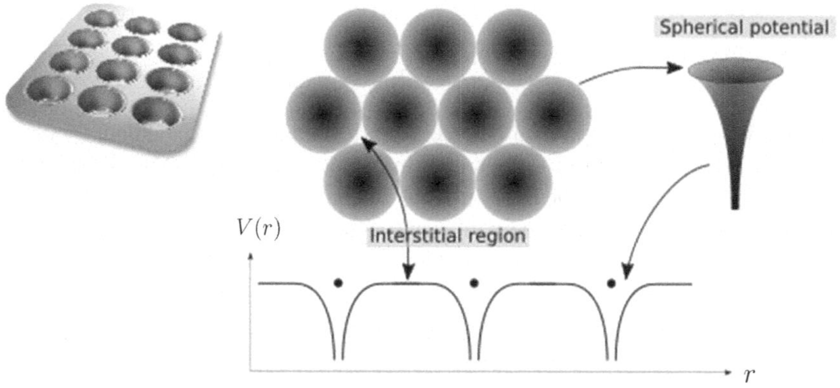

Fig. 1.8 Representations of the muffin-tin potential

1.4.2 The Muffin-Tin Approximation

In Sects. 1.2 and 1.3, we have supposed the potential to be spherically symmetric. Now we make a further approximation, we take the potential of the system as a *muffin-tin potential*: in addition of being spherically symmetric at atomic sites, the potential is constant in the interstitial region. This is sketched in Fig. 1.8.

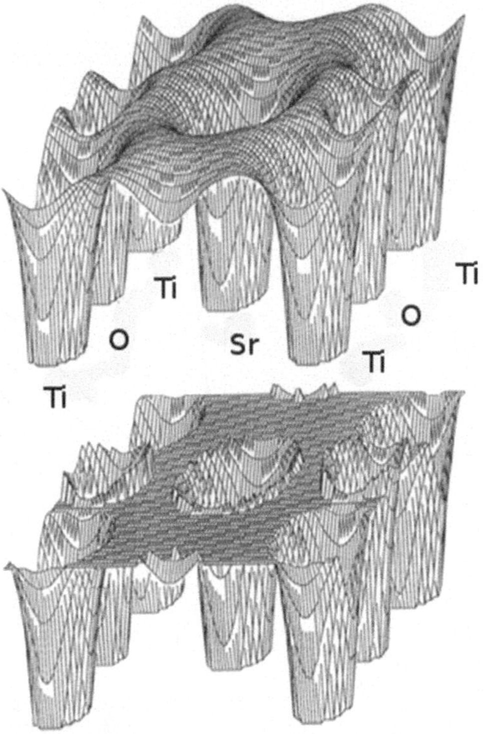

Fig. 1.9 Comparison between the full potential (top view) and the muffin-tin approximation to the potential (bottom view) of $SrTiO_3$ [24]

Figure 1.9 gives a comparison between the full potential, as calculated with an *ab initio* code and the muffin-tin potential in the case of $SrTiO_3$ [24]. We see from this comparison that it is in the interstitial region that the approximation is the worst. Therefore, the muffin-tin approximation will essentially be valid for close packed systems, or for high kinetic energies (where the electron is not sensitive to the fine details of the potential as it is essentially scattered by the core).

1.4.3 The Transition Operator of the System

We turn now to the calculation of the transition operator T of the system, associated to potential V. We denote $\bar{T}_i$ the individual transition of operator corresponding to the scattering by potential $\bar{V}_i$ *alone*. Following (1.32), they can be expressed as

$$\bar{T}_i = \bar{V}_i + \bar{V}_i \bar{G}_i \bar{V}_i \ . \tag{1.62}$$

Here again we drop the $\pm$ in order to simplify the notation. $\bar{G}_i$ is the propagator in the presence of $\bar{V}_i$ only and is given by the ith individual Dyson equation

$$\bar{G}_i = G_0 + G_0 \bar{T}_i G_0 . \tag{1.63}$$

Contrarily to the potentials, the individual $\bar{T}_i$ do not add to form the system transition operator T. Indeed, from the Dyson equation in the form (1.32), we can write

$$T = \sum_i \bar{\mathfrak{T}}_i , \tag{1.64}$$

with

$$\bar{\mathfrak{T}}_i = \bar{V}_i + \bar{V}_i G V , \tag{1.65}$$

which differs from (1.62) by the quantities G and V which take into account the effect of *all* the atoms.

In order to obtain an additive formulation, we must therefore correct the individual $\bar{T}_i$ by the so-called *distortion operator* $\bar{W}_i$ [25] which takes care of the corrections due to the presence of the rest of the atoms occuring in G and V and write

$$T = \sum_i \bar{T}_i \ \bar{W}_i . \tag{1.66}$$

Because $\bar{W}_i$ takes into account the distortion to $\bar{T}_i$ due to all the atoms of the system, we can expand it in terms of a sum over all the atoms j. Combining the two sums, we have the formulation

$$T = \sum_{i,j} \bar{\tau}^{ji} . \tag{1.67}$$

It is then straightforward to show from (1.32) that the quantity $\bar{\tau}^{ji}$ must be expressed as

$$\bar{\tau}^{ji} = \bar{V}_j \delta_{ij} + \bar{V}_j G \bar{V}_i , \tag{1.68}$$

which satisfies

$$\bar{\tau}^{ji\,\pm} = \left[\bar{\tau}^{ij\,\mp} \right]^{\dagger} , \tag{1.69}$$

as the potentials are Hermitian.

It is related to the origin-centered τ^{ji} through

$$\bar{\tau}^{ji} = \mathcal{T}(\boldsymbol{R}_j) \tau^{ji} \mathcal{T}(-\boldsymbol{R}_i) . \tag{1.70}$$

This expression is very rewarding as it gives us a direct physical interpretation for this operator $\bar{\tau}^{ji}$. Indeed, when acting on an electron state, we see that we have first scattering by atom i, then propagation throughout the whole system (G) and finally scattering by atom j. In other words, $\bar{\tau}^{ji}$ describes all the possible ways the electron can travel to go from i to j. For this reason, it is called the *scattering path operator*. It was first introduced in nuclear physics (without being given a name) by Faddeev [26], and later rediscovered and named in condensed matter physics by Györffy [27]. We can now recast our whole formalism into the scattering path operator approach.

Note that the additive operator $\bar{\mathfrak{T}}_j$ defined in (1.65) can be expressed as

$$\bar{\mathfrak{T}}_j = \sum_i \bar{\tau}^{ji} . \tag{1.71}$$

This operator allows us to derive the optical theorem satisfied by the scattering path operator

$$\bar{\tau}^{ji\,+} - \bar{\tau}^{ij\,+\,\dagger} = -\frac{\mathrm{i}\pi}{k} \bar{\mathfrak{T}}_j^{\pm} \bar{\mathfrak{T}}_i^{\pm\,\dagger} , \tag{1.72}$$

which, for $i = j$ simplifies to

$$\Im \left[\bar{\tau}^{ii\,(+)} \right] = -\frac{\pi}{2k} \bar{\mathfrak{T}}_i^{\pm} \bar{\mathfrak{T}}_i^{\pm\,\dagger} . \tag{1.73}$$

This result makes the connection between x-ray absorption and angular-resolved photoemission. Indeed, upon excitation of a core state $|\phi_c\rangle$ on an atom labelled 0 by an incoming photon, multiplying (1.73) by the photon-electron excitation operator O on the right and by its Hermitian conjugate on the left, we obtain

$$\langle \phi_c | O^\dagger \Im \left[\bar{\tau}^{00\,(+)} \right] O | \phi_c \rangle = -\frac{k}{2\pi} \int \left| \langle \boldsymbol{k} | \bar{\mathfrak{T}}_0^{-\,\dagger} O | \phi_c \rangle \right|^2 d\hat{\boldsymbol{k}} . \tag{1.74}$$

The left-hand side corresponds to the x-ray absorption spectroscopy (XAS) cross-section while in the right-hand side, $\left| \langle \boldsymbol{k} | \bar{\mathfrak{T}}_0^{-\,\dagger} O | \phi_c \rangle \right|^2$ is proportional to the angular-resolved photoemission differential cross-section (also called *photoelectron diffraction* when modulations of a given feature in the photoemission spectrum are studied as a function of energy or of the exit angle). Therefore, the optical theorem tells us that the XAS cross-section is nothing else than the integral of the photoelectron diffraction (PED) differential cross-section over all the directions of the outgoing photoelectron. $\bar{\mathfrak{T}}_0^{-} |\boldsymbol{k}\rangle$ is the so-called time-reversed LEED state commonly used to describe the final state in the photoemission process. Expression (1.74) can be verified numerically. This is shown in Fig. 1.10 for a 23-atom MgO(001) cluster.

Note also that the full Green's function $G(\boldsymbol{r}', \boldsymbol{r}; k^2)$ can be expanded into partial waves, like the free electron Green's function. This expansion is [19]

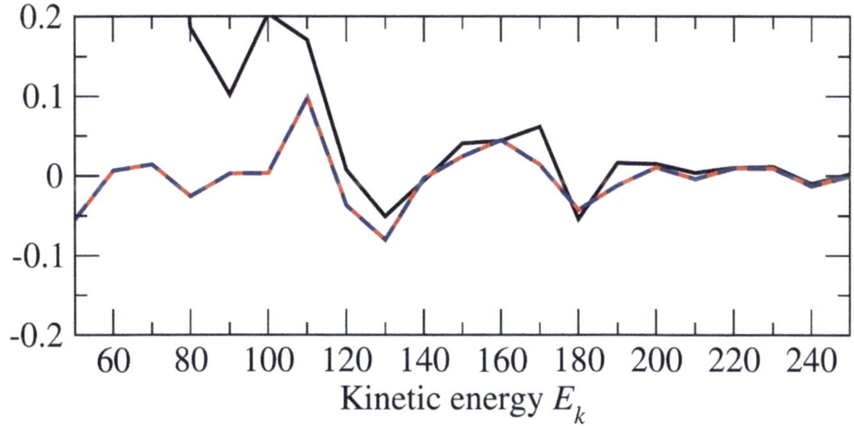

Fig. 1.10 Comparison of the XAS cross-section and the cross-section of PED integrated over 2030 directions for an 23-atom MgO(001) cluster. Black line: PED series expansion up to order 5, red line: PED matrix inversion and dashed blue line: XAS matrix inversion. The energies are expressed in eV (Reproduced from the MsSpec user's guide [28])

$$\begin{aligned} G(\boldsymbol{r}', \boldsymbol{r}; k^2) = \frac{\pi}{2} & \sum_{L_i, L_j} \left[\tilde{\mathcal{R}}_{L_j}^{j\,-}(\boldsymbol{r}_j) \right]^* \tau^{ji}_{L_j L_i} \tilde{\mathcal{R}}_{L_i}^{i\,+}(\boldsymbol{r}_j) \\ & + k \sum_{L_i} \left[\tilde{\mathcal{R}}_{L_i}^{i\,-}(\boldsymbol{r}_{i<}) \right]^* \tilde{\mathcal{J}}^i_{L_i}(\boldsymbol{r}_{i>})\, \delta_{ij} \; . \end{aligned} \tag{1.75}$$

Here, $\tilde{\mathcal{R}}_{L_i}^{i+}(\boldsymbol{r}_i)$ is the *regular* solution of the Schrödinger equation *inside the range* of potential V_i that matches $t_{\ell_i}^{-1} \mathrm{i}^{\ell_i} \left[j_{\ell_i}(kr_i) + \mathrm{i} t_{\ell_i} \, h_{\ell_i}^{(1)}(kr_i) \right] Y_{L_i}(\hat{\boldsymbol{r}}_i)$ *outside the range*. $\tilde{\mathcal{J}}^i_{L_i}(\boldsymbol{r}_i)$ is the corresponding *irregular* solution inside V_i. It matches $\mathrm{i}^{\ell_i} j_{\ell_i}(kr_i) Y_{L_i}(\hat{\boldsymbol{r}}_i)$. The geometry corresponding to this equation is given in Fig. 1.7. Expansion (1.75) gives an alternative way to derive Lehmann spectral representation (1.27) to compute the Green's function of the system.

We can now summarize the multiple scattering framework that will allow us to solve the Schrödinger equation of the problem by

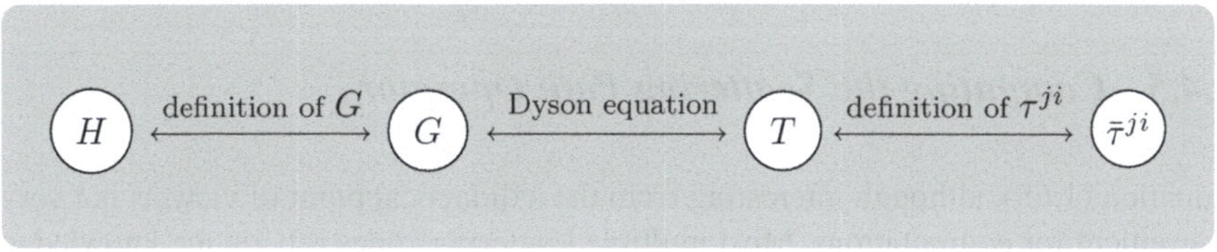

1.4.4 Normalization Issues

Expression (1.75) differs from the result derived originally in the seminal paper by Faulkner and Stocks [29]. This is due to the fact that they chose a different normalization for the spherical wave basis functions — see (1.42). In their approach (FS), as in many other papers in the literature, they use the definition

$$t_\ell^{FS} = -\frac{1}{k} \sin \delta_\ell \, e^{i\delta_\ell} \;, \tag{1.76}$$

with their wave functions related to ours by

$$\tilde{\mathcal{R}}_L^{FS}(\boldsymbol{r}) = -k \, \tilde{\mathcal{R}}_L(\boldsymbol{r}) \tag{1.77}$$

$$\tilde{\mathcal{J}}_L^{FS}(\boldsymbol{r}) = \tilde{\mathcal{J}}_L(\boldsymbol{r}) \tag{1.78}$$

If we make the substitution and take into account the fact that our scattering path operator matrix elements $\tau_{L_jL_i}^{ji}$ scale like $-2k/\pi$ times our t_ℓ and theirs like $-1/k$ times t_ℓ, we recover exactly their expression

$$\begin{aligned} G(\boldsymbol{r}', \boldsymbol{r}; k^2) = &\sum_{L_i, L_j} \left[\tilde{\mathcal{R}}_{L_j}^{j\,FS\,-}(\boldsymbol{r}_j)\right]^* \tau_{L_jL_i}^{ji\,FS} \tilde{\mathcal{R}}_{L_i}^{i\,FS\,+}(\boldsymbol{r}_j) \\ &-\sum_{L_i} \left[\tilde{\mathcal{R}}_{L_i}^{i\,FS\,-}(\boldsymbol{r}_{i<})\right]^* \tilde{\mathcal{J}}_{L_i}^{i\,FS}(\boldsymbol{r}_{i>}) \, \delta_{ij} \;. \end{aligned} \tag{1.79}$$

Note also that in certain articles, the so-called *normalization to one state per Rydberg* is used. It corresponds to radial wave functions of the form [30]

$$R_\ell(r) \;=\; \sqrt{\frac{k}{\pi}} \, t_\ell^{-1} \left[j_\ell(kr) + i t_\ell h_\ell^{(1)}(kr) \right] \tag{1.80}$$

More information on normalization issues in scattering theory can be found in [31].

1.4.5 Computing the Scattering Path Operator

Equation (1.68), although interesting from the a didactical point of view, is not very convenient for computations. Most multiple scattering codes rely on the knowledge of matrix elements of the individual T_i, which are one of the building blocks of this approach. Therefore, for practical calculations, we will use the more convenient *equation of motion* of the scattering path operator given by [19, 26]

$$\bar{\tau}^{ji} = \bar{T}_j \delta_{ij} + \sum_{k \neq j} \bar{T}_j G_0 \bar{\tau}^{ki} \tag{1.81a}$$

$$= \bar{T}_j \delta_{ij} + \sum_{k \neq i} \bar{\tau}^{jk} G_0 \bar{T}_i \ . \tag{1.81b}$$

This equation of motion is particularly interesting as it provides us directly with two practical schemes to compute the scattering path operator from the individual T_i:

1. Factorization:
   We can consider the individual T_i as matrix elements of a **T** matrix and likewise for the propagators, when expressed in the angular momentum representation. The equation of motion can then be seen as a *matrix equation* which can be factorized as

$$\tau = (\mathbf{T}^{-1} - \mathbf{G_0})^{-1} = \mathbf{T}(\mathbf{I} - \mathbf{G_0}\mathbf{T})^{-1} = (\mathbf{I} - \mathbf{T}\mathbf{G_0})^{-1}\mathbf{T} \ . \tag{1.82}$$

   It is always exact from the scattering point of view. However, as it involves to fill up a matrix and invert it, its use is limited by the size of the matrix. Indeed, for a $N_m \times N_m$ matrix, the storage will scale like N_m^2 and the computing time, if the matrix is not sparse which is generally the case, like N_m^3. Here, $N_m = N(\ell_{max} + 1)^2$, where N is the number of atoms in the cluster.
2. Iteration:
   Upon the replacement of $\bar{\tau}^{ki}$ by its own equation of motion in (1.81), we obtain

$$\bar{\tau}^{ji} = \bar{T}_j \delta_{ij} + \bar{T}_j G_0 \bar{T}_i + \sum_{k \neq i,j} \bar{T}_j G_0 \bar{T}_k G_0 \bar{T}_i + \cdots \tag{1.83}$$

   This is the so-called (Watson) *multiple scattering series expansion* [32]. It is a *perturbative expansion* and therefore its validity is subject to convergence issues [33, 34]. More specifically, it converges only if the spectral radius $\rho(\mathbf{G_0T})$ of the kernel matrix $\mathbf{G_0T}$ in (1.82), defined by

$$\rho(\mathbf{G_0T}) = \max_i |\lambda_i| \ , \tag{1.84}$$

   where λ_i is the ith eigenvalue of the kernel matrix, is *strictly lower than 1*.
   In practice, it should be restricted to higher energies (typically larger than $\approx$ 100 eV) [34]. In addition, it has to be truncated at a given scattering order.

The solution to (1.82) and (1.83) only coincide if (1.83) is convergent.

A third approach, called *correlation expansion* [35] or *n-body expansion* [36], is discussed in Chap. 8.

The multiple scattering series expansion of the scattering path operator involves the summation over pathway terms of the form $\cdots \bar{T}_j G_0 \bar{T}_k G_0 \bar{T}_i \cdots$, or equivalently

$$\cdots \boxed{\mathcal{T}(\boldsymbol{R}_j)T_j\mathcal{T}(-\boldsymbol{R}_j)}\boxed{G_0}\boxed{\mathcal{T}(\boldsymbol{R}_k)T_k\mathcal{T}(-\boldsymbol{R}_k)}\boxed{G_0}\boxed{\mathcal{T}(\boldsymbol{R}_i)T_i\mathcal{T}(-\boldsymbol{R}_i)} \cdots$$

We can group them differently as

$$\boxed{\cdots\mathcal{T}(\boldsymbol{R}_j)}\boxed{T_j}\underbrace{\boxed{\mathcal{T}(-\boldsymbol{R}_j)G_0\mathcal{T}(\boldsymbol{R}_k)}}_{G_0^{jk}}\boxed{T_k}\underbrace{\boxed{\mathcal{T}(-\boldsymbol{R}_k)G_0\mathcal{T}(\boldsymbol{R}_i)}}_{G_0^{ki}}\boxed{T_i}\boxed{\mathcal{T}(-\boldsymbol{R}_i)\cdots} .$$

The underbraced quantities are the matrix elements of the free electron propagator between two atoms. In electronic structure models, they are usually called *KKR structure constants* as they depend only on the positions of the atoms. They are the matrix elements of the matrix $\mathbf{G_0}$ introduced in (1.82).

Expressed into the spherical wave basis defined by (1.42), and assuming that all operators are computed at the same reduced energy k^2, the only two ingredients of our multiple scattering framework are given by [19]

$$\langle L_j|T^+|L_i\rangle = -\frac{2k}{\pi} \sin\delta_{\ell_i}\, \mathrm{e}^{\mathrm{i}\delta_{\ell_i}}\, \delta_{L_jL_i} \tag{1.85}$$

$$\langle L_j|\mathcal{T}(-\boldsymbol{R}_{ji})G_0^+|L_i\rangle = -\frac{2\mathrm{i}\pi^2}{k} \sum_L \mathrm{i}^\ell h_\ell^{(1)}(kR_{ji})\, Y_L(\hat{\boldsymbol{R}}_{ji})\, G(L_jL|L_i) \,, \tag{1.86}$$

where $\boldsymbol{R}_{ji} = \boldsymbol{R}_j - \boldsymbol{R}_i$ and the Gaunt coefficient $G(L_jL|L_i)$ is defined by [37]

$$G(L_jL|L_i) = \int Y_{L_j}(\hat{\boldsymbol{r}})Y_L(\hat{\boldsymbol{r}})Y^*_{L_i}(\hat{\boldsymbol{r}})\,\mathrm{d}\hat{\boldsymbol{r}} \,. \tag{1.87}$$

Note that here, in order to simplify the notations, we have replaced $|kL\rangle$ by $|L\rangle$, assuming that all states and operators correspond to the same wave number k.

Another important quantity in the multiple scattering description of spectroscopies is the *multiple scattering amplitude*. It arises each time we need the amplitude of an electron incoming on an atom, or when we want to calculate the amplitude of an electron originating from a given atom when it reaches the detector. It is a matrix element of the distorted transition operator $\bar{\mathfrak{T}}_i$, which represents the scattering of the electron by atom i in the presence of all the other atoms. Using (1.59)a, (1.71) and the fact that $\langle \boldsymbol{k}|L\rangle = 1/k\; Y_L(\hat{\boldsymbol{k}})$, we have

$$\langle \bar{L}_i|\bar{\mathfrak{T}}_i^{\pm}|\boldsymbol{k}\rangle = \frac{1}{k}\, B_{L_i}^{i\,(\pm)}(\boldsymbol{k}) \,, \tag{1.88}$$

with the multiple scattering amplitudes $B_{L_i}^{i\,\pm}(\boldsymbol{k})$ given by

$$B_{L_i}^{i\,(\pm)}(\boldsymbol{k}) = \sum_{j,L_j} \tau_{L_iL_j}^{ij\,\pm}\, Y^*_{L_j}(\hat{\boldsymbol{k}})\, \mathrm{e}^{\mathrm{i}\boldsymbol{k}\cdot\boldsymbol{R}_i} \,. \tag{1.89}$$

$|\bar{L}_i\rangle = \mathcal{T}(\boldsymbol{R}_i)|L_i\rangle$ is the spherical wave state with respect to atom i.

1.5 Expression of the Cross-Sections

1.5.1 General Expression

Let us write now V_I the interaction potential underlying the physical process, so that we keep V for the potential of the sample. V_I can be for example the photon-electron interaction or a Coulomb interaction. Under this interaction, the system described by $H = H_0 + V_I = H_S + H_P + V_I$, where H_P is the Hamiltonian of the particle and H_S that of the sample in the absence of any interaction. We denote $\mathcal{E}$ the eigenvalues of H_0, which are a sum of the eigenvalues of H_P and that of H_S and $|\Phi\rangle$ the corresponding eigenstates (product of one eigenstate of H_P by one eigenstate of H_S.

Using the expression of the transition probability (1.36), and generalizing for the case of possible multiple outgoing electrons (see Chap. 9 for such an example), we can derive the differential cross-section as [17, 39]

$$\frac{\mathrm{d}^n\sigma}{\underbrace{\mathrm{d}E_1\,\mathrm{d}\Omega_1\ \cdots}_{n \text{ terms}}} = \frac{1}{I_0}\frac{2\pi}{\hbar}\left|T_{fi}\right|^2 \varrho\left(\mathcal{E}_f\right) , \tag{1.90}$$

where I_0 is the flux of the incoming beam and $\varrho\left(\mathcal{E}_f\right)$ is the density of states of the detected particle(s) at energy $\mathcal{E}_f$. Ω_m is the solid angle into which particle m is detected (see Fig. 1.11). The T operator matrix element is given by

$$T_{fi} = \langle\Phi_f|T_I|\Phi_i\rangle , \tag{1.91}$$

where T_I is the transition operator corresponding to potential V_I. As discussed in Sect. 1.3.3, the transition operator for our system can be written as

$$T_I = V_I + V_I G(\mathcal{E}_i) V_I , \tag{1.92}$$

where $G(\mathcal{E}_i)$ is the full propagator. As we do not know it, we can use the Dyson expansion (1.25) to approximate it

$$T_I = V_I + V_I G_0(\mathcal{E}_i) V_I + \cdots \tag{1.93}$$

If we keep only the first term V_I, we obtain the so-called *Fermi's golden rule No 2* [40]. This is the approximation usually made to compute the photoemission cross-section. For spectroscopies such as Resonant Elastic X-ray Spectroscopy (REXS), or spectroscopies involving two outgoing electrons detected in coincidence, such as Auger PhotoElectron Coincidence Spectroscopy (APECS) or $(e, 2e)$ spectroscopy (see Chap. 9), we need to go to second order which is Fermi's golden rule No 1 [41].

The flux and the density of states for photons and electrons are given in Table 1.1 [17].

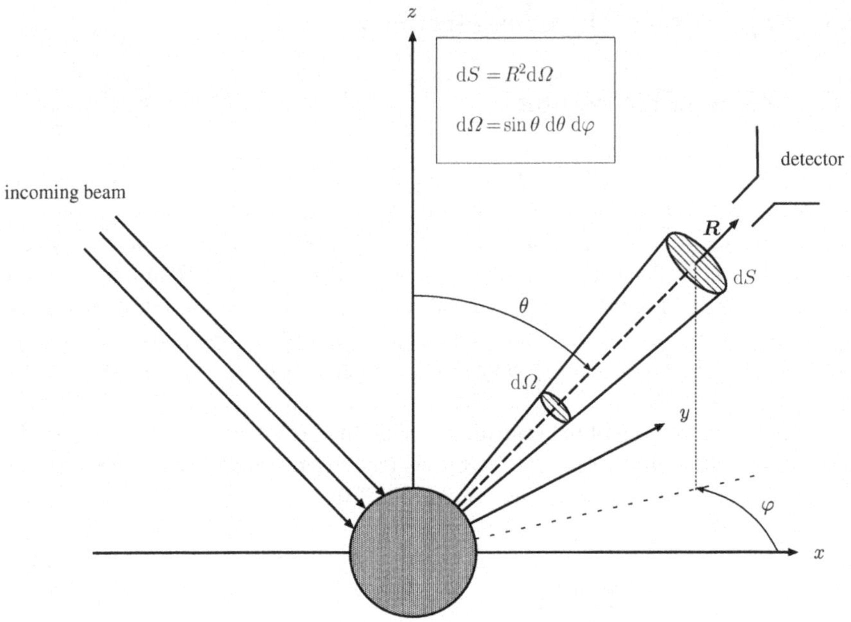

Fig. 1.11 Geometry for the expression of the cross-section in the case of a single outgoing electron

Table 1.1 Values of the flux and density of states for photons and electrons [17]

Particle	Flux	Density of states
Photons	$\frac{c}{\mathcal{V}^a}$	$\frac{\mathcal{V}}{(2\pi)^3}\frac{(\hbar\omega_q)^2}{(\hbar c)^3}$
Electrons	$\frac{1}{\mathcal{V}}\frac{\hbar}{m}k_i$	$\frac{\mathcal{V}}{(2\pi)^3}\frac{m}{\hbar^2}k_f$

$^a\mathcal{V}$ is the box normalization volume, normally taken as $(2\pi)^3$

1.5.2 *Cross-Section for Some Spectroscopies*

We will not make any demonstration here, but just give the resulting expression. We refer to the literature for the exact derivation of these cross-sections.

1.5.2.1 Photoelectron Diffraction (PED)

This is an angular-resolved photoemission experiment where a core level peak is generally monitored as a function of the escape direction. In this case, the cross-section writes, within the dipole approximation for the photon-electron interaction, as [19]

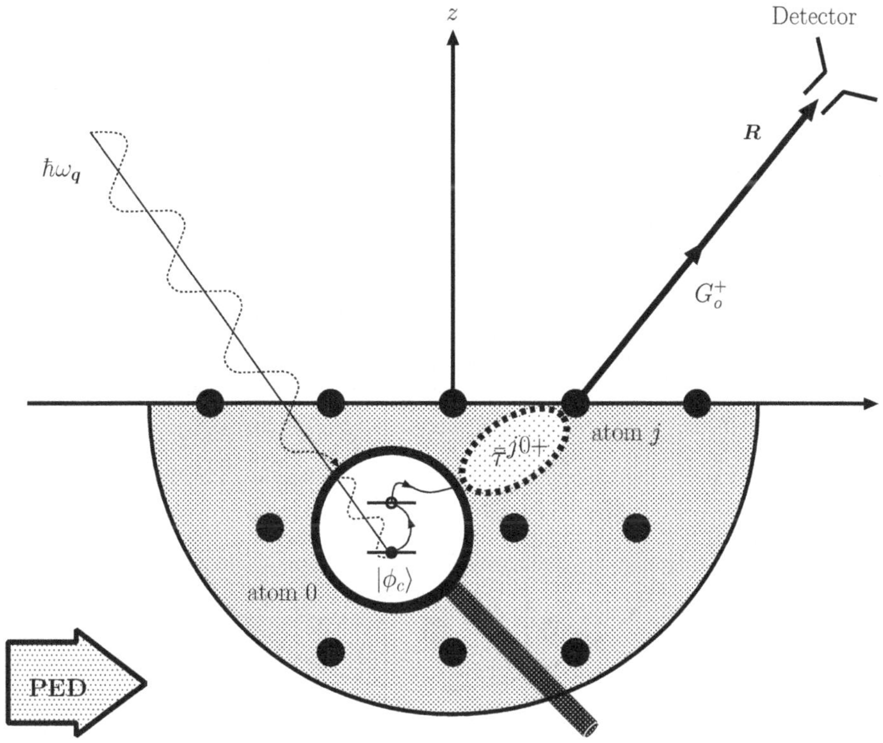

Fig. 1.12 Pictorial representation of the photoelectron diffraction process

$$\frac{\mathrm{d}\sigma}{\mathrm{d}\hat{\boldsymbol{k}}} = 4\pi^2\alpha k\frac{m\omega_{\boldsymbol{q}}}{\hbar}\sum_0\sum_{m_c}\left|\langle\boldsymbol{k}|\sum_j \bar{\tau}^{j0}\tilde{\bar{\Omega}}_0^{(-)\dagger}\overline{(\hat{\boldsymbol{e}}_{\boldsymbol{q}}\cdot\boldsymbol{r})}_0\mathcal{T}(\boldsymbol{R}_0)|\phi_c\rangle\right|^2 . \tag{1.94}$$

α is the fine structure constant, $\hat{\boldsymbol{e}}_{\boldsymbol{q}}$ the polarization of the incoming photon beam of energy $\hbar\omega_{\boldsymbol{q}}$, and the sums are over all the absorbing atoms 0 and the azimuthal quantum number m_c of the core state $|\phi_c\rangle$. Here, we have replaced the $\bar{\mathfrak{T}}_0^{-\dagger}$ in (1.74) by its equivalent form as a sum over the scattering path operators. $\tilde{\bar{\Omega}}_0^{(-)\dagger}$ is the so-called *renormalized Møller wave operator* [19] that takes cares of the matching between the excitation part and the multiple scattering part. $\overline{(\hat{\boldsymbol{e}}_{\boldsymbol{q}}\cdot\boldsymbol{r})}_0$ is the dipole operator with respect to atom 0 (see (1.61) for its expression in terms of the standard dipole operator and translation operators). A pictorial representation of the photoelectron diffraction process with the different operators involved in the description of this process is shown in Fig. 1.12.

A workable expression is obtained by inserting the closure relation of the spherical wave basis whenever necessary. We find [19]

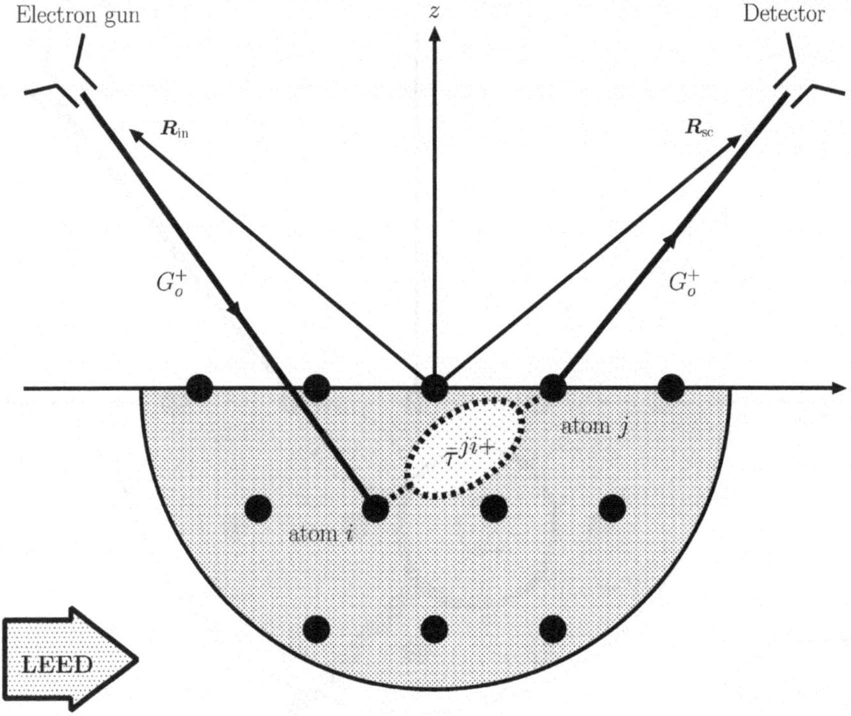

Fig. 1.13 Pictorial representation of the LEED process

$$\frac{d\sigma}{d\hat{\boldsymbol{k}}} = 8\pi^2\alpha\frac{m\omega_{\boldsymbol{q}}}{\hbar k}\sum_{0}\sum_{m_c}\left|\sum_{L_0} M^{00}_{L_0 L_c}\sum_{j,L_J}\bar{\tau}^{j0+}_{L_j L_0}\,Y_{L_j}(\hat{\boldsymbol{k}})\,\mathrm{e}^{-\mathrm{i}\boldsymbol{k}\cdot\boldsymbol{R}_j}\right|^2 , \tag{1.95}$$

where $M^{00}_{L_0 L_c}$ is the matrix element describing the excitation of the photoelectron.

1.5.2.2 Low-Energy Electron Diffraction (LEED)

The LEED cross-section is probably the simplest possible cross-section. A sketch of the experiment together with the operators involved is given Fig. 1.13. This cross-section is simply

$$\frac{d\sigma}{d\hat{\boldsymbol{k}}_{\text{sc}}} = 4\pi^4\left|\langle\boldsymbol{k}_{\text{sc}}|\sum_{i,j}\bar{\tau}^{ji+}|\boldsymbol{k}_{\text{in}}\rangle\right|^2 . \tag{1.96}$$

In a spherical wave representation, it gives [17]

$$\frac{\mathrm{d}\sigma}{\mathrm{d}\hat{\boldsymbol{k}}_{\mathrm{sc}}} = \frac{4\pi^4}{k^4} \left| \sum_{i,L_i} \sum_{j,L_j} \mathrm{e}^{-\mathrm{i}\boldsymbol{k}_{\mathrm{sc}}\cdot\boldsymbol{R}_j}\, Y_{L_j}(\hat{\boldsymbol{k}}_{\mathrm{sc}})\, \bar{\tau}^{ji+}_{L_j L_i}\, \mathrm{e}^{\mathrm{i}\boldsymbol{k}_{\mathrm{in}}\cdot\boldsymbol{R}_i}\, Y^*_{L_i}(\hat{\boldsymbol{k}}_{\mathrm{in}}) \right|^2 . \tag{1.97}$$

References

1. T. Ikebe, Arch. Rat. Mech. Anal. **5**(1) (1960); T. Ikebe, Pacific. J. Math. **15**(511) (1965)
2. A. Bohm, *Quantum Mechanics: Foundations and Applications*, 3rd edn. (Springer, Berlin, 1993), p. 433
3. C. Cohen-Tannoudji, B. Diu, F. Laloë, *Quantum Mechanics* (Wiley-VCH, New York, 1977), vol. 2, p. 910
4. L.S. Rodberg, R.M. Thaler, *Introduction to the Quantum Theory of Scattering* (Academic Press, New York, 1967), p. 6
5. C. Cohen-Tannoudji, B. Diu, F. Laloë, *Quantum Mechanics* (Wiley-VCH, New York, 1977), vol. 1, p. 662
6. M. Abramowitz, I.A. Stegun, *Handbook of Mathematical Functions* (Dover, New York, 1965), p. 437
7. C. Cohen-Tannoudji, B. Diu, F. Laloë, *Quantum Mechanics* (Wiley-VCH, New York, 1977), vol. 1, pp. 930–931
8. C.J. Joachain, *Quantum Collision Theory* (North-Holland, Amsterdam, 1983), p. 67
9. M.-L. Xu, J.J. Barton, M.A. Van Hove, Phys. Rev. B **39**, 8275 (1989)
10. C.J. Joachain, *Quantum Collision Theory* (North-Holland, Amsterdam, 1983), p. 79
11. C. Cohen-Tannoudji, B. Diu, F. Laloë, *Quantum Mechanics* (Wiley-VCH, New York, 1977), vol. 2, p. 917
12. L.S. Rodberg, R.M. Thaler, *Introduction to the Quantum Theory of Scattering* (Academic Press, New York, 1967), p. 114
13. W.E. Boyce, R.C. DiPrima, *Elementary Differential Equations and Boundary Value Problems*, 7th edn. (Wiley, New York, 2001), pp. 211–212
14. R.G. Newton, *Scattering Theory of Waves and Particles*, 2nd edn. (Springer, Heidelberg, 2008), p. 182
15. R.G. Newton, *Scattering Theory of Waves and Particles*, 2nd edn. (Springer, Heidelberg, 2008), p. 177
16. L.S. Rodberg, R.M. Thaler, *Introduction to the Quantum Theory of Scattering* (Academic Press, New York, 1967), p. 179
17. D. Sébilleau, R. Gunnella, Z.-Y. Wu, S. Di Matteo, C.R. Natoli, Topical review. J. Phys. Condens. Matter **18**, R175–R230 (2006)
18. L.S. Rodberg, R.M. Thaler, *Introduction to the Quantum Theory of Scattering* (Academic Press, New York, 1967), p. 184
19. D. Sébilleau, Phys. Rev. B **61**, 14167 (2000)
20. R.G. Newton, *Scattering Theory of Waves and Particles*, 2nd edn. (Springer, Heidelberg, 2008), p. 155
21. J.R. Taylor, *Scattering Theory* (Wiley, New York, 1972), pp. 28–35
22. H.A. Bethe, R. Jackiw, *Intermediate Quantum Mechanics*, 3rd edn. (Menlo Park, Benjamin/Cummings, 1986), p. 310
23. M. Danos, L.C. Maximon, J. Math. Phys. **6**, 766 (1965)
24. K. Schwarz, in *Selected Topics in Applications of Quantum Mechanics*, Intech, 2015, ed. by Mohammad Reza Pahlavani, pp. 275–310
25. P.C. Tandy, R.M. Thaler, Phys. Rev. C **22**, 2321 (1980)

26. L.D. Faddeev, in *Mathematical Aspects of the Three-body Problem in the Quantum Scattering Theory*, Israel Program for Scientific Translation, Jerusalem, 1965, p. 12
27. B.L. Györffy, in *Fondamenti di Fisica dello Stato Solido*, (Scuola Nazionale di Struttura della Materia, 1971); B.L. Györffy, M.J. Stott, in *Band Structure Spectroscopy of Metals and Alloys*, Academic Press, London, 1973, pp. 385–403 ed. by D.J. Fabian, L.M. Watson
28. D. Sébilleau, C.R. Natoli, M. Gavaza, H. Zhao, F. Da Pieve, K. Hatada, Comput. Phys. Commun. **182**, 2567–2579 (2011). for the first version. The latest version of the MsSpec package can be obtained from the author of the chapter
29. J.S. Faulkner, G.M. Stocks, Phys. Rev. B **8**, 3222 (1980)
30. T.A. Tyson, K.O. Hodgson, C.R. Natoli, M. Benfatto, Phys. Rev. B **46**, 5997 (1992)
31. M.A. Morrison, A.N. Feldt, Am. J. Phys. **75**, 67 (2007)
32. K.M. Watson, Phys. Rev. **89**, 575 (1953)
33. C.R. Natoli, M. Benfatto, J. Phys. (France) C **8**, 11 (1986)
34. D. Sébilleau, C.R. Natoli, J. Phys. Conf. Ser. **190**, 012002 (2009)
35. H.-F. Zhao, D. Sébilleau, Z.-Y. Wu, J. Phys. Condens. Matter **20**, 275241 (2008)
36. A. Filipponi, A. Di Cicco, C.R. Natoli, Phys. Rev. B **52**, 15122 (1995)
37. D. Sébilleau, J. Phys. A: Math. Gen. **31**, 7157–7168 (1998)
38. D.S. MacMillan, E.F. Redish, Phys. Rev. C **33**, 804 (1986)
39. L.S. Rodberg, R.M. Thaler, *Introduction to the Quantum Theory of Scattering* (Academic Press, New York, 1967), p. 196
40. L.I. Schiff, *Quantum Mechanics* (McGraw-Hill/Kogakusha, 1968), p. 314
41. L.I. Schiff, *Quantum Mechanics* (McGraw-Hill/Kogakusha, 1968), p. 336

Chapter 2
Generating Phase-Shifts and Radial Integrals for Multiple Scattering Codes

Calogero R. Natoli and Didier Sébilleau

Abstract A brief derivation of the cross section in the independent particle approximation for some of the spectroscopies treated in the MSSPEC program package is presented . We solve the related Schrödinger equation with a complex energy-dependent effective potential in the framework of multiple scattering theory to write the cross-section for photoemission and photoabsorption in a physically transparent way that provides insight in their interpretation and analysis. Relativistic corrections are also implemented. In order to be able to apply this theory to a wide variety of systems we use a kind of *all-purpose* optical potential, depending only on the local density of the system under investigation, and discuss its merits and drawbacks. A Green's function approach is shown to be necessary to write the photoabsorption cross section in the case of complex potential.

2.1 Introduction

The purpose of this chapter is to provide an introductory description of the steps necessary to interpret and analyze the most common x-ray spectroscopies, notably core-level photoemission, photoabsorption and photon scattering. Actually the same scattering solution of the Schrödinger equation (SE) can also be used for other spectroscopies, like LEED (Low Energy Electron Diffraction) and EELS (Electron Energy Loss Spectroscopy), where the incoming probe is a beam of electrons, so that their interpretation bears a close resemblance to the photon spectroscopies and can be described in the same terms. The unifying framework is provided by multiple scattering theory (MST) which we use to solve the Lippmann-Schwinger equation (LSE) associated to the SE. This derivation generalizes that given by one of us (DS)

C. Natoli (✉)
LNF-INFN, Via E. Fermi 40, 00044 Frascati, Italy
e-mail: calogero.natoli@lnf.infn.it

R. Sébilleau (✉)
Institut de Physique de Rennes (IPR), Univ Rennes, CNRS, UMR 6251,
F-35000 Rennes, France
e-mail: didier.sebilleau@univ-rennes1.fr

D. Sébilleau et al. (eds.), *Multiple Scattering Theory for Spectroscopies*,
Springer Proceedings in Physics 204,
https://doi.org/10.1007/978-3-319-73811-6_2

in Chap. 1 in terms of scattering path and projection operators, in that the local scattering potentials are not spherically symmetric. We use this solution to write down the photoemission and photoabsorption cross sections, illustrating their physical meaning in the framework of MST. Having in mind that the potential in the effective SE can be complex we reformulate the photoabsorption cross section in terms of the Green's function (GF) of the system. For real potential we recover the expression for the cross section written in terms of the scattering wave function, solution of the SE, while for complex potential this is no longer valid and we need to use the expression involving the imaginary part of the GF. In this reformulation we take advantage of the fact that MST provides an explicit expression for the GF that is expedient to the analysis of the photoabsorption spectroscopy. As a particular application we illustrate how in the case of complex potential the concept of mean free path arises in the general term of the MS series. We show how to construct, in the muffin-tin (MT) approximation, an *all-purpose* effective potential, sum of a Coulomb part and a (complex) local density exchange-correlation part easily applicable to a wide variety of systems. Finally we describe how to introduce relativistic corrections in the solution of the radial SE inside the MT atomic spheres to generate corresponding atomic t-matrices and atomic cross sections, which are the main ingredients, together with the KKR structure factors, of any MST. They serve as an input to various MS programs like MsSpec, MXAN, GNXAS.

2.2 Derivation of the Cross-Section for Various Spectroscopies

For the benefit of the reader, we derive here the expression of the cross-section for various spectroscopies dealt with in the program package MsSpec [1], restricting ourselves to those processes where the impinging and/or the detected particles are either electrons or photons. This derivation partly overlaps and partly complements the one given by one of us (DS) in Chap. 1. For more details the reader is referred to [2].

The total Hamiltonian of the particle impinging on the matter system is given by

$$H = H_P + H_M + V_I = H_0 + V_I \ , \tag{2.1}$$

where H_P describes the kinetic energy of the projectile, H_M the matter system and V_I their interaction. In terms of the eigenstates $|\Phi\rangle$ of the *free* Hamiltonian H_0, scattering theory [3] tells us that the transition probability per unit time between an initial state Φ_i with energy E_i and a final state Φ_f with energy E_f is given by

$$W_{i\to f} = \frac{2\pi}{\hbar} \left|\langle\Phi_f|T_I(E_f)|\Phi_i\rangle\right|^2 \varrho(E_f) \ , \tag{2.2}$$

where $E_i = E_f$ for a real interaction, $\varrho(E_f)$ is the density of the final states (in general determined by the detected particle) and $T_I(E)$ is the transition operator associated with the interaction potential V_I.

The equation defining T_I is

$$T_I(E) = V_I + V_I\, G(E)\, V_I \ , \tag{2.3}$$

where $G(E)$ is the Green's function of the total Hamiltonian H, given by [3]

$$G^+(E) \ = \ \lim_{\varepsilon \to 0^+} \frac{1}{E - H + \mathrm{i}\varepsilon} \ . \tag{2.4}$$

Even though (2.3) is difficult to solve, its advantage lies in the fact that, when the interaction potential V_I is *small* (in a sense that can be rigorously defined), one can expand T_I in series of the perturbation V_I, so that to the lowest order

$$T_I^{(1)}(E) \approx V_I \ , \tag{2.5}$$

whereas to the second order

$$T_I^{(2)}(E) \approx V_I + V_I\, G_0(E)\, V_I \ , \tag{2.6}$$

where $G_0(E)$ is obtained from (2.4) by replacing H with H_0.

The cross section is obtained by dividing the transition probability per unit time by the incoming flux of the incident particle

$$I_o = \frac{v}{V} \ , \tag{2.7}$$

where v is the velocity of the projectile and V the quantization volume.

2.2.1 *Cross Section for Incoming Photons*

In this case the interaction potential V_I is given by

$$V_I \ = \sum_i \left(\frac{e}{m} \boldsymbol{A}(\boldsymbol{r}_i) \cdot \boldsymbol{p}_i + \frac{e^2}{2m}\, \boldsymbol{A}^2(\boldsymbol{r}_i) \right) \ , \tag{2.8}$$

where, in second quantization,

$$\boldsymbol{A}(\boldsymbol{r}) \ = \sum_{\boldsymbol{q},\hat{\boldsymbol{e}}_q} \sqrt{\frac{\hbar}{2\varepsilon_0 \omega_{\boldsymbol{q}} V}} \left[a_{\boldsymbol{q},\hat{\boldsymbol{e}}_q} \mathrm{e}^{\mathrm{i}\boldsymbol{q}\cdot\boldsymbol{r}} \hat{\boldsymbol{e}}_{\boldsymbol{q}} + a^{\dagger}_{\boldsymbol{q},\hat{\boldsymbol{e}}_q} \mathrm{e}^{-\mathrm{i}\boldsymbol{q}\cdot\boldsymbol{r}} \hat{\boldsymbol{e}}^{*}_{\boldsymbol{q}} \right] \ , \tag{2.9}$$

is the vector potential of the incoming radiation and $\boldsymbol{p}$ is the momentum of the electron in the matter system, with customary meaning of the other symbols [2]. In the following we shall omit the sum over all the electrons, since we shall work in the independent particle approximation. How to reduce the many-body problem to an effective one-particle problem is outside the scope of this article and the reader is referred to [2].

2.2.1.1 Photoabsorption and Photoemission Cross Section

In the process of photoemission the incoming photon is absorbed and an electron of the matter system is ejected into the continuum and detected along a certain direction $\hat{\boldsymbol{k}}$. By integrating over all the emission angles one counts the number of electron holes created, obtaining the photoabsorption cross section. We shall henceforth limit ourselves to photoemission from atomic core states.

In both cases, since the incoming photon is absorbed, we need T_I only to first order in V_I and consider only the destruction part of the photon operator in (2.9), so that

$$T_I^{\mathrm{ph}} = \frac{e}{m} \sum_{\boldsymbol{q},\hat{\boldsymbol{e}}_q} \sqrt{\frac{\hbar}{2\varepsilon_0 \omega_{\boldsymbol{q}} V}}\, (\hat{\boldsymbol{e}}_q \cdot \boldsymbol{p})\, a_{\boldsymbol{q},\hat{\boldsymbol{e}}_q}\, \mathrm{e}^{\mathrm{i}\,\boldsymbol{q}\cdot\boldsymbol{r}} \,. \tag{2.10}$$

The cross-section $\mathrm{d}\sigma/\mathrm{d}\hat{\boldsymbol{k}}$ for emission of an electron along the direction $\hat{\boldsymbol{k}}$ is given by

$$\frac{\mathrm{d}\sigma}{\mathrm{d}\hat{\mathbf{k}}}\, I_o = \frac{\mathrm{d}W_{i\to f}}{\mathrm{d}\hat{\mathbf{k}}} \,, \tag{2.11}$$

where $I_o = c/V$ is the incident photon flux and $\mathrm{d}W_{i\to f}/\mathrm{d}\hat{\boldsymbol{k}}$ is the transition probability per unit time and solid angle given by (2.2), in which $\varrho(E_f)$ is now the density of free final states for an electron ejected along $\hat{\boldsymbol{k}}$ at the energy E_f. This latter is given by the Einstein relation $E_f = \hbar\omega_{\boldsymbol{q}} - I_c$, where I_c is the core ionization potential. Moreover, $|\Phi_i\rangle$ is the system initial state, the tensor product of a photon state $a^{\dagger}_{\boldsymbol{q},\hat{\boldsymbol{e}}_q}|0\rangle$ times a matter electronic ground state $|\phi^e_g\rangle$, whereas $|\Phi_f\rangle$ is the product of the photon vacuum state $|0\rangle$ times an electronic final state $|\psi^-_{\hat{\boldsymbol{k}}}\rangle$, the time-reversed solution of the Schrödinger equation with scattering boundary conditions at infinity [2, 4] and plane-wave normalization.

Inserting (2.10) into (2.2) we find

$$\frac{\mathrm{d}\sigma}{\mathrm{d}\hat{\boldsymbol{k}}} = 4\pi^2\alpha\, \frac{\hbar}{m^2\,\omega_{\boldsymbol{q}}}\, \varrho(E_f)\, \left| \langle \psi^-_{\hat{\boldsymbol{k}}} | \hat{\boldsymbol{e}}_q \cdot \boldsymbol{p}\, \mathrm{e}^{\mathrm{i}\,\boldsymbol{q}\cdot\boldsymbol{r}} | \phi_i \rangle \right|^2 \,, \tag{2.12}$$

where $\alpha = e^2/(4\pi\varepsilon_0\hbar c)$ is the fine structure constant, a dimensionless coefficient whose numerical value is $\approx 1/137$.

When in the integration volume of the core state one has $qr \ll 1$, we can neglect the spatial dispersion of the exponential factor in this expression and use the well known relation

$$\langle \phi_f | \hat{\boldsymbol{e}}_q \cdot \boldsymbol{p} | \phi_i \rangle = \mathrm{i}m\,\omega_q\,\langle \phi_f | \hat{\boldsymbol{e}}_q \cdot \boldsymbol{r} | \phi_i \rangle \,, \tag{2.13}$$

since $E_f - E_i = \hbar\omega_q$. This is the so called *length form* of the transition matrix element [5], valid when $|\phi_{i,f}\rangle$ are eigenstates of the *same* Hamiltonian. Therefore (2.12) can be written as

$$\frac{\mathrm{d}\sigma}{\mathrm{d}\hat{\boldsymbol{k}}} = 4\pi^2\alpha\,\hbar\omega_q\,\varrho(E_f)\,\left|\langle \psi^-_{\hat{\boldsymbol{k}}} | \hat{\boldsymbol{e}}_q \cdot \boldsymbol{r} | \phi_i \rangle\right|^2 . \tag{2.14}$$

The density of free photoelectron final states per unit energy and solid angle is easily found to be (see Table 1.1 of Chap. 1)

$$\varrho(E_f) = \frac{V}{(2\pi)^3}\,\frac{\mathrm{d}^3\boldsymbol{k}}{\mathrm{d}E_f\,\mathrm{d}\hat{\boldsymbol{k}}} = \frac{V}{(2\pi\hbar)^3}\,\hbar k\, m\,. \tag{2.15}$$

It is customary to incorporate this factor into the normalization of the photoelectron wavefunction passing from a plane wave normalization $|\psi^-_{\hat{\boldsymbol{k}}}\rangle$ to a normalization to one state per unit energy interval $|\phi^-_{\hat{\boldsymbol{k}}}\rangle$. Using atomic units for lengths and Rydberg units of energy ($2m/\hbar^2 \to 1$) we have

$$|\phi^-_{\hat{\boldsymbol{k}}}\rangle = \frac{1}{4\pi}\sqrt{\frac{k}{\pi}}\,|\psi^-_{\hat{\boldsymbol{k}}}\rangle\,. \tag{2.16}$$

The absorption cross section is found by integrating (2.14) over all emission angles at the same final photoelectron energy

$$\sigma_{\mathrm{abs}}(\omega) = \int \frac{\mathrm{d}\sigma}{\mathrm{d}\hat{\boldsymbol{k}}}\,\mathrm{d}\hat{\boldsymbol{k}} = 4\pi^2\alpha\,\hbar\omega_q \int \mathrm{d}\hat{\boldsymbol{k}}\,\left|\langle \phi^-_{\hat{\boldsymbol{k}}} | \hat{\boldsymbol{e}}_q \cdot \boldsymbol{r} | \phi_i \rangle\right|^2 . \tag{2.17}$$

2.2.1.2 Photon Scattering Cross Section

Photon scattering is a second order process in which a photon is absorbed and another emitted. Typical processes are Thomson and resonant scattering. The transition operator should be of second order in the vector potential (2.9), so that from (2.6) we have

$$T_I^{(2)} = \left(\frac{e}{m}\right)^2\left[\frac{m}{2}\boldsymbol{A}\cdot\boldsymbol{A} + \boldsymbol{A}\cdot\boldsymbol{p}\,G_0(E_f)\,\boldsymbol{A}\cdot\boldsymbol{p}\right] + O\left(\frac{e^3}{m^3}\right)\,, \tag{2.18}$$

dropping the first order term $e/m\ \boldsymbol{A}\cdot\boldsymbol{p}$. The first term $\boldsymbol{A}\cdot\boldsymbol{A}$ gives rise to Thomson scattering (a) whereby the matrix element is

$$\langle \Phi_f | T^{(2)}_{I\,(a)} | \Phi_i \rangle = \frac{e^2}{2m} \frac{\hbar}{2\varepsilon_0 V} \frac{1}{\sqrt{\omega_{\boldsymbol{q}_f} \omega_{\boldsymbol{q}_i}}} \langle \Phi_f | (\hat{\boldsymbol{e}}^*_{\boldsymbol{q}_f} \cdot \hat{\boldsymbol{e}}_{\boldsymbol{q}_i}) \mathrm{e}^{\mathrm{i}(\boldsymbol{q}_i - \boldsymbol{q}_f)\cdot \boldsymbol{r}}$$

$$\times \left[a_{\boldsymbol{q}_i \hat{\boldsymbol{e}}_{\boldsymbol{q}_i}} a^\dagger_{\boldsymbol{q}_f \hat{\boldsymbol{e}}_{\boldsymbol{q}_f}} + a^\dagger_{\boldsymbol{q}_f \hat{\boldsymbol{e}}_{\boldsymbol{q}_f}} a_{\boldsymbol{q}_i \hat{\boldsymbol{e}}_{\boldsymbol{q}_i}} \right] | \Phi_i \rangle \,.$$

The second term gives rise to resonant (b) and non resonant scattering (c), according to whether the incoming photon is absorbed before the scattered photon is emitted or vice-versa. The relative matrix elements are

$$\begin{cases} \langle \Phi_f | T^{(2)}_{I\,(b)} | \Phi_i \rangle = N \displaystyle\sum_n^{E_n > E_F} \frac{\langle \phi_f | \hat{\boldsymbol{e}}^*_{\boldsymbol{q}_f} \cdot \boldsymbol{p}\, \mathrm{e}^{-\mathrm{i}\boldsymbol{q}_f \cdot \boldsymbol{r}} | \phi_n \rangle \langle \phi_n | \hat{\boldsymbol{e}}_{\boldsymbol{q}_i} \cdot \boldsymbol{p}\, \mathrm{e}^{\mathrm{i}\boldsymbol{q}_i \cdot \boldsymbol{r}} | \phi_i \rangle}{E_i - E_n + \hbar\omega_{\boldsymbol{q}_i} + \mathrm{i}\varepsilon} \\ \langle \Phi_f | T^{(2)}_{I\,(c)} | \Phi_i \rangle = N \displaystyle\sum_n^{E_n > E_F} \frac{\langle \phi_f | \hat{\boldsymbol{e}}_{\boldsymbol{q}_i} \cdot \boldsymbol{p}\, \mathrm{e}^{\mathrm{i}\boldsymbol{q}_i \cdot \boldsymbol{r}} | \phi_n \rangle \langle \phi_n | \hat{\boldsymbol{e}}^*_{\boldsymbol{q}_f} \cdot \boldsymbol{p}\, \mathrm{e}^{-\mathrm{i}\boldsymbol{q}_f \cdot \boldsymbol{r}} | \phi_i \rangle}{E_i - E_n - \hbar\omega_{\boldsymbol{q}_f} + \mathrm{i}\varepsilon} \\ N = \left(\dfrac{e}{m}\right)^2 \dfrac{\hbar}{2\varepsilon_0 V} \dfrac{1}{\sqrt{\omega_{\boldsymbol{q}_f} \omega_{\boldsymbol{q}_i}}} \,. \end{cases}$$

The corresponding scattering cross section is found by dividing these matrix elements by the incoming photon flux c/V and multiplying by the photon final density of states

$$\varrho^{\mathrm{ph}}(\hbar\omega_{\boldsymbol{q}_f}) = \frac{V}{(2\pi)^3} \frac{(\hbar\omega_{\boldsymbol{q}_f})^2}{\hbar^3 c^3} \,.$$

We find, putting $\boldsymbol{Q} = \boldsymbol{q}_f - \boldsymbol{q}_i$,

$$\frac{\mathrm{d}\sigma}{\mathrm{d}\Omega} = r_0^2 \frac{\omega_{\boldsymbol{q}_f}}{\omega_{\boldsymbol{q}_i}} \Bigg| \hat{\boldsymbol{e}}^*_{\boldsymbol{q}_f} \cdot \hat{\boldsymbol{e}}_{\boldsymbol{q}_i} \langle \phi_f | \mathrm{e}^{-\mathrm{i}\boldsymbol{Q}\cdot \boldsymbol{r}} | \phi_i \rangle$$

$$\frac{1}{m} \sum_n^{E_n > E_F} \left[\frac{\langle \phi_f | \hat{\boldsymbol{e}}^*_{\boldsymbol{q}_f} \cdot \boldsymbol{p}\, \mathrm{e}^{-\mathrm{i}\boldsymbol{q}_f \cdot \boldsymbol{r}} | \phi_n \rangle \langle \phi_n | \hat{\boldsymbol{e}}_{\boldsymbol{q}_i} \cdot \boldsymbol{p}\, \mathrm{e}^{\mathrm{i}\boldsymbol{q}_i \cdot \boldsymbol{r}} | \phi_i \rangle}{E_i - E_n + \hbar\omega_{\boldsymbol{q}_i} + \mathrm{i}\varepsilon} \right.$$

$$\left. + \frac{\langle \phi_f | \hat{\boldsymbol{e}}_{\boldsymbol{q}_i} \cdot \boldsymbol{p}\, \mathrm{e}^{\mathrm{i}\boldsymbol{q}_i \cdot \boldsymbol{r}} | \phi_n \rangle \langle \phi_n | \hat{\boldsymbol{e}}^*_{\boldsymbol{q}_f} \cdot \boldsymbol{p}\, \mathrm{e}^{-\mathrm{i}\boldsymbol{q}_f \cdot \boldsymbol{r}} | \phi_i \rangle}{E_i - E_n - \hbar\omega_{\boldsymbol{q}_f} + \mathrm{i}\varepsilon} \right] \Bigg|^2 \,,$$

where $r_0 = e^2/(4\pi\varepsilon_0 mc^2)$ is the classical electron radius.

In passing from the many-body matter states to the one-particle we have taken into account the Pauli principle, whereby transitions to already occupied states are forbidden ($E_n > E_F$, where E_F is the Fermi energy). An elegant way for performing the sum over the intermediate states in terms of the Green's function of the system is given in [6].

2.2.2 Cross Section for Incoming Electrons

When the incoming probe is a beam of electrons we have different kinds of spectroscopies. In LEED (Low Energy Electron Diffraction), the scattering is elastic (no energy-loss for the scattered electrons), whereas in EELS the detected electrons have a different energy from the incoming beam. We refer the reader to Chap. 1 for LEED and Chap. 24 for the derivation of the corresponding cross sections and discussions. Here we briefly sketch the derivation of the cross section for EELS in a way which is complementary to that used in Chap. 24 and is suitable for our purposes to illustrate the calculations of transition matrix elements.

The expression for the cross section is given by ([2], p. 214, (114))

$$\frac{d\sigma}{d\hat{\boldsymbol{k}}_f} = 4\pi \frac{k_f}{k_i} \sum_n \left| \langle \Phi_n \psi_f^- | V_c | \Phi_0 \psi_i^+ \rangle_d \pm \langle \psi_f^- \Phi_n | V_c | \Phi_0 \psi_i^+ \rangle_e \right|^2 , \tag{2.19}$$

where the plus (minus) sign applies if the two electrons are in a triplet (singlet) state. Here $V_c = \sum_{ij} 2/|r_i - r_j|$ is the Coulomb interaction, Φ_0 and Φ_n are the ground state and the excited state of the system, whereas ψ_i^+ is the scattering wave function of the incident electron, and ψ_f^- is the time reversal scattering wave function of the final (outgoing) scattered electron. If Φ_0 and Φ_n are Slater Determinants (SD), then we can take $\Phi_0 = \phi_c$ (the initial core hole state) and $\Phi_n = \phi_\epsilon$, where ϵ is the energy of the secondary excited electron given by $E_{\text{exc}} = E_{\text{inc}} - E_{\text{sct}} - I_c$, I_c being the ionization energy of the core state. In this case we have (in Ryd units)

$$\begin{aligned}
\langle \phi_\epsilon \psi_f^- | V_c | \Phi_0 \psi_i^+ \rangle_d &= \int d\boldsymbol{r} d\boldsymbol{r}' \phi_\epsilon(\boldsymbol{r}) [\psi_f^-(\boldsymbol{r}')]^* \frac{2}{|\boldsymbol{r} - \boldsymbol{r}'|} \phi_c(\boldsymbol{r}) \psi_i^+(\boldsymbol{r}') \\
\langle \psi_f^- \phi_\epsilon | V_c | \Phi_0 \psi_i^+ \rangle_e &= \int d\boldsymbol{r} d\boldsymbol{r}' \phi_\epsilon(\boldsymbol{r}') [\psi_f^-(\boldsymbol{r})]^* \frac{2}{|\boldsymbol{r} - \boldsymbol{r}'|} \phi_c(\boldsymbol{r}) \psi_i^+(\boldsymbol{r}') ,
\end{aligned} \tag{2.20}$$

for the direct and exchange matrix elements.

When the secondary (excited) electron is not observed (the usual case), one has to sum over all the emission directions of the intermediate state ϕ_ϵ so that, making use of the spectral representation of the Green's function for this electron, the EELS cross section can be written as

$$\begin{aligned}
\frac{d\sigma}{d\hat{\boldsymbol{k}}_f} = -4 \frac{k_f}{k_i} \int d\boldsymbol{r}_1 d\boldsymbol{r}_2 \int d\boldsymbol{r}'_1 d\boldsymbol{r}'_2 [\psi_i^+(\boldsymbol{r}'_2)]^* V'(|\boldsymbol{r}'_1 - \boldsymbol{r}'_2|) \psi_f^-(\boldsymbol{r}'_2) \times \\
\{\phi_c(\boldsymbol{r}'_1) \Im[G(\boldsymbol{r}'_1, \boldsymbol{r}_1; E_{\text{exc}})] \phi_c(\boldsymbol{r}_1)\} \times \\
\psi_i^+(\boldsymbol{r}_2) V'(|\boldsymbol{r}_1 - \boldsymbol{r}_2|) [\psi_f^-(\boldsymbol{r}_2)]^* ,
\end{aligned} \tag{2.21}$$

where $V'(|\boldsymbol{r}_1 - \boldsymbol{r}_2|) \equiv (I - P_{12}) V_c \equiv (I - P_{12}) 2/(|\boldsymbol{r}_1 - \boldsymbol{r}_2|)$ and P_{12} is the permutation of the elements 1 and 2. Similarly for the primed variables. In keeping with the definitions in (2.20), the permutation operator acts only on the indices of the same type (primed or unprimed) of the GF and the final state ψ_f^-. We show in Sect. 2.4.4 that in (2.21) one can apply the operator $\Im$ in front of the integral.

2.3 Multiple Scattering Theory

The method for solving the Schrödinger equation both for bound and continuum states, necessary to calculate the various spectroscopic response functions, is based on multiple scattering theory (MST). In its essence MST is a technique for solving a linear partial differential equation over a region of space with certain boundary conditions. It is implemented by dividing the space into non-overlapping domains Ω_j (cells), solving the differential equation separately in each of the cells and then assembling together the partial solutions into a global solution that is continuous and smooth across the whole region and satisfies the given boundary conditions.

After the reduction of the many-body problem to an effective one particle problem, the calculation of the photoemission (photoabsorption) cross section requires the solution of the effective SE

$$\left[\nabla^2 + E - V_{\text{eff}}(\boldsymbol{r};\, E)\right] \psi_{\boldsymbol{k}}(\boldsymbol{r}) = 0\,, \tag{2.22}$$

with incoming wave boundary conditions [4] (see also Sect. 1.3.5 of Chap. 1)

$$\psi_{\boldsymbol{k}}^-(\boldsymbol{r}) \simeq \mathrm{e}^{\mathrm{i}\boldsymbol{k}\cdot\boldsymbol{r}} + f^*(\hat{\boldsymbol{r}};\, -\boldsymbol{k})\frac{\mathrm{e}^{-\mathrm{i}kr}}{r}\,. \tag{2.23}$$

Here and in the following we shall use atomic units for lengths and Rydberg units for energies. The normalization to one state per Rydberg can be restored at the end by using the relation (2.16). The effective potential $V_{\text{eff}}(\boldsymbol{r};\, E)$ (henceforth written simply $V(\boldsymbol{r};\, E)$) might be complex and energy dependent. For convenience we shall work with outgoing waves and then apply the time reversal to calculate the photoemission cross section. Neglecting spin, this amount to take the complex conjugate of the wave function.

For the solution of (2.22) it is expedient to introduce the free GF $G_0^+(\boldsymbol{r}-\boldsymbol{r}')$ obeying the equation

$$\left[\nabla^2 + E\right] G_0^+(\boldsymbol{r}-\boldsymbol{r}') = \delta(\boldsymbol{r}-\boldsymbol{r}')\,, \tag{2.24}$$

so that (2.22) together with the complex conjugate of the boundary condition (2.23) is equivalent to the Lippmann-Schwinger equation

$$\psi_{\boldsymbol{k}}(\boldsymbol{r}) = \mathrm{e}^{\mathrm{i}\boldsymbol{k}\cdot\boldsymbol{r}} + \int G_0^+(\boldsymbol{r}-\boldsymbol{r}')\, V(\boldsymbol{r}';\, E)\, \psi_{\boldsymbol{k}}(\boldsymbol{r}')\, \mathrm{d}^3 r'\,. \tag{2.25}$$

Before proceeding with the solution of (2.25) we need to define the scattering matrix of the potential $V(\boldsymbol{r})$ in response to an incident spherical wave $J_L(\boldsymbol{r}; k) = j_\ell(kr)Y_L(\hat{\boldsymbol{r}})$, where $j_\ell(kr)$ is the spherical Bessel function and for short we put $L \equiv \ell, m$. In the following we shall use *real* spherical harmonics $Y_L(\hat{\boldsymbol{r}})$.

To this purpose we remember the expansions already introduced in Chap. 1

$$
\begin{aligned}
\mathrm{e}^{\mathrm{i}\boldsymbol{k}\cdot\boldsymbol{r}} &= 4\pi \sum_L \mathrm{i}^\ell Y_L(\hat{\boldsymbol{k}}) J_L(\boldsymbol{r}; k) = \sum_L A_L(\hat{\boldsymbol{k}}) J_L(\boldsymbol{r}; k) \\
G_0^+(\boldsymbol{r} - \boldsymbol{r}'; E) &= \sum_L J_L(\boldsymbol{r}; k) \tilde{H}_L^+(\boldsymbol{r}'; k) \; (r < r') \\
&= \sum_L J_L(\boldsymbol{r}'; k) \tilde{H}_L^+(\boldsymbol{r}; k) \; (r > r') \,, \qquad (2.26)
\end{aligned}
$$

where $\tilde{H}_L^+(\boldsymbol{r}'; k) = -\mathrm{i}\, h_\ell^+(kr) Y_L(\hat{\boldsymbol{r}})$, $h_\ell^+(kr)$ being the Hankel function of the first kind (outgoing wave behavior at great distances), following the definition of [3] (Chap. 2, (2.20) and footnote). It can also be referred as $h_\ell^{(1)}(kr)$.

Introducing the partial wave components of $\psi_{\mathbf{k}}(r)$ through the relation

$$
\psi_{\boldsymbol{k}}(\boldsymbol{r}) \equiv \psi(\boldsymbol{r}; \boldsymbol{k}) = \sum_L A_L(\hat{\boldsymbol{k}}) \psi_L(\boldsymbol{r}; \boldsymbol{k}) \,, \qquad (2.27)
$$

we find from (2.25)

$$
\psi_L(\boldsymbol{r}; \boldsymbol{k}) = J_L(\boldsymbol{r}; k) + \int G_0^+(\boldsymbol{r} - \boldsymbol{r}')\, V(\boldsymbol{r}'; E)\, \psi_L(\boldsymbol{r}'; \boldsymbol{k})\, \mathrm{d}^3 r' \,. \qquad (2.28)
$$

Using the third of (2.26), the asymptotic behavior of $\psi_L(r; \mathbf{k})$ at great distances from the scattering center is found to be

$$
\begin{aligned}
\psi_L(\boldsymbol{r}; \boldsymbol{k}) &= J_L(\boldsymbol{r}; k) + \sum_{L'} \tilde{H}_{L'}^+(\boldsymbol{r}; k) \int J_{L'}(\boldsymbol{r}'; k)\, V(\boldsymbol{r}')\, \psi_L(\boldsymbol{r}'; \boldsymbol{k})\, \mathrm{d}^3 r' \\
&= J_L(\boldsymbol{r}; k) + \sum_{L'} \tilde{H}_{L'}^+(\boldsymbol{r}; k)\, T_{L'L} \,, \qquad (2.29)
\end{aligned}
$$

putting

$$
T_{L'L} = \int J_{L'}(\boldsymbol{r}'; k)\, V(\boldsymbol{r}')\, \psi_L(\boldsymbol{r}'; \boldsymbol{k})\, \mathrm{d}^3 r' \,. \qquad (2.30)
$$

This quantity defines the amplitude of the wave function scattered into a spherical wave of angular momentum L' in response to an exciting partial wave of angular momentum L. This is the generalization of (1.16) in Chap. 1 to the non-MT case and coincides with it for radial potentials ($V(\boldsymbol{r}) \equiv V(r)$) except for a factor i (imaginary unit).

In order to solve (2.25) with the method of multiple scattering, we introduce a partition of the potential that follows that of the space, so that $V(\boldsymbol{r};\, E) = \sum_j v_j(\boldsymbol{r};\, E)$, where $v_j(\boldsymbol{r};\, E)$ coincides with $V(\boldsymbol{r};\, E)$ within the cell Ω_j and is zero outside. We then introduce local scattering solutions

$$\psi_L(\boldsymbol{r}_j;\, \boldsymbol{k}) = J_L(\boldsymbol{r}_j;\, k) + \int_{\Omega_j} G_0^+(\boldsymbol{r}_j - \boldsymbol{r}'_j;\, k)\, v_j(\boldsymbol{r}'_j)\, \psi_L(\boldsymbol{r}'_j;\, \boldsymbol{k})\, \mathrm{d}^3 r'_j \,. \tag{2.31}$$

These solutions are local in the sense that they are defined only inside the bounding sphere of cell Ω_j, which is centered at $\boldsymbol{R}_j$ so that $\boldsymbol{r}_j = \boldsymbol{r} - \boldsymbol{R}_j$.

Then the global solution of (2.25) can be expressed locally as

$$\psi(\boldsymbol{r}_j;\, \boldsymbol{k}) = \sum_L C_L^j(\boldsymbol{k}) \psi_L(\boldsymbol{r}_j;\, k) \,. \tag{2.32}$$

We now write (2.25) referred to a generic cell Ω_i. We find

$$\begin{aligned}\psi(\boldsymbol{r}_i;\, \boldsymbol{k}) &= \mathrm{e}^{\mathrm{i}\boldsymbol{k}\cdot\boldsymbol{r}_i}\, \mathrm{e}^{\mathrm{i}\boldsymbol{k}\cdot\boldsymbol{R}_i} \\ &+ \int_{\Omega_i} G_0^+(\boldsymbol{r}_i - \boldsymbol{r}'_i;\, k)\, v_i(\boldsymbol{r}'_i)\, \psi(\boldsymbol{r}'_i;\, \boldsymbol{k})\, \mathrm{d}^3 r'_i \\ &+ \sum_{j \neq i} \int_{\Omega_j} G_0^+(\boldsymbol{r}_j - \boldsymbol{r}'_j;\, k)\, v_j(\boldsymbol{r}'_j)\, \psi(\boldsymbol{r}'_j;\, \boldsymbol{k})\, \mathrm{d}^3 r'_j \,.\end{aligned}$$

Inserting here the relation (2.32) and remembering (2.31) we obtain an equation determining the unknown coefficients C_L^j

$$\begin{aligned}\sum_L C_L^i(\boldsymbol{k})\, J_L(\boldsymbol{r}_i;\, k) = \mathrm{e}^{\mathrm{i}\boldsymbol{k}\cdot\boldsymbol{r}_i}\, \mathrm{e}^{\mathrm{i}\boldsymbol{k}\cdot\boldsymbol{R}_i} + \sum_{j \neq i} \int_{\Omega_j} G_0^+(\boldsymbol{r}_i - \boldsymbol{r}'_j;\, k)\, v_j(\boldsymbol{r}'_j) \\ \times \sum_L C_L^j(\boldsymbol{k}) \psi_L(\boldsymbol{r}'_j;\, k)\, \mathrm{d}^3 r'_j \,.\end{aligned} \tag{2.33}$$

We can simplify this relation by introducing the two center expansion of the free GF $G_0^+(\boldsymbol{r} - \boldsymbol{r}';\, k)$ whereby

$$G_0^+(\boldsymbol{r} - \boldsymbol{r}';\, k) = \sum_{LL'} J_L(\boldsymbol{r}_i;\, k)\, G_{LL'}^{ij}\, J_L(\boldsymbol{r}_j;\, k) \quad (j \neq i) \,, \tag{2.34}$$

where $G_{LL'}^{ij}$ are the KKR structure factors given by

$$G_{LL'}^{ij} = 4\pi \sum_{L''} G(L, L';\, L'')\, \mathrm{i}^{l-l'+l''}\, \tilde{H}_{L''}^+(\boldsymbol{R}_{ij};\, k) \,, \tag{2.35}$$

in terms of the Gaunt coefficients $G(L, L'; L'') = \int Y_L(\Omega)Y_{L'}(\Omega)Y_{L''}(\Omega)\mathrm{d}\Omega$. The relation (2.34) is absolutely convergent only if $R_{ij} \equiv |\boldsymbol{R}_i - \boldsymbol{R}_j| > r_i + r_j$, a condition that might not be always satisfied for two contiguous cells Ω_i and Ω_j, when r_i and r_j are within the bounding spheres of the respective cells. We observe however that the relation (2.33) is valid whatever $\boldsymbol{r}_i$ in cell Ω_i and therefore also for $\boldsymbol{r}_i \to 0$. In this limit, exploiting the fact that the coefficients C^j_L do not depend on $\boldsymbol{r}_j$, we can safely use (2.34) and obtain for them the following simple equation

$$C^i_L(\boldsymbol{k}) = I^i_L(\boldsymbol{k}) + \sum_{j\neq i}\sum_{L'L''} G^{ij}_{LL'}\, T^j_{L'L''}\, C^j_{L''}(\boldsymbol{k})\ , \tag{2.36}$$

where we have defined the cell T-matrix $T^j_{L'L} = \int J_{L'}(\boldsymbol{r}'; k)\, v_j(\boldsymbol{r}')\, \psi_L(\boldsymbol{r}'; k)\, \mathrm{d}^3r'$ according to (2.30) and $I^i_L(\boldsymbol{k}) = 4\pi\, \mathrm{i}^\ell Y_L(\hat{\boldsymbol{k}})\mathrm{e}^{\mathrm{i}\boldsymbol{k}\cdot\boldsymbol{R}_i}$. Accordingly the cell partition should only satisfy the weaker condition that R_{ij} should be greater than the radius of the bounding sphere of the cell Ω_i (or Ω_j).

Summarizing, we have transformed the Lippmann–Schwinger equation into an algebraic equation for the coefficients $C^j_L(\boldsymbol{k})$ where the only ingredients are the structure factors $G^{ij}_{LL'}$ and the cell T-matrices $T^j_{L'L}$, obtained in terms of the local solutions (2.31). This is the famous separation of structure and dynamics peculiar to MST. We refer the reader to [7] for the numerical generation of these solutions and the calculation of $T^j_{L'L}$ in the general case of non spherically symmetric potentials.

2.3.1 Expression of Cross Sections in MST

We are now in a position to express the cross sections (2.14) and (2.17) in terms of the solution (2.32) given by MST.

To this purpose we introduce new expansion coefficients $B^j_L(\boldsymbol{k})$ defined by

$$B^j_L(\boldsymbol{k}) = \sum_{L'} T^j_{LL'} C^j_{L'}(\boldsymbol{k})\ , \tag{2.37}$$

and new local basis functions given by

$$\Phi_L(\boldsymbol{r}_j; k) = \sum_{L'} [\mathbf{T}^j]^{-1}_{L'L}\, \psi_{L'}(\boldsymbol{r}_j; k)\ , \tag{2.38}$$

so that

$$\psi^+_{\boldsymbol{k}}(\boldsymbol{r}_j) = \sum_{L} B^j_L(\boldsymbol{k})\Phi_L(\boldsymbol{r}_j; k)\ . \tag{2.39}$$

On the basis of (2.36) the new coefficients satisfy

$$\sum_{L'} (\mathbf{T}^i)^{-1}_{LL'} B^i_{L'}(\boldsymbol{k}) - \sum_{j,L'} (1-\delta_{ij})\, G^{ij}_{LL'} B^j_{L'}(\boldsymbol{k}) = I^i_L(\boldsymbol{k}) \,, \tag{2.40}$$

from which we derive the solution

$$B^i_L(\boldsymbol{k}) = \sum_{jL'} \tau^{ij}_{LL'} I^j_{L'}(\boldsymbol{k}) = \sqrt{\frac{k}{\pi}} \sum_{jL'} \tau^{ij}_{LL'}\, \mathrm{i}^{\ell'}\, Y_{L'}(\hat{\boldsymbol{k}})\, \mathrm{e}^{\mathrm{i}\boldsymbol{k}\cdot\boldsymbol{R}_j} \,, \tag{2.41}$$

in terms of the inverse of the MS matrix $(\mathbf{T}^{-1} - \mathbf{G_0})$

$$\tau = (\mathbf{T}^{-1} - \mathbf{G_0})^{-1} \,, \tag{2.42}$$

known as scattering path operator. Notice that we have restored the normalization to one state per Rydberg by multiplying $I^i_L(\boldsymbol{k})$ by the factor in (2.16).

The advantage of introducing the coefficients $B^i_L(\boldsymbol{k})$ lies in the fact that they are scattering amplitudes that in case of real potentials satisfy the relation

$$\int \mathrm{d}\hat{\boldsymbol{k}}\; B^i_L(\boldsymbol{k}) \left[B^j_{L'}(\boldsymbol{k})\right]^* = -\frac{1}{\pi} \Im \left[\tau^{ij}_{LL'}\right] \,, \tag{2.43}$$

which is a kind of generalized optical theorem. This relation is very important, since it establishes the connection between the angle-integrated photoemission and the photoabsorption cross section [7].

Using (2.39), due to the localization of the inital core state at site c, we can write the photoemission cross section (2.14) as

$$\frac{\mathrm{d}\sigma}{\mathrm{d}\hat{\boldsymbol{k}}} = 4\pi^2\, \alpha\, \hbar\omega_q \sum_{m_c\sigma_c} \left| \sum_L M_{L_c\sigma_c L}\, [B^c_L(\boldsymbol{k})]^* \right|^2 \,, \tag{2.44}$$

where

$$M_{L_c\sigma_c L}(E) = \int_{\Omega_c} \mathrm{d}\boldsymbol{r}_c\, \phi^c_{L_c\sigma_c}(\boldsymbol{r}_c)\, (\hat{\boldsymbol{e}}_q \cdot \boldsymbol{r}_c)\, \Phi_L(\boldsymbol{r}_c; k) \,, \tag{2.45}$$

In the presence of spin-orbit coupling in the initial core state one should change the notation $m_c\sigma_c$ to j^z_c for a given split edge j_c. Henceforth we shall neglect altogether the sum over spin, which will make a factor of two in front of the cross section in non magnetic systems.

In MST the quantity $\tau^{ij}_{LL'}$ is a scattering amplitude for propagation of the excited photoelectron from site i to site j, starting with angular momentum L around site i and arriving with L' around site j. On the basis of the solution (2.41) for the amplitude $B^c_L(\boldsymbol{k})$ the expression (2.44) substantiates our intuitive representation of the photoemission process as one in which the photoelectron is created at site c with a probability amplitude given by (2.45), propagates from this site to any site j to escape toward the detector with an angular distribution $Y_{L'}(\hat{\boldsymbol{k}})$ determined by its angular momentum L' around site j [2]. Since this site can coincide with the photoabsorber site c, both closed and open scattering paths contribute to the cross section.

For the total absorption cross section we get

$$\sigma_{\text{tot}}(\omega_q) = \int d\hat{\boldsymbol{k}} \, \frac{d\sigma}{d\hat{\boldsymbol{k}}} = 8\pi^2 \, \alpha \, \hbar\omega_q \sum_{m_c} \int d\hat{\boldsymbol{k}} \left| \sum_L M_{L_c L}(E) \, B_L^c(\boldsymbol{k}) \right|^2$$

$$= -8\pi \, \alpha \, \hbar\omega_q \sum_{\substack{LL' \\ m_c}} M_{L_c L}(E) \Im \left[\tau_{LL'}^{cc} \right] M_{L_c L'}(E) \, , \quad (2.46)$$

the second step being valid only in the case of real potentials due to the relation (2.43). This relation indicates that now only closed paths contribute to the process, the photoabsorber acting simultaneously as the source and the detector of the excited photoelectron. In this way the energy modulations observed in the cross section originate from the interference between the outgoing and the incoming photoelectron wave at the photoabsorbing site.

2.3.2 *The Green's Function Approach to Photoabsorption: Real Potential*

The Green's function of the effective SE (2.22) is defined as

$$\left[\nabla^2 + E - V_{\text{eff}}(\boldsymbol{r}; E) \right] G^+(\boldsymbol{r} - \boldsymbol{r}'; E) = \delta(\boldsymbol{r} - \boldsymbol{r}') \, . \quad (2.47)$$

It is an analytic function of E in the complex plane if one assumes the analyticity of the energy dependence of the effective potential in the whole plane.

A solution of (2.47) analytical in the upper energy plane is given by the so-called spectral representation in terms of all the eigenstates $\psi_n(\boldsymbol{r}; E)$ of the Schrödinger operator (SO) at energy E_n:

$$G^+(\boldsymbol{r}, \boldsymbol{r}'; E) = \lim_{\varepsilon \to 0^+} \sum_n \frac{\psi_n(\boldsymbol{r}; E) \, \psi_n(\boldsymbol{r}'; E)}{E - E_n + \mathrm{i}\,\varepsilon} \, . \quad (2.48)$$

Here the index n runs over all eigenstates, discrete and continuum, of the SO. We shall omit henceforth the additional dependence of the state ψ on the energy E via that of $V_{\text{eff}}(\boldsymbol{r}; E)$, since it will not affect the following manipulations on the GF.

From (2.48) it follows that

$$\Im \left[G(\boldsymbol{r}, \boldsymbol{r}'; E) \right] = -\frac{1}{\pi} \sum_n \psi_n(\boldsymbol{r}) \, \psi_n(\boldsymbol{r}') \, \delta(E - E_n) \, . \quad (2.49)$$

Remembering that

$$\sigma_{\text{abs}}(\omega_q) = 4\pi^2\,\alpha\,\hbar\omega_q \sum_f \left|\langle\Phi_f|\hat{\boldsymbol{e}}_q\cdot\boldsymbol{r}|\Phi_i\rangle\right|^2\,\delta(E_i - E_f + \hbar\omega_q)$$

we find that $\sigma_{\text{abs}}(\omega_q)$ can also be expressed as

$$\sigma_{\text{tot}}(\omega_q) = -8\pi\alpha\,\hbar\omega_q \sum_{m_c} \Im \int\!\!\int_{\Omega_c} \langle\phi^c_{L_c}(\boldsymbol{r})|\hat{\boldsymbol{e}}_q\cdot\boldsymbol{r}|G(\boldsymbol{r},\boldsymbol{r}';E)|\hat{\boldsymbol{e}}_q\cdot\boldsymbol{r}'|\phi^c_{L_c}(\boldsymbol{r}')\rangle \mathrm{d}\boldsymbol{r}\mathrm{d}\boldsymbol{r}'\,, \tag{2.50}$$

dropping again the sum over σ_c.

One of the advantages of MST is that one can write an explicit form of the full GF in terms of the scattering path operator as

$$G(\boldsymbol{r}_i,\boldsymbol{r}'_j;E) = \sum_{LL'} \Phi^i_L(\boldsymbol{r}_i;k)\tau^{ij}_{LL'}\Phi^j_L(\boldsymbol{r}_j;k) - \delta_{ij}\sum_L \Phi^i_L(\boldsymbol{r}_<;k)\Lambda^i_L(\boldsymbol{r}'_>;k)\,, \tag{2.51}$$

where $\Phi^j_L(\boldsymbol{r}_j;k)$ is the same local function as defined in (2.38) and $\Lambda^i_L(\boldsymbol{r}_>;k)$ is the irregular solution of the SE inside cell Ω_i matching smoothly to $J_L(\boldsymbol{r}_i;k)$ on the bounding sphere of the cell [7]. Notice that the singular part of the GF is real if the potential is real, so that its imaginary part is zero. In this case, by inserting this form of the GF into (2.50) one recovers (2.46).

2.3.3 The Green's Function Approach to Photoabsorption: Complex Potential

A realistic treatment of photoemission process cannot leave out of consideration the fact that the emitted photoelectron can loose its energy either by plasmon creation and/or inelastic scattering with other electrons in its propagation throughout the system (extrinsic processes) or via shake-up, shake-off excitations into the system itself (intrinsic processes).

In order to treat this problem, one can start from a many-channel formulation of the many-body process and then eliminate all the channels except the one of interest, usually the completely relaxed channel where the photoelectron has its maximum energy given by the Einstein relation [2]. This reduction process leads to an effective complex optical potential $V_{\text{eff}}(\boldsymbol{r};E)$ that describes the effect of the eliminated channels. Except in particular cases, their effect results in a smoothing of the spectral features of the main channel. A concomitant effect of the complex potential is to give a finite mean free path to the excited photoelectron given approximately by $\lambda = k/\Im[V_{\text{eff}}]$. In this way photoemission and photoabsorption become local structural techniques.

With a complex potential one can still use (2.44) to calculate the photoemission spectrum. However the integrated spectrum is no longer equal to the photoabsorption spectrum because this latter is an all inclusive measurement that counts all processes, elastic and inelastic, whereas in photoemission one measures a particular channel selected by the final kinetic energy of the detected photoelectron. In other words, the first line of (2.46) is no longer valid and is not equal to the second line.

However the second line can still be used to calculate the photoabsorption cross section in case of complex potential, if one remembers that in scattering theory with complex potential the optical theorem relates the imaginary part of the forward scattering amplitude to the sum of the elastic plus inelastic total cross sections (Chap. 7 of [3]).

Reference [2] shows how to express the cross section in terms of the one-particle GF with complex potential. The result is that we can still use (2.50) multiplied by a modulation function $|S_0(\omega)|^2$ that takes into account the overlap factor between the initial and final many-body states and describes the effect of the creation of the core hole on the spectrum. It is usually approximated by a numerical constant of the order 0.8–0.9.

It is expedient in this case to use an alternative form of the GF (2.51) obtained by using the relation

$$\psi_L(\boldsymbol{r}_j;\boldsymbol{k}) = J_L(\boldsymbol{r}_j;k) + \sum_{L'} \tilde{H}^+_{L'}(\boldsymbol{r}_j;k)\, T^j_{L'L} \,,$$

valid at the bounding sphere of cell Ω_j (use (2.29)).

By integrating backward the SE toward the center we find, remembering (2.38),

$$\sum_{L'} \Phi_{L'}(\boldsymbol{r}_j;\boldsymbol{k})\, T^j_{L'L} = \Lambda_L(\boldsymbol{r}_j;k) + \sum_{L'} \tilde{\Psi}^+_{L'}(\boldsymbol{r}_j;k)\, T^j_{L'L} \,, \tag{2.52}$$

where $\tilde{\Psi}^+_{L'}(\boldsymbol{r}_j;k)$ is the irregular solution of the SE matching smoothly to $\tilde{H}^+_{L'}(\boldsymbol{r}_j;k)$ at the boundary of the cell.

Inserting this relation into (2.51) we find

$$\begin{aligned} G(\boldsymbol{r}_i,\boldsymbol{r}'_j;E) &= \sum_{LL'} \Phi^i_L(\boldsymbol{r}_i;k)[\tau^{ij}_{LL'} - \delta_{ij}\, T^i_{LL'}]\Phi^j_{L'}(\boldsymbol{r}_j;k) \\ &+ \delta_{ij} \sum_{LL'} \Phi^i_L(\boldsymbol{r}_<;k) T^i_{LL'} \tilde{\Psi}^i_{L'}(\boldsymbol{r}'_>;k) \,. \end{aligned} \tag{2.53}$$

This form of the GF makes explicit the separation between the spectral contribution given by the environment (first term) and the photoabsorber (second term).

In fact, insertion into (2.50) gives

$$\sigma_{\mathrm{abs}}(\omega_{\boldsymbol{q}}) = \sigma_{\mathrm{env}}(\omega_{\boldsymbol{q}}) + \sigma_{\mathrm{at}}(\omega_{\boldsymbol{q}}) \tag{2.54}$$

where

$$\sigma_{\rm at}(\omega_q) = -8\pi\,\alpha\,\hbar\omega_q \sum_{LL'} \Im\left[\tilde{M}_{L_c;LL'}\,T^c_{LL'}\right]\,, \tag{2.55}$$

defining the new atomic absorption matrix element as

$$\tilde{M}_{L_c;LL'} = \iint_{\Omega_c} \mathrm{d}\boldsymbol{r}\,\mathrm{d}\boldsymbol{r}'\langle \phi^c_{L_c}(\boldsymbol{r})|\hat{\boldsymbol{e}}_q\cdot\boldsymbol{r}|\Phi^i_L(\boldsymbol{r}_<;k)\tilde{\Psi}^i_{L'}(\boldsymbol{r}'_>;k)|\hat{\boldsymbol{e}}_q\cdot\boldsymbol{r}'|\phi^c_{L_c}(\boldsymbol{r}')\rangle\,, \tag{2.56}$$

and

$$\begin{aligned}\sigma_{\rm env}(\omega_q) &= -8\pi\,\alpha\,\hbar\omega_q \sum_{m_cLL'} \Im\left[M_{L_cL}(E)(\tau^{cc}_{LL'} - T^c_{LL'})\,M_{L_cL'}(E)\right]\\ &= -8\pi\,\alpha\,\hbar\omega_q \sum_{m_cLL'} \Im\left[M_{L_cL}(E)\sum_{n=2}^{\infty}[(\mathbf{TG_0})^n\,\mathbf{T}]^{cc}_{LL'}\,M_{L_cL'}(E)\right]\,. \end{aligned}\tag{2.57}$$

The second line of this equation follows from the expansion of the scattering path operator in series of the matrix $\mathbf{TG_0}$

$$\tau^{ij}_{LL'} = \left[\left(\mathbf{T}^{-1} - \mathbf{G_0}\right)^{-1}\right]^{ij}_{LL'} = \left[\sum_{n=0}^{\infty}(\mathbf{TG_0})^n\,\mathbf{T}\right]^{ij}_{LL'} = \left[\sum_{n=0}^{\infty}\mathbf{T}\,(\mathbf{G_0T})^n\right]^{ij}_{LL'}\,, \tag{2.58}$$

which is absolutely convergent if the spectral radius (maximum eigenvalue of $\mathbf{TG_0}$) $\rho(\mathbf{TG_0}) < 1$, otherwise it is only conditionally (or asymptotically) convergent [2, 8]. The expansion in (2.57) starts from $n = 2$ since $[\mathbf{G_0}]^{cc}_{LL'} = 0$. Due to the meaning of $\mathbf{T}$ (site scattering amplitude) and $\mathbf{G_0}$ (free intersite spherical wave propagator) the matrix power $[(\mathbf{TG_0})^n]^{cc}_{LL'}$ defines closed multiple scattering paths beginning and ending at the photoabsorber. Therefore, with due precautions, one can almost always analyze the energy modulations in terms of MS paths.

Finally we want to point out that in the case of complex potential the atomic cross section does not factor out from the environment cross section, so that the theoretical signal to be compared with experimental energy modulations is

$$\chi(\omega_q) = \frac{\sigma_{\rm env}(\omega_q)}{\sigma_{\rm at}(\omega_q)}\,. \tag{2.59}$$

2.3.3.1 General Term of MST with Complex Potential: The Mean Free Path

It is instructive to see how the concept of mean free path arises in the general term of the multiple scattering series. We discuss this point assuming for simplicity the muffin-tin (MT) approximation for the cell potential. In this case we have

$[\mathbf{T}]_{LL'} = t_\ell = \mathrm{e}^{\mathrm{i}\,\delta_\ell} \sin\delta_\ell/k$ where $\delta = \delta_1 + \mathrm{i}\,\delta_2$ is the (complex) phase shift of the atomic potential (we shall omit henceforth the index ℓ since not necessary to the following argument).

The general term of MS series is given by

$$[(\mathbf{TG})^n]^{cc}_{LL'} = \sum_{ij\ldots k} \sum_{L_1L_2\ldots L_n} t^c_\ell\, G^{ci}_{LL_1}\, t^i_{\ell_1}\, G^{ij}_{L_1L_2}\, t^j_{\ell_2} \ldots . G^{kc}_{L_nL'} \,. \tag{2.60}$$

Since the propagator $G^{ij}_{LL'}$ is proportional to k we shall work with the dimensionless quantity kt. We find

$$kt = \mathrm{e}^{\mathrm{i}\delta} \sin\delta = \mathrm{e}^{-2\delta_2}\, \mathrm{e}^{\mathrm{i}\delta_1} \sin\delta_1 + \frac{\mathrm{i}}{2}\,[1 - \mathrm{e}^{-2\,\delta_2}]\,, \tag{2.61}$$

so that, assuming $\delta_2 \ll 1$ (which is usually the case),

$$\Im\,[kt] = |kt|^2 + \frac{\mathrm{i}}{4}\,[1 - \mathrm{e}^{-4\,\delta_2}] \approx |kt|^2 + \mathrm{i}\,\delta_2\,. \tag{2.62}$$

As anticipated, according to electron-atom scattering theory with complex potential [3], $\Im[t]$ represents the total cross section (elastic plus inelastic), while $|t|^2$ gives the elastic cross section (no energy loss for the impinging electron).

It is therefore clear that taking the imaginary part of (2.60) will describe all processes (elastic and inelastic) undergone by the photoelecton in the particular path, whereas the coherent elastic process will be obtained by taking for all kt factors the first term $\mathrm{e}^{-2\delta_2}\, \mathrm{e}^{\mathrm{i}\delta_1} \sin\delta_1$ of (2.61). This is the same factor as for real potential, except for the presence of the damping factor $\mathrm{e}^{-2\delta_2}$.

In order to evaluate this factor for the general MS term, we illustrate the derivation for three sites c, i, j, since the extension of the argument to the most general case will then be straightforward.

Remembering the approximate expression for the propagator [9][1]

$$G^{ij}_{LL'} \approx -\mathrm{i}\,k\, Y_L(\hat{\boldsymbol{R}}_{ij})\, Y_{L'}(\hat{\boldsymbol{R}}_{ij})\, \frac{\mathrm{e}^{\mathrm{i}\,kR_{ij}}}{k\,R_{ij}}\,, \tag{2.63}$$

where $k = \sqrt{E - V_I} \equiv k_1 + \mathrm{i}\,k_2$, V_I being the interstitial potential, and omitting the factor $-\mathrm{i}$ irrelevant to the argument, for three sites the damping is given by the exponential factor with exponent

$$-\,2\delta^c_2 - k_2R_{ci} - 2\delta^i_2 - k_2R_{ij} - 2\delta^j_2 - k_2R_{jc}\,. \tag{2.64}$$

[1]This approximation can be easily derived from the definition (2.35) by putting $\mathrm{i}^{\ell''}\,\tilde{H}^+_{L''}(\boldsymbol{R}_{ij};k) \approx -\mathrm{i}k\mathrm{e}^{\mathrm{i}\,kR_{ij}}/(k\,R_{ij})\,Y_{L''}(\hat{\boldsymbol{R}}_{ij})$ and closing the sum over L'' by using the relation $\sum_{L''} G(L, L'; L'')\, Y_{L''}(\hat{\boldsymbol{R}}) = Y_L(\hat{\boldsymbol{R}})\, Y_{L'}(\hat{\boldsymbol{R}})$.

In the WKB approximation of the phase shift we find

$$\delta^j = \int_0^{R_j} [E - v_j(r)]^{1/2} \mathrm{d}r - kR_j \ , \tag{2.65}$$

where R_j is the MT radius of the potential at site j.

Therefore, putting $R_{\text{path}} = R_{oi} + R_{ij} + R_{jo}$, the damping factor is given by

$$\mathrm{e}^{-\kappa_2 R_{\text{path}}} \ , \tag{2.66}$$

where

$$\kappa_2 = \frac{1}{R_{\text{path}}} \Im \left[\int_{\text{path}} [E - v(r)]^{1/2} \mathrm{d}r \right] . \tag{2.67}$$

The mean free path is accordingly

$$\lambda = \frac{1}{2\kappa_2} \ , \tag{2.68}$$

consistent with the fact that $\mathrm{e}^{-\kappa_2 R_{\text{path}}}$ is an attenuation factor for an amplitude of propagation.

A further simplification comes from observing that $[E - v_1(r)] \gg v_2(r)$. Therefore

$$[E - v(r)]^{1/2} \approx [E - v_1(r)]^{1/2} + \frac{\mathrm{i}}{2} \frac{v_2(r)}{[E - v_1(r)]^{1/2}} \ ,$$

so that

$$\kappa_2 = \frac{1}{2R_{\text{path}}} \int_{\text{path}} \frac{v_2(r)}{[E - v_1(r)]^{1/2}} \mathrm{d}r \leq \frac{1}{2k_1 R_{\text{path}}} \int_{\text{path}} v_2(r) \mathrm{d}r \ , \tag{2.69}$$

implying that, in atomic units,

$$\lambda = \frac{k_1}{\overline{v}_2} \ , \tag{2.70}$$

where $\overline{v}_2 = \int_{\text{path}} v_2(r)\mathrm{d}r / R_{\text{path}}$ is the average of the complex part of the potential over the path. The core hole half width $\Gamma_h/2$ can also be considered a complex potential that should be added to $\overline{v}_2$ in (2.70) to obtain the total mean free path.

Due to the presence of the damping of the photoelectronic wave, in absorption spectra only atoms within a sphere centered on the emitter with radius equal to the mean free path contribute to the structural signal, since only closed paths are possible. In contrast, in photoelectron diffraction spectroscopy this radius equals twice the mean free path, since the cross-section is proportional to the square of a scattering amplitude with interference between different contributions (direct atomic versus propagation processes), as clear from (2.44).

2.4 An *All-Purpose* Optical Potential

Constructing an effective ad hoc potential for any system to investigate would be time-consuming and not very practical. In order to cope in a simple way with the wide variety of systems encountered in practical applications one makes the *ansatz* that the optical potential is a functional of the density $\varrho(\boldsymbol{r})$ of the system under consideration. This assumption is akin to the local density approximation made in Density Functional Theory (DFT) for the exchange-correlation potential (the Coulomb part of the potential is already a functional of the density). There is a difference, however, since in our case the potential regards an excited state and not the ground state. Experience has shown that the Hedin-Lundqvist exchange-correlation potential [10] for excited states constitutes a good candidate for this kind of *universal* optical potential.

Moreover, partitioning the space in non overlapping space-filling domains in an automatic way is not straightforward for arbitrary atomic locations. This difficulty motivates the introduction of the muffin-tin (MT) approximation, whereby one draws a sphere around each atomic position, inside which the potential $v_i(r_i)$ is spherically averaged, whereas it is approximated by a constant suitably chosen in the interstitial region between the spheres. Empty spheres to minimize the interstitial volume can also be drawn.

Obviously these approximations have a price. They are reasonably good for photoelectron energies E greater than $\approx$50 eV, less good at lower energies.

2.4.1 *The Construction of the Muffin-Tin Potential*

Given a cluster of atoms, the Coulomb potential (in Rydberg units of energy) is given by

$$V_c(r) = \underbrace{-\sum_j \frac{2Z_j}{|\boldsymbol{r} - \boldsymbol{R}_j|}}_{\text{nuclear part}} + \underbrace{2\int \mathrm{d}\boldsymbol{r}' \frac{\varrho_{\text{tot}}(\boldsymbol{r}')}{|\boldsymbol{r} - \boldsymbol{r}'|}}_{\text{electron part}}, \tag{2.71}$$

where $\boldsymbol{R}_j$ and Z_j denote the position and the charge of the jth atomic nucleus and $\varrho_{\text{tot}}(\boldsymbol{r})$ is the total electronic charge density of the system.

We construct the total density by superimposing spherically symmetric self-consistent atomic charge densities that are generated by currently available atomic programs [11]. The resulting cluster density is not self-consistent, but it is a good approximation to it. In order to make this superposition we have to expand a spherically symmetric function referred to one center j around another center i and take only the $L = 0$ component. This method was proposed by Löwdin [12] and later utilized by Mattheiss [13]. If $\varrho(r_j)$ is the radial charge density around center j, normalized such that $\int_0^\infty \varrho(r_j)\, r_j^2\, \mathrm{d}r_j = Z_j$, then the component $\varrho_{L=0}(r)$ of the charge density at distance r from the center i is given by

$$\varrho_{L=0}(r) = \frac{1}{2R\,r}\int_{|R-r|}^{R+r} \varrho_j(r_j) r_j \mathrm{d}r_j \,, \tag{2.72}$$

where R is the distance between the two centers. Therefore, the total overlapped, spherically averaged, charge density around center i, due to all its neighbors, is given by

$$\varrho^i_{\text{tot}}(\boldsymbol{r}) = \varrho^i(\boldsymbol{r}) + \sum_j \frac{1}{2R_{ij}\,r}\int_{|R_{ij}-r|}^{R_{ij}+r} \varrho_j(r_j) r_j \mathrm{d}r_j \,, \tag{2.73}$$

where R_{ij} is the distance between the two centers i and j and $\varrho^i(\boldsymbol{r})$ is the charge density of the atom at site i, which would be zero if one were dealing with an empty sphere.

Similarly, due to the linearity of the Poisson equation in the charge density, the total Coulomb potential around center i, generated by the charge density (2.72) is given by

$$V^i_{\text{tot}}(\boldsymbol{r}) = V^i(\boldsymbol{r}) + \sum_j \frac{1}{2R_{ij}\,r}\int_{|R_{ij}-r|}^{R_{ij}+r} V_j(r_j) r_j \mathrm{d}r_j \,, \tag{2.74}$$

where $V_j(r)$ is the atomic potential generated by the charge density $\varrho_j(r_j)$ of the atom at site j:

$$V_j(r) = \frac{2}{r}\int_0^r r_j^2 \varrho(r_j)\mathrm{d}r_j + 2\int_r^\infty \varrho(r_j) r_j \mathrm{d}r_j \,. \tag{2.75}$$

In the same spirit we can calculate the constant interstitial potential $\overline{V}_c$ by taking the average of the total Coulomb potential in the interstitial region $\Delta\Omega$. To this purpose we define a center of the cluster as the center of all the atomic charges, draw a sphere of radius R_o of minimum volume encircling the cluster and then expanding the total potential around this center. Then we easily find

$$\overline{V}_c = \frac{1}{\Delta\Omega}\int_{\Delta\Omega} V_{\text{tot}}(r)\mathrm{d}r = \frac{1}{\Delta\Omega}\left[\int_0^{R_o} V_c^o(r) r^2 \mathrm{d}r - \sum_j \int_0^{R_j} V_c^j(r_j) r_j^2 \mathrm{d}r_j\right] , \tag{2.76}$$

where R_j is the radius of the MT sphere at site j. Similarly, the interstitial charge density $\overline{\varrho}_{\text{int}}$, necessary to find the interstitial exchange-correlation part, is given by

$$\overline{\varrho}_{\text{int}} = \frac{1}{\Delta\Omega}\int_{\Delta\Omega} \varrho(\boldsymbol{r})\mathrm{d}\boldsymbol{r} = \frac{1}{\Delta\Omega}\left[\int_0^{R_o} \varrho^o_{\text{tot}}(r) r^2 \mathrm{d}r - \sum_j \int_0^{R_j} \varrho^j(r_j) r_j^2 \mathrm{d}r_j\right] , \tag{2.77}$$

In order to determine the MT radii we can follow one of the two prescriptions present in the literature: one given by Norman [14] and the other by Wille et al. [15].

In the Norman prescription, a Norman radius of each site i is determined such that

$$\int_0^{R_i^{\mathrm{Norm}}} \varrho^i(r_i) r_i^2 \mathrm{d}r_i = Z_i \; . \tag{2.78}$$

Then, given two nearest-neighbor sites i and j, the touching sphere MT radius is given by

$$R_i = R_{ij} \frac{R_i^{\mathrm{Norm}}}{R_i^{\mathrm{Norm}} + R_j^{\mathrm{Norm}}} \; , \tag{2.79}$$

and similarly for R_j, which implies that $R_i/R_j = R_i^{\mathrm{Norm}}/R_j^{\mathrm{Norm}}$. Empirically this prescription works for systems with covalent bonds. An overlap of about 10–15% of these radii is also allowed.

On the other hand, in the Wille et al. prescription the choice of the MT radii is made in such a way that the potential discontinuity at the boundary of the MT spheres are minimized. This prescription seems to work with ionic compounds. In any case discontinuities are always present in the MT partition of space causing unphysical photoelectron scattering, although this effect can be minimized by varying the interstitial potential and overlap factors. However this situation is unsatisfactory and has been solved by Full Potential MST [7].

2.4.2 *The Construction of the Exchange-Correlation Potential*

In a many-body approach, the elimination of all the channels different from the one carrying the structural information would lead to a very complicated and unpractical optical potential [2], since it shoud describe all processes, intrinsic and extrinsic. The intrinsic processes are those inelastic events ensuing the sudden creation of the core-hole, while the extrinsic processes are inelastic events excited by the photoelectron in its propagation through the system. These latter can be described in terms of the self-energy of the photoelectron, for which there is a whole panoply of methods of calculations. In the framework of many-body theory, the GW approximation to the calculation of the self-energy is the method of choice. However, even with this restriction, a calculation adapted to each particular system is not practically viable.

In the spirit of the local density approximation one makes the assumption that for any system the photoelectron self-energy is given by that of the homogeneous interacting electron gas calculated at the local density of the system under study. Therefore we can define a kind of *universal* exchange-correlation potential as

$$\Sigma(\boldsymbol{r}; E) \approx \Sigma_h[p(\boldsymbol{r}), E - V_c(\boldsymbol{r}); \varrho(\boldsymbol{r})] \; , \tag{2.80}$$

where, as anticipated, $\Sigma_h[p, \omega; \varrho]$ is the self-energy of an electron in an homogeneous interacting electron gas with momentum $p(\boldsymbol{r})$, energy $\omega = E - V_c(\boldsymbol{r})$ and density $\varrho = \varrho(\boldsymbol{r})$, the local density of the actual physical system. Since $E - V_c(\boldsymbol{r}) \approx p^2(\boldsymbol{r})$, neglecting the small exchange and correlation correction, we can define an exchange-correlation potential for the photoelectron in our system as

$$V_{\text{exc}}(\boldsymbol{r}; E) \simeq \Sigma_h[p(\boldsymbol{r}), p^2(\boldsymbol{r}); \varrho(\boldsymbol{r})] \,, \tag{2.81}$$

where the local momentum $p(\boldsymbol{r})$ is defined as

$$p^2(r) + \Sigma_h[p(\boldsymbol{r}), p^2(\boldsymbol{r}); \varrho(\boldsymbol{r})] = k^2 + k_F^2(\boldsymbol{r}) + \Sigma_h[k_F, k_F^2; \varrho(\boldsymbol{r})] \,. \tag{2.82}$$

Here $k^2 = E$ is the photoelectron kinetic energy measured from the Fermi level in the system and $k_F^2(\boldsymbol{r}) = [3\pi^2\varrho(\boldsymbol{r})]^{2/3}$ is the local Fermi energy. Usually the small difference between the two self-energies is ignored and one takes $p^2(\boldsymbol{r}) \approx k^2 + k_F^2(\boldsymbol{r})$.

Introducing the Wigner parameter $r_s = [3/(4\pi\varrho)]^{1/3}$, which measures the average inter electronic distance in an electron gas at density ϱ, and the constant $\beta = [4/9\pi]^{1/3} \approx 0.52$, we have

$$k_F = (3\pi^2\varrho)^{1/3} = \frac{1}{\beta r_s} \,. \tag{2.83}$$

Then, according to [16], the GW expression for the electron self-energy in the interacting electron gas is given by

$$\Sigma_h(p, \omega) = \frac{\mathrm{i}}{(2\pi)^4} \int \mathrm{e}^{\mathrm{i}\omega'\delta} \frac{V(\boldsymbol{p}')}{\varepsilon(\boldsymbol{p}', \omega')} G_0(\boldsymbol{p} + \boldsymbol{p}', \omega + \omega') \mathrm{d}\boldsymbol{p}' \mathrm{d}\omega' \,, \tag{2.84}$$

corresponding to the self-energy of a test electron interacting with the charge fluctuations of the medium. Here

$$G_0(\boldsymbol{p}, \omega) = \frac{1}{\omega - p^2 + \mathrm{i}\,\mathrm{sign}(\omega - \epsilon_F)} \,, \tag{2.85}$$

is the momentum space propagator of the test electron and

$$\frac{1}{\varepsilon(\boldsymbol{p}, \omega)} = 1 + \frac{\omega_p^2}{\omega^2 - \omega_1^2(p)} \,, \tag{2.86}$$

is the frequency and momentum dependent dielectric function of the electron gas in the *plasmon-pole approximation* [16] that screens the bare Coulomb interaction . In this last equation

$$\omega_1^2(p) = \omega_p^2 + \epsilon_F^2 \left[\frac{4p^2}{3k_F^2} + \frac{p^4}{k_F^4} \right] \,, \tag{2.87}$$

is the momentum dispersion of the plasmon pole, given at zero momentum by

$$\omega_p = 4\epsilon_F \left[\frac{\beta r_s}{3\pi}\right]^{1/2} = \frac{41.7}{[r_s(\text{au})]^{3/2}}\,\text{eV}\,. \tag{2.88}$$

We now have all the ingredients to calculate the self-energy (2.84) and we refer the reader to the Appendix B of [10] for an explicit calculation. Even though this potential was initially devised by Hedin and Lundqvist to describe exchange and correlation corrections to the Hartree potential due to the valence charge, we follow Lee and Beni [17] to extend the validily of (2.84) to the atomic region.

By way of illustration Fig. 2.1 presents the energy dependence of the real and imaginary part of the self-energy (2.84) at $r_s = 2.88$ au, corresponding to a density of $\varrho = 0.01\,\text{au}^{-3}$. At this value the plasmon energy is $\omega_p = 9.64$ eV and the local Fermi energy is $k_F^2 = 6.04$ eV. We assume that the Fermi level in the system is at −9.5 eV, from which the kinetic energy of the photoelectron E and the plasmon energy are counted. We see that the imaginary part of $\Sigma_h(p, p^2)$ begins to be different from zero for values E slightly greater then ω_p [10]. At the same energy value the real part has a dip that comes from the screened exchange part of the potential [10]. The sudden onset of the photoelectron damping and the dip in the exchange part are rather unphysical, since the plasmon pole approximation of the dielectric function neglects the contribution of particle-hole excitations. A way to improve on this situation is indicated in [18] where a scheme on a many-pole approximation of the dielectric function, via the loss function, is implemented. This approach eliminates the unphysical behavior of $\Sigma_h(p, p^2)$.

Figure 2.1 also indicates the constant value of the Slater exchange $3\alpha k_F/\pi = -6.23$ eV, at a typical value of $\alpha = 0.72$. We see that the exchange-correlation part of $\Sigma_h(p, p^2)$ decreases with energy as the inverse of $\sqrt{E}$ [10], in keeping with the fact that the Pauli principle is less and less effective the higher the photoelectron kinetic energy, due to the descreasing overlap of its wave function with the occupied states of the system. This behaviour is born out by the observation of the EXAFS maxima, whose position is not reproduced by a constant exchange but is well predicted by the HL energy-dependent exchange [19].

For comparison, in Fig. 2.2 we also give $\Sigma_h(p, p^2)$ at $r_s = 1.34$ au, corresponding to a density of $\varrho = 0.1\,\text{au}^{-3}$. The Fermi level lies again at −9.5 eV. We see that the plasmon energy is now much higher ($\omega_p = 30.49$ eV) whereas the Slater exchange $3\alpha k_F/\pi = -13.42$ eV is much lower.

2.4.3 Generating Phase Shifts and Atomic Cross Sections

Based on Sect. 2.4, the effective potential is given by $V_{\text{eff}}(\boldsymbol{r}; E) = V_c(\boldsymbol{r}) + V_{\text{exc}}(\boldsymbol{r}; E)$. After reduction to the MT form, in each atomic (or empty) sphere we have to solve for each partial wave ℓ a radial SE of the type

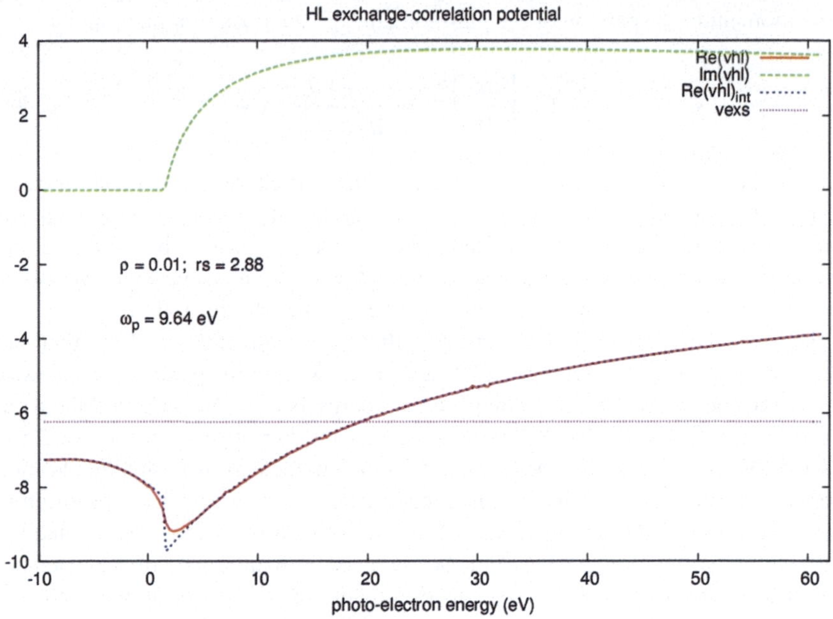

Fig. 2.1 Behavior of the real and imaginary part of the HL exchange-correlation potential at $r_s = 2.88$ au: Re(vhl) = real part, Im(vhl), imaginary part, (both calculated according to the formulas of Appendix B of [10]), $\mathrm{Re(vhl)}_{\mathrm{int}}$, real part (calculated according the interpolation scheme of [19]), vexs $= 3\alpha k_F/\pi$, constant Slater exchange with $\alpha = 0.72$

$$\left[\frac{\mathrm{d}^2}{\mathrm{d}r^2} + \frac{2}{r}\frac{\mathrm{d}}{\mathrm{d}r} + E - \frac{\ell(\ell+1)}{r^2} - V(r;\, E)\right] R_\ell(r) = 0\,, \tag{2.89}$$

where, for brevity, we have indicated by $V(r;\, E)$ the spherically averaged $V_{\mathrm{eff}}(\boldsymbol{r};\, E)$ inside the sphere under consideration.

In reality this is a non relativistic potential. Relativistic corrections can be introduced by the method followed in [20], whereby one starts from the radial Dirac equation with potential $V(r;\, E)$ by eliminating the lower component of the wave function $g_\ell(r)$ in favor of the upper component $f_\ell(r)$. Defining

$$B(r;\, E) = \{1 + \alpha^2[E - V(r;\, E)]\}^{-1}\,, \tag{2.90}$$

where again α is the fine structure constant, and putting

$$f_\ell(r) = \frac{u_\ell(r)}{r} = \frac{1}{r}\,\frac{q_\ell(r)}{B^{1/2}(r;\, E)}\,, \tag{2.91}$$

one finds that $q_\ell(r)$ obeys a pseudo-SE of the type

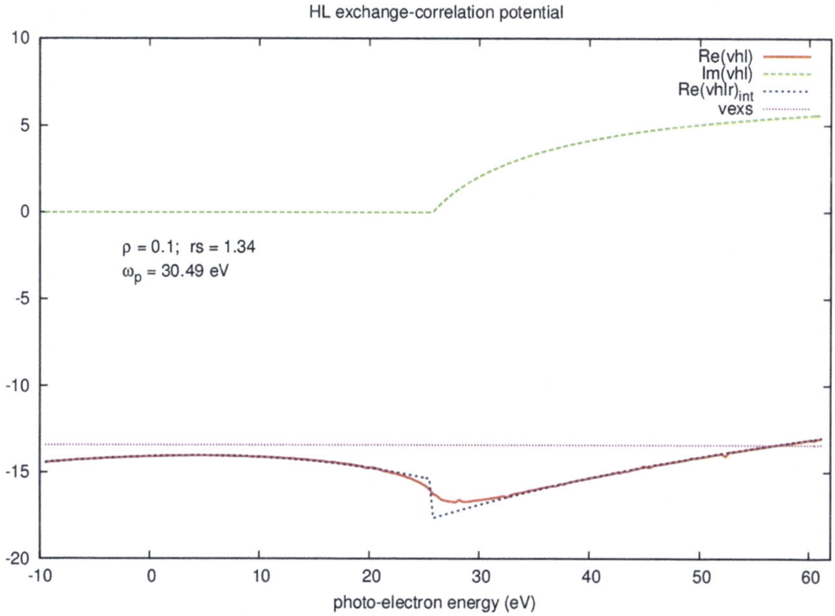

Fig. 2.2 Behavior of the real and imaginary part of the HL exchange-correlation potential at $r_s = 1.34$ au: same legend as for Fig. 2.1

$$\frac{\mathrm{d}^2}{\mathrm{d}r^2} q_\ell(r) = G(r; E)\, q_\ell(r) \,, \tag{2.92}$$

where

$$\begin{aligned} G(r) &= V(r; E) + \frac{\ell(\ell+1)}{r^2} - E - \frac{\alpha^2}{4}[E - V(r; E)] \\ &+ \frac{\alpha^2}{8} B(r; E) \left\{ \frac{1}{r} \frac{\mathrm{d}^2}{\mathrm{d}r^2} [r\, V(r; E)] + \frac{3\alpha^2}{8} B(r; E) \left[\frac{\mathrm{d}}{\mathrm{d}r} V(r; E) \right]^2 \right\} \\ &+ \frac{\alpha^2}{4} B(r; E) \frac{1}{r} \frac{\mathrm{d}}{\mathrm{d}r} [V(r; E)]\, 2\boldsymbol{\ell} \cdot \boldsymbol{s} \,. \end{aligned} \tag{2.93}$$

Throughout E is the photoelectron energy without its rest mass. The last term is the spin orbit interaction such that

$$\begin{aligned} 1 + 2\boldsymbol{\ell} \cdot \boldsymbol{s} &= \ell + 1 \quad \text{for } j = \ell + \frac{1}{2} \\ &= -\ell \qquad \text{for } j = \ell - \frac{1}{2} \,, \end{aligned} \tag{2.94}$$

where $\boldsymbol{j} = \boldsymbol{\ell} + \boldsymbol{s}$. Equation (2.92) can be solved numerically by the Numerov method with the starting behavior at the origin, given by the indicial equation,

$$\lim_{r \to 0} q_\ell(r) \sim r^{1/2+[\kappa^2-(Z\alpha)^2]^{1/2}} , \tag{2.95}$$

where $-\kappa$ is the eigenvalue of $1 + 2\boldsymbol{\ell} \cdot \boldsymbol{s}$ (see (2.94)) and Z is the atomic number of the atom inside the sphere (equal to zero if empty).

Specializing (2.29) to the spherical case, we must match smoothly the solution of the (pseudo-)SE $R_\ell(r)(f_\ell(r))$ to $t_\ell\, j_\ell(kr) - \mathrm{i}kh^+_\ell(kr)$ at the MT radius R_s of the sphere so that

$$t_\ell = \frac{\mathrm{i}}{k} \frac{[R_\ell(r)\, j'_\ell(kr) - R'_\ell(r)\, j_\ell(kr)]}{[R_\ell(r)\,(h^+)'_\ell(kr) - R'_\ell(r)\, h^+_\ell(kr)]}\Bigg|_{r=R_s} , \tag{2.96}$$

where the prime indicates derivative with respect to r. Similarly for $f_\ell(r)$ if the relativistic solution is used. If the spin-orbit interaction is considered, the index ℓ in this equation should be replaced by j_ℓ.

Moreover, by inward integration of the radial equations (2.89) and (2.92), we can also generate the irregular solutions $\tilde{R}_\ell(r)$ that matches smoothly to $-\mathrm{i}h^+_\ell(kr)$ at the MT radius. Then the atomic cross section (2.55), using t_ℓ and (2.56), turns out to be proportional to

$$\sum_L t_\ell \left\{ \int_0^{R_s} \mathrm{d}\boldsymbol{r}'\ \phi_c(\boldsymbol{r}')T(\boldsymbol{r}')\tilde{R}_L(\boldsymbol{r}') \int_0^{r'} \mathrm{d}\boldsymbol{r}\ R_L(\boldsymbol{r})T(\boldsymbol{r})\phi_c(\boldsymbol{r}) + \right.$$
$$\left. \int_0^{R_s} \mathrm{d}\boldsymbol{r}'\ \phi_c(\boldsymbol{r}')T(\boldsymbol{r}')R_L(\boldsymbol{r}') \int_{r'}^{R_s} \mathrm{d}\boldsymbol{r}\ \phi_c(\boldsymbol{r})T(\boldsymbol{r})\tilde{R}_L(\boldsymbol{r}) \right\} , \tag{2.97}$$

where $T(\boldsymbol{r}) = \hat{\boldsymbol{e}}_q \cdot \boldsymbol{r}$ is the dipole transition operator, $\phi_c(\boldsymbol{r}) = R^c_\ell(r)\, Y_{L_c}(\hat{\boldsymbol{r}})$, $R_L(\boldsymbol{r}) = R_\ell(r)\, Y_L(\hat{\boldsymbol{r}})$ and $\tilde{R}_L(\boldsymbol{r}) = \tilde{R}_\ell(r)\, Y_L(\hat{\boldsymbol{r}})$, with obvious modifications if one uses the relativistic wave functions.

2.4.4 Calculating EELS Matrix Elements

In calculating EELS matrix elements, we follow two different strategies according to whether we calculate Coulomb matrix elements as in (2.20) corresponding to an experiment of the type ($e \to 2e$), where both electrons in the final state are analyzed, or we are interested in an all-inclusive experiment ($e \to e$), in which case we have to calculate an expression like the one in (2.21).

In the first case, in order to calculate the integrals in (2.20) we need to use an expansion formula for the Coulomb interaction. Without loss of generality we introduce a damped Coulomb interaction so that

$$\frac{2\exp(-\kappa|\boldsymbol{r}-\boldsymbol{r}'|)}{|\boldsymbol{r}-\boldsymbol{r}'|} = -8\pi\kappa\sum_L I_L(\boldsymbol{r}_<)K_L(\boldsymbol{r}_>)(-1)^\ell\ , \tag{2.98}$$

which reduces to the usual Coulomb interaction for $\kappa \to 0$. Here

$$I_L(\boldsymbol{r}) = i_\ell(\kappa r)Y_L(\hat{\boldsymbol{r}});\ \ K_L(\boldsymbol{r}) = k_\ell(\kappa r)Y_L(\hat{\boldsymbol{r}})\ , \tag{2.99}$$

where $i_\ell(\kappa r)$ and $k_\ell(\kappa r)$ are modified spherical Bessel and Hankel functions (the analytical continuation of the usual functions $j_\ell(kr)$ and $h_\ell(kr)$ with $k = \mathrm{i}\kappa$). This relation is the analytical continuation to complex k of the same expression given in (2.26). The damping constant κ describes the screening of the Coulomb interaction of the incoming electron by the other electrons in the system. In a metal the electron screening gives rise to a Coulomb interaction of the Yukawa type, like in (2.98), with $\kappa = (12/\pi)^{1/3}\ (r_s)^{-1/2}\mathrm{au}^{-1}$ in the Thomas-Fermi approximation. For semiconductors one can define a sort of Thomas-Fermi spatial dielectric screening $\bar{\varepsilon}(r)$ that gives rise to a screened Coulomb interaction of the type $V_c = -2/(r\,\bar{\varepsilon}(r))$ (see [21] and references therein) and approximate $\bar{\varepsilon}(r)$ as a constant. In both cases one can use the expansion (2.99).

The analytical continuation of the usual spherical wave propagator with positive k given by (2.35) leads to the following two site expansion

$$\frac{\exp(-\kappa|\boldsymbol{r}-\boldsymbol{r}'|)}{|\boldsymbol{r}-\boldsymbol{r}'|} = -\kappa\sum_L I_L(\boldsymbol{r}_i)G^{ij}_{LL'}I_{L'}(\boldsymbol{r}'_j)\ , \tag{2.100}$$

where the factor of $4\pi(-1)^\ell$ is incorporated in the definition of $G^{ij}_{LL'}$. Therefore an integral of the type

$$\int \mathrm{d}\boldsymbol{r}\mathrm{d}\boldsymbol{r}'\frac{\varrho(\boldsymbol{r})\varrho'(\boldsymbol{r}')}{|\boldsymbol{r}-\boldsymbol{r}'|}\ , \tag{2.101}$$

where the functions $\varrho(\boldsymbol{r})$ and $\varrho'(\boldsymbol{r}')$ are represented locally, as in multiple scattering theory, can be written as

$$\begin{aligned}\int \mathrm{d}\boldsymbol{r}\mathrm{d}\boldsymbol{r}'\frac{\varrho(\boldsymbol{r})\varrho'(\boldsymbol{r}')}{|\boldsymbol{r}-\boldsymbol{r}'|} &= \sum_{i\neq j}\sum_{LL'}\int_{\Omega_i}\mathrm{d}\boldsymbol{r}_i\ \varrho(\boldsymbol{r}_i)I_L(\boldsymbol{r}_i)G^{ij}_{LL'}\int_{\Omega_j}\mathrm{d}\boldsymbol{r}_j\ \varrho'(\boldsymbol{r}_j)I_L(\boldsymbol{r}_j)\\ &+\sum_i\sum_L\int_0^{R_i^s}\mathrm{d}\boldsymbol{r}_i\varrho(\boldsymbol{r}_i)\left\{K_L(\boldsymbol{r}_i)\int_0^{r_i}\mathrm{d}\boldsymbol{r}'_i\ \varrho'(\boldsymbol{r}'_i)I_L(\boldsymbol{r}'_i)\right.\\ &\left.+\ I_L(\boldsymbol{r}_i)\int_{r_i}^{R_i^s}\mathrm{d}\boldsymbol{r}'_i\ \varrho'(\boldsymbol{r}'_i)K_L(\boldsymbol{r}'_i)\right\}\ ,\end{aligned} \tag{2.102}$$

where $\Omega_{i,j}$ denote the various cells (MT spheres) in which the space has been partitioned.

Now from (2.20) we see that for the direct integral $\varrho(\boldsymbol{r})$ should be identified with $\phi_\epsilon(\boldsymbol{r})\phi_c(\boldsymbol{r})$ and $\varrho(\boldsymbol{r}')$ with $\psi_i^+(\boldsymbol{r}')\psi_f^-(\boldsymbol{r}')$, while for the exchange integral $\varrho(\boldsymbol{r})$ should be identified with $\psi_f^-(\boldsymbol{r})\phi_c(\boldsymbol{r})$ and $\varrho(\boldsymbol{r}')$ with $\phi_\epsilon(\boldsymbol{r}')\psi_i^+(\boldsymbol{r}')$. From MST we know that at site i

$$\psi_i^+(\boldsymbol{r}_i) = \sum_L B_L^i(\boldsymbol{k}_i) R_\ell(r_i) Y_L(\hat{\boldsymbol{r}}_i)$$

$$\psi_f^-(\boldsymbol{r}_i) = \sum_L \left[B_L^i(\boldsymbol{k}_f)\right]^* R_\ell(r_i) Y_L(\hat{\boldsymbol{r}}_i)\,, \tag{2.103}$$

so that at site i we can write in general

$$\varrho(\boldsymbol{r}_i) = \sum_{\Lambda\Lambda'} \sum_{\Lambda''} \varrho_{\Lambda\Lambda'}(r_i) G(\Lambda, \Lambda', \Lambda'') Y_{\Lambda''}(\hat{\boldsymbol{r}}_i) \tag{2.104}$$

where $G(L, L', L'')$ is the usual Gaunt coefficient. Inserting this expression and the corresponding one for $\varrho(\boldsymbol{r}')$ into (2.102), one can perform the angular integrals with the aid of the Gaunt coefficients and be left only with the radial integrals, which are calculated in the subroutine *radialx_eels* of the program *phagen_scf.f*. Notice also that whenever $\phi_c(\boldsymbol{r})$ appears, the integral should be limited only to the cell (sphere) of the photoabsorber.

The calculation of (2.21) proceeds along the following lines. Applying the permutation operator we get four terms, a direct term, two mixed terms and a full exchange term, according to whether no permutation, only one or both are applied to the respective variables.

We consider first the direct term (no permutation) in (2.21). In order to make the expression of the cross section in (2.21) resemble the usual one for electric absorption we define the effective transition operator

$$T_d(\boldsymbol{r}_1) \equiv \int \mathrm{d}\boldsymbol{r}_2 \psi_i^+(\boldsymbol{r}_2) V_c(|\boldsymbol{r}_1 - \boldsymbol{r}_2|)[\psi_f^-(\boldsymbol{r}_2)]^*\,, \tag{2.105}$$

so that the EELS cross section can be written as the imaginary part of

$$\iint \mathrm{d}\boldsymbol{r}_1 \mathrm{d}\boldsymbol{r}_1'\; \phi_c(\boldsymbol{r}_1')[T_d(\boldsymbol{r}_1')]^* G(\boldsymbol{r}_1', \boldsymbol{r}_1; E_{\text{exc}}) T_d(\boldsymbol{r}_1)\phi_c(\boldsymbol{r}_1)\,. \tag{2.106}$$

We now specialize the expression of the Green's function (2.53) to the MT case (as in (2.97)), whereby

$$G(\boldsymbol{r}_i, \boldsymbol{r}_j'; E_{\text{exc}}) = \sum_{LL'} R_L(\boldsymbol{r}_i)[\tau_{LL'}^{ij} - t_\ell^i \delta_{LL'}\delta_{ij}] R_{L'}(\boldsymbol{r}_j')(1 - \delta_{ij})$$

$$+ \delta_{ij} \sum_L R_L(\boldsymbol{r}_i^<)\, t_\ell\, \overline{R}_L(\boldsymbol{r}_i^>)\,. \tag{2.107}$$

We see that the regular part leads to radial matrix elements of the type already introduced in (2.102), whereas the irregular part gives rise to an expression of the type

$$\sum_L \iint \mathrm{d}\boldsymbol{r}\mathrm{d}\boldsymbol{r}' \; \phi_c(\boldsymbol{r}')[T_d(\boldsymbol{r}')]^* \left[R_L(\boldsymbol{r})\overline{R}_L(\boldsymbol{r}')\Theta(r'-r) + R_L(\boldsymbol{r}')\overline{R}_L(\boldsymbol{r})\Theta(r-r')\right] T_d(\boldsymbol{r})\phi_c(\boldsymbol{r}) \;, \tag{2.108}$$

which lead to the same integral of (2.97), where now the transition operator $T_d(\boldsymbol{r})$ is given by (2.105). Remembering the expansion (2.98) and putting for short $\varrho(\boldsymbol{r}) = \psi_i^+(\boldsymbol{r})[\psi_f^-(\boldsymbol{r})]^*$ we can write this transition operator as

$$T_d(\boldsymbol{r}) = -8\pi\kappa \sum_L (-1)^\ell \left\{ I_L(\boldsymbol{r}) \int_r^{R_b} \mathrm{d}\boldsymbol{r}' \; K_L(\boldsymbol{r}')\varrho(\boldsymbol{r}') + K_L(\boldsymbol{r}) \int_0^r \mathrm{d}\boldsymbol{r}' \; I_L(\boldsymbol{r}')\varrho(\boldsymbol{r}') \right\} \;. \tag{2.109}$$

Considering the full exchange term, i.e. permuting both primed and unprimed variables, we get

$$\frac{\mathrm{d}\sigma_e}{\mathrm{d}\hat{\boldsymbol{k}}_f} = -4\frac{k_f}{k_i}\Im \left\{ \int \mathrm{d}\boldsymbol{r}_1\mathrm{d}\boldsymbol{r}_2 \int \mathrm{d}\boldsymbol{r}_1'\mathrm{d}\boldsymbol{r}_2' \left[\psi_i^+(\boldsymbol{r}_2')\right]^* V_c(|\boldsymbol{r}_1' - \boldsymbol{r}_2'|)\psi_f^-(\boldsymbol{r}_1') \phi_c(\boldsymbol{r}_1')G(\boldsymbol{r}_2', \boldsymbol{r}_2; E_{\mathrm{exc}})\phi_c(\boldsymbol{r}_1) \psi_i^+(\boldsymbol{r}_2)V_c(|\boldsymbol{r}_1 - \boldsymbol{r}_2|)\left[\psi_f^-(\boldsymbol{r}_1)\right]^* \right\} \;. \tag{2.110}$$

Introducing the exchange transition operator

$$T_e(\boldsymbol{r}_2) \equiv \int \mathrm{d}\boldsymbol{r}_1 \phi_c(\boldsymbol{r}_1)V_c(|\boldsymbol{r}_1 - \boldsymbol{r}_2|)\left[\psi_f^-(\boldsymbol{r}_1)\right]^* \;, \tag{2.111}$$

we can write (2.110) as

$$-\Im \left\{ \iint \mathrm{d}\boldsymbol{r}_1\mathrm{d}\boldsymbol{r}_1' \; \left[\psi_i^+(\boldsymbol{r}_1')\right]^* \left[T_e(\boldsymbol{r}_1')\right]^* G(\boldsymbol{r}_1', \boldsymbol{r}_1; E_{\mathrm{exc}})T_e(\boldsymbol{r}_1)\psi_i^+(\boldsymbol{r}_1) \right\} \;. \tag{2.112}$$

Similarly, when operating only on the primed variables, we get the mixed term

$$\iint \mathrm{d}\boldsymbol{r}_1\mathrm{d}\boldsymbol{r}_1' \; \left[\psi_i^+(\boldsymbol{r}_1')\right]^* \left[T_e(\boldsymbol{r}_1')\right]^* G(\boldsymbol{r}_1', \boldsymbol{r}_1; E_{\mathrm{exc}})T_d(\boldsymbol{r}_1)\phi_c(\boldsymbol{r}_1) \;, \tag{2.113}$$

whereas, by operating only on the unprimed variables, we get the other mixed term

$$\iint \mathrm{d}\boldsymbol{r}_1 \mathrm{d}\boldsymbol{r}'_1 \; \phi_c(\boldsymbol{r}'_1) \left[T_d(\boldsymbol{r}'_1)\right]^* G(\boldsymbol{r}'_1, \boldsymbol{r}_1; E_{\text{exc}}) T_e(\boldsymbol{r}_1) \psi_i^+(\boldsymbol{r}_1) \, . \tag{2.114}$$

In the expression of the total cross section, the direct and full exchange term come with a positive sign, whereas the mixed terms come with a negative sign. These latter should be considered together before taking the imaginary part. For what concerns the singular part of the GF, these last three expressions can be easily calculated on the basis of (2.97) and (2.109), *mutatis mutandis*. The regular part does not pose particular problems.

All EELS integrals in the program for the regular and irregular part of the Green's function are calculated following the formulas presented in this section.

It remains to show that in (2.21) one can apply the operator $\Im$ in front of the integral. To this purpose we note that the scattering part of the GF in (2.107) is invariant under interchange of $\boldsymbol{r}$ and $\boldsymbol{r}'$, provided real spherical harmonics are used, since in this case the scattering path operator τ and the site t-matrix are symmetric in the site and angular momentum indices. This is also true for the singular part, as apparent from the expression in square brakets in (2.108). Moreover we note that all the various terms in the EELS cross section can be written as

$$\left\{ \iint \mathrm{d}\boldsymbol{r}_1 \mathrm{d}\boldsymbol{r}'_1 \; \left[A(\boldsymbol{r}'_1)\right]^* \Im \left[G(\boldsymbol{r}'_1, r_1; E_{\text{exc}})\right] A(\boldsymbol{r}_1) \right\} \, , \tag{2.115}$$

provided the two mixed exchange terms are lamped together.

By interchanging the integration coordinates in the integrand and taking into account the invariance property of the GF under this operation, we find that this expression is equal to its complex conjugate

$$\left\{ \iint \mathrm{d}\boldsymbol{r}_1 \mathrm{d}\boldsymbol{r}'_1 \; A(\boldsymbol{r}'_1) \Im \left[G(\boldsymbol{r}'_1, \boldsymbol{r}_1; E_{\text{exc}})\right] \left[A(\boldsymbol{r}_1)\right]^* \right\} \, . \tag{2.116}$$

We can therefore move the $\Im$ sign in front of the integral.

References

1. D. Sébilleau, C.R. Natoli, M. Gavaza, H. Zhao, F. Da Pieve, K. Hatada, Comp. Phys. Commun. **182**, 2567–2579 (2011)
2. D. Sébilleau, R. Gunnella, Z.-Y. Wu, S. Di Matteo, C.R. Natoli, J. Phys.: Condens. Matter **18**, R175–R230 (2006)
3. L.F. Canto, M.S. Hussein, *Scattering Theory of Molecules. Atoms and Nuclei* (World Scientific, Singapore, 2013)
4. G. Breit, H. Bethe, Phys. Rev. **93**, 888 (1954)
5. B.H. Bransden, C.J. Joachain, *Physics of Atoms and Molecules* (Longman, London, 1983)
6. R.V. Vedrinskii, V.L. Kraizman, A.A. Novakovich, VSh Machavariani, J. Phys.: Condens. Matter **4**, 6155–6169 (1992)
7. K. Hatada, K. Hayakawa, M. Benfatto, C.R. Natoli, J. Phys.: Condens. Matter **22**, 185501 (24pp) (2010)

8. D. Sébilleau, C.R. Natoli, J. Phys.: Conf. Ser. **190**, 012002 (2009)
9. C.R. Natoli, M. Benfatto, S. Doniach, Phys. Rev. A **34**, 4682–4694 (1986)
10. C.R. Natoli, M. Benfatto, S. Della Longa, K. Hatada, J. Synchrotron Rad. **10**, 26–42 (2003)
11. A.L. Ankudinov, S.I. Zabinsky, J.J. Rehr, Comp. Phys. Commun. **98**, 359–364 (1996)
12. P. Löwdin, Adv. Phys. **5**, 3–172 (1956)
13. L.F. Mattheiss, Phys. Rev. **133**, A1399–A1403 (1964)
14. J.G. Norman, Mol. Phys. **81**, 1191–1198 (1974)
15. I.T. Wille, P.J. Durham, P.A. Sterne, J. Phys. (Paris) **47**(C8), 43–45 (1986)
16. L. Hedin, S. Lundqvist, Solid State Phys. **23**, 1–181 (1969)
17. P.A. Lee, G. Beni, Phys. Rev. B **15**, 2862–2883 (1977)
18. J.J. Kas, A.P. Sorini, M.P. Prange, L.W. Cambell, J.A. Soininen, Phys. Rev. B **76**, 195116 (2007)
19. J. Mustre de Leon, J.J. Rehr, S.I. Zabinsky, R.C. Albers, Phys. Rev. B **44**, 4146 (1991)
20. J.H. Wood, A.M. Boring, Phys. Rev. B **18**, 2701 (1978)
21. R. Resta, Phys. Rev. B **16**, 2717–2722 (1977)

Chapter 3
Real Space Full Potential Multiple Scattering Theory

Keisuke Hatada and Calogero R. Natoli

Abstract We show how to implement a Full Potential Multiple Scattering (FPMS) code based on a real-space FPMS theory valid for both continuum and bound states, under conditions for space partitioning that are less restrictive than those applied so far. This theory is free from the need to expand cell shape functions in spherical harmonics or to use rectangular matrices. Tests of the program show that it is able to reproduce with very good accuracy known solutions of the Schrödinger equation. Applications to the spectroscopy of low dimensional systems, such as one-dimensional (1D) chain like systems, 2D layered systems and 3D diamond structure systems, where the Muffin-Tin approximation is known to give very poor results, show a remarkable improvement toward the agreement with experiments. The default mode of the code uses superimposed atomic charge densities, which works satisfactorily in most of the applications, but with help of the ES2MS interface, incorporated in the program, one can also use self-consistent charge densities derived from the VASP program. The program is also incorporated in the photoelectron diffraction code MSSPEC and parallelized for energy point.

3.1 Introduction

Multiple Scattering (MS) theory is and has been one of the techniques of election for solving the Schrödinger Equation (SE) due to its suggestive description of the electronic structure of solids and spectroscopic response functions, which appeals to our physical intuition. For its implementation, one partitions the space into non-overlapping domains (cells), solves the differential equation separately in each of

K. Hatada (✉)
Department Chemie, Ludwig-Maximilians-Universität München,
Butenandtstr. 5-13, 81377 Munich, Bavaria, Germany
e-mail: keisuke.hatada@cup.uni-muenchen.de

C. R. Natoli
LNF-INFN, Via E. Fermi 40, 00044 Frascati, Italy
e-mail: calogero.natoli@lnf.infn.it

D. Sébilleau et al. (eds.), *Multiple Scattering Theory for Spectroscopies*,
Springer Proceedings in Physics 204,
https://doi.org/10.1007/978-3-319-73811-6_3

the cells and then assembles together the partial solutions into a global solution that is continuous and smooth across the whole space and satisfies the given boundary conditions. The theory was proposed originally by Korringa and by Kohn and Rostoker (KKR) as a viable method for calculating the electronic structure of solids [1, 2] and was later extended to the calculation of bound states of polyatomic molecules by Slater and Johnson [3] and continuum states by Dill and Dehmer [4].

A characteristic feature of the method is the complete separation between the dynamical aspect of the system under consideration, represented by the cell scattering power, from the structural aspect of the problem, that reflects the geometrical position of the atoms in space. Another advantage of the theory is that one can write an explicit form of the Green's function (GF) (the solution of the SE with a delta-like source term) which is essential for the description of many properties of the systems under investigation.

For ease of computation, traditionally the KKR method has been implemented within the so-called muffin-tin (MT) approximation, in which the potential is spherically averaged inside non overlapping spheres (usually containing a physical atom) and takes a constant value in the interstitial region. However it is known that this approximation is only good for closed-packed and works poorly for covalently bonded and low dimensional systems, like surfaces, sparse (and/or) layered systems and diamond like structures [5–8]. Moreover, the introduction of empty spheres to reduce the interstitial volume does not mitigate the problem [9]. Although in some cases this introduction improves the calculations for X-ray Absorption Fine Structure (XAFS) and density of states (DOS) in bulk systems, where an angle-integrated feature is probed, it still generates spurious peaks in angle resolved low-energy photoelectron diffraction spectra, due to the unphysical diffraction caused by the potential discontinuities between the physical and artificial scatterers.

Due to poor performance of the MT approximation, both for bound and continuum states, investigations to overcome this approach started quite early. In their pioneering work [10] Williams and van Morgan reformulated the MS theory for arbitrary local potentials by partitioning the space with space filling truncated cells and applied with success the method to a model of crystalline silicon for which exact numerical solutions were available. The introduction of empty cells (EC) was found necessary for the diamond type lattice of silicon in order to represent adequately the potential in regions of substantial anisotropy and to satisfy some geometrical constraint imposed by the re-expansion of the free GF around two sites.

They showed that the practical implementation of the method did not imply large increases in computation with respect to the MT approach. The only point of difference was the calculation of the single site scattering power ($T_{LL'}$ matrix, no more diagonal in the angular momentum indexes), for which they developed the variable phase method to solve the SE for the truncated cell potential. However the implications of the truncation of the angular momentum expansion necessary in the numerical implementation of the method, posing convergence problems, were not analyzed and remained unanswered.

Many features of the full potential MS (FPMS) theory have limited its applications. The need to expand the truncated potential (or, equivalently, the cell shape function) in spherical harmonics (SH) (giving rise to the well known Gibbs phenomenon, like in Fourier expansion), the need to converge internal sums arising from the re-expansion of the free GF around two sites, entailing the use of rectangular matrices, the geometrical restrictions on the space partitioning cells induced by this re-expansion, the solution of a fairly complicated system of coupled differential equations to determine the local (cell) solutions, based on the phase function method, the question of the angular momentum convergence of the whole theory have all contributed to the slow progress of the theory. In the few cases in which the FPMS theory has been applied, the general attitude has been the empirical approach used in [10, 11]. The reader is referred to the book of Gonis and Butler [12] (and references therein) for a discussion on these points.

In [7] we have presented a derivation of a real-space FPMS theory, valid both for continuum and bound states, that is free from the drawbacks described above (in particular the need to expand cell shape functions in SH and the introduction of rectangular matrices) under conditions for space partitioning that are less restrictive than those previously applied. This approach eliminates the inconveniences of multiple scattering theory formulated in the muffin-tin approximation (MTA), while preserving its ease and simplicity of application. We have implemented the FPMS code based on this theory.

3.2 Multiple Scattering Theory

As anticipated in the introduction, MS theory is a method for solving the SE with an effective (optionally complex for scattering states) potential $V(\boldsymbol{r})$

$$\left[\nabla^2 + E - V(\boldsymbol{r})\right] \psi_{\boldsymbol{k}}(\boldsymbol{r}) = 0 \,, \tag{3.1}$$

with incoming wave boundary conditions [13] in the case of photoemission (emission of an electron of wave-vector $\boldsymbol{k}$ into a continuum state) or photoabsorption (obtained by integrating over all photoemission angles). When dealing with bound states one instead imposes the usual boundary condition that $\lim_{r\to\infty} \psi(\boldsymbol{r}) = 0$, dropping the unnecessary index $\boldsymbol{k}$. We use throughout Rydberg units for energies and atomic units for lengths.

Through the introduction of the free GF $G_0^+(\boldsymbol{r} - \boldsymbol{r}')$ obeying the equation

$$\left[\nabla^2 + E\right] G_0^+(\boldsymbol{r} - \boldsymbol{r}') = \delta(\boldsymbol{r} - \boldsymbol{r}') \,, \tag{3.2}$$

one can transform the differential equation with boundary conditions into an integral problem known as the Lippmann–Schwinger equation (LSE)

$$\psi_{\boldsymbol{k}}(\boldsymbol{r}) = \mathrm{e}^{\mathrm{i}\boldsymbol{k}\cdot\boldsymbol{r}} + \int G_0^+(\boldsymbol{r} - \boldsymbol{r}')\, V(\boldsymbol{r}')\, \psi_{\boldsymbol{k}}(\boldsymbol{r}')\, \mathrm{d}^3 r' \,, \tag{3.3}$$

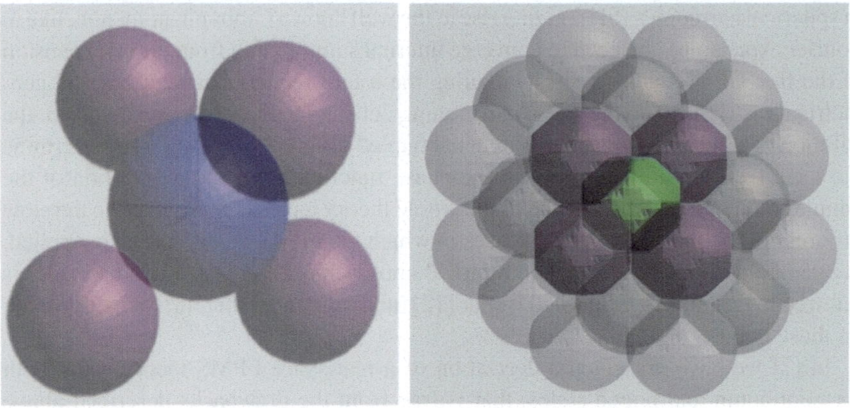

Fig. 3.1 Partition of space in non overlapping cells compared with a MT partition

where the inhomogeneous term $e^{i\boldsymbol{k}\cdot\boldsymbol{r}}$ is set to zero when dealing with bound states, in which case $k = i\sqrt{|E|}$ in the analytical continuation of G_0^+.

In Chap. 2 of this book a method of solution of this equation in the framework of MS theory has been given. We present here an alternative method that will allow us to highlight some additional features of MS equations.

In order to solve (3.3), one introduces a partition of the space into non overlapping domains Ω_j, called cells, and at the same time defines a partition of the potential that follows that of the space, so that $V(\boldsymbol{r}) = \sum_j v_j(\boldsymbol{r})$, where $v_j(\boldsymbol{r})$ coincides with $V(\boldsymbol{r})$ within the cell Ω_j and is zero outside. We make the assumption that the potential is short-ranged, in the sense that it behaves as $1/r^{2+\varepsilon}$ with positive ε at great distances. We also assume that a finite neighborhood around the origin of each cell lies in the domain of the cell. Figure 3.1 illustrates the partition of space in the case of a five atoms in tetrahedral coordination compared with a MT partition. The cells are BCC Voronoi polyhedra. Empty cells (EC) are introduced in order to cover the regions of space in which the potential is not negligible.

We start from the following identity involving surface integrals

$$\sum_{j=1}^{N} \int_{S_j} \left[G_0^+(\boldsymbol{r}-\boldsymbol{r}')\nabla\psi_{\boldsymbol{k}}(\boldsymbol{r}') - \psi_{\boldsymbol{k}}(\boldsymbol{r}')\nabla G_0^+(\boldsymbol{r}-\boldsymbol{r}')\right]\cdot \boldsymbol{n}_j \, \mathrm{d}\sigma_j'$$

$$= \int_{S_o} \left[G_0^+(\boldsymbol{r}-\boldsymbol{r}')\nabla\psi_{\boldsymbol{k}}(\boldsymbol{r}') - \psi_{\boldsymbol{k}}(\boldsymbol{r}')\nabla G_0^+(\boldsymbol{r}-\boldsymbol{r}')\right]\cdot \boldsymbol{n}_o \, \mathrm{d}\sigma_o' \quad , \qquad (3.4)$$

valid for all $\boldsymbol{r}$ lying in the neighborhood of the origin of each cell. Here $\Omega_o = \sum_j \Omega_j$, with surface S_o. We choose the number of cells N in such a way as to cover the region of space beyond which the potential reaches its asymptotic behavior.

To proceed further, we remember the following expansion of the GF

$$
\begin{aligned}
G_0^+(\boldsymbol{r}-\boldsymbol{r}';E) &= \sum_L J_L(\boldsymbol{r};k)\tilde{H}_L^+(\boldsymbol{r}';k)\ (r<r') \\
&= \sum_L J_L(\boldsymbol{r}';k)\tilde{H}_L^+(\boldsymbol{r};k)\ (r>r'),
\end{aligned}
\tag{3.5}
$$

where $J_L^+(\boldsymbol{r}';k) = j_\ell(kr)\,Y_L(\hat{\boldsymbol{r}})$, $\tilde{H}_L^+(\boldsymbol{r}';k) = -\mathrm{i}\,h_\ell^+(kr)Y_L(\hat{\boldsymbol{r}})$, $j_\ell(kr)$ and $h_\ell^+(kr)$ being respectively the Bessel and Hankel function of the first kind (outgoing wave behavior at great distances), following the definition of [14] (Chap. 2, (2.20) and footnote).

We now introduce the functions $\Phi_L(\boldsymbol{r})$ which are local solutions of the SE behaving as J_L in the limit $r \to 0$. They form a complete set of basis functions such that the global scattering wave function can be expanded inside the cell region as [15]

$$
\psi_{\boldsymbol{k}}(\boldsymbol{r}) = \sum_L A_L(\boldsymbol{k})\Phi_L(\boldsymbol{r}) = \sum_{LL'} A_L(\boldsymbol{k})R_{L'L}(r)Y_{L'}(\hat{\boldsymbol{r}}).
\tag{3.6}
$$

Using the expansions (3.5) referred to the origin of cell i located at $\boldsymbol{R}_i$, so that $G_0^+(\boldsymbol{r}_i - \boldsymbol{r}_i') = \sum_L J_L(\boldsymbol{r}_{i<})\tilde{H}_L^+(\boldsymbol{r}_{i>})$, taking $\boldsymbol{r}_i = \boldsymbol{r} - \boldsymbol{R}_i$ near the origin, we readily arrive at the MST compatibility equations for the amplitudes $A_L^i(\boldsymbol{k})$ in cell i:

$$
\sum_{jL'} H_{LL'}^{ij} A_{L'}^j(\boldsymbol{k}) = Y_L(\hat{\boldsymbol{k}})\mathrm{e}^{\mathrm{i}\boldsymbol{k}\cdot\boldsymbol{R}_i}(k/\pi)^{1/2} = I_L^i(\boldsymbol{k}),
\tag{3.7}
$$

where

$$
H_{LL'}^{ij} = \int_{S_j} [\,\tilde{H}_L^+(\boldsymbol{r}_i)\nabla\Phi_{L'}(\boldsymbol{r}_j) - \Phi_{L'}(\boldsymbol{r}_j)\nabla\tilde{H}_L^+(\boldsymbol{r}_i)\,]\cdot\boldsymbol{n}_j\,\mathrm{d}\sigma_j.
\tag{3.8}
$$

The term $I_L^i(\boldsymbol{k}) = \mathrm{i}^\ell Y_L(\boldsymbol{k})\,\mathrm{e}^{\mathrm{i}\boldsymbol{k}\cdot\boldsymbol{R}_{io}}\sqrt{\frac{k}{\pi}}$ comes from the surface integral over S_o in the rhs of (3.4), taking into account the asymptotic form of $\psi_{\boldsymbol{k}}(\boldsymbol{r}')$ in this region when normalized to one state per Rydberg (we refer for details to the Appendix A of [16]).

The usual derivation of the MS equations now proceeds by re-expanding $\tilde{H}_L^+(\boldsymbol{r}_i)$ around cell j in order to calculate the surface integrals over the various surfaces S_j, when $j \neq i$, using the relation

$$
\tilde{H}_L^+(\boldsymbol{r}_i) = \sum_{L'} G_{LL'}^{ij} J_{L'}(\boldsymbol{r}_j) \qquad (R_{ij} > r_j),
\tag{3.9}
$$

where $G_{LL'}^{ij}$ are the usual free electron propagator in the site and angular momentum basis (KKR real space structure factors) and $R_{ij} = |\boldsymbol{R}_i - \boldsymbol{R}_j|$. Unfortunately this re-expansion introduces a further expansion parameter L' into the theory (with related convergence problems) that is actually unnecessary, as we are going to show.

In fact, provided $R_{ij} > r_j$, one can show that the re-expansion (3.9) is absolutely and uniformly convergent in the angular domain [7]. In order to take advantage of this property, we impose the restriction that the partition of the space in cells satisfies the requirement that the shortest inter-cell vector R_{ij} joining the origins of the nearest-neighbor cells i and j, is larger than any intracell vector r_i or r_j, when r is inside cell i or j. If necessary, empty cells can be introduced to satisfy this requirement.

We observe now that the integrals over the surfaces of the various cells j can be calculated over the surfaces of the corresponding bounding spheres (with radius R_b^j) by application of the Green's theorem, since the potential is taken to be zero outside the domain of the cell. Due to the uniform convergence of the re-expansion (3.9) we obtain the following relations by integrating term by term

$$\int_{S_j} Y_{L'}(\hat{\boldsymbol{r}}_j)\, \tilde{H}_L^+(\boldsymbol{r}_i)\, \mathrm{d}\sigma_j = (R_b^j)^2\, G_{LL'}^{ij}\, j_{\ell'}(kR_b^j)$$

$$\int_{S_j} Y_{L'}(\hat{\boldsymbol{r}}_j)\, \nabla_j\, \tilde{H}_L^+(\boldsymbol{r}_i) \cdot \boldsymbol{n}_j\, \mathrm{d}\sigma_j = (R_b^j)^2\, G_{LL'}^{ij}\, \frac{\mathrm{d}}{\mathrm{d}R_b^j} \left[j_{\ell'}(kR_b^j) \right] , \tag{3.10}$$

which are exact for all L, L', as confirmed also by numerical tests. By inserting in (3.7) the expression for the basis functions expanded in spherical harmonics $\Phi_L(\boldsymbol{r}) = \sum_{L'} R_{L'L}(r) Y_{L'}(\hat{\boldsymbol{r}})$, we finally obtain

$$\sum_{L'} E_{LL'}^i A_{L'}^i(\boldsymbol{k}) - \sum_{j,L',L''}^{j \neq i} G_{LL''}^{ij} S_{L''L'}^j A_{L'}^j(\boldsymbol{k}) = I_L^i(\boldsymbol{k}) , \tag{3.11}$$

where, defining $W[f, g] = fg' - gf'$,

$$E_{LL'} = R_b^2\, W[-\mathrm{i}kh_\ell^+, R_{LL'}]; \qquad S_{LL'} = R_b^2\, W[j_\ell, R_{LL'}] , \tag{3.12}$$

which reduce to diagonal matrices for MT potentials, since in this case $R_{LL'}(r) = R_\ell(r)\delta_{LL'}$. Equation (3.11) looks formally similar to the usual MSE. However we notice that the sum over L'' runs over the angular momentum components of the basis functions and is not affected by convergence constraints related to the re-expansion (3.9). Therefore the only angular momentum indexes appearing in (3.11) are those of the radial functions $R_{LL'}$ in the expansion (3.6).

We now note that the truncation value for both indexes is the same and corresponds to the classical relation $\ell_{\max} = kR_b^j$, where R_b^j is the radius of the bounding sphere of the cell at site j. This is true for the index L, which is a reminder that the basis function Φ_L behaves like $j_\ell(kr)Y_L$ near the origin. Due to the properties of the spherical Bessel functions, when $\ell \gg kR_b^j$, Φ_L becomes very small inside the cell, decreasing like $[(2\ell + 1)!!]^{-1}$. Therefore its weight in the expansion (3.6) will be negligible. The other index L', as will be clear from the following (3.17), measures the response of the potential inside the cell to an incident wave J_L of angular momentum L into an outgoing wave of angular momentum L'. Due to the same argument as above, familiar to scattering theory, the scattering matrix $T^j_{L'L}$ will decrease like $[(2\ell+1)!!(2\ell'+1)!!]^{-1}$ (see Appendix B of [7] for $\ell, \ell' \gg kR^j_b$). As a consequence $\mathbf{E}^j$ and $\mathbf{S}^j$ can be considered to be square matrices.

To proceed further, we introduce new expansion coefficients defined by

$$C^i_L(\boldsymbol{k}) = \sum_{L'} E^i_{LL'} A^i_{L'}(\boldsymbol{k}) \,, \tag{3.13}$$

implying that we use new basis functions given by

$$\Psi_L(\boldsymbol{r}_i) = \sum_{L'} [\mathbf{E}]^{-1}_{L'L} \, \Phi_{L'}(\boldsymbol{r}_i) \,, \tag{3.14}$$

so that

$$\sum_L C^i_L(\boldsymbol{k}) \, \Psi_L(\boldsymbol{r}_i) = \sum_L A_L(\boldsymbol{k}) \Phi_L(\boldsymbol{r}_i) \,. \tag{3.15}$$

Then (3.11) can be rewritten as

$$C^i_L(\boldsymbol{k}) = I^i_L(\boldsymbol{k}) + \sum_{j \neq i} \sum_{L'L''} G^{ij}_{LL'} \, T^j_{L'L''} \, C^j_{L''}(\boldsymbol{k}) \,, \tag{3.16}$$

putting

$$T^j_{LL'} = \sum_{L''} S^j_{LL''} \, [\mathbf{E}^j]^{-1}_{L''L'} \,. \tag{3.17}$$

The comparison of (3.16) with (2.36) of Chap. 2 shows that $\Psi_L(\boldsymbol{r}_i)$ is a local scattering function, so that the cell T-matrix is given by the relation (3.17) in terms of the local solutions

$$\Phi_L(\boldsymbol{r}) = \sum_{L'} R_{L'L}(r) Y_{L'}(\hat{\boldsymbol{r}}) \,. \tag{3.18}$$

A more direct way to derive (3.17) is to remember the definition of the cell T-matrix $T^j_{L'L}$ as the scattering amplitude into an outgoing spherical wave of angular momentum L' in response to an exciting spherical wave of angular momentum L. This can be expressed in terms of the local scattering wave function $\Psi_L(\boldsymbol{r}_j)$ obeying the LSE (3.3) for one cell j. Omitting for simplicity the index j in $\boldsymbol{r}_j$, we have

$$\Psi_L(\boldsymbol{r}) = J_L(\boldsymbol{r};k) + \int G^+_0(\boldsymbol{r}-\boldsymbol{r}') \, v_j(\boldsymbol{r}') \, \Psi_L(\boldsymbol{r}') \, \mathrm{d}^3 r' \,, \tag{3.19}$$

where $J_L(\boldsymbol{r};k)$ is the spherical wave component of angular momentum L of the incident plane wave $\mathrm{e}^{i\boldsymbol{k}\cdot\boldsymbol{r}} = 4\pi\sum_L \mathrm{i}^\ell Y_L(\hat{\boldsymbol{k}})J_L(\boldsymbol{r};k)$. Then using (3.5) in the limit $r \to \infty$, we obtain the relation

$$\Psi_L(\boldsymbol{r}) = J_L(\boldsymbol{r};k) + \sum_{L'} \tilde{H}^+_{L'}(\boldsymbol{r};k)\, T^j_{L'L} \,, \tag{3.20}$$

putting $T^j_{L'L} = \int J_{L'}(\boldsymbol{r}';k)\, v_j(\boldsymbol{r}')\, \Psi_L(\boldsymbol{r}';k)\, \mathrm{d}^3r'$. From this expression, using (3.14) and transforming the volume integral into a surface integral by application of Green's theorem, we derive (3.17).

Summarizing, in order to generate a solution of the LSE, we need to calculate the local solutions $\Phi_L(\boldsymbol{r})$ and solve (3.16) for the coefficients C^i_L.

3.2.1 The Local Solutions

One of the key ingredients of our approach to FPMS [7] is the new scheme to generate local basis functions for the truncated potential. Starting from the SE written in polar coordinates, after elimination of the radial first derivative, the local solution $P_L(\boldsymbol{r}) = r\ \Phi_L(\boldsymbol{r})$ satisfies the equation

$$\left[\frac{\mathrm{d}^2}{\mathrm{d}r^2} + E - V(r,\hat{\boldsymbol{r}})\right] P_L(r,\hat{\boldsymbol{r}}) = \frac{1}{r^2}\tilde{L}^2 P_L(r,\hat{\boldsymbol{r}}) \,, \tag{3.21}$$

where $\tilde{L}^2$ is the angular momentum operator, whose action on $P_L(r,\hat{\boldsymbol{r}})$ can be calculated in terms of the expansion $\Phi_L(\boldsymbol{r}) = \sum_{L'} R_{L'L}(r)Y_{L'}(\hat{\boldsymbol{r}})$ as:

$$\tilde{L}^2 P_L(r,\hat{\boldsymbol{r}}) = \sum_{L'} \ell'(\ell'+1) r R_{L'L}(r) Y_{L'}(\hat{\boldsymbol{r}}) \,, \tag{3.22}$$

and we have called $V(r,\hat{\boldsymbol{r}})$ the truncated cell potential, which coincides with the true potential inside the cell and is zero outside. We remind that the index L of $P_L(\boldsymbol{r})$ is reminiscent of its behaviour at the origin: $\lim_{r\to 0} P_L(\boldsymbol{r}) \sim J_L(\boldsymbol{r})$. Its expansion in SH does not pose convergence problems, since it is continuous with its first derivative; however we do not expand the potential, but we use an extension of the Numerov's method to solve it. In fact (3.21) in the variable r looks like a second order radial equation with an inhomogeneous term. Accordingly, putting $f^L_{i,j} = P_L(r_i,\hat{\boldsymbol{r}}_j)$ and dropping for simplicity the index L, the associated three point recursion relation in Numerov approach is

$$A_{i+1,j} f_{i+1,j} - B_{i,j} f_{i,j} + A_{i-1,j} f_{i-1,j} = g_{i,j} - \frac{h^6}{240} f^{vi}_{i,j} \,, \tag{3.23}$$

where,

$$
\begin{aligned}
A_{i,j} &= 1 - \frac{h^2}{12} v_{i,j} \\
B_{i,j} &= 2 + \frac{5h^2}{6} v_{i,j} = 12 - 10A_{i,j} \\
v_{i,j} &= V(r_i, \hat{\boldsymbol{r}}_j) - E \\
g_{i,j} &= \frac{h^2}{12}[q_{i+1,j} + 10q_{i,j} + q_{i-1,j}] \\
q_{i,j} &= \frac{1}{r_i^2} \sum_{L'} \ell'(\ell'+1) r_i R_{L'L}(r_i) Y_{L'}(\hat{\boldsymbol{r}}_j) \,.
\end{aligned} \tag{3.24}
$$

Here i is the index of the radial mesh points and j the index of the angular points on a Lebedev surface grid [17]. Accordingly, $r_i R_{L'L}(r_i) = \sum_j w_j P_L(r_i, \hat{\boldsymbol{r}}_j) Y_{L'}(\hat{\boldsymbol{r}}_j)$, where w_j is the weight function for angular integration associated with the chosen grid. The number of surface points N_{Leb} is given by $N_{\text{Leb}} \approx (2\ell_{\max} + 1)^2/3$ as a function of the maximum angular momentum used [18], taking into account that one has to integrate the product of two spherical harmonics.

As it is, we cannot use (3.23) to find $f_{i+1,j}$ by iteration, from the knowledge of $f_{i,j}$ and $f_{i-1,j}$ at all the angular points, since the *inhomogeneous* term $q_{i+1,j}$ is not expressible in terms of $f_{i+1,j}$ due to the last line of (3.24), calculated at the radial mesh point $i + 1$.

In order to eliminate this point from the expression of $g_{i,j}$, we observe that

$$
\begin{aligned}
g_{i,j} &= \frac{h^2}{12} \left[q_{i+1,j} + 10q_i + q_{i-1,j} \right] \\
&= \frac{h^2}{12} \left[\frac{q_{i+1,j} - 2q_i + q_{i-1,j}}{h^2} h^2 + 12 q_{i,j} \right] .
\end{aligned} \tag{3.25}
$$

The second order central difference is given by [19]

$$
q_{i+1} - 2q_i + q_{i-1} = h^2 q_i'' + \frac{h^4}{12} q_i^{iv} + \frac{h^6}{360} q_i^{vi} + \frac{h^8}{20160} q_i^{viii} + \cdots \tag{3.26}
$$

so that

$$
g_{i,j} \sim \frac{h^2}{12} \left[\left(q_{i,j}'' + \frac{h^2}{12} q_{i,j}^{iv} \right) h^2 + 12 q_{i,j} \right] , \tag{3.27}
$$

omitting errors of order h^6 and higher.

Now for the second derivative $q_{i,j}''$ we use the backward formula [19]

$$
q_{i,j}'' = \frac{q_{i,j} - 2q_{i-1,j} + q_{i-2,j}}{h^2} + h q_{i,j}''' - \frac{7h^2}{12} q_{i,j}^{iv} , \tag{3.28}
$$

to avoid the contribution of the point $i+1$. Inserting (3.28) into (3.27)

$$g_{i,j} \sim \frac{h^2}{12}\left[13q_{i,j} - 2q_{i-1,j} + q_{i-2,j}\right] + \frac{h^5}{12}q'''_{i,j} - \frac{h^6}{24}q^{iv}_{i,j}\,, \tag{3.29}$$

which is the formula we wanted to arrive at. Therefore our modified Numerov procedure becomes:

$$A_{i+1,j}f_{i+1,j} - B_{i,j}f_{i,j} + A_{i-1}f_{i-1,j} = g_{i,j} + \frac{h^5}{12}q'''_{i,j}\,, \tag{3.30}$$

where,

$$\begin{aligned} A_{i,j} &= 1 - \frac{h^2}{12}v_{i,j} \\ B_{i,j} &= 2 + \frac{5h^2}{6}v_{i,j} = 12 - 10A_{i,j} \\ g_{i,j} &= \frac{h^2}{12}\left[13q_{i,j} - 2q_{i-1,j} + q_{i-2,j}\right]\,, \end{aligned} \tag{3.31}$$

which now needs three backward points to start.

The appearance of the third r derivative of q'''_i in (3.30), which is strictly infinite at the step point, does not cause practical problems. Although not necessary, one can always assume a smoothing of the potential at the cell boundary *à la Becke* [20], reducing at the same time the mesh h, so that the error at that particular step point is negligible.

In this way, at the cost of a larger error $O(h^5)$ compared to the original Numerov formula and the introduction of a further backward point (three points i, $i-1$ and $i-2$ are now involved in (3.31)), the three-dimensional discretized equation can be solved along the radial direction for all angles in an onion-like way, provided the expansion (3.22) is performed at each new radial mesh point to calculate $q_{i,j}$. We use a log-linear mesh $\rho = \alpha r + \beta \ln r$, to reduce numerical errors around the origin and the bounding sphere [21]. In reality, in the program we divide the radial mesh into two regions, the first one ranging from the origin to the radius of the sphere inscribed in the cell (usually the MT radius), the second one from this radius to the bounding radius of the cell. The above method is used only in this second region, which crosses the discontinuities of the truncated potential, whereas in the first region one can safely expand the potential in spherical harmonics, due to the continuity of its first derivatives and the fact that only few multipoles are appreciable. Here we can project (3.21) onto $Y_L(\hat{\boldsymbol{r}})$ obtaining

$$\sum_{L''}\left[\left(-\frac{\mathrm{d}^2}{\mathrm{d}r^2} + \frac{\ell(\ell+1)}{r^2} - E\right)\delta_{LL''} + V_{LL''}(r)\right]X_{L''L'}(r)\,, \tag{3.32}$$

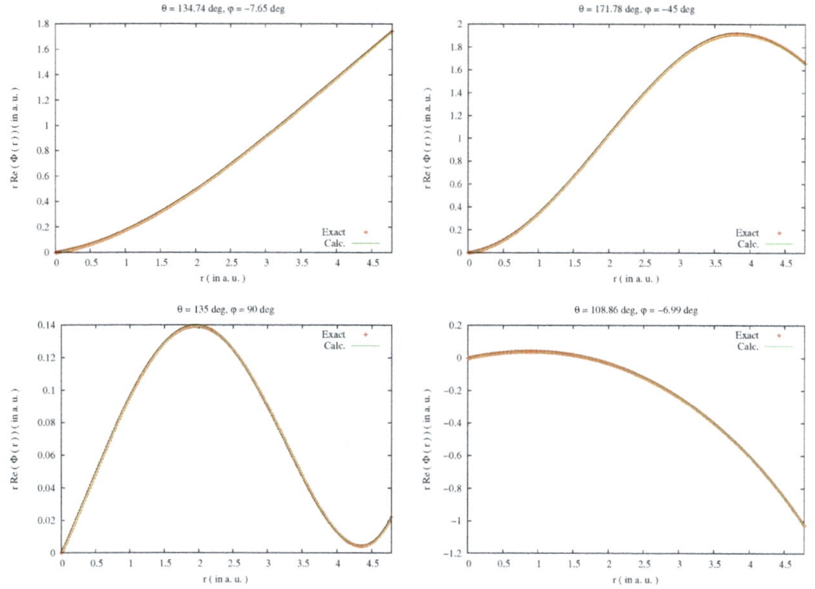

Fig. 3.2 Comparison between the real part of the analytical solution of the truncated potential well described in the text with the one generated by the 3D Numerov method at four different polar angles

where $X_{L\,L'}(r) = r\,R_{L\,L'}(r)$. This equation is then solved by the matrix Numerov method with a Gaussian elimination procedure. We refer for details to [7]. This method of generating $P_L(\boldsymbol{r})$ is simple, fast, efficient, valid for any shape of the cell and reduces to the minimum the number of SH in the expansion of the scattering wavefunction. The cell T-matrix is then calculated according to (3.17).

We have checked that the method works by comparing with known solutions of the SE [7]. As a further example, Fig. 3.2 shows the comparison between the analytical and the numerical solution for certain directions in the special case of the truncated potential $V(x, y, z) = a\,\theta(|x| - R_c) + b\,\theta(|y| - R_c) + c\,\theta(|z| - R_c)$, where θ is the step function, $R_c = 3.78\,\text{au} = 2.0\,\text{Å}$ and $a = -0.05,\ b = -0.1,\ c = -0.15\,\text{Ryd}$, for an energy $E = 0.3$ Ryd (for the solution see [22]). For this comparison we used an $\ell_{\max} = 7$ and a number of surface points on the Lebedev grid equal to 266. For other comparisons we refer the reader to [7].

Summarizing, contrary to past approaches, we have avoided the double series expansion of the free GF around two centers, so that the angular momentum indexes LL' are the same as those of the T-matrix $T^j_{LL'}$ and originate from the function $R_{L'L}(r)$ in (3.6). As a consequence, the MS matrix $(\mathbf{I} - \mathbf{G}_0\mathbf{T})$ can be considered as square and we show in the following section that in the $\ell \to \infty$ its inverse exists, providing a firm ground for the use of FPMS theory as a viable method for electronic structure calculations and spectroscopic response functions, with the ease and versatility of the corresponding MT theory. The truncation parameter in the SH expansion is given by the classical relation $\ell_{max} = kR_b$, where k is the electron wavevector and R_b is the radius of the bounding sphere of the scattering cell.

We refer to [7] for details on how to use an outer sphere region in order to avoid introducing too many empty cells to cover the region of space where the asymptotic behavior of the scattering wave function is not yet reached or to use a constant potential (instead of zero) outside the cell in order to accelerate the L-convergence. The final result obviously does not depend on this constant.

3.2.2 *The L-Convergence of Full Potential Multiple Scattering Theory*

The inversion of the MS matrix in (3.16) is usually done by series expansion, whereby

$$(\mathbf{I} - \mathbf{G_0 T})^{-1} = \sum_n (\mathbf{G_0 T})^n \, . \tag{3.33}$$

Due to the meaning of the matrix $\left[G^{ij}_{LL'}\right]$, describing a free particle propagation of spherical waves from site i to site j, and of $T^i_{LL'}\delta_{ij}$, giving the scattering amplitude of the potential at site i, a photoemission process is usually seen as a propagation of the excited photoelectron from site to site, intercalated by scattering events due the atomic potentials. Hence the name of the theory.

However, while this series is absolutely convergent for nonoverlapping MT spheres, provided the spectral radius of the matrix $\mathbf{G_0 T}$ (i.e. the modulus of its highest eigenvalue) is less than one [23], it is known to diverge for the case of space-filling cells due to the asymptotic relation (see Appendix B of [7])

$$|G_{\ell\ell} T_{\ell\ell}| \approx R_b \left(\frac{2R_b}{R_{ij}} \right)^{2\ell+1} , \tag{3.34}$$

since in this case $2R_b > R_{ij}$, at least for nearest neighbors.

Due to this asymptotic behavior there is a widespread belief that in the case of space-filling cells the limiting procedure of inverting exactly an ℓ truncated MS matrix does not converge for $\ell \to \infty$. We want to show here that this is not the case, provided a slight modification of the free propagator G_0 is adopted.

To this purpose, let us start by solving the LSE (3.3) using the theory of integral equations, before applying the method of MST. We cannot apply Fredholm theory [24] to the solution of (3.3) as it stands, since the *kernel* for this integral equation

$$\mathbf{K}(\boldsymbol{r}', \boldsymbol{r}) = -\frac{1}{4\pi} \frac{e^{ik\cdot|\boldsymbol{r}'-\boldsymbol{r}|}}{|\boldsymbol{r}'-\boldsymbol{r}|} V(\boldsymbol{r}) \, , \tag{3.35}$$

is such that

$$\operatorname{Tr}(\mathbf{K}^\dagger\mathbf{K}) = \iint \mathrm{d}\boldsymbol{r}\mathrm{d}\boldsymbol{r}'\, \mathbf{K}^*(\boldsymbol{r}',\boldsymbol{r})\,\mathbf{K}(\boldsymbol{r}',\boldsymbol{r})$$
$$= \left(\frac{1}{4\pi}\right)^2 \iint \mathrm{d}\boldsymbol{r}\mathrm{d}\boldsymbol{r}'\, \frac{V^2(\boldsymbol{r})}{|\boldsymbol{r}'-\boldsymbol{r}|^2}\,, \tag{3.36}$$

and obviously diverges.

However, following the expedient of Sect. 10.3, p. 280 of [25], we can multiply the LSE (3.3) by $|V(\boldsymbol{r}')|^{1/2}$, writing at the same time $V(\boldsymbol{r}) = |V(\boldsymbol{r})|v(\boldsymbol{r})$, where $v(\boldsymbol{r})$ is a sign factor, equal to $+1$ where the potential is positive and to -1 where it is negative. Then the modified equation reads

$$\psi_s(\boldsymbol{r}';\boldsymbol{k}) \equiv |V(\boldsymbol{r}')|^{1/2}\,\psi(\boldsymbol{r}';\boldsymbol{k})$$
$$= |V(\boldsymbol{r}')|^{1/2}\,\mathrm{e}^{\mathrm{i}\boldsymbol{k}\cdot\boldsymbol{r}'}$$
$$+\,|V(\boldsymbol{r}')|^{1/2}\int G_0^+(\boldsymbol{r}'-\boldsymbol{r};k)\,|V(\boldsymbol{r})|^{1/2}\,v(\boldsymbol{r})\,\psi_s(\boldsymbol{r};\boldsymbol{k})\,\mathrm{d}^3r\,. \tag{3.37}$$

whose kernel is

$$\mathbf{K}_s(\boldsymbol{r}',\boldsymbol{r}) = -\frac{1}{4\pi}\,|V(\boldsymbol{r}')|^{1/2}\,\frac{\mathrm{e}^{\mathrm{i}\boldsymbol{k}\cdot|\boldsymbol{r}'-\boldsymbol{r}|}}{|\boldsymbol{r}'-\boldsymbol{r}|}\,|V(\boldsymbol{r})|^{1/2}\,v(\boldsymbol{r})\,, \tag{3.38}$$

Consequently

$$\operatorname{Tr}(\mathbf{K}_s^\dagger\mathbf{K}_s) = \iint \mathrm{d}\boldsymbol{r}\mathrm{d}\boldsymbol{r}'\, \mathbf{K}_s^*(\boldsymbol{r}',\boldsymbol{r})\,\mathbf{K}_s(\boldsymbol{r}',\boldsymbol{r})$$
$$= \left(\frac{1}{4\pi}\right)^2 \iint \mathrm{d}\boldsymbol{r}\mathrm{d}\boldsymbol{r}'\, \frac{V(\boldsymbol{r})\,|V(\boldsymbol{r}')|}{|\boldsymbol{r}'-\boldsymbol{r}|^2}$$
$$\leq \left(\frac{1}{4\pi}\right)^2 \iint \mathrm{d}\boldsymbol{r}\mathrm{d}\boldsymbol{r}'\, \frac{|V(\boldsymbol{r})|\,|V(\boldsymbol{r}')|}{|\boldsymbol{r}'-\boldsymbol{r}|^2}\,, \tag{3.39}$$

is finite for a large class of potentials (including the molecular ones). The kernel $\mathbf{K}_s$ is therefore of the Hilbert–Schmidt type and Fredholm theory for $\mathcal{L}_2$-kernels can be applied. We can then obtain the solution of (3.3) simply by dividing $\psi_s(\boldsymbol{r}';\boldsymbol{k})$ by $|V(\boldsymbol{r}')|^{1/2}$, except at points for which $|V(\boldsymbol{r}')|^{1/2} = 0$, where it can be defined by continuity.

Now, the application of MST to the LSE of (3.19) for the scattering wave functions $\Psi_L(\boldsymbol{r})$ transforms this equation into the infinite set of algebraic equations (3.16) for the coefficients $C_L(\boldsymbol{k})$, which in matrix form can be written as

$$(\mathbf{I}-\mathbf{K})\,\mathbf{C} = \mathbf{A}\,. \tag{3.40}$$

putting $\mathbf{K} = \mathbf{G_0T}$ and calling $\mathbf{A}$ the vector $I_L^i(\boldsymbol{k})$.

The matrix $\mathbf{K}$ here is not an operator of the Hilbert–Schmidt type, since Tr $(\mathbf{K}^\dagger \mathbf{K})$ diverges, due to (3.34) and in keeping with (3.36). However, following the procedure used above in passing from (3.3) to (3.37), we multiply both sides of (3.40) by $\mathbf{T}^{1/2}$ obtaining the new equation

$$(\mathbf{I} - \mathbf{K}_s)\,\mathbf{C}' = \mathbf{A}' \,, \tag{3.41}$$

where $\mathbf{C}' = \mathbf{T}^{1/2}\mathbf{C}$, $\mathbf{A}' = \mathbf{T}^{1/2}\mathbf{A}$ and

$$\mathbf{K}_s = \mathbf{T}^{1/2}\mathbf{G}_0\mathbf{T}^{1/2} \,. \tag{3.42}$$

In order to calculate the square root of the matrix $\mathbf{T}$, we first find a similarity transformation $\mathbf{S}$ such that $\mathbf{\Lambda} = \mathbf{S}\mathbf{T}\mathbf{S}^{-1}$ is diagonal. Then we put $\mathbf{T}^{1/2} = \mathbf{S}^{-1}\mathbf{\Lambda}^{1/2}\mathbf{S}$ so that $\mathbf{T}^{1/2}\mathbf{T}^{1/2} = \mathbf{S}^{-1}\mathbf{\Lambda}\mathbf{S} = \mathbf{T}$. These operations with the infinite matrix $\mathbf{T}$ can be safely performed since Tr $(\mathbf{T}^\dagger\mathbf{T}) < \infty$, as can be seen from the asymptotic behavior of its matrix element in Appendix B, (B10) of [7]. Consequently the limiting procedure of truncating the matrices, performing the various operations and then taking the limit to infinite dimensions, is well defined.

By virtue of (3.39), one can show that the kernel $\mathbf{K}_s$ is such that Tr $(\mathbf{K}_s^\dagger\mathbf{K}_s)$ is finite, provided a slight modification of the free propagator $\mathbf{G}_0$ is used (we refer to Appendix G of [7] for details). Following this modification, in order to calculate the matrix elements of $\mathbf{K}_s$, one should use the displaced cell approach [26], according to which

$$[\mathbf{K}_s]^{ij}_{LL'} = \sum_{\underline{\Lambda}} \left\{ \sum_{\Lambda\Lambda'} (T^i_{L\Lambda})^{1/2} \; J_{\Lambda\underline{\Lambda}}(\boldsymbol{b}) G_{\underline{\Lambda}\Lambda'}(\boldsymbol{R}_{ij} + \boldsymbol{b}) \; \left(T^i_{\Lambda' L'}\right)^{1/2} \right\} \,, \tag{3.43}$$

provided that the vector $\boldsymbol{b}$ is such that $|\boldsymbol{R}_{ij} + \boldsymbol{b}| > R^i_b + R^j_b$ and the sums inside the curly brackets is performed first. Here $J_{\Lambda\Lambda'}(\boldsymbol{b})$ is the usual translation operator in MST [26]. For $\boldsymbol{b} = 0$ this equation reduces to the customary one, since $J_{\Lambda\Lambda'}(\mathbf{0}) = \delta_{\Lambda\Lambda'}$. As is well known (see [24]), Tr $(\mathbf{K}_s^\dagger\mathbf{K}_s) < \infty$ is the condition for the existence of the determinant $|\mathbf{I} - \mathbf{K}_s|$ necessary to define its inverse, since by Hadamard's inequality, for any finite $L_{\max}$, one has

$$|\mathbf{I} - \mathbf{K}_s|^2 \le \Pi_L^{L_{\max}} \left(1 + \sum_{L'}^{L_{\max}} |\,[\mathbf{K}_s]_{LL'}\,|^2 \right) \,, \tag{3.44}$$

and in the limit $\ell_{\max} \to \infty$ the infinite product will converge if $\sum_{LL'} |\,[\mathbf{K}_s]_{LL'}\,|^2 \equiv$ Tr $(\mathbf{K}_s^\dagger\mathbf{K}_s) \le N < \infty$ [27].

This means that the sequency of inverse matrices obtained by inverting the finite matrices $\mathbf{I} - \mathbf{K}_s$ truncated at a certain $\ell_{\max}$ converges absolutely in the limit $\ell_{\max} \to \infty$. Once $\mathbf{C}'$ is obtained, the original problem is solved by putting $\mathbf{C} = \mathbf{T}^{-1/2}\mathbf{C}'$. Moreover the scattering path operator $\boldsymbol{\tau} = (\mathbf{I} - \mathbf{G}_0\mathbf{T})^{-1}$ is given by

$$\tau = \mathbf{T}^{1/2}\,(\mathbf{I} - \mathbf{K}_s)^{-1}\,\mathbf{T}^{1/2}\,. \tag{3.45}$$

It is worth noting here that the customary way of inverting the matrix $\mathbf{I} - \mathbf{K}_s$ by series expansion, using the formal relation

$$(\mathbf{I} - \mathbf{K}_s)^{-1} = \sum_n (\mathbf{K}_s)^n\,. \tag{3.46}$$

is not always viable. In fact, even if the kernel $\mathbf{K}_s$ is of the Hilbert- -Schmidt type but Tr $(\mathbf{K}_s^\dagger \mathbf{K}_s) \geq 1$, the series in (3.46) diverges, whereas the process of truncating and taking the inverse always converges. On the contrary, the series $\sum_n \mathbf{K}^n$ is always divergent, since Tr $(\mathbf{K}^\dagger \mathbf{K})$ is infinite.

In practical numerical applications one does not have to worry about modifying the structure constants as indicated by (3.43) since, for the cell geometries ordinarily encountered in the applications (see the restrictions described in Sect. 3.2 soon after (3.9)), ℓ-convergence in the ℓ-truncation procedure of the MS matrix shows up much earlier than what predicted by the onset of divergence in (3.34), written with the unmodified structure constants $G^{ij}_{LL'}$. An example of this behavior was found in [5] for the $GeCl_4$ molecule. Similar results were found for other compounds. This behavior is probably indicative of the fact that the sequence of the inverse $(\mathbf{I} - \mathbf{G_0 T})^{-1}$, truncated as a function of $\ell_{\max}$, is asymptotic in character, as is the expansion (3.33). All this justifies the empirical attitude of many authors of MST to truncate the MS matrix (or the MS series) to a certain $\ell_{\max}$ and then check if the results change by increasing this value by one or two.

3.2.3 Construction of the Green's Function in MST

One of the major advantages of MST is the possibility to write the Green's function of the system, solution of the SE with a source term

$$(\nabla^2 + E - V(\boldsymbol{r}))\,G(\boldsymbol{r}, \boldsymbol{r}'; E) = \delta(\boldsymbol{r} - \boldsymbol{r}')\,. \tag{3.47}$$

in an explicit form. This is important both for writing down spectroscopic response functions (see [16]) and for the calculation of ground state properties through contour integration in the complex energy plane (see e.g. [11] and references therein).

In the derivation of the explicit expression for the GF, we shall follow the method used in [28] for the MT case, generalizing it to the full potential case.

First of all, we need an expression for the GF in the case of an isolated cell. If we write

$$G^+(\boldsymbol{r} - \boldsymbol{r}') = \sum_{LL'} Y_L(\hat{\boldsymbol{r}}) g_{LL'}(r, r') Y_{L'}(\hat{\boldsymbol{r}}')\,, \tag{3.48}$$

the radial part $g_{LL'}(r, r')$ of the GF is solution of the equation

$$\sum_{L''} D_{L'L''}(r)\, g_{L''L}(r, r') = \frac{1}{r^2}\delta(r - r')\delta_{LL'} \,, \tag{3.49}$$

where the operator $D_{L'L''}(r)$ is given by

$$D_{L'L''}(r) = \left[\left(D(r) + E - \frac{\ell(\ell+1)}{r^2}\right)\delta_{L'L''} - V_{L'L''}(r)\right]$$
$$D(r) = \frac{1}{r^2}\frac{\mathrm{d}}{\mathrm{d}r}r^2\frac{\mathrm{d}}{\mathrm{d}r} \,. \tag{3.50}$$

By projecting the single site regular solution (3.18) of the SE onto the SH $Y_L(\hat{\boldsymbol{r}})$ we easily see that

$$\sum_{L''} D_{L'L''}(r)\, R_{L''L}(r) = 0 \,. \tag{3.51}$$

We also introduce the irregular solution $\tilde{H}_{L'L}(r)$ that matches smoothly to $\tilde{h}_L(r)\delta_{LL'}$ at the bounding sphere of the cell, so that

$$\sum_{L''} D_{L'L''}(r)\, \tilde{H}_{L''L}(r) = 0 \,. \tag{3.52}$$

From (3.51) and (3.52) we easily derive that

$$\frac{\mathrm{d}}{\mathrm{d}r}\sum_{L''}\left(r^2 W[\tilde{H}_{L''L}, R_{L''L'}]\right) = 0 \,. \tag{3.53}$$

In analogy with the general method of second order equations with a δ-function source term, we try

$$g_{L'L}(r, r') = \sum_{\Lambda\Lambda'} R_{\Lambda\Lambda'}(r_<) c_{\Lambda' L} \tilde{H}_{\Lambda L'}(r_>) \,, \tag{3.54}$$

where $r_<$ ($r_>$) is the lesser (greater) of r, r', and determine the coefficients $c_{\Lambda\Lambda'}$ from the condition that the jump of the first derivatives in r and r' of $g_{L'L}(r, r')$ when $r = r'$ be $1/r^2\delta_{LL'}$. In other words we require that

$$\left[\frac{\mathrm{d}}{\mathrm{d}r} g_{L'L}(r, r') - \frac{\mathrm{d}}{\mathrm{d}r'} g_{L'L}(r, r')\right]\Bigg|_{r=r'} = \sum_{\Lambda\Lambda'} c_{\Lambda' L} W[\tilde{H}_{\Lambda' L'}, R_{\Lambda\Lambda'}]|_{r=r'} = \frac{1}{r^2}\delta_{LL'}. \tag{3.55}$$

Due to (3.53) we can calculate this expression at the bounding sphere of the cell with radius R_b, so that

$$\sum_{\Lambda} R_b^2 \, W[\tilde{h}_{L'}, R_{L'\Lambda}] c_{\Lambda L} = \delta_{LL'} \; . \tag{3.56}$$

Therefore $c_{LL'} = [\mathbf{E}]^{-1}_{LL'}$ due to (3.12). Hence from (3.48), in view of (3.14), we find

$$G^+(\boldsymbol{r} - \boldsymbol{r}') = \sum_L \Psi_L(\boldsymbol{r}_<) \, \tilde{\mathcal{H}}_L(\boldsymbol{r}_>) \; . \tag{3.57}$$

putting $\tilde{\mathcal{H}}_L(\boldsymbol{r}) = \sum_{L'} \tilde{H}_{L'L}(r) \, Y_{L'}(\hat{\boldsymbol{r}})$. This expression gives the GF for a single cell. In order to find the GF for a cluster, we assume that

$$G^+(\boldsymbol{r}_i + \boldsymbol{R}_i, \boldsymbol{r}_j + \boldsymbol{R}_j) = \sum_{LL'} \Psi_L(\boldsymbol{r}_i) \mathcal{G}^{ij}_{LL'} \Psi_{L'}(\boldsymbol{r}_j) + \delta_{ij} \sum_L \Psi_L(\boldsymbol{r}_<) \tilde{\mathcal{H}}_L(\boldsymbol{r}_>) \; , \tag{3.58}$$

and try to determine the matrix elements $\mathcal{G}^{ij}_{LL'}$ starting from the identity

$$\begin{aligned}
&\sum_{j=1}^{N} \int_{S_j} \left[G_0^+(\boldsymbol{r} - \boldsymbol{r}_1) \nabla_{\boldsymbol{r}_1} G^+(\boldsymbol{r}_1 - \boldsymbol{r}') - \nabla_{\boldsymbol{r}_1} G_0^+(\boldsymbol{r} - \boldsymbol{r}_1) G^+(\boldsymbol{r}_1 - \boldsymbol{r}') \right] \cdot \boldsymbol{n}_j \, \mathrm{d}\sigma_j \\
&= \int_{S_o} \left[G_0^+(\boldsymbol{r} - \boldsymbol{r}_1) \nabla_{\boldsymbol{r}_1} G^+(\boldsymbol{r}_1 - \boldsymbol{r}') - \nabla_{\boldsymbol{r}_1} G_0^+(\boldsymbol{r} - \boldsymbol{r}_1) G^+(\boldsymbol{r}_1 - \boldsymbol{r}') \right] \cdot \boldsymbol{n}_o \, \mathrm{d}\sigma_o \, ,
\end{aligned} \tag{3.59}$$

where the surface integration is over the variable $\boldsymbol{r}_1$. This identity can also be obtained on the basis of (3.2) and (3.47) for the free and cluster GF. It is similar to (3.4) when one replaces the wave function $\psi_k(\boldsymbol{r})$ with the cluster GF (3.47). For simplicity, we neglect again the contribution of the outer surface S_o, imagining to cover the space where the potential is substantially different from zero by N cells.

We have two cases:

(a) $\boldsymbol{r}$ and $\boldsymbol{r}'$ inside cell i
(b) $\boldsymbol{r}$ inside cell i and $\boldsymbol{r}'$ inside cell k $(i \neq k)$

In case (a) we take $\boldsymbol{r}$ and $\boldsymbol{r}'$ very near the origin of cell i. Then, using (3.5) and (3.58), we find

$$\begin{aligned}
&\sum_{LL'} \sum_{L''} J_L(\boldsymbol{r}_i) \, \Psi_{L'}(\boldsymbol{r}'_i) \quad \times \\
&\left\{ \sum_{j \neq i} \int_{S_j} [\, \tilde{H}_L^+(\boldsymbol{r}_i) \nabla \Psi_{L''}(\boldsymbol{r}_j) - \Psi_{L''}(\boldsymbol{r}_j) \nabla \tilde{H}_L^+(\boldsymbol{r}_i) \,] \cdot \boldsymbol{n}_j \, \mathrm{d}\sigma_j] \, \mathcal{G}^{jn}_{L''L'} + \right. \\
&\left. \int_{S_i} [\, \tilde{H}_L^+(\boldsymbol{r}_i) \nabla \Psi_{L''}(\boldsymbol{r}_i) - \Psi_{L''}(\boldsymbol{r}_i) \nabla \tilde{H}_L^+(\boldsymbol{r}_i) \,] \cdot \boldsymbol{n}_j \, \mathrm{d}\sigma_j] \, \mathcal{G}^{nn}_{L''L'} \right\} = 0 \; .
\end{aligned} \tag{3.60}$$

We now observe that in this equation we can move the integral from the surface S_j to the corresponding bounding sphere with radius R_j by Green's theorem so that $\Psi_L(\boldsymbol{r}_j)$ is given by (3.20). Then, using (3.9) to re-expand $\tilde{H}_L^+(\boldsymbol{r}_i)$ around another center j, remembering that

$$\int_{S_j} [\, \tilde{H}_{L'}^+(\boldsymbol{r}_j) \nabla J_L(\boldsymbol{r}_j) - J_L(\boldsymbol{r}_j) \nabla \tilde{H}_{L'}^+(\boldsymbol{r}_j) \,] \cdot \boldsymbol{n}_j \, \mathrm{d}\sigma_j = -\delta_{LL'} \,, \tag{3.61}$$

and putting to zero the expression inside curly brackets, we obtain the condition

$$\sum_j \sum_{\Lambda\Lambda'} G_{L\Lambda}^{ij} \, T_{\Lambda\Lambda'}^{j} \, \mathcal{G}_{\Lambda' L'}^{ji} - \mathcal{G}_{LL'}^{ii} = 0 \,. \tag{3.62}$$

Notice that the KKR structure factors $G_{LL'}^{ij}$ are not to be confused with the full GF factors $\mathcal{G}_{LL'}^{ij}$.

In case (b) we take $\boldsymbol{r}$ and $\boldsymbol{r}'$ very near the origin of cells i and k respectively, so that we have

$$\begin{aligned} &\sum_{LL'} \sum_{L''} J_L(\boldsymbol{r}_i) \, \Psi_{L'}(\boldsymbol{r}'_k) \times \\ &\quad \Big\{ \sum_{j \neq i} \int_{S_j} [\, \tilde{H}_L^+(\boldsymbol{r}_i) \nabla \Psi_{L''}(\boldsymbol{r}_j) - \Psi_{L''}(\boldsymbol{r}_j) \nabla \tilde{H}_L^+(\boldsymbol{r}_i) \,] \cdot \boldsymbol{n}_j \, \mathrm{d}\sigma_j] \, \mathcal{G}_{L''L'}^{jk} + \\ &\qquad \int_{S_i} [\, \tilde{H}_L^+(\boldsymbol{r}_i) \nabla \Psi_{L''}(\boldsymbol{r}_i) - \Psi_{L''}(\boldsymbol{r}_i) \nabla \tilde{H}_L^+(\boldsymbol{r}_i) \,] \cdot \boldsymbol{n}_i \, \mathrm{d}\sigma_i] \, \mathcal{G}_{L''L'}^{ik} + \\ &\qquad \int_{S_k} [\, \tilde{H}_L^+(\boldsymbol{r}_i) \nabla \tilde{\mathcal{H}}_{L''}(\boldsymbol{r}_k) - \tilde{\mathcal{H}}_{L''}(\boldsymbol{r}_k) \nabla \tilde{H}_L^+(\boldsymbol{r}_i) \,] \cdot \boldsymbol{n}_k \, \mathrm{d}\sigma_k] \Big\} = 0 \,. \end{aligned} \tag{3.63}$$

Using the same procedure as above we find the condition

$$\sum_j \sum_{\Lambda\Lambda'} G_{L\Lambda}^{ij} \, T_{\Lambda\Lambda'}^{j} \, \mathcal{G}_{\Lambda' L'}^{jk} - \mathcal{G}_{LL'}^{ik} + G_{LL'}^{ik} = 0 \,. \tag{3.64}$$

Putting together (3.62) and (3.64) we find that the matrix $\mathcal{G}_{LL'}^{ik}$ satisfies the Dyson equation

$$\mathcal{G}_{LL'}^{ik} = G_{LL'}^{ik} \, (1 - \delta_{ik}) + \sum_j \sum_{\Lambda\Lambda'} G_{L\Lambda}^{ij} \, T_{\Lambda\Lambda'}^{j} \, \mathcal{G}_{\Lambda' L'}^{jk} \,. \tag{3.65}$$

This equation constitutes the extension of the customary relation for the MT case to the full potential (FP) case.

3.2.4 Spectroscopic Response Functions

In Chap. 2 it has been shown that the photon absorption cross-section from a core state (ignoring for simplicity spin-orbit coupling) can be written in terms of the GF (3.58), even in case of complex absorptive potential, as

$$\sigma_{\rm tot}(\omega_q) = -4\pi\alpha\,\hbar\omega_q \sum_{m_c\sigma_c} \Im \int \langle \phi^c_{L_c}(\boldsymbol{r}_o)|\boldsymbol{e}_q\cdot\boldsymbol{r}|G(\boldsymbol{r},\boldsymbol{r}';E)|\boldsymbol{e}_q\cdot\boldsymbol{r}'|\phi^c_{L_c}(\boldsymbol{r}_o)\rangle\,\mathrm{d}\boldsymbol{r}\mathrm{d}\boldsymbol{r}', \tag{3.66}$$

where $\hbar\omega_q$ is the impinging photon energy, o is the site of the photoabsorber and $\alpha \approx 1/137$ is the fine structure constant. From the expression (3.58) we obtain two terms: the singular part provides the atomic absorption, whereas the regular part gives the contribution of the environment of the absorber. This is very handy in case of structural analysis.

The photoemission cross-section is better written in terms of new amplitudes

$$B^j_L(\boldsymbol{k}) = \sum_{L'} T^j_{LL'} C^j_{L'}(\boldsymbol{k})\,, \tag{3.67}$$

which imply another set of local basis functions

$$\psi(\boldsymbol{r}_j) = \sum_L B^j_L(\boldsymbol{k})\Psi_L(\boldsymbol{r}_j)\,. \tag{3.68}$$

It is easily seen that, on the basis of (3.16), the new amplitudes satisfy the equation

$$\sum_{jL'} \left[\mathbf{T}^{-1} - \mathbf{G_0}\right]^{ij}_{LL'} B^j_{L'}(\boldsymbol{k}) = I^i_L(\boldsymbol{k})\,, \tag{3.69}$$

whereby, writing $\tilde{\boldsymbol{\tau}} = \mathbf{T}\,(\mathbf{I} - \mathbf{G_0}\mathbf{T})^{-1}$, we find the solution

$$B^o_L(\boldsymbol{k}) = \sum_{jL'} \tilde{\tau}^{oj}_{LL'} \mathrm{i}^{\ell'} Y_{L'}(\hat{\boldsymbol{k}}) \mathrm{e}^{\mathrm{i}\boldsymbol{k}\cdot\boldsymbol{R}_{jo}} (k/\pi)^{1/2}\,. \tag{3.70}$$

Then the photoemission cross-section is given by

$$\frac{\mathrm{d}\sigma}{\mathrm{d}\hat{\boldsymbol{k}}} = 4\pi^2\,\alpha\,\hbar\omega_q \sum_{m_c\sigma_c} \left| \sum_L M_{L_cL}\,[B^o_L(\boldsymbol{k})]^* \right|^2\,, \tag{3.71}$$

where M_{L_cL} is the dipole matrix element between the initial core state and the local basis function (3.68).

These are the formulas for photoemission and photoabsorption in the case of FP. For their physical interpretation and other spectroscopies we refer to Chap. 2 of this book.

3.3 The Program

3.3.1 Features and Capabilities

The FPMS code [30] focuses on the calculations of XANES spectra, but can also calculate projected DOS and Resonant X-ray elastic Scattering [31]. It incorporates a part of ES2MS code [32], which is an interface to use charge densities and potentials generated by electronic structure codes, notably LMTO [9] and VASP [33]. FPMS is incorporated into the MXAN code [34] (the so-called FP- MXAN code), to perform structural fitting of XANES spectra without the need of the MT approximation [35]. FPMS has an option to print out T-matrices to feed the input of the MsSpec code [36], so that one can perform full potential calculations for photoelectron diffraction. Point group symmetries can be specified so as to reduce the computational cost considerably. Truncation of cells may be checked with an interactive animation based on OpenGL library [37].

3.3.2 Requirements

The code is platform independent, it may run on Linux, Windows and Mac OS X. The prepared executable is provided only for the serial mode, and it is stand alone, no additional programs or libraries are required. Instead, the parallel version must be compiled by users on their own platform. For the compilation, one needs Fortran 2003 compilers, OpenGL, MPI [38] and LAPACK [39] and BLAS [40].

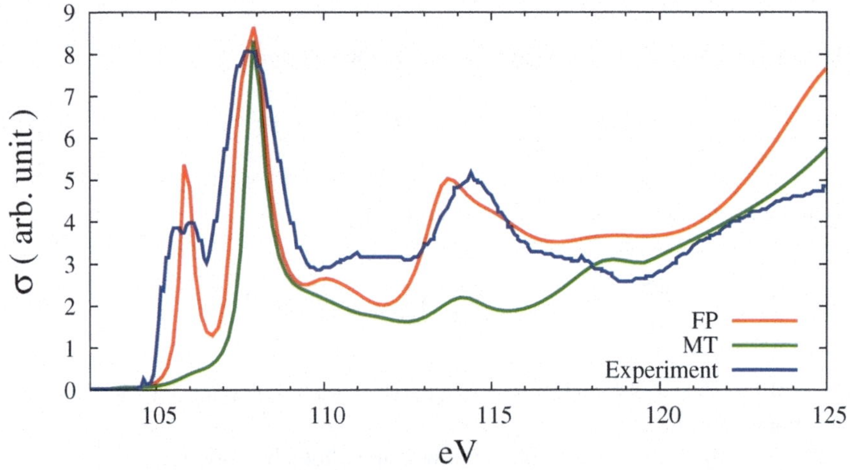

Fig. 3.3 $L_{2,3}$-edges unpolarized absorption cross-section for α-quartz, showing the comparison between the MT and FP calculations against the experimental data [6]

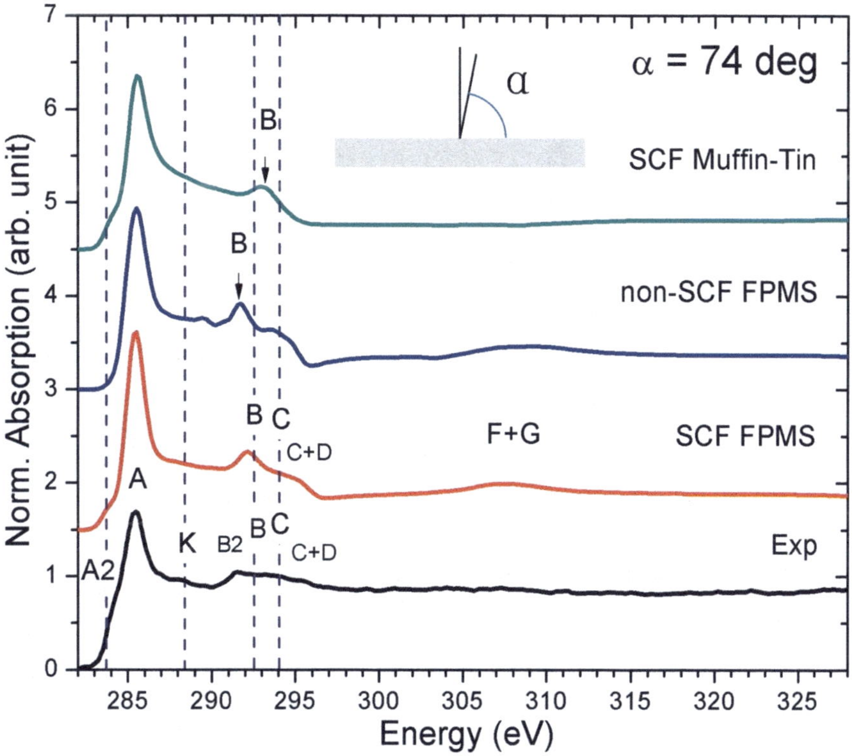

Fig. 3.4 Comparison [8] between calculated polarized C K-edge XANES of graphene with a cluster of radius 30 Å and the experimental data from [29]. α is the angle between electric field and the xy plane. Since $\alpha = 74°$, $\sigma \approx 0.924\sigma_z + 0.076\sigma_x$. Dashed vertical lines show the experimental peak positions. SCF-FP, non-SCF-FP and SCF-MT. (non-)SCF represents the results of calculations using (non-)self-consistent potential in the Full Potential (FP) and Muffin-Tin (MT) version of multiple scattering theory

3.4 MT Versus FP Calculations

For electronic and structural studies of materials, it is important to go beyond the MT approximation, especially for systems with open structures like layers or diamond structure.

Figure 3.3 shows the comparison between the MT and FP calculations against the experimental data, obtained by an electron energy loss (EELS) technique [6]. It is known that EELS spectra, in the limit of small momentum transfer and high energy of the incoming beam, can be described as an absorption spectrum in the dipole approximation with polarization given by the momentum transfer vector. We see that MT calculation gives a very poor result.

In Fig. 3.4 [8], the experimental absorption spectrum [29] for graphene is shown along with the present calculations done for different potential approximations. The

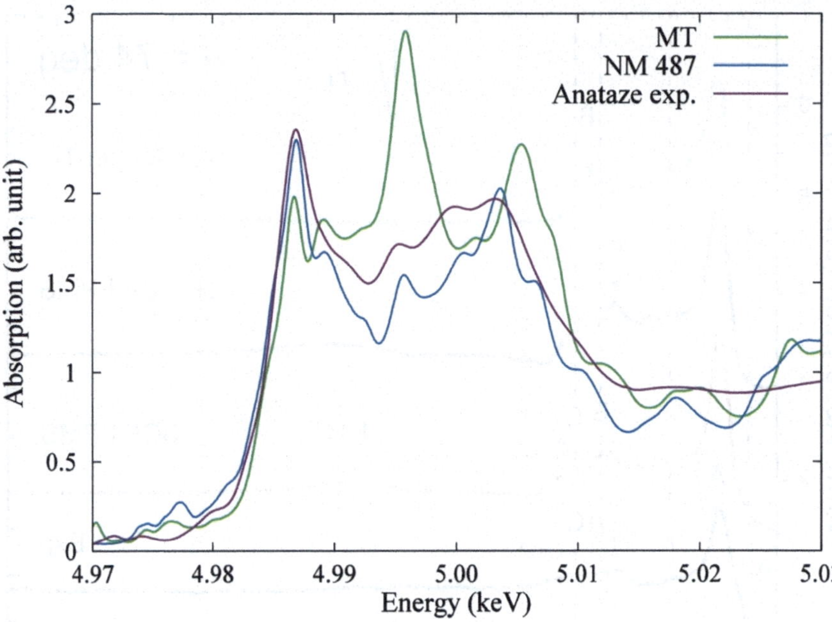

Fig. 3.5 Ti K-edge absorption in TiO_2 (Anatase phase)

SCF potential has been transferred from VASP code. The size of radius of the cluster is 30 Å, indeed rather large, due to the need to describe the focusing effect of the chain like structure of Carbon atoms. As in similar cases, the graphene layer is covered by layers of EC from both sides. It is obvious that the FP calculations (non-SCF-FP or SCF- FP) agree much better with experiment than the MT calculation (SCF-MT). The differences between non- SCF-FP and SCF-FP spectra are small, indicating that self-consistency affects XANES much less than FP corrections.

As a last example we show in Fig. 3.5 the absoprtion spectrum of TiO_2 in the Anatase phase, calculated in the MT and FP mode for a cluster of 487 atoms, in comparison with the experimental spectrum taken from [41]. Here too, we note the dramatic improvement brought in by the FP approach.

Finally, it instructive to show that in the case of close-packed structures the MT approximation gives comparable results with FP programs. Figure 3.6 shows the result of a calculation of the local density of states for a cluster of 459 Cupper atoms with a radius of 11 Å calculated in the MT approximation, compared with the same quantity calculated with FPMS, VASP and FP- ASA programs. As anticipated, the curves are very similar.

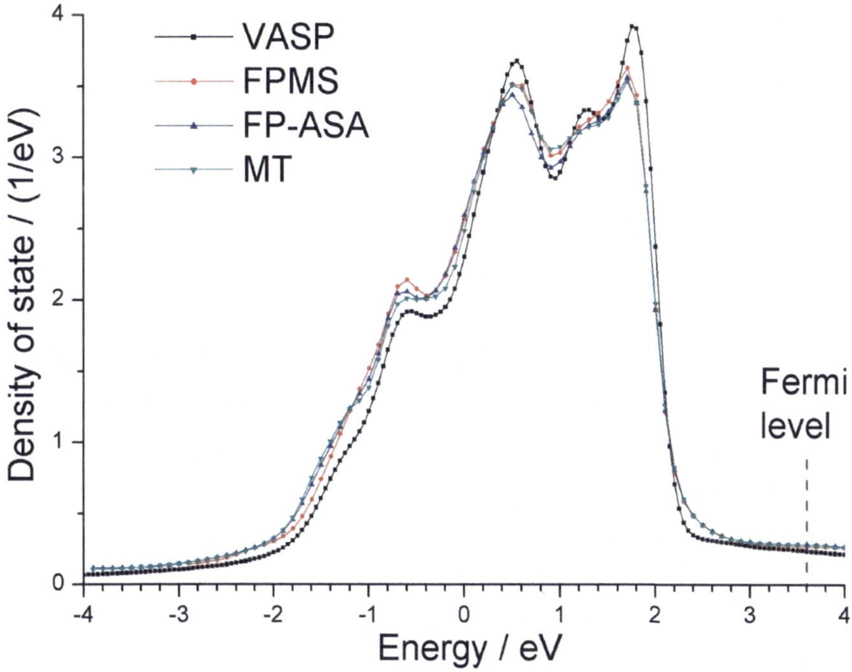

Fig. 3.6 Cupper local density of states

3.5 Future Perspectives

3.5.1 *Optimization*

In the last version of FPMS, we introduced a new way of partitioning the MS matrix for inversion. On an Intel XEON E5-2650 V2 2.6G 8C CPU the computation time per energy point for a Cu cluster of 233 atoms with $\ell_{max} = 6$ was 203 s without partitioning, reducing to 13.5 with partition, with a gain of a factor of 15. We expect an even larger gain for larger clusters.

3.5.2 *Other Spectroscopies*

Now UPS (Ultraviolet Photoemission Spectroscopy) code is under development using FPMS. Implementations for STM (Scanning Tunneling Microscope) and BEEM (Ballistic Emission Electron Microscope) are in progress as well.

References

1. J. Korringa, Physica **13**, 392–400 (1947)
2. W. Kohn, N. Rostoker, Phys. Rev. **94**, 1111–1120 (1954)
3. J.C. Slater, K.H. Johnson, Phys. Rev. B **5**, 844–853 (1972)
4. D. Dill, J.L. Dehmer, J. Chem. Phys. **61**, 692–699 (1974)
5. K. Hatada, K. Hayakawa, M. Benfatto, C.R. Natoli, Phys. Rev. B **76**, 060102R1–4 (2007)
6. K. Hatada, K. Hayakawa, M. Benfatto, C.R. Natoli, J. Phys.: Condens. Matter **21**, 104206 (2009)
7. K. Hatada, K. Hayakawa, M. Benfatto, C.R. Natoli, J. Phys.: Condens. Matter **22**, 185501 (2010)
8. J. Xu, P. Krüger, C.R. Natoli, K. Hayakawa, Z. Wu, K. Hatada, Phys. Rev. B **92**, 125408–1–11 (2015)
9. O.K. Andersen, Phys. Rev. B **12**, 3060–3083 (1975)
10. A.R. Williams, J.W. van Morgan, J. Phys. C: Solid State Phys. **7**, 37–60 (1974)
11. N. Papanikolau, R. Zeller, P.H. Dederich, J. Phys.: Condens. Matter **14**, 2799–2823 (2002)
12. A. Gonis, W.H. Butler, *Multiple Scattering in Solids* (Springer Science & Business Media, New York, 2012)
13. G. Breit, H.A. Bethe, Phys. Rev. **93**, 888–890 (1954)
14. L.F. Canto, M.S. Hussein, *Scattering Theory of Molecules, Atoms and Nuclei* (World Scientific, Singapore, 2013)
15. W.H. Butler, A. Gonis, X.G. Zhang, Phys. Rev. B **45**, 11527–11541 (1992)
16. D. Sebilleau, R. Gunnella, Z.-Y. Wu, S. Di Matteo, C.R. Natoli, J. Phys.: Condens. Matter **18**, R175–R230 (2006)
17. V.I. Lebedev, Comput. Math. Math. Phys. **15**, 44–51 (1975)
18. X.G. Wang, T. Carrington Jr., J. Theor. Comput. Chem. **4**, 599–608 (2003)
19. M. Abramowitz, I.N. Stegun, *Handbook of Mathematical Functions with Formulas, Graphs, and Mathematical Tables* (U.S. Goverment Printing Office, Washington, 1972)
20. A.D. Becke, J. Chem. Phys. **88**, 2547–2553 (1988)
21. V.F. Brastev, *Atomic Wavefunctions* (Nauka, Moscow, 1966)
22. S. Flügge, *Practical Quantum Mechanics*, Springer study edn. (Springer, Berlin, 1974), problem 22, p. 42
23. M. Benfatto, C.R. Natoli, A. Bianconi, J. Garcia, A. Marcelli, M. Fanfoni, I. Davoli, Phys. Rev. B **34**, 5774–5781 (1986)
24. F.G. Tricomi, *Integral Equations* (Courier Dover Publications, New York, 1985)
25. R.G. Newton, *Scattering Theory of Waves and Particles*, 2nd edn. (Courier Dover Publications, New York, 2002)
26. A. Gonis, W.H. Butler, *Multiple Scattering in Solids* (Springer, New York, 2000), and references therein
27. E.T. Whittaker, G.N. Watson, *A Course of Modern Analysis* (Cambridge University Press, Cambridge, 1965)
28. R.V. Vedrinskii, A.A. Novakovich, Phys. Met. Metallogr. **39**, 7–15 (1975)
29. D. Pacilé, M. Papagno, A.F. Rodríguez, M. Grioni, L. Papagno, Ç.Ö. Girit, J.C. Meyer, G.E. Begtrup, A. Zettl, Phys. Rev. Lett. **101**, 066806 (2008)
30. K. Hatada, J. Xu, K. Hayakawa, C.R. Natoli (2005), https://www.lnf.infn.it/theory/CondensedMatter/fpms.html
31. G. Subías, J. Herrero-Martín, J. García, J. Blasco, C. Mazzoli, K. Hatada, S. Di Matteo, C.R. Natoli, Phys. Rev. B **75**, 235101 (2007)
32. X. Junqing, C.R. Natoli, P. Krüger, K. Hayakawa, D. Sébilleau, L. Song, K. Hatada, Comput. Phys. Commun. **203**, 331–338 (2016)
33. G. Kresse, D. Joubert, Phys. Rev. B **59**, 1758–1775 (1999)
34. K. Hayakawa, K. Hatada, S. Della Longa, P. D'Angelo, M. Benfatto, AIP Conf. Proc. **882**, 111–113 (2007)

35. K. Hatada, K. Hayakawa, unpublished 2011
36. D. Sébilleau, C. Natoli, G.M. Gavaza, H. Zhao, F. Da Pieve, K. Hatada, Comput. Phys. Commun. **182**, 2567–2579 (2011)
37. OpenGL, Opengl 4.0 specification (2013), http://www.opengl.org/registry/doc/glspec40.core.20100311.pdf
38. Message P Forum, Mpi: a message-passing interface standard, Technical report, Knoxville, TN, USA (1994)
39. E. Anderson, Z. Bai, J. Dongarra, A. Greenbaum, A. McKenney, J. Du Croz, S. Hammerling, J. Demmel, C. Bischof, D. Sorensen, in *Proceedings of the 1990 ACM/IEEE Conference on Supercomputing, Supercomputing'90, Los Alamitos, CA, USA* (IEEE Computer Society Press, 1990), pp. 2–11
40. L.S. Blackford, J. Demmel, J. Dongarra, I. Duff, S. Hammarling, G. Henry, M. Heroux, L. Kaufman, A. Lumsdaine, A. Petitet, R. Pozo, K. Remington, R.C. Whaley, ACM Trans. Math. Softw. **28**, 135–151 (2002)
41. G. Rossi, M. Calizzi, V. Di Cintio, S. Magkos, L. Amidani, L. Pasquini, F. Boscherini, J. Phys. Chem. C **120**, 7457–7466 (2016)

Chapter 4
KKR Green's Function Method in Reciprocal and Real Space

Ján Minár, Ondřej Šipr, Jürgen Braun and Hubert Ebert

Abstract The Korringa–Kohn–Rostoker (KKR) method is a very flexible band structure technique which is based on the multiple scattering formalism. In contrast to many other band structure methods, which are based on a representation of the electronic structure in terms of Bloch wave functions, the KKR method represents the properties of solids in terms of Green's functions. In this chapter we demonstrate the flexibility of the KKR method as a tool to describe spectroscopic aspects such as x-ray absorption spectra theory and one-step model of photoemission.

4.1 Introduction to the KKR Green's Function Method

4.1.1 General Features

Multiple scattering formalism, as it is in detail formally discussed in the first chapter of this book, is also the basis of the Korringa–Kohn–Rostoker Green's function (KKR) band structure method. Over the years many developments of this method have been performed and are recently reviewed in detail by Ebert et al. [1]. Many methods for solving the Schrödinger equation expand the wave functions in a suitable basis set and use the Rayleigh–Ritz variational principle. Examples are the pseudopotential methods which use the most simple and convenient set, i.e. plane waves, and the modern linear band-structure methods [2] which use numerical basis functions constructed from solutions of the Schrödinger equation for the potential inside an

J. Minár (✉)
New Technologies-Research Center, University of West Bohemia, Univerzitní 8, 306 14 Plzeň, Czech Republic
e-mail: jminar@ntc.zcu.cz

O. Šipr
Institute of Physics, Czech Academy of Sciences, Cukrovarnicka 10, 162 53 Praha 6, Czech Republic

J. Braun · H. Ebert
Department Chemie, Ludwig-Maximilians-Universität München, Butenandtstraße 11, 81377 Munich, Bavaria, Germany

D. Sébilleau et al. (eds.), *Multiple Scattering Theory for Spectroscopies*, Springer Proceedings in Physics 204,
https://doi.org/10.1007/978-3-319-73811-6_4

atomic cell centered at a lattice site. A disadvantage of variational methods is that for very accurate results the number of basis functions rapidly increases unless they are really well chosen for the physical system under consideration.

An alternative to variational methods is the KKR method, which uses multiple scattering theory to provide an exact solution of the Schrödinger equation. This is particularly appropriate when considering unoccupied states which can have arbitrarily large energies and which are very important for the spectroscopy applications discussed in this chapter. Unlike most band structure methods, that provide the electronic wave functions and energy eigenvalues for a crystal, the KKR method is aiming at a calculation of the single-particle Green's function. In fact, knowledge of the Green's function is sufficient to calculate in a very elegant and efficient way all relevant electronic properties of a crystal such as charge and spin densities and to determine various spectroscopic properties. The application of multiple scattering theory for the electronic structure problem in solids was first suggested by Korringa [3] and, independently a bit later, by Kohn and Rostoker [4]. The starting point of this original version of the KKR method, is the Lippmann–Schwinger equation for Bloch states that involve the free-electron Green's function, i.e. the Schrödinger equation is formulated as an integral equation. For the ansatz of the Bloch wave function a minimal basis set is used that is constructed from energy and angular momentum dependent partial waves leading to an eigenvalue problem with a correspondingly low dimensionality. Since its introduction the KKR method has been continuously further developed leading to a wide regime of applications. Apparently, it was first realized by Beeby that the KKR or multiple scattering formalism gives direct access to the electronic Green's function for a considered system [6]. The corresponding formal developments led finally to the KKR-Green's function (KKR-GF) method. The last decades brought further developments for the KKR-GF theory, among others by Faulkner [5, 7], Faulkner and Stocks [8], Györffy and Stocks [9] for the non–relativistic case, and by Onodera and Okazaki [10], Strange et al. [11], Weinberger [12], Strange [13] and Ebert [14] for the relativistic case. Among the outstanding features of the KKR-GF method one has to emphasize that the Green's function is transparently related to single-site scattering and to structural quantities, and that there are no preconditions with regard to the arrangement of the atoms, as e.g. spatial periodicity. One can therefore separate the electronic structure calculations into two parts. The first one supplies the solution corresponding to single-site scattering, while the second one is dependent solely on the geometry of the system considered. Moreover, the Green's function of any perturbed system can be connected to the Green's function of a corresponding reference system without perturbation by means of the Dyson equation. These properties of the KKR-GF method virtually predestinate it for the treatment of defect systems such as e.g. vacancies and impurities [15], but also surfaces and interfaces [16]. A further important field of application is disordered alloys, for which the KKR-GF Green's function method in conjunction with the coherent potential approximation (CPA) [17] constitutes the best available single-site method of calculating their electronic structure [18].

In the following sections we pick up two areas of applications of the KKR-GF method in spectroscopy. In particular, we will focus on x-ray absorption as well as on angular resolved photoemission.

4.1.2 Treatment of Disorder

One of the outstanding features of the KKR-GF method is that it provides the one-electron Green's function of a system directly without making use of Bloch's theorem. Because of this, the KKR-GF method allows one to deal with substitutional diluted and concentrated alloys when combined with the CPA [17, 72]. Within this combined approach (KKR-CPA) the propagation of an electron in a disordered alloy is regarded as a sequence of scattering processes due to a random distribution of scatterers on a lattice. The necessary average over all configurations of the atoms on the lattice is taken by introducing an auxiliary effective CPA medium. As it illustrated by Fig. 4.1, this medium is determined by demanding that embedding of a component A or B of an alloy A_xB_{1-x} should in the average lead to no additional scattering.

Within the CPA the configurationally averaged electronic properties of a disordered alloy are represented by the auxiliary ordered CPA-medium. Making use of the KKR formalism this can be described in turn by a corresponding single-site t-matrix t^{CPA} and site-diagonal scattering path operator τ^{CPA}. In case of a binary system A_xB_{1-x} for example t^{CPA} and τ^{CPA} for the CPA medium are determined by the following so called CPA-condition (cf. Fig. 4.1):

$$x\tau^{\mathrm{A}} + (1-x)\tau^{\mathrm{B}} = \tau^{\mathrm{CPA}} , \tag{4.1}$$

where the angular momentum subscripts were omitted. In this equation the site-diagonal component-projected scattering path operator τ^{A} given by the expression

$$\tau^{\mathrm{A}} = \tau^{\mathrm{CPA}} \left[1 + \left(t_{\mathrm{A}}^{-1} - t_{\mathrm{CPA}}^{-1}\right) \tau^{\mathrm{CPA}}\right]^{-1} , \tag{4.2}$$

represents the scattering properties of an A atom embedded in the CPA medium, where t^{A} is the single-site t-matrix of this component. The scattering properties of the component B embedded in the CPA medium is described by an analogous equation. Obviously, the coupled set of equations for τ^{CPA} and t^{CPA} have to be solved self-consistently. This is usually achieved iteratively by means of the so-called CPA cycle.

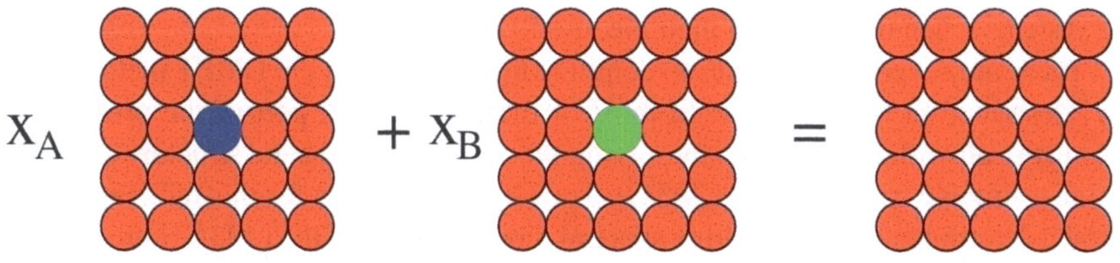

Fig. 4.1 The major idea of the CPA: an auxiliary CPA medium that is meant to represent the average over all possible configurations of a disordered substitutional alloy A_xB_{1-x} is constructed by the requirement that embedding of an A or B atom does not give rise to additional scattering with respect to the CPA medium

4.1.3 Many-Body Effects: LSDA+DMFT Within the KKR Formalism

The spectrum of one-particle excitations of a system of correlated electrons in a solid is of central interest in condensed-matter physics. The theoretical understanding of the excitation spectrum poses a long-standing and not yet generally solved problem. Within the independent-electron approximation the spectrum is simply given in terms of the one-particle eigenenergies of the Hamiltonian. Analogously, it is widely accepted to interpret a measured photoemission spectrum by referring to the results of band-structure calculations that are based on density functional theory (DFT) and the local spin-density approximation (LSDA) [40, 41, 73]. Such an interpretation is questionable since there is actually no direct correspondence between the Kohn–Sham eigenenergies and the one-particle excitations of the system [71, 73]. For an in principle correct description of the excitation energies, the LSDA exchange-correlation potential has to be supplemented by a non-local, complex and energy-dependent self-energy which leads to the Dyson equation [73, 74] instead of the Schrödinger- or Dirac-type equation within the non- or fully relativistic Kohn–Sham scheme.

To deal with this situation we have developed a generalized approach by accounting for electronic correlations beyond the LSDA. A general non-local, site-diagonal, complex and energy-dependent self-energy Σ^{DMFT} [75–77], has been included in the fully relativistic Korringa–Kohn–Rostoker multiple scattering theory [63]. In the following we shortly review an implementation of the LSDA+dynamical mean-field theory (DMFT) formalism within the full potential fully relativistic multiple scattering KKR method [63, 78]. The implementation is completely self consistent with respect to the charge density as well as to the self-energy. As the DMFT method itself has already been described in several reviews, we focus here on those aspects which are specific for the KKR framework and the charge self-consistency.

The central idea of the implementation of the LSDA+DMFT within the KKR method is to account for the corresponding non-local, site-diagonal, complex and energy-dependent self-energy Σ^{DMFT} when solving the single site Schrödinger (or Dirac) equation and this way to account for it via the numerical basis functions used. As a consequence one can exploit directly all advantageous features of the standard KKR Green's function method when performing LSDA+DMFT type calculations. This implies in particular that one can account for correlation effects for a wide range of systems when dealing with their electronic structure.

There are several approaches to combine the LSDA with the DMFT method [76, 77] described in the literature. In most of these the LSDA part of the problem is usually dealt with in a first step by solving the corresponding band structure problem using a suitable basis set (e.g. LMTO). The Green's function for this reference level is then determined by the resulting spectral representation of the LSDA Hamiltonian. Solving in a second step the DMFT part of the electronic structure problem, the resulting local self-energy Σ^{DMFT} can be used together with the local Green's function to calculate a new charge density and a corresponding effective LSDA potential. However, to a achieve a coherent combination of LSDA with DMFT in line with spectral density functional theory [74] one has to solve the following Dyson equation in a self-consistent manner:

$$\begin{aligned}G(\boldsymbol{r},\boldsymbol{r}',E) &= G_0(\boldsymbol{r},\boldsymbol{r}',E)\\ &\quad+\int d^3r''\int d^3r''' G_0(\boldsymbol{r},\boldsymbol{r}'',E)\,[V_{\text{eff}}(\boldsymbol{r}'')\delta(\boldsymbol{r}''-\boldsymbol{r}''')\\ &\quad+\Sigma(\boldsymbol{r}'',\boldsymbol{r}''',E)]\,G(\boldsymbol{r}''',\boldsymbol{r}',E)\,,\end{aligned} \tag{4.3}$$

with $G_0(\boldsymbol{r},\boldsymbol{r}'',E)$ being the Green's function for the chosen free electron reference system. Furthermore, the effective potential is defined as $V_{\text{eff}}(\boldsymbol{r}) = [\overline{V}_{\text{eff}}(\boldsymbol{r}) + \beta\boldsymbol{\sigma}\cdot\boldsymbol{B}_{\text{eff}}(\boldsymbol{r})]$, with $\overline{V}_{\text{eff}}(\boldsymbol{r})$ its spin-independent potential and $\boldsymbol{B}_{\text{eff}}(\boldsymbol{r})$ the corresponding spin-dependent part or magnetic field [11]. Within the relativistic formulation used here, the matrices β and α_k are the standard Dirac matrices where the latter ones can be expressed by $\alpha_k = \sigma_x \otimes \sigma_k$ by means of the 2×2 Pauli matrices σ_k $(k = x, y, z)$.

The KKR-GF method offers the possibility to solve equation (4.3) in a very efficient way. With the system decomposed into atomic regions, usually this means into Wigner–Seitz-cells, and making use of the fact that Σ^{DMFT} is an on-site quantity a straight forward solution of the equation can be achieved via the standard KKR technique. In practice, this implies that one first solves the corresponding so-called single site scattering problem to obtain the wave function $\Psi(\boldsymbol{r})$ inside an atomic cell and the associated single site scattering t-matrix. Within relativistic spin density functional theory [79, 80] the corresponding single site Dirac equation is given by:

$$\left[\frac{\hbar}{i}c\,\boldsymbol{\alpha}\cdot\nabla + \beta mc^2 + V_{\text{eff}}(\boldsymbol{r})\right]\Psi(\boldsymbol{r}) + \int d^3r'\Sigma(\boldsymbol{r},\boldsymbol{r}',E)\Psi(\boldsymbol{r}') = E\Psi(\boldsymbol{r})\,, \tag{4.4}$$

where $\Psi(\boldsymbol{r})$ is an energy-dependent four-component spinor function for a given energy E. To solve this equation the following ansatz $\Psi = \sum_\Lambda \Psi_\Lambda$ for the wave function is used. In practice, application of the DMFT scheme is restricted to correlated d- or f- orbitals. In line with this, the self-energy $\Sigma(\boldsymbol{r},\boldsymbol{r}',E)$ has to be projected onto a localized set of orbitals $\phi^n_\Lambda(\boldsymbol{r})$. The corresponding self-energy matrix $\Sigma_{\Lambda\Lambda'}(E)$ is then delivered by a suitable DMFT solver.

It is worth to note that even if the spherical muffin-tin or atomic sphere approximation to the shape of the effective potential is used, the components of (4.4) associated with different Λ's are coupled in the same way as it is in a full-potential calculation. This implies that the implementation of the LSDA+DMFT approach on the basis of the KKR-GF requires the use of its full-potential version. Solving the set of coupled equations for the wave functions $\Psi(\boldsymbol{r})$ gives direct access to the associated single site t-matrix by matching the wave functions to the Hankel and Bessel functions as free electron solutions outside the atomic cell. When dealing this way with the single-site problem, one accounts obviously for the entire complexity of the underlying complex non-local self-energy of the LSDA+DMFT approach. Because of this, the resulting regular and irregular scattering wave functions $Z_\Lambda(\boldsymbol{r},E)$ and $J_\Lambda(\boldsymbol{r},E)$, respectively, as well as the corresponding single-site t-matrix carry the full information on the underlying LSDA+DMFT Hamiltonian. This means in particular that in contrast to other LSDA+DMFT implementations that use basis functions derived from the LSDA-part of the Hamiltonian, the effect of the self-energy is now also incorporated into the wave functions Ψ. This becomes relevant, for example, when dealing with disordered alloys on the basis of the CPA or when calculating the total energy of a system or matrix elements for spectroscopy.

With the single-site t-matrix made available, the next step of a KKR-type calculation is to deal with the multiple scattering problem. This task implies in general to find the corresponding scattering path operator τ [81]. Technically, this step is completely independent from the DMFT part of the electronic structure problem. Accordingly, the expression for the retarded site diagonal Green's function $G(\boldsymbol{r}, \boldsymbol{r}', E)$ as specified in (4.10) above [8, 82] can be used. Knowing the Green's function $G(\boldsymbol{r}, \boldsymbol{r}', E)$ makes it possible to derive all electronic properties of interest as e.g. the charge density in a straight manner with the calculated Green's function $G(\boldsymbol{r}, \boldsymbol{r}', E)$ accounting for all effects of the self-energy Σ^{DMFT}.

In order to provide the bath Green's function that is needed as input for the DMFT solver, the corresponding localized Green's function matrix is calculated by projecting the Green's function given by (4.10) according to the selection of correlated atomic orbitals. This is achieved by a projection of $G(\boldsymbol{r}, \boldsymbol{r}', E)$ with respect to a localized set of orbitals $\phi_\Lambda(\boldsymbol{r})$:

$$G_{\Lambda\Lambda'}(E) = \int \mathrm{d}^3 r \int \mathrm{d}^3 r\,' \phi_\Lambda(\boldsymbol{r}) G(\boldsymbol{r}, \boldsymbol{r}', E) \phi_{\Lambda'}(\boldsymbol{r}'). \tag{4.5}$$

A natural choice for these localized orbitals or projection functions $\phi_\Lambda(\boldsymbol{r})$ are regular solutions of the Kohn–Sham–Dirac single-site equations that are normalized to unity and also used for a representation of the self energy Σ (see below). In the case of transition metal systems in general only the d-d sub-block of the structural Green's function matrix $\underline{G}^{nn}(E)$ has to be considered. Apart from these restrictions, the choice of the functions $\phi_\Lambda(\boldsymbol{r})$ is rather arbitrary provided they form a complete set of functions. This implies that a localized basis set may be set up for a given reference energy E_{ref} (taken to be at the center of gravity of the occupied part of the d-band for the cases discussed here), with the spin-dependent part of the potential set to zero. As a consequence one has to suppress all further coupling to the other ℓ-channels imposed by the crystal symmetry in the case of full-potential calculations.

A flow diagram that describes the resulting KKR-GF based self consistency cycle of the LSDA+DMFT scheme is given in Fig. 4.2. For this, the set of regular (Z) and irregular (J) solutions to the single-site problem that account for the LSDA potential as well as for the DMFT self energy Σ is calculated via (4.4). Using the corresponding t-matrix the multiple scattering problem is solved in terms of the scattering path operator τ. With this available the KKR Green's function is constructed based on (4.10). As a next step the projected Green's function matrix is constructed according to (4.5) providing this way the necessary input to deal with the many-body problem. The way the mutual connection between the charge density and the self-energy is implemented deserves special attention. To start with, it should be stressed that the

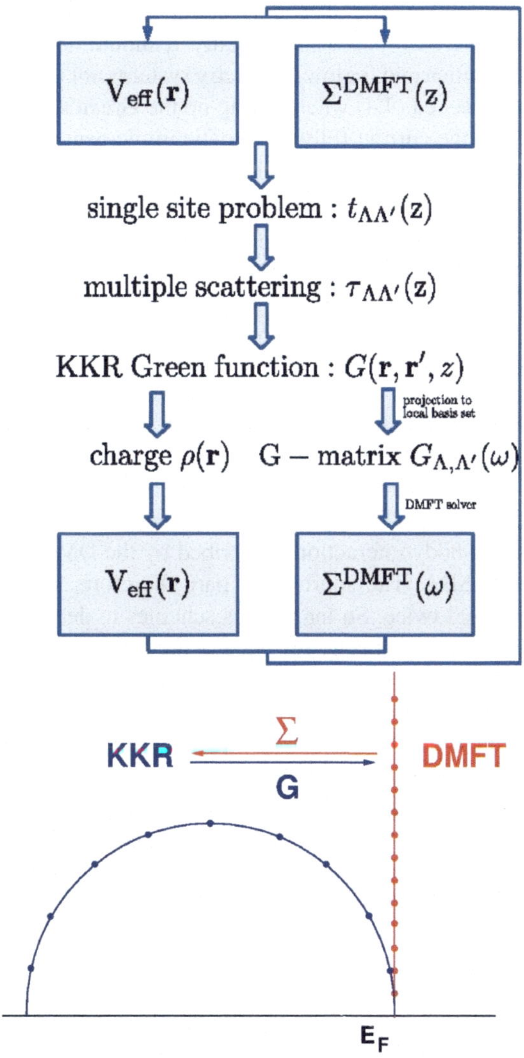

Fig. 4.2 *Top*: flow diagram for a KKR-based LSDA+DMFT calculation. *Bottom*: representation of the various involved energy paths. The *blue semicircle* represents the complex energy path with energies z used by the KKR-GF method to calculate the charge density. After the corresponding bath Green's function G has been determined, it is analytically extrapolated to the straight line perpendicular to the real energy axis (*red*) to calculate the self-energy Σ^{DMFT} using the DMFT impurity solver. Afterwards, Σ^{DMFT} is analytically extrapolated back to the *semicircle* energy path used by the KKR-GF

definition of the Green's function is not restricted to real energies E but applies also for arbitrary complex energies z. The fact that the Green's function $G(\boldsymbol{r}, \boldsymbol{r}', z)$ is analytical throughout the complex plane away from the real axis [83] allows, in particular, to perform the necessary energy integration for the set up of the charge density on an arbitrary contour in the complex energy plane [84]. Using a semi circular path around 30 energy mesh points are sufficient in general. On the other hand many schemes to deal with the DMFT problem use a Matsubara-type mesh of frequencies providing the self-energy Σ^{DMFT} for this mesh. As a consequence, it is necessary to apply a suitable technique for an analytical continuation i.e. to transform Σ^{DMFT} from Matsubara frequencies ω to the energy contour used within the KKR-GF based step. Finally, it should be stressed that Σ^{DMFT} is in general not Hermitian and for low symmetry systems not only right but also left handed solutions one needed (4.4) when setting up the Green's function $G(\boldsymbol{r}, \boldsymbol{r}', E)$ [85].

In the current fully relativistic implementation, the many-body problem is dealt with by relying on perturbation theory. One can use either the SPTF solver (spin-polarized T-matrix+FLEX) [86] or spin-polarized T-matrix [87] solver for $T = 0\,K$. However, any DMFT solver which supplies the self-energy $\Sigma(E)$ can be included when dealing with the many-body part of the electronic structure problem. To transfer the self-energy from the Matsubara energy path to the complex energy path used to set up the updated KKR Green's function the Padé analytical continuation is used once more. In this context, the key role is played by the scattering path operator $\tau^{nn}_{\Lambda\Lambda'}(E)$, that is used to generate the new single particle Green's function and from this the new charge and potential in each SCF iteration.

Finally, the double counting correction H_{DC} has to be considered. This problem is definitely one of the main challenges for first-principle calculations within the LSDA+DMFT. The double counting correction stems from the fact that some many-body interactions described by the DMFT formalism are already included in the LSDA. The appropriate part, therefore, has to be subtracted to avoid it being counted twice. So far various schemes to deal with the double counting correction problem have been suggested in the literature [77]. The most simple scheme in this context corresponds to the static LSDA+U scheme and has been used in our implementation. In practice, the double counting correction to the self-energy is applied at the step when the many-body problem is solved. This implies that first the complete static part of the self-energy coming from the solver is removed and then the LSDA+U-like static part is added [88]. From many applications to pure transition metals as well as their metallic compounds or alloys the so called around the mean field (AMF) double counting correction was found to be most appropriate [64, 66, 88, 89]. It should be stressed, that so far it was not possible to derive an exact and simple analytical expression for the double counting correction. On the other hand, the combined GW+DMFT scheme [90] allows in principle to deal with the double counting problem in an exact way. It is therefore important to perform direct comparison of LSDA+DMFT based results to experimental data coming for example from angle-resolved photoemission (ARPES) measurements to test the adequacy of the chosen double counting correction scheme. However, to be able to select among the various suggestions for the correction H_{DC} in a most reliable way it is advantageous to calculate not only the bare spectral function, i.e. $\Im\, G$, but to calculate the photoemission spectra in a most accurate way. Actually, on the basis of the one-step model of photoemission it could be convincingly demonstrated that the AMF suggestion for H_{DC} is appropriate for transition metal systems [64, 66, 89].

The fact that the DMFT is a mean field theory relying on an effective medium (bath) reminds to another successful use of the mean field formalism in the KKR, namely, the CPA method for dealing with disordered systems. Indeed, a similar philosophy is applied in both cases. Thus, combining the DMFT and KKR-CPA methods is a relatively straightforward task, because both schemes are formulated in their standard version on a single-site level. On the DMFT side this means that any correlation concerning the occupation of neighboring sites for example due to short range order is ignored and the self-energy Σ is accordingly assumed to be on-site only. The extension of the CPA scheme for disordered alloys to account simultaneously for many-body correlation effects can straightforwardly be done within the KKR+GF. This is achieved by incorporating the local multi-orbital and energy dependent self-energies Σ_A^{DMFT} and Σ_B^{DMFT} of the components directly into the corresponding single-site matrices t_A and t_B, respectively, when solving the component specific single site problem (4.4) [63]. As a consequence, all quantities related to the Green's function, as for example the charge density, represent the impact of the electronic correlations beyond the LSDA level. The combination of the CPA with the LSDA+DMFT scheme proved to be a rather robust and powerful technique [62, 63, 88, 91]. It is worth noting that the combination of the KKR-CPA for disordered alloys and the DMFT scheme is based on the same arguments as given by Drchal et al. [92] who did corresponding work earlier using the LMTO Green's function method for alloys [16].

We conclude by noting that the solution of the Dyson equation for a given self-energy allows to deduce a *raw* photoemission spectrum. However, to allow a reliable interpretation of experimental data, however, it is indispensable to incorporate all matrix element effects which may have a dramatic influence on the spectrum. For a direct comparison to experiment, in particular the full wave-vector and energy dependence of the transition-matrix elements has to be taken care within such calculations. The impact of these matrix element effects are well-known to be important but nevertheless often ignored. On the one hand side, they result from multiple-scattering processes which dominate the electron dynamics in particular in the low-energy regime of typically 1–200 eV [39]. At least equally important is the inclusion of selection rules and of the excitation cross section or oscillator strength via the transition-matrix elements that are completely ignored when working with the raw spectrum. In fact, one can say that the main task of photoemission theory is to provide a bridge between the raw spectrum obtained by any LSDA+DMFT based electronic structure calculation and experiment. In this context, the one-step model of photoemission [37–39] proved to be the most successful and flexible theoretical approach available providing a coherent description of all relevant aspects of the electronic structure and of the photo emission experiment itself.

4.2 Applications of KKR-Green Function Formalism in the Spectroscopy

4.2.1 X-ray Absorption: Formalism

If an x–ray beam with intensity I_0 passes through a sample of thickness d, its attenuation is determined by

$$I = I_0 \exp[-\mu^{\hat{e}_q}(\omega_q) d] \ . \tag{4.6}$$

The absorption coefficient $\mu^{\hat{e}_q}(\omega_q)$ depends on the energy $\hbar\omega_q$, on the polarization vector $\hat{\boldsymbol{e}}_q$ and also on the wave vector $\boldsymbol{q}$ ($q = \omega/c$) of the incoming radiation. The x-ray absorption coefficient $\mu^{\hat{e}_q}(\omega)$ can be expressed via the Fermi's golden rule as [19]:

$$\mu^{\hat{e}_q}(\omega_q) \propto \sum_i \sum_f^{\text{unocc}} \left| \langle \Psi_f \,|\, \hat{X}_{\hat{e}_q} \,|\, \Psi_i \rangle \right|^2 \delta(E_f - E_i - \hbar\omega_q) \,. \tag{4.7}$$

The sum here spans all initial core states $|\,\Psi_i\,\rangle$ with energies E_i that are involved in the x-ray absorption process and all empty final states $|\,\Psi_f\,\rangle$ whose energy E_f is above E_F.

The operator $\hat{X}_{\hat{e}_q}$ in (4.7) describes the interaction between electrons and photons with polarization vector $\hat{\boldsymbol{e}}_q$ and wave vector $\boldsymbol{q}$. In the relativistic formalism $\hat{X}_{\hat{e}_q}$ is given by [19, 20]:

$$\hat{X}_{\hat{e}_q} = e\boldsymbol{\alpha} \cdot \boldsymbol{A}_{\hat{e}_q} \,, \tag{4.8}$$

where the classical vector potential $\boldsymbol{A}_{\hat{e}_q}$ represents the photon field and $ec\boldsymbol{\alpha}$ is the operator of the electron current. The vector $\boldsymbol{\alpha}$ stands for the Dirac matrices [20]. In most cases, one can employ the electric dipole approximation when dealing with $\hat{X}_{\hat{e}_q}$. Going beyond the dipole approximation is straightforward [19].

When dealing with ordered solids, Bloch wavefunctions can be substituted for the final states $|\,\Psi_f\,\rangle$ in (4.7). Suitable all-electron band-structure method can then be applied to calculate the absorption coefficient $\mu^{\hat{e}_q}(\hbar\omega_q)$ [21]. Another approach consists in representing the final states $|\,\Psi_f\,\rangle$ using the Green's function $G^+(E)$ to arrive at [22, 23]:

$$\mu^{\hat{e}_q}(\hbar\omega_q) \propto \sum_i^{\text{occ}} \langle \Psi_i \,|\, \hat{X}^{\dagger}_{\hat{e}_q} \, \Im\left[G^+(E_i + \hbar\omega_q)\right] \hat{X}_{\hat{e}_q} \,|\, \Psi_i \rangle \, \delta(E_i + \hbar\omega_q - E_F) \,. \tag{4.9}$$

This equation can be further transformed if the Green's function is expressed within the multiple scattering framework as [8]:

$$\begin{aligned} G^+(\boldsymbol{r}_n, \boldsymbol{r}_{n'}{}', E) = & \sum_{\Lambda\Lambda'} Z^n_\Lambda(\boldsymbol{r}, E) \tau^{nn'}_{\Lambda\Lambda'}(E) Z^{n'\times}_{\Lambda'}(\boldsymbol{r}\,', E) \\ & - \sum_{\Lambda} \big[\, Z^n_\Lambda(\boldsymbol{r}, E) J^{n\times}_\Lambda(\boldsymbol{r}\,', E) \Theta(r' - r) \\ & + J^n_\Lambda(\boldsymbol{r}, E) Z^{n\times}_\Lambda(\boldsymbol{r}\,', E) \Theta(r - r') \big] \, \delta_{nn'} \,. \end{aligned} \tag{4.10}$$

The matrix $\tau^{nn'}$ is the so-called scattering path operator. It transforms an incoming electron wave arriving at the atomic site n' into an outgoing electron wave leaving site n while accounting for all possible scattering events that may take place in between (see (1.68) of Chap. 1 of this book). Because the initial core states $|\Psi_i\rangle$ occurring in (4.9) are localized, only site-diagonal terms of the scattering path operator τ^{nn}

are needed; the site label n identifies the photoabsorbing atom. If one restricts to a non-relativistic description in (4.10), the functions Z_Λ and J_Λ represent regular and irregular solutions to the Schrödinger equation but normalized as required by the scattering theory [8]. The combined subscript $\Lambda = (\ell, m_\ell, m_s)$ then stands the orbital (ℓ) and magnetic (m_ℓ, m_s) quantum numbers. However, in this chapter the relativistic formalism is mostly employed, meaning that Λ usually represents a set of relativistic quantum numbers $\Lambda = (\kappa, \mu)$, where κ and μ stand for the spin-orbit and magnetic quantum numbers, respectively.

When dealing with finite systems, the $\tau^{nn'}$ matrix can be evaluated by inverting the so called *KKR-matrix* according to $\boldsymbol{\tau}(E) = [\mathbf{t}(E)^{-1} - \mathbf{G}_0(E)]^{-1}$, where $\mathbf{t}$ is the single-site t-matrix and the matrix $\mathbf{G}_0$ represents the real space structure constants. (The upright bold font indicates matrices with respect to site and spin-angular (Λ) subscripts.) One speaks of calculations in the real space in this case. Alternatively, when dealing with a translationally periodic system such as a crystal, the equation can be solved by employing the Fourier transformation. Then one speaks of calculations in the reciprocal space.

In this section the absorption coefficient is mostly calculated using an equation which can be obtained by inserting the expression (4.10) for the Green's function into (4.9), namely,

$$\mu^{\hat{e}_q}(\hbar\omega_q) \sim \sum_i^{\text{occ}} \sum_{\Lambda\Lambda'} M_{\Lambda i}^{\hat{e}_q *}(E_f) \Im \left[\tau_{\Lambda\Lambda'}^{nn}(E_f)\right] M_{\Lambda' i}^{\hat{e}_q}(E_f) \,, \tag{4.11}$$

where $E_f = E_i + \hbar\omega_q$. The matrix elements $M_{\Lambda i}^{\hat{e}_q}(E_f)$ are given by:

$$M_{\Lambda i}^{\hat{e}_q}(E_f) = \langle Z_\Lambda(E_f)|\hat{X}_{\hat{e}_q}|\Psi_i\rangle \,. \tag{4.12}$$

Equation (4.11) presents quite a general expression for evaluating the absorption coefficient, applicable in many circumstances. For example, it can be employed for systems which do not have translational periodicity, such as alloys, surfaces, adsorbates or defects.

It has been assumed in (4.11) that the final states energies are restricted to the real axis. If one uses complex energy formalism to describe the effects of relaxation or self-energy, an additional term appears that comes from the second term in (4.10). An explicit form for this atomic-like term can be found, e.g., in [19].

An important role in applications of spectroscopy is nowadays played by magnetic dichroic effects. Generally, magnetic dichroism is the difference between two spectra for a magnetized sample recorded for different polarizations of the incoming radiation. Magnetic dichroism results from a combined influence of spin polarization and spin-orbit coupling. The computational scheme has to be amended to describe this. The most consistent and robust way in this respect is to treat everything within a fully relativistic formalism. Let us restrict the discussion to the case of x-ray magnetic circular dichroism (XMCD) in polar geometry. If we assume that the wave vector of the incoming radiation $\boldsymbol{q}$ points along the z direction, the polarization vector of the radiation can be written as

$$\boldsymbol{a}_+ = -\frac{1}{\sqrt{2}}\begin{pmatrix}1\\ i\\ 0\end{pmatrix} \quad \boldsymbol{a}_- = \frac{1}{\sqrt{2}}\begin{pmatrix}1\\ -i\\ 0\end{pmatrix} . \tag{4.13}$$

We use the convention that the first vector describes light which is left circularly polarized (positive helicity) and the second vector in (4.13) describes light which is right circularly polarized (negative helicity). It should be emphasized here that the particular choice of the geometry (coordinate system) does not harm the generality of the formalism, any other polarization of the incoming radiation can be treated analogously.

If a fully relativistic formalism is employed, the functions Z_Λ and J_Λ in (4.10) are solutions of the Dirac equation. Because the potential is spin-dependent, the symmetry of the Dirac Hamiltonian is reduced and it is not possible to assign to the functions Z_Λ and J_Λ a pure spin-angular character Λ [24, 25]. Rather, they can be viewed as superpositions of functions $Z_{\Lambda'\Lambda}$ and $J_{\Lambda'\Lambda}$ with character Λ', such as $Z_\Lambda = \sum_{\Lambda'} Z_{\Lambda'\Lambda}$. If the initial core states Ψ_i are also treated within in a fully relativistic framework [26], one can account for all possible sources of the magnetic dichroism. The form of the transition matrix elements $M^{\hat{e}_q}_{\Lambda i}$ is given in [19], together with corresponding discussion.

4.2.2 *X-Ray Absorption and X-Ray Magnetic Circular Dichroism of Clusters*

In this section we will introduce several illustrative examples how calculations of x-ray absorption spectra via the KKR-GF formalism as given by (4.9) can be used for structural analysis. We will deal with situations when multiple scattering is important. Our focus is thus not on the extended x-ray absorption fine structure (EXAFS) region but rather on the x-ray absorption near edge structure (XANES) region. We will present also some cases where analysis of XANES and EXAFS complement each other. The calculations presented below have been performed within the real space formulation of KKR-GF equations, i.e., the τ matrix in (4.9) is evaluated for a finite set of atoms (scattering centers), directly in real space.

4.2.2.1 XANES and XMCD of Free Fe Clusters

One of the areas where the real-space formulation of KKR-GF equations comes handy is spectroscopy of clusters. We will illustrate few concepts on the study of XANES and x-ray magnetic circular dichroism (XMCD) spectra of free Fe clusters

Table 4.1 Dependence of magnetic properties of free Fe clusters on the cluster size. Number of atoms is in the first column, average d components of the spin and orbital magnetic moments are in the second and the third columns and the average number of holes in the d band is in the last column [94]. Magnetic moments are in μ_B

Size	$\mu_{\text{spin}}^{(d)}$	$\mu_{\text{orb}}^{(d)}$	$n_h^{(d)}$
9	2.84	0.209	3.36
15	2.54	0.071	3.15
27	2.82	0.125	3.40
51	2.62	0.075	3.36
59	2.67	0.063	3.35
65	2.65	0.075	3.38
89	2.68	0.068	3.48
Bulk	2.37	0.055	3.44

[93]. Interest in these systems is motivated mainly by their magnetic properties: magnetic moments of small systems are larger than magnetic moments of corresponding bulk materials (when related to the same amount of atoms of given type). Some magnetic properties of free Fe clusters as calculated via the KKR-GF formalism are summarized in Table 4.1 [94].

Comparison of theoretical Fe $L_{2,3}$-edge XANES spectra for few free clusters and for the crystal is shown in Fig. 4.3. The clusters are assumed to have a bcc-like geometry as if cut from the bulk, so we investigate pure effect of the finite size of the system. Spectra for clusters comprising 9, 15, 51, and 89 atoms are shown. Spectra of clusters were obtained by superposing spectra originating from the individual atoms of the cluster. Furthermore, common normalization was achieved by dividing the spectra by the number of atoms in each cluster [93].

First thing to notice in Fig. 4.3 is that there is little significant variation of the spectra with cluster size, except for some fine structure following the L_3 white line for the 9- and 15-atom clusters (at about $E \approx 2$–6 eV). These features could be markers of truly discrete states appearing below the vacuum level. The height of the vacuum level above E_F varies from 5 eV for the 9-atom cluster to 8 eV for the 89-atom cluster [94, 95]. Formally, the continuous character of theoretical x-ray absorption spectra in the region between E_F and the vacuum is a consequence of the Green's function formalism, in particular, of a small imaginary component which is included in the energy. If the cluster size increases, the number of discrete levels increases as well, smoothening thus the (overlapping) spectral peaks and resulting in a spectral *band*. Generally, it is evident from Fig. 4.3 that white lines for small clusters are sharper than white lines for large clusters. The vacuum edge itself does not seem to give rise to a distinct feature in the spectrum.

Another characteristic feature worth mentioning is the absence of the small bulge around 8 eV in spectra of clusters. This feature appears in calculated XANES of bulk Fe and has been associated with a Van Hove singularity. Its occurrence is linked to the translational periodicity, meaning that it can appear only for large systems. In

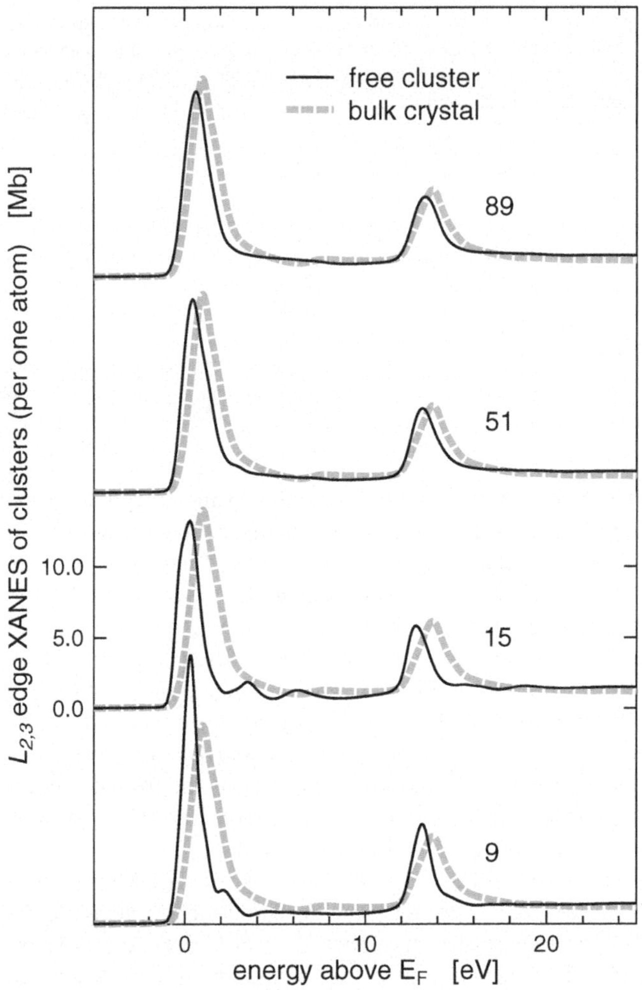

Fig. 4.3 Theoretical $L_{2,3}$-edge XANES of free Fe clusters (*thin full line*) together with XANES of a crystal (*thick dashed line*). The number of atoms in the cluster is indicated at each spectrum. Note that the curve representing the crystal spectrum is identical for each of the four sub-graphs. Figure reproduced from [93]

particular, such feature is present in a spectrum generated by the central atom of a cluster which contains more than ~100 atoms [96].

As one of the strong motivations for research on clusters is magnetism, it is useful to have a look at the XMCD spectra. This is done in Fig. 4.4. It follows from the calculations that the shape of XMCD for the clusters and for the bulk is generally similar: spectra are dominated by two main peaks which are split by the spin-orbit splitting. The intensities of the peaks systematically decrease if the clusters increase. Another trend concerns the widths of the peaks: small clusters give rise to narrower

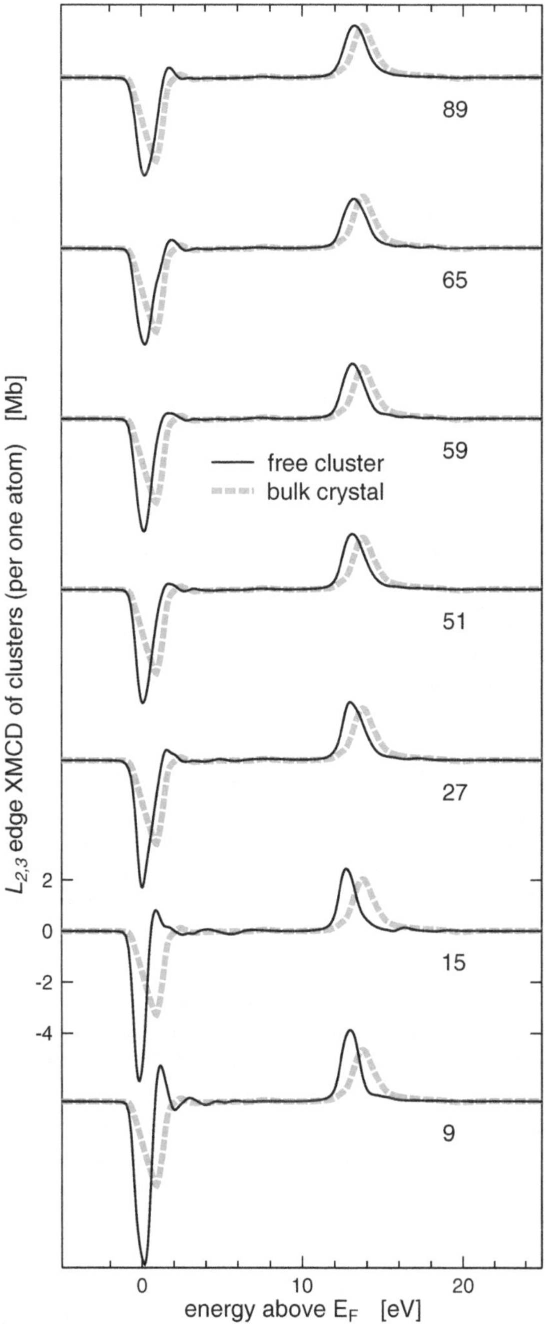

Fig. 4.4 Theoretical $L_{2,3}$-edge XMCD of free Fe clusters (*thin full line*) together with XMCD of a crystal (*thick dashed line*). The numbers of atoms in the clusters are also given. Figure reproduced from [93]

XMCD peaks than large clusters. The trends in XMCD can be seen as counterparts to analogous trends in XANES; they can be traced to a higher localization of *d* electrons in the smaller clusters.

The intensities of XMCD peaks decrease with increasing cluster size nearly monotonously. This trend reflects the changing character of *d* electron states (from atomic-like to bulk-like). On the other hand, the areas of $L_{2,3}$-edge XMCD peaks do not follow a simple trend. This should not be surprising because these areas reflect values of magnetic moments of cluster atoms [97–99] and these values oscillate with the cluster size (see Table 4.1).

X-ray absorption spectra (XAS) for small charged clusters were measured by Hirsch et al. [100]. The theoretical x-ray absorption spectra shown in Fig. 4.3 contain more structure and exhibit narrower lines than the measured spectra. This difference is probably associated with differences in the geometric structure of clusters investigated by experiment and by theory: For calculations, clusters with bulk-like bcc structure were employed whereas in experiment it is more likely that other structures with lower symmetry were involved. For these low symmetry clusters, more inequivalent atoms are present. This gives rise to a multitude of closely separated levels and subsequent smearing of originally sharp spectral features. XMCD spectra of free charged Fe clusters are also available [101]. Again, experimental spectra contain less sharp features than the calculations. The general trend that XAS and XMCD spectra of clusters exhibit more features than spectra of bulk Fe is, nevertheless, present both in theory and in experiment.

4.2.2.2 Rh Clusters Supported by Ag(100)

Magnetism of Rh clusters has been a controversial topic, with contradictory reports on presence or absence of magnetism in thin Rh films and/or small Rh clusters [102–105]. A combined experimental and theoretical study of Rh clusters on Ag(100) demonstrated that magnetism of Rh clusters is strongly related to the geometry of the clusters and also to the interaction of the clusters with the support [106].

Free Rh clusters are magnetic, as it follows from the experiment [104] as well as from calculation [107–111]. To illustrate this, we present in Table 4.2 average magnetic moments of atoms in Rh clusters of different sizes, together with magnetization energies per atom (defined as differences of total energies obtained if the cluster is either magnetic or non-magnetic). One can see that magnetic moments per atom as well as magnetization energies mostly decrease with cluster size — even though this decrease is not monotonic. Magnetic and non-magnetic states are practically degenerate for Rh cluster of 68 atoms. If the cluster contains 135-atoms, its ground state is non-magnetic [111].

For supported Rh clusters, intermixing of the cluster atoms with the substrate and formation of a surface alloy appears to be an important factor which suppresses the magnetism. Combining experiment and theory helps to understand how the magnetic moment is formed for Rh clusters if they are prepared by buffer-layer-assisted growth in a Xe matrix. To begin with, the KKR-GF calculations suggest that if a Rh_4 cluster

Table 4.2 Magnetic moments μ_{spin} and μ_{orb} (in μ_B) and magnetization energies (in mRy) per atom for free Rh clusters of 13–68 atoms [111]

N	μ_{spin}	μ_{orb}	ΔE_{mag}
13	1.423	0.108	–11.74
14	1.258	0.147	–2.88
19	0.771	0.110	–2.34
38	0.696	0.040	–0.61
43	0.202	0.009	–0.15
55	0.526	0.037	–0.66
68	0.062	0.009	0.00

is located atop the Ag(100) surface, its average spin magnetic moment (per Rh atom) is $\mu_{\text{spin}} = 0.92\mu_B$ [106]. However, if it gets embedded into the surface or even buried beneath it, it is nonmagnetic.

Experimental XMCD study [106] demonstrates that if Rh clusters are prepared on a Xe matrix, they are magnetic. However, if Rh is deposited directly on the Ag(100) surface at low temperature, it shows no magnetism – in contrast to theoretical predictions that Rh clusters should be magnetic. To address this issue, spectra of Rh-Ag systems in different geometries are calculated and compared to measurements. The experimental XANES for a 0.1-monolayer amount of Rh deposited directly on Ag(100) (see Fig. 4.5a) is very different from spectrum of equivalent amount of Rh on the Xe buffer layer. In the former case the diffuse peak is not present and instead a hump on the high-energy sides of the $M_{3,2}$ peaks appears. Moreover, the resonances are shifted to higher energies. This can be understood if one accounts for the alloying processes (bottom panel of Fig. 4.5b). Here one compares theoretical XAS for a flat Rh_4 cluster in three different positions: (i) on the top of the Ag(100) surface, (ii) embedded within the surface, and (iii) submerged down to the subsurface layer (i.e., located below the surface). In real samples the shoulder at the high-energy side of the M_3 edge would be smeared-out due to superposition of spectra originating from clusters of different shapes and sizes, so the resulting spectra would be similar to what was measured. One concludes that the suppression of Rh magnetism observed experimentally during the buffer-layer-assisted growth is mainly caused by the increase of the cluster size and not by the cluster-substrate interactions. Nevertheless, hybridization is important for Rh atoms if they are deposited directly on Ag(100) surface, without the buffer layer; in that case, magnetic moments are quenched [106].

4.2.3 Modeling the Structure of Glasses

When using x-ray absorption spectroscopy for structural studies, one usually makes use of the EXAFS range where multiple scattering effects are not very important. Consequently, the extraction of structural data from the spectra is relatively straightforward. Analysis of XANES is quite complicated in this respect. In some situations,

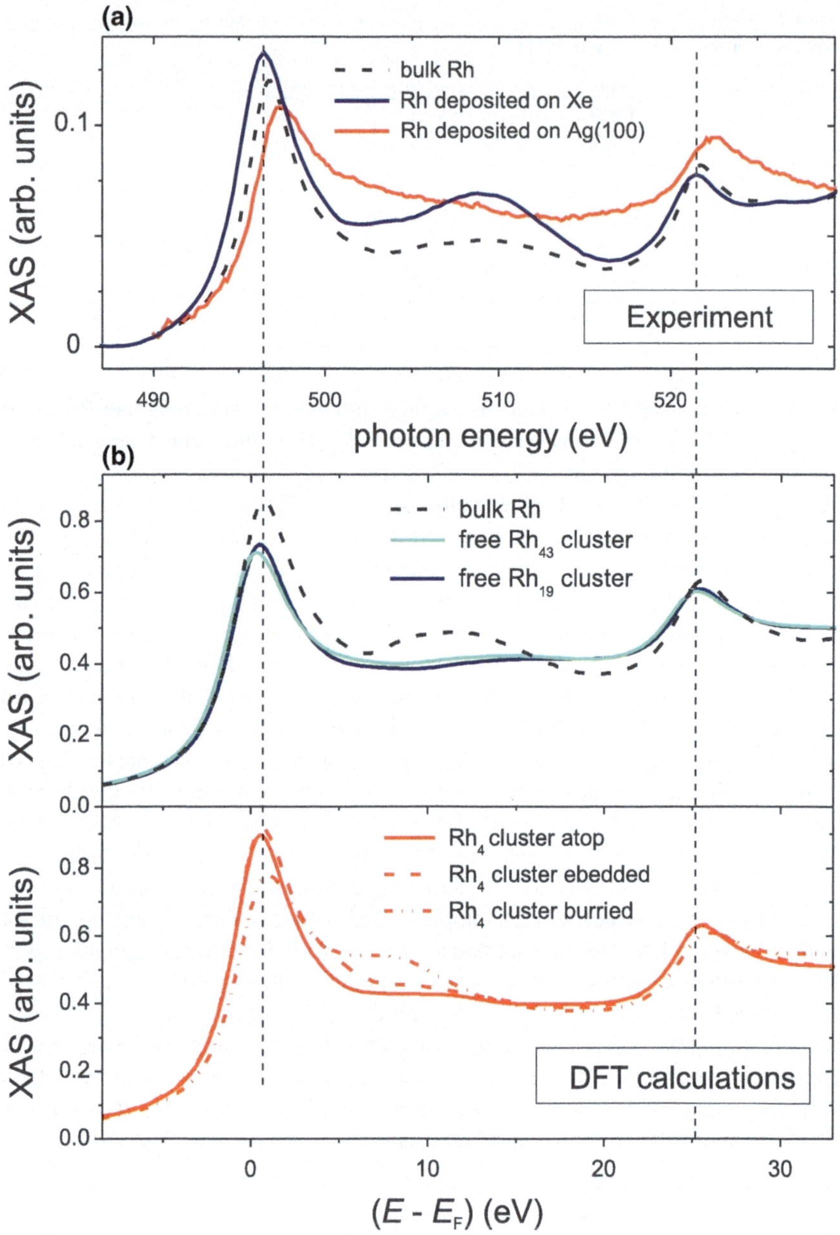

Fig. 4.5 **a** Experimental Rh $M_{3,2}$ XANES of Rh clusters deposited on a Xe buffer layer on Ag(100) and of Rh clusters deposited directly on Ag(100). Spectrum of bulk Rh is shown for comparison. **b** Theoretical XANES: (*upper panel*) free spherically-shaped Rh clusters of 19 and 43 atoms together with Rh crystal, (*lower panel*) flat two-dimensional Rh clusters of four atoms located either on top of Ag(100) surface or embedded into it or buried below it. Figure reproduced from [106]

however, getting involved in it is necessary. The cases introduced in Sect. 4.2.2 above concerned situations where recording EXAFS is not possible because the material is very diluted so one does not have sufficient intensity of the signal to isolate small EXAFS oscillations. In this part we will deal with other situations where it is useful to go beyond the single-scattering EXAFS analysis.

There are several reasons why the views on the geometry provided by XANES and by EXAFS could be different. First, XANES is quite sensitive to multiple-scattering contributions which are usually not so important for EXAFS. XANES thus reflects also the bond angles. Second, as the scattering amplitude depends on the electron energy differently for different atomic types, XANES may depend differently than EXAFS on the chemical composition of the nearest neighborhood of the photoabsorber. Finally, XANES features will be less damped by the disorder, so the signals which originate from scattering by atoms in the second and further coordination shells can be more important in XANES than in EXAFS [112–114]. One expects that this will become significant for studies of systems with a strong static disorder – such as glasses.

4.2.3.1 Local Geometry Around Ag Atoms in Ag-B-O Glasses

As an illustration, we present a study of possible structural configurations of Ag atoms in silver borate glasses $Ag_2O \cdot nB_2O_3$. Borate glasses in general attracted attention because of their varied optical, electrochemical or magnetic properties. Silver borate glasses in particular aroused interest due to their high ionic conductivity, which promises potential applications in electrochemistry (solid electrolytes). Optimizing technologically interesting properties requires understanding the structure at the atomic level. Specifically, a detailed information about the local environment of the moving ions is desirable [115].

The local geometry around Ag atoms has been studied by various approaches both for binary ($Ag_2O \cdot nB_2O_3$) and ternary silver borate glasses (e.g., mixed with AgI). Based on the experimental results, a model for the structure around Ag was initially proposed. EXAFS experiments suggested that the Ag-O distance should be close to 2.3 Å and that the coordination number of Ag should be low (around two) [116, 117]. On the other hand, neutron scattering experiments in connection with a reverse Monte Carlo analysis of the data presented evidence for a different model, where several different local geometries around Ag would co-exist, with the average coordination number higher than what was deduced from the EXAFS [118]. This indicated gaps in the knowledge of the local geometry around Ag in $Ag_2O \cdot nB_2O_3$ glasses. Therefore, a thorough XANES analysis was performed [119–121].

First, one has to find a good starting point. This can be achieved by comparing theoretical spectra obtained for few generic polyhedra which simulate the nearest neighborhood of Ag atoms. Inspired by EXAFS results [116], one can start with one Ag atom surrounded by few O atoms at the distance of 2.27 Å. The number of nearest oxygens is varied between two and eight. We show several Ag K-edge spectra of basic units in Fig. 4.6. Apparently, the crucial task for the modeling is to reproduce

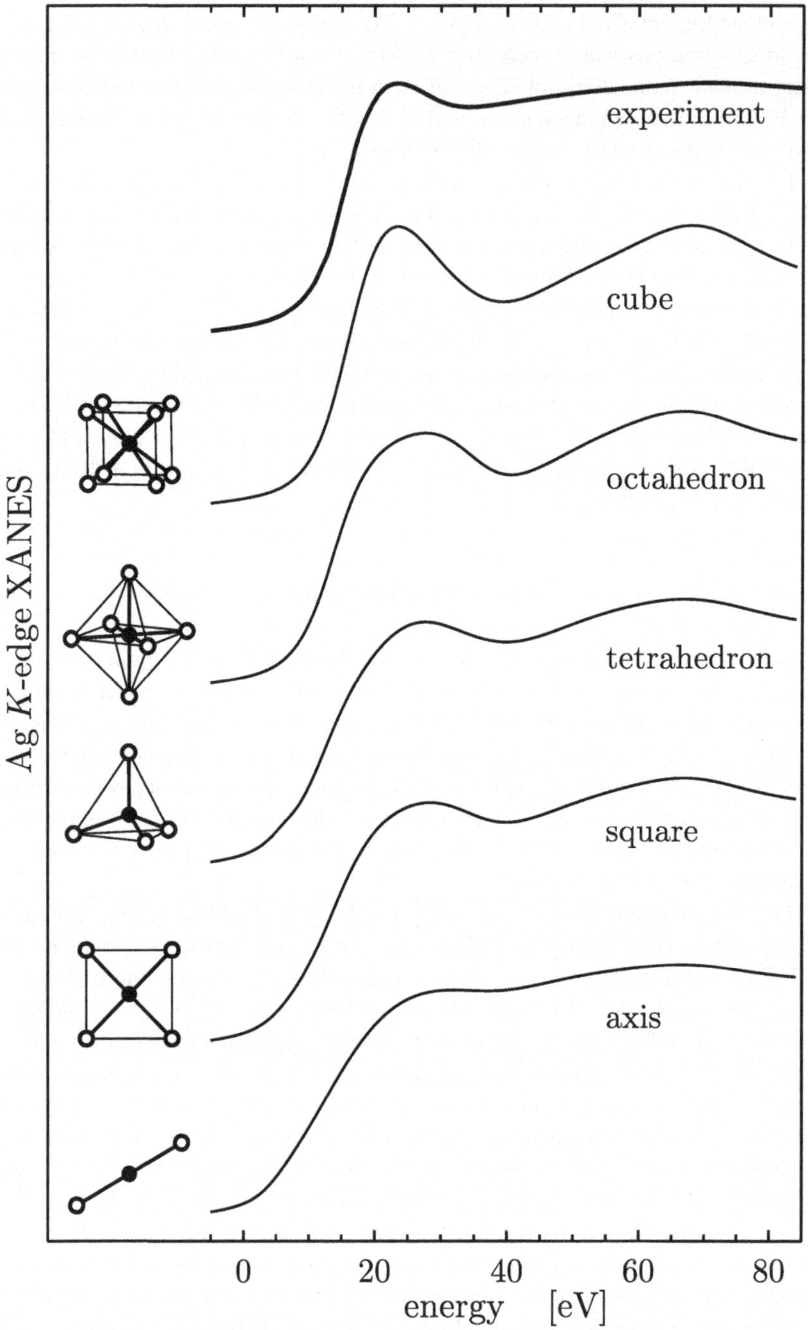

Fig. 4.6 Model Ag *K*-edge spectra calculated for five generic Ag-centered polyhedra. Spectral curves are identified by names and schematical diagrams of respective polyhedra (*full circles* depict Ag atoms, *open circles* depict O atoms). Spectrum measured for silver diborate glass $Ag_2O{\cdot}2B_2O_3$ is included for comparison. Figure reproduced from [120]

the first spectral peak at about 22 eV. In most cases, it is too wide and/or too low. The model with eight O atoms, with cubic order which resemble the nearest neighborhood in a bcc crystal, seems to be the most plausible [120]. The general picture provided by the data of Fig. 4.6 remains even if we randomly add some more distant atoms (to simulate the effect of further coordination shells which would be quite disordered in a glass). We conclude that the most promising candidate for a further improvement of the structural model of $Ag_2O{\cdot}nB_2O_3$ glasses is the eight-atoms cube-like polyhedron. Further study can focus just on this model and elaborates it in more details.

The simple cube-based model is clearly too symmetric to describe a real situation. One should consider that atoms in the nearest Ag neighborhood will have different radial distances and different chemical types. This implies that one cannot reproduce the measured Ag K-edge XANES of Ag-B-O glasses by employing just a single geometric configuration. To obtain an acceptable agreement between model spectrum and experiment, it is necessary to incorporate large structural disorder present in glasses via the multi-configurational approach. Conventional best-fitting methods would fail in this situation. Therefore an alternative approach has to be employed. Namely, we construct several cube-based semi-ordered structural models, each comprising several individual configurations, and explore how the calculated XAS changes if some characteristics of the models such as radial distances, coordination numbers and chemical compositions of atoms in the nearest and next nearest neighborhoods are varied.

The multi-configurational approach promoted here thus implies that each model structure comprises a *set of individual geometric configurations* or clusters, as shown in Fig. 4.7. Every cluster consists of a central Ag atom and of additional eight atoms which are located on the diagonals of the cube – close to cube vertices but not necessarily exactly in them. These "corner atoms" are further divided into two groups. The nearest N_1 atoms create a more rigid *first shell*; these atoms are always oxygens, all at the same distance R_1 from the Ag atom in the center. The rest of the $(8\text{-}N_1)$ corner atoms create a more diffuse *second shell*. We assume that K_B of them are borons and the remaining are oxygens. The second shell is radially smeared, the radial distances

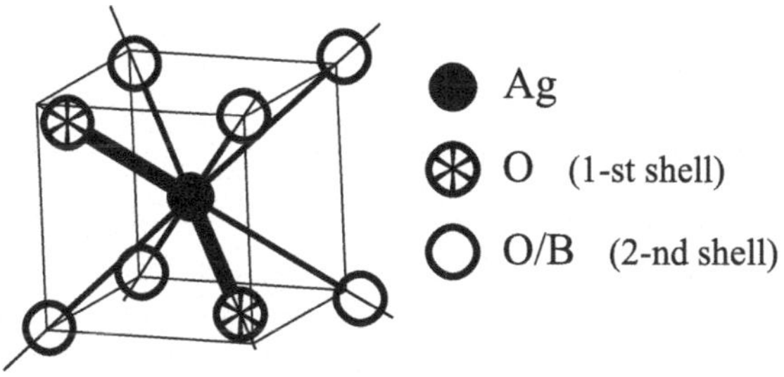

Fig. 4.7 Model of local environment around Ag in $Ag_2O{\cdot}nB_2O_3$ glasses based on a cubic structure. Positions of the central Ag atom and of the first-shell O atoms (marked by *crosses* and *thicker bonds*) are fixed for all configurations which form a particular set. Radial distances and chemical types of atoms in the second shell (depicted via *empty circles* and *thinner bonds*) are different for different configurations. Figure reproduced from [120]

span the 2.5–3.0 Å range equidistantly. To present a particular example, if the second shell is formed by five atoms, each of them has a different radial distance from the set 2.5, 2.625, 2.75, 2.875 and 3.0 Å. Imposing these conditions allows to generate many different geometric configurations with the same first shell geometry and with a fixed number of B atoms K_B. We categorize individual configurations in such a way that configurations with the same positions of first-shell atoms belong to the same *set*. Configurations differing only by arrangement of the second-shell atoms thus belong to the same set. Each such set defines a structural model.

We consider models with the first-shell coordination numbers $N_1 = 2, 3$, or 4, with the first-shell distances $R_1 = 2.0, 2.2$, or 2.4 Å and with $K_B = 0, 2$, or 4 borons in each cluster. Note that the parameters N_1, R_1, and K_B still do not specify the model uniquely because there are several non-equivalent ways to distribute the first-shell atoms among the eight bond directions. Ag K-edge XANES was calculated for each of the models by averaging theoretical spectra for all configurations which are associated with the particular model (set). Theoretical spectra for structural models specified by $N_1 = 4$ oxygens located in the first shell in a specific way are shown in Fig. 4.8 [120]. The arrangement of first-shell atoms is depicted by the diagram in the

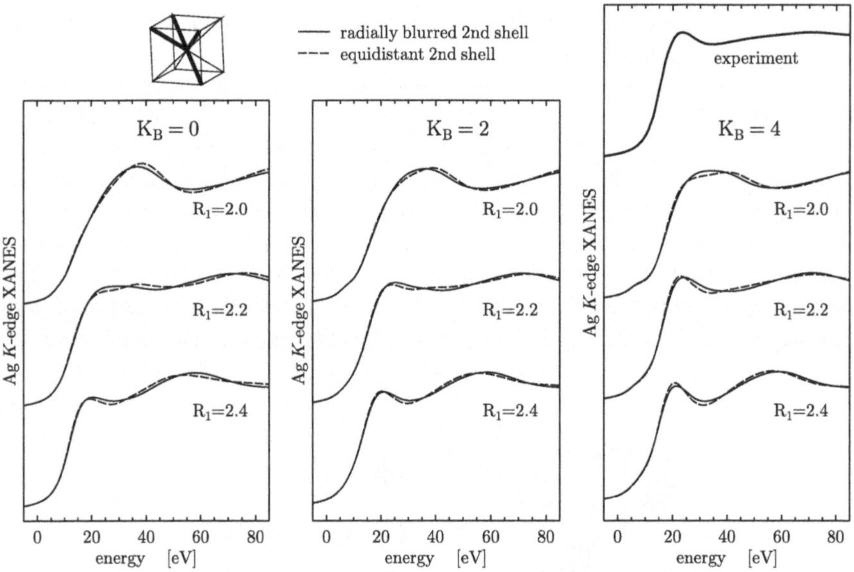

Fig. 4.8 Theoretical Ag K-edge XANES of $Ag_2O{\cdot}nB_2O_3$ glasses for models based on four oxygen atoms in the first shell. Positions of atoms in the first-shell are shown in the *inset above the left panel* (directions of first-shell bonds are shown via *thick lines*). Numbers of B atoms located in the second shell K_B vary from $K_B = 0$ (*left panel*) to $K_B = 2$ (*middle panel*) to $K_B = 4$ (*right panel*). Distances between O atoms in the first-shell and the Ag atom in the center vary from $R_1 = 2.0$ Å (*uppermost graphs*) to $R_1 = 2.2$ Å (*middle graphs*) to $R_1 = 2.4$ Å (*lowermost graphs*). Spectra for corresponding models which have all the (8-N_1) second-shell atoms at the same distance (2.75 Å) are shown by *dashed curves*. Measured K-edge XANES of $Ag_2O{\cdot}2B_2O_3$ glass is shown at the *top of the right panel*. Figure reproduced from [120]

inset. The plot shows how the averaged spectra evolve if R_1 or K_B are varied. The effect of smearing of the second shell can be seen as well: spectra for corresponding models with equidistant second shells are shown via dashed lines for comparison.

Whether a structural model is plausible or not should be assessed according to the ability of the model to reproduce essential spectral features seen in the experiment. In our case this includes the relatively sharp first peak followed by a second flat shoulder; the distance between these peaks is 41–47 eV. It turns out that the crucial test is whether a particular structural model gives rise to the first peak. The best agreement is obtained for model with $R_1 = 2.2$ Å [120]. A too short Ag-O distance ($R_1 = 2.0$ Å) results in unsatisfactory XANES curves. If some second-shell O atoms are substituted with B atoms, the results improve. Using $K_B = 4$ borons leads to better results than using $K_B = 2$ borons. Smearing the distances in the second shell over the whole 2.5–3.0 Å interval suppresses some redundant spectral oscillations. It also results in a small increase of the distance between both peaks, improving further the agreement with experiment.

Exploiting the K-edge of Ag in Ag-B-O glasses for structural analysis is quite complicated because of the large number of O and B atoms which surround Ag at distances below 3 Å. For glasses, this appears to be a general situation, especially if one is interested in the local structure around the glass network modifier (as is the case of Ag). In these cases analyzing XANES spectra can help in selecting most plausible models for local structure.

More reliable information about local structure could be got by analyzing both XANES and EXAFS signals together. Specifically in the case of Ag-B-O glasses, one could use the Ag-related radial distribution function (RDF) obtained from EXAFS as a guide for a multi-configuration analysis of XANES. It should be noted that a conventional EXAFS analysis based on Fourier filtering and cumulant expansion would not lead to reliable results for the structure of $Ag_2O \cdot nB_2O_3$ glasses due to a strong disorder around Ag ions. Instead, one could use a method based on a direct inversion of the EXAFS formula. Such procedure does not require a priori assumptions about the shape of the RDF [122]. Application of this method to Ag K-edge XANES of g-$Ag_2O \cdot nB_2O_3$ leads to the RDF shown in Fig. 4.9 [121]. Unfortunately, the procedure used to get the RDF shown in Fig. 4.9 does not allow for specifying chemical types of atoms in Ag neighborhood, because boron and oxygen are too close in the periodic table and their scattering properties are too similar in the EXAFS photoelectron energy range. This problem can be circumvented by employing XANES analysis: One can calculate the Ag K-edge XANES via the multi-configuration approach, using the RDF obtained from EXAFS analysis. When generating individual configurations employed in the XANES calculations, one requires that the average distribution of Ag–X distances (X = O, B) is the same as the EXAFS-derived RDF shown in Fig. 4.9. For technical reasons, the desired distribution is modeled by superposing two Gaussian RDF's.

In this way one can generate several sets of geometric configurations or *structural models* which differ one from another by the average number of B atoms K_B. As noted above, the RDF was written in terms of superposition of two Gaussians; the B atoms were put preferentially into the more distant of them. Only in case that

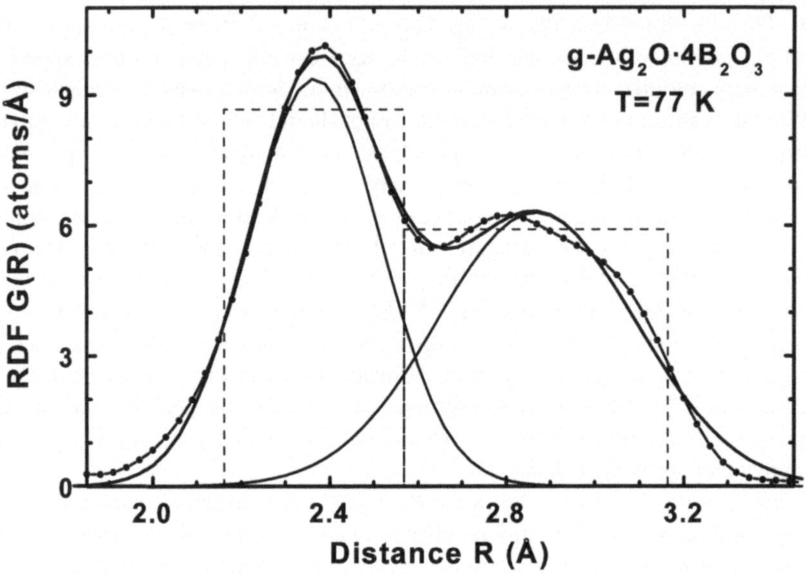

Fig. 4.9 Radial distribution function (RDF) for atoms around Ag in $Ag_2O{\cdot}4B_2O_3$ glass as estimated by best-fitting the EXAFS spectrum (*full line with dots*). Also shown is the approximation of this RDF by a superposition of two Gaussians (*full lines without dots*) and by a semi-uniform distribution (*dashed lines*). Figure reproduced from [121]

$K_B \geq 4$, some B atoms had to be placed also to the first shell. In the study of Kuzmin et al. [121] each model is represented by 200 configurations. Consistently with the multi-configuration approach, XANES for each structural model was obtained by averaging spectra obtained for all configuration that form the respective structural model (set of configurations).

Theoretical XANES obtained for the structural models has to be compared with experiment; this is done in Fig. 4.10. Calculated XANES clearly depends on the average number of B atoms K_B. For small K_B, the first peak appearing at about 20 eV has too low intensity in comparison with experiment. If the number of B atoms increases, theoretical XANES becomes similar to the experimental spectrum at about $K_B = 4$. Further increase of K_B leads to decrease of the amplitude of the first spectral peak and to an appearance of a false pre-peak at 8 eV.

A more quantitative comparison of calculated and experimental XANES signals one can achieved by using the R^2-factor,

$$R^2 = 100 \, \frac{\int dE \, \left[Y_{\text{the}}(E) - Y_{\text{exp}}(E) \right]^2}{\int dE \, \left[Y_{\text{exp}}(E) \right]^2} \, ,$$

with Y_{the} and Y_{exp} being the theoretical and experimental spectral intensities. The R^2-factors evaluated for each of the spectral curves shown in Fig. 4.10 are presented

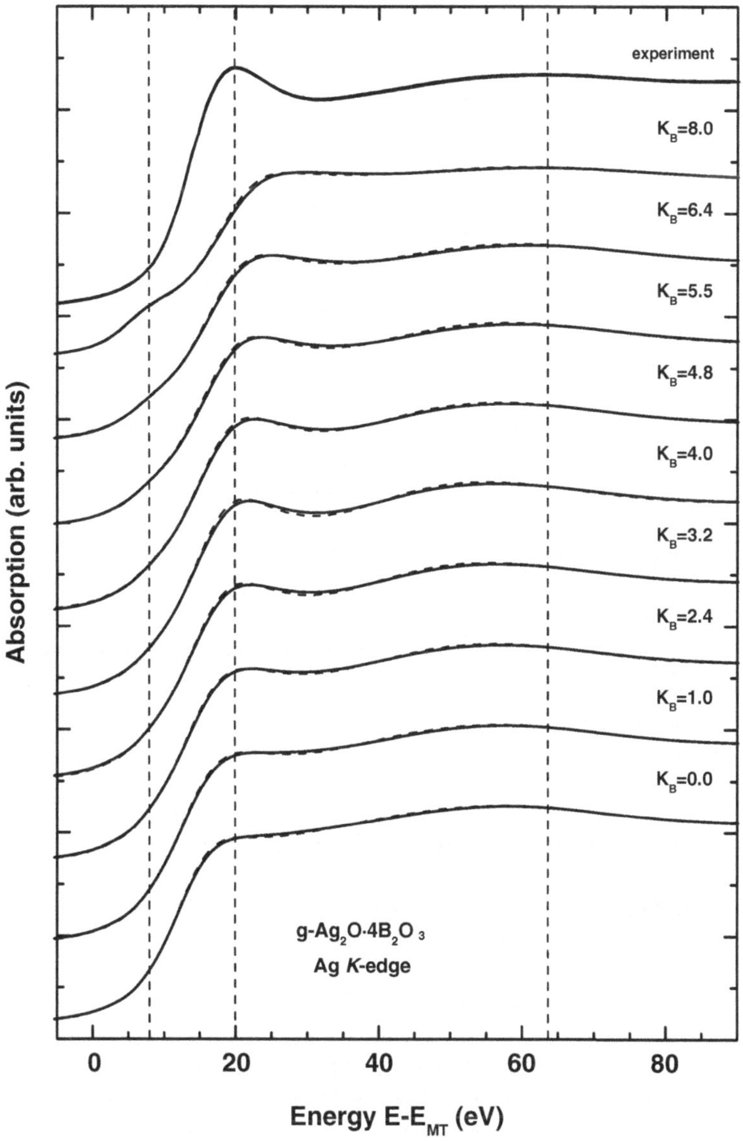

Fig. 4.10 Calculated Ag K-edge spectrum for $Ag_2O{\cdot}4B_2O_3$ glass. The spectra were calculated for structural models generated consistently with either double Gaussian (*full lines*) or semi-uniform (*dashed lines*) RDF's (see Fig. 4.9). Structural models are determined by average numbers of B atoms K_B (the rest of the first-shell atoms are oxygens). *Vertical lines* serve as guides for eye. Figure reproduced from [121]

Table 4.3 Quantification of the differences between calculated and measureed spectra for g-$Ag_2O{\cdot}4B_2O_3$ via the R^2-factor. This table accompanies the Fig. 4.10

Number of borons	R^2
0.0	0.2999
1.0	0.2708
2.4	0.2389
3.2	0.2112
4.0	0.1860
4.8	0.1849
5.5	0.1867
6.4	0.2084
8.0	0.3276

in Table 4.3. It can be seen that most plausible models are those containing 4.0–5.5 boron atoms, consistently with what was inferred based on Fig. 4.10.

Let us conclude this part by observing that by exploiting complementarity of EXAFS and XANES techniques one is able to find a more complete and reliable structural model than would be possible if only EXAFS or only XANES was used. Calculations of XANES in the real space within the KKR-GF formalism forms an important ingredient of this approach.

In the examples presented in this section, the inclusion of the disorder was achieved by directed averaging over many configurations. An alternative approach might rely on effective averaging over atomic positions via the CPA formalism (see Chap. 21 of this book).

4.2.3.2 Local Geometry Around B Atoms in Ag-B-O Glasses

There is a difference between the roles which Ag and B atoms have in $Ag_2O{\cdot}nB_2O_3$ glasses. Namely, B atoms are directly incorporated into the B-O network, forming rings of BO_3 and BO_4 units. They are thus network formers. On the other hand, Ag atoms mostly occupy the voids between the B-O rings, they are just network modifiers. A very important question related to the structure of borate glasses is the ratio between the numbers of BO_3 and BO_4 units in the borate network.

Conventionally, the ratio of B atoms occurring in BO_3 and BO_4 units has been determined by analyzing nuclear magnetic resonance (NMR) spectra of ^{11}B [123]. Having an independent method of measuring this ratio would be, nevertheless, quite useful. One of the reasons is that NMR spectroscopy cannot be used if the concentration of B atoms is very small (for example, in materials where boron is used as dopant). Therefore, there is a need for alternative ways of determining the ratio of three-fold and four-fold coordinated B atoms.

The ability of XANES to distinguish between BO_3 and BO_4 sites can be tested on boron-containing minerals. Their structure is quite complicated so it can be considered as a good approximation for the local structure of glasses. Let us start by examining whether spectra generated by three-fold and by four-fold coordinated B are sufficiently different one from another and whether this difference can be reproduced by calculations. Theoretical and experimental B *K*-edge XANES spectra for systems which contain only BO_3 units are presented in the left graphs of Fig. 4.11, spectra for systems with only BO_4 units are presented in the right graphs of Fig. 4.11 [124]. To test the robustness of theoretical results, the calculations were done both for non-self-consistent and self-consistent potentials, with the 1*s* core hole included within the relaxed and screened model [125].

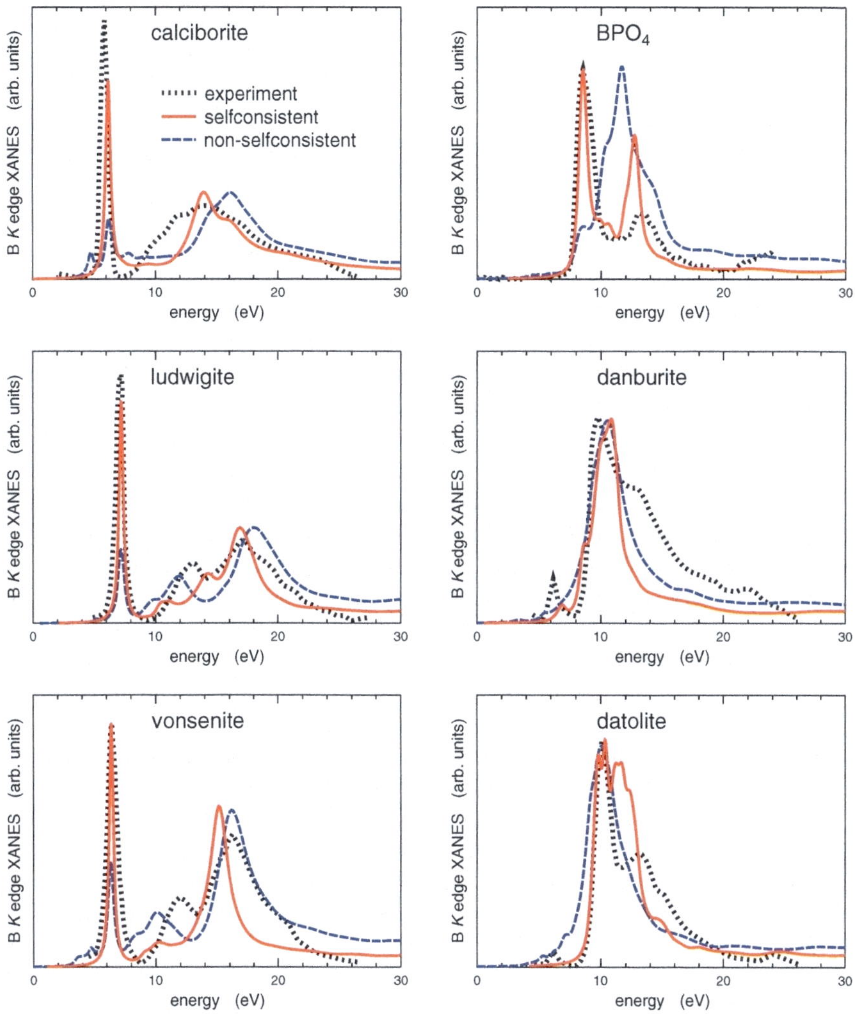

Fig. 4.11 B *K*-edge XANES of selected boron-contaning minerals calculated for either selfconsistent (*solid lines*) or non-selfconsistent (*broken lines*) potentials, together with experimental spectra (*dotted lines*). Left panels show spectra for systems with B in BO_3, right panels show spectra for systems with B in BO_4. Figure reproduced from [124]

Figure 4.11 shows that involving self-consistent potentials mostly does not have a major impact on the calculated XANES in comparison with non-self-consistent potentials. Particularly, the difference between the general spectral shapes for minerals with BO_3 and with BO_4 is reproduced sufficiently well for both potentials. The biggest impact of using a self-consistent potential can be observed in the pre-edge region (in case of BO_3-containing systems): the pre-peak intensity is properly reproduced only for a self-consistent potential. The situation for BPO_4 is specific: without using a self-consistent potential, the B K-edge XANES cannot be reproduced even as concerns its gross shape. It should be noted that the small pre-peak appearing in the experimental XANES of danburite and datolite (cf. Fig. 4.11) has been attributed to a small amount of BO_3 units formed in the material due to radiation damage [126, 127].

The objective of the research is to find out whether the main distinction between the B K-edge XANES of mineral containing BO_3 and BO_4 is robust. It is evident that the differences in theoretical B K-edge XANES spectra induced by using self-consistent or non-self-consistent potential are smaller than differences that would result from changing the local geometry. The conclusions drawn from minerals thus can be taken over for studying glasses.

A more targeted test of the sensitivity of XANES towards distinction between BO_3 and BO_4 units can be done on minerals which contain both BO_3 and BO_4 units. The focus of such a study should not be on a detailed analysis of XANES of a single compound but rather on observing spectral features which would be common to all boron sites with three-fold or four-fold coordinations. That requires an analysis of a large number of spectra. Figure 4.12 displays theoretical spectra of ten minerals which contain three-fold- as well as four-fold-coordinated B atoms [128]. To separate the possible effect of the long-range order present in crystalline minerals but not in glasses, the calculations are done both for large clusters of more than hundred of atoms and for very small clusters comprising just the basic BO_3 and BO_4 units. To provide a complex view, the spectra are shown for each boron site; this includes 13 curves corresponding to three-fold coordinated boron and 26 curves corresponding to four-fold coordinated boron. As an aid to the eye, spectra averaged over all sites in BO_3 units and in BO_4 units are shown via thick lines.

Apparently, the spectra shown in Fig. 4.12 can be categorized into two groups according to the coordination of the photoabsorbing B atom (similarly as in Fig. 4.11). Characteristic shapes of spectra generated at BO_3 and BO_4 sites are well-separated. The gross shape of B K-edge XANES is clearly dominated by short-range order – averaged spectra of small clusters are very similar to averaged spectra of large clusters. Conclusions based on spectra of minerals thus may be transferred to glasses.

Further quantitative analysis of results presented in Fig. 4.12 shows that the area of the main peak labelled as B depends only on the number of nearest oxygens (it does not depend on compound, distances, middle-range order etc.) It appears therefore that the ratio of areas of peaks labelled as C and B could serve as a criterion for estimating the ratio of BO_3 and BO_4 units in borate minerals and in borate glasses alike.

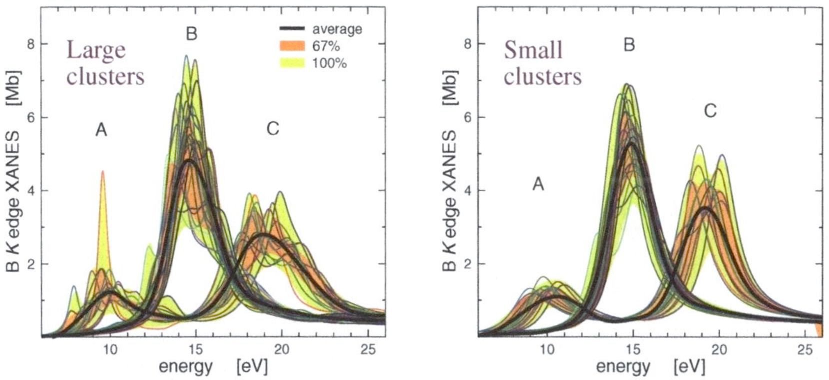

Fig. 4.12 Spectra calculated at all boron sites in ten different B-contaning minerals. *Thin lines* represent spectra at individual sites, *thick lines* show averages over all spectra at sites with BO_3 or with BO_4 coordination. Areas within which 100% or 67% of curves associated with given coordination are contained are also marked. The *left panel* shows spectra calculated for large clusters of about 140 atoms, the *right panel* shows spectra calculated for small clusters of 4–5 atoms

4.2.4 Interdiffusion at Interface: Interplay Between Electronic and Real Structure

In this part we will deal with Au $L_{2,3}$-edge x-ray magnetic circular dichroism (XMCD) of Co/Au multilayers. Experimental Au $L_{2,3}$-edge XANES and XMCD spectra are shown by thick dotted lines in Fig. 4.13 [129]. This experiment is remarkable because it shows that Au atoms at the Co/Au interface have appreciable magnetic moments, despite the fact Au atoms are hard to polarize. Let us start by comparing these experimental spectra with theoretical spectra obtained for Au atoms at clean (sharp) Co/Au interface (full lines in Fig. 4.13) [130]. It is obvious from this comparison that the system studied by Wilhelm et al. [129] cannot be modeled by a multilayer with a clean interface: the calculated L_2 XMCD peak has the same sign as the L_3 peak, which strongly contradicts the experiment. However, the experimental XMCD signal can be reproduced if one allows for some interdiffusion at the Co/Au interface. Two models of Co-Au interdiffusion were considered. In the first model, the interdiffusion spans two layers and the concentration of Au atoms in the interface layers is 70 and 30%. In the second model, the interdiffusion spans four layers and the Au concentration varies across the interface as 70, 55, 45, and 30%. Theoretical Au $L_{2,3}$-edge spectra obtained for these models are also included in Fig. 4.13. One can see that allowing for the interdiffusion improves the agreement between theory and experiment considerably. In particular, the model with interdiffusion spanning two layers yields XMCD peak intensities very similar to what was measured [130].

Interdiffusion at the Co/Au interface is something one would not expect in the beginning because Au and Co are mutually non-miscible. Nevertheless, for systems prepared by various non-equilibrium procedures such as evaporating and sputtering

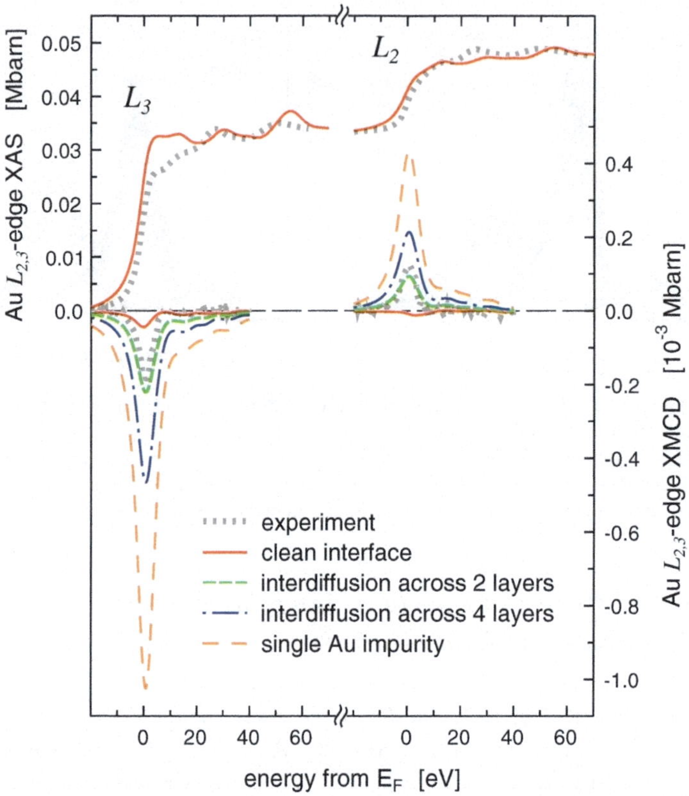

Fig. 4.13 Calculated and measured Au $L_{2,3}$-edge XANES and XMCD of $Co_{11}Au_4$ multilayer. Experimental data are from Wilhelm et al. [129]. The XANES and XMCD signals are shown for a clean interface and for two model cases with interdiffusion at the interface. Theoretical XMCD for an Au atom embedded in bulk Co is shown for comparison. Note that XANES curves calculated for both models of interdiffusion nearly coincide (in contrast to XMCD curves). Figure reproduced from [130]

some interdiffusion may be present. It should be noted in this regard that interdiffusion in Co/Au multilayers was assumed earlier when interpreting the dependence of magnetic anisotropy energy on annealing conditions [131–133].

The fact that the best agreement between theoretical and experimental XMCD signal is obtained for a specific model with interdiffusion across two layers should not be interpreted as implying that other interdiffusion models are not acceptable. It is quite possible that similar spectra could be obtained also for different distributions of Au and Co atoms across the interface. Moreover, one has to keep in mind that the structural model considered in the study [130] is still quite limited and that calculations which rely on the LDA also have their limitations. Nevertheless, the tendency to increase the heights of XMCD peaks if Au/Co interface interdiffusion is allowed for is obvious.

Interestingly, the changes in XMCD spectra which occur due to interdiffusion are not accompanied by corresponding changes in calculated XANES (see Fig. 4.13). This is in accordance with the idea that XANES is determined mostly by positions of atoms around the photoabsorber: one can plausibly assume that interdiffusion will change the chemical type of atoms close to the interface but not their positions. It seems that the difference in scattering amplitudes of Co and Au atoms is too small to give rise to significant changes in calculated XANES. On the other hand, magnetic moments of Au atoms are changed considerably via interdiffusion and this leads to significant changes in the XMCD.

4.2.5 Doped Materials

One of the areas where analysis of XANES spectra may bring useful structural information is research on doped materials because here it might be very difficult to get an EXAFS signal of sufficient intensity. Multiple-scattering analysis may be very helpful in these circumstances. To demonstrate this, we present few illustrative results for transition-metal-doped ZnO. The origin of magnetism in wurtzite w-ZnO doped by transition metal atoms is still subject of debate [134]. Despite numerous studies of structure and/or magnetism of doped ZnO, it is difficult to obtain a coherent picture [135–141]. One of the reasons for this is that it is difficult to investigate magnetism and structure simultaneously *on the same footing*. XANES and XMCD spectroscopy offer interesting possibilities in this respect.

One of the key question to answer is where the dopant is located. The potential of XANES modeling within the real-space KKR-GF formalism can be illustrated by the study of Co-doped wurtzite ZnO by Ney et al. [142]. Comparison of experimental and theoretical Co K edge x-ray linear dichroism (XLD) provides a convincing proof that Co atoms are located in Zn-substitutional sites (Fig. 4.14). A similar situation was found for Cu-doped ZnO pellets [143]. On the other hand, it appears that for Cu-doped ZnO thin films most of the Cu atoms are not built-in into the host lattice but occur rather in CuO-like coordinations [141]. A proper analysis thus has to be done for each system.

To assess the reliability of structural analysis based on XANES of doped ZnO, one needs information about the structural sensitivity of the spectra for the case of interest. Therefore, as another example, we present calculations of Cu K-edge XANES for Cu in a number of structures: Cu at Zn-substitutional site (Cu_{Zn}) in w-ZnO, Cu at interstitial site (Cu_i) in w-ZnO, monoclinic CuO, and also a hypothetic Cu metal with the structure of w-ZnO [144]. Apart from XANES averaged over all spatial orientations (as in polycrystalline sample), we evaluated also the XLD by subtracting the spectra for two perpendicular orientations of the polarization vector of the incoming radiation. The outcome is presented in Fig. 4.15. One can see that having Cu in different environments leads to different XAS and especially different XLD spectra. Employing the XLD thus makes it possible to distinguish reliably between Cu_{Zn} and Cu_i positions and also between CuO and Cu clustering.

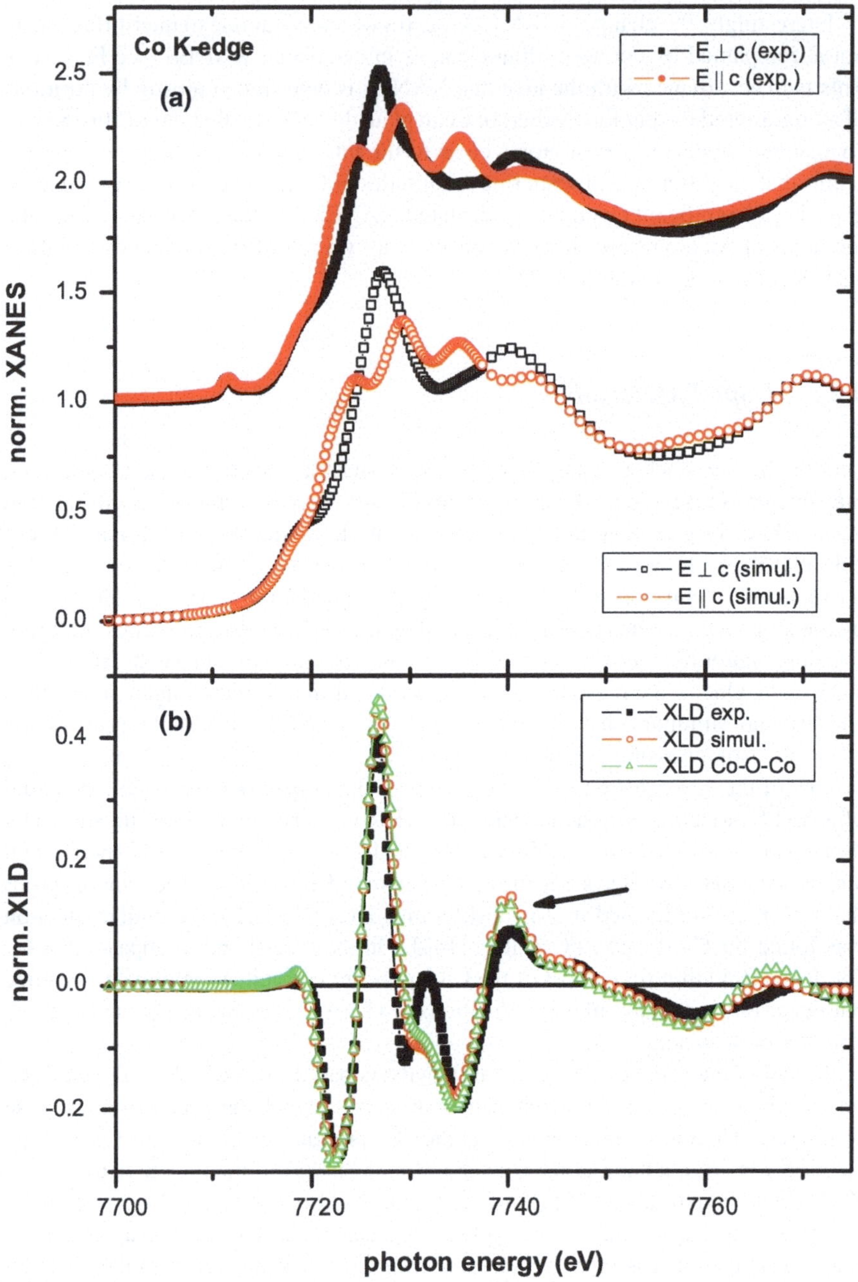

Fig. 4.14 Measured and calculated XANES (*upper panel*) and XLD (*lower panel*) at the Co *K* edge of Co-doped ZnO. A Zn-substitutional Co position was assumed. Figure reproduced from [142]

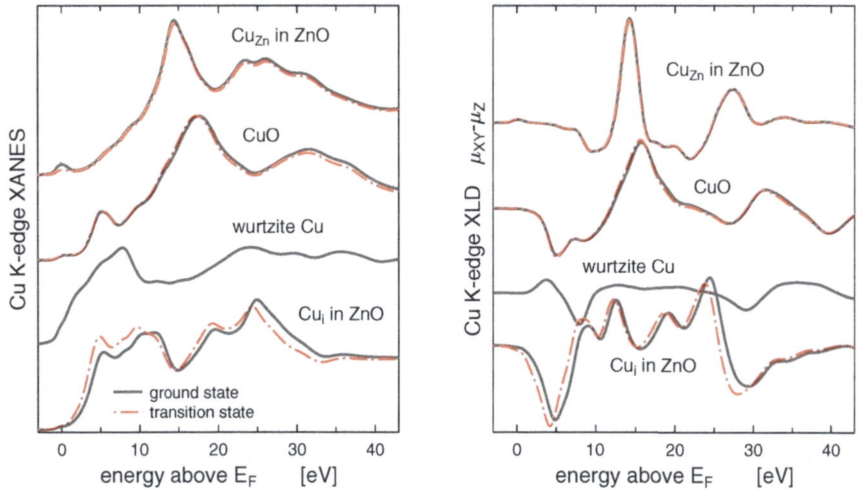

Fig. 4.15 Comparison of Cu K-edge spectra (XAS in the *left panel*, XLD in the *right panel*) for Cu in different environments: substitutional Cu_{Zn} in w-ZnO, monoclinic CuO, hypothetic Cu metal with structure of w-ZnO and interstitial Cu_i in w-ZnO. Results obtained for a ground state potential are shown via *full grey lines*, results obtained with a core hole included within the transition state approximation are shown via *dash-dotted lines*. Figure reproduced from [144]

Table 4.4 Magnetic moments (in μ_B) for Cu_{Zn} in ZnO if there are no vacancies and if vacancies are present in the *ab* plane — either in the first or in the second shell of oxygen atoms [144]

	μ_{spin}(tot)	μ_{spin}(Cu)	μ_{orb}(Cu)
No vacancy	1.70	0.59	0.14
Vacancy in 1st shell	0.54	0.18	0.06
Vacancy in 2nd shell	1.18	0.52	0.30

Another example how analysis of polarized XANES can facilitate studying structure of materials is given in Sect. 22.3 of this book.

One of the main motivations for research on doped ZnO is magnetism. Theoretical studies suggest that whether the dopant in ZnO is magnetic or not depends on its location. It was found that Cu atom in interstitial position in ZnO is nonmagnetic, while Cu atom in a substitutional position is magnetic [143]. As some studies suggest that oxygen vacancies may be important [137], it is interesting to calculate magnetic moments for a substitutional impurity Cu_{Zn} in ZnO if vacancies are included: the oxygen vacancy V_O was located in the *ab* wurtzite plane, in the first or in the second shell of O atoms (Cu–V_O distances were 1.98 Å or 3.80 Å) [144]. Theoretical results for μ_{spin} and μ_{orb} at the Cu sites are summarized in Table 4.4. The table presents also the spin magnetic moment for the whole system, to demonstrate that the polarization cloud around Cu must be quite extended.

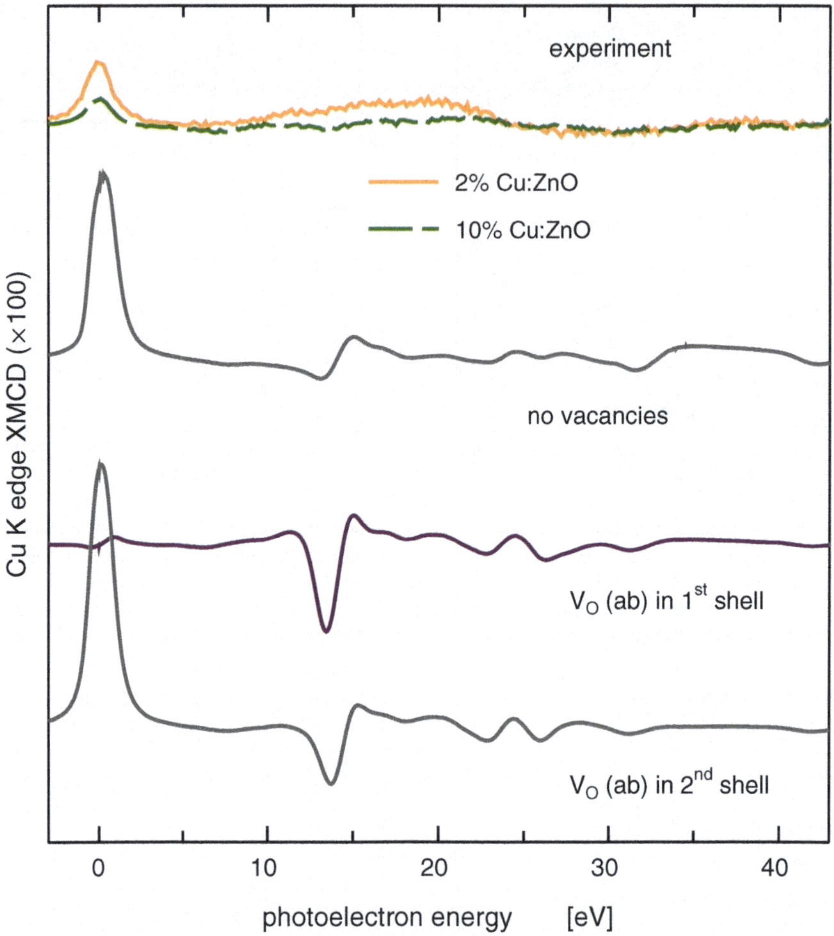

Fig. 4.16 Theoretical Cu K-edge XMCD for substitutional Cu_{Zn} in ZnO either without or with oxygen vacancies in the ab plane. Calculations were done without the core hole, with magnetization $M \| c$. Measured Cu K-edge XMCD spectrum for $Zn_{0.98}Cu_{0.02}O$ and $Zn_{0.90}Cu_{0.10}O$ pellets is shown at the top. Figure reproduced from [144]

To study the influence of vacancies further, one can calculate Cu K-edge XMCD for substitutional Cu_{Zn} either with vacancies or without vacancies. The results are shown in Fig. 4.16 [144], together with Cu K edge XMCD measured for paramagnetic Cu-doped ZnO pellets [143]. If the vacancy is next to Cu_{Zn}, the sharp XMCD peak in the pre-edge region is nearly suppressed. If the vacancy is further from Cu_{Zn}, it leads to changes in the *fine XMCD structure* in the energy region which corresponds to the main XANES peak. A particularly interesting feature is the case of an oxygen vacancy in the first shell. There is small non-zero magnetic moment on Cu for this case but the XMCD signal does not exhibit the strong pre-edge signal, clearly shown in the theory for other geometries (and also observed in the experiment for Cu-doped

ZnO pellets [143]). This strong sensitivity of the Cu K-edge XMCD to the presence of vacancies can serve as yet another example how XMCD spectroscopy, together with proper theoretical support, could be useful for structural studies.

4.2.6 Angular Resolved Photoemission

4.2.6.1 One-Step Model of Photoemission

Angle resolved photoemission (ARPES) and bremsstrahlung isochromat spectroscopy (ARBIS) are experimental techniques for the direct determination of the electronic structure of materials [27, 28]. Within this techniques, we have direct and very accurate access to measure the band dispersion and the electronic structure around the Fermi level with high resolution. During the last years many technical improvements led to an increase of the resolution of ARPES down to the meV regime. Among others, these developments includes the use of the synchrotron and laser photon sources and improvements of the detector side (e.g. spin resolution). On the theory side, the photoemission many-body theory has been developed 50 years ago [29–33]. Based on these formulations, Berglund and Spicer [34] derived so called three-step model (3SM) of photoemission (PE), which was the first and very simplified version of a one-electron approximation for the photocurrent, In this model the photocurrent is divided into 3 incoherent steps: the excitation of the photoelectron (PhEl), its transport through the bulk and its escape into the vacuum. Self-energy corrections, which describe among others damping processes and shifts in the quasiparticle spectrum, are neglected. This means that the final and initial states in the ARPES process are considered to be Bloch-waves with an infinite lifetime. The assumption of an infinite lifetime can not account for transitions into evanescent gap states. It means, exponentially decaying states into the bulk. This assumption for the initial state also does not allow to describe PE spectra that includes surface states.

To overcome the limitations of the 3SM, a dynamic description was proposed first for the final state by Liebsch [35] and Spanjaard et al. [36]. Later on multiple scattering effects were included for the initial as well as final states by Pendry and coworkers [37, 38]. This allows to include self-energy effects on equal footing. Within the Pendry's one-step approach (1SM) to ARPES [37, 39] the retarded one-electron Green's function for the initial state is calculated within the DFT [40]. Here the electronic correlation effects are in most cases considered in by the local (spin) density approximation (L(S)DA) [41, 42]. Lifetime effects in the initial state as well as final states are modeled by an imaginary potential term V_0. The time reversed spin-polarized low-energy electron diffraction (SPLEED) state [39, 43] is the proper final states for the PE process. The imaginary part $V_{0,f}$ here simulates the inelastic mean free path (IMFP) leading to the decaying amplitude of the high-energy PhEl state inside the bulk [37].

After the 1SM has been formulated, it has been generalised in many ways. For example, the quantitative description of spin-orbit induced dichroism was worked out by several groups [43–50]. Furthermore, the so called full-potential formulation of PE led to the accurate description for complex surface systems [43, 49, 51] (see Chap. 3 of this book). The treatment of disordered systems has been first proposed by Durham et al. [46, 52]. Nowadays, the 1SM allows to consider photon energies in a wide range from a few eV to more than 12 keV [53–60], for arbitrarily ordered [61] and disordered systems [62] at finite temperatures, and including in addition strong correlation effects within the DMFT [63–68].

The idea of the 1SM is to treat the actual excitation, the transport as well as the escape into the vacuum [34] as a single quantum-mechanically coherent process. This includes all multiple-scattering events. By describing the final and initial states within the fully relativistic layer-dependent KKR method [1, 69, 70], it is possible to treat ARPES from complex layered materials like thin films and multilayers. As mentioned above, within the 1SM of PE, the ARPES is described by Pendry's formula [37]:

$$I^{\mathrm{PES}} \propto \Im \, \langle E_f, \boldsymbol{k}_{\|} | G_2^+ \hat{X} G_1^+ \hat{X}^\dagger G_2^- | E_f, \boldsymbol{k}_{\|} \rangle \, . \tag{4.14}$$

This equation is derived Fermi's golden rule for the transition probability per unit time [43, 71] within the *sudden approximation*. Here the many body interaction of the PhEl with the rest of excited system is neglected. The initial state is described by the layer KKR Green's function G_1^+. Treating an angle-, spin- and energy-resolved, experiment the state of the PhEl at the detector can be formulated as $|E_f, \boldsymbol{k}_{\|}\rangle$, where $\boldsymbol{k}_{\|}$ is the component of the wave vector parallel to the surface, and E_f is the kinetic energy of the PhEl. The spin state of the PhEl is included in $|E_f, \boldsymbol{k}_{\|}\rangle$ as a four-component Dirac spinor. The advanced Green's function G_2^- in (4.14) describes the final-state e.g., the scattering properties of the surface at E_f. By $|\Psi_f\rangle = G_2^- |E_f, \boldsymbol{k}_{\|}\rangle$ all multiple-scattering events are formally included. We treat the final state within SPLEED theory by a single plane wave $|E_f, \boldsymbol{k}_{\|}\rangle$ penetrating onto the crystal surface. The $\hat{X}$ is the dipole operator of electron-photon interaction and is shown in its relativistic form in (4.8). However in ARPES, due to the numerical stability it is usefull to have a different form of this operator, i.e. the so called ∇V form. This is derived by making use of anticommutator and commutator with Dirac Hamiltonian analogously to the nonrelativistic case [19].

4.2.6.2 Correlation Effects in Transition Metals

The most detailed description of the band structure of correlated matter can be seen by angle- and spin-resolved valence band PE. Here we show various examples of ARPES calculations done within the 1SM. These examples will demonstrate the need to treat matrix elements in ARPES calculations in order to get a quantitative understanding of the experimental data.

The following examples concern the ferromagnetic (FM) transition metals like Fe and Ni as prototypes to understand magnetism and electronic correlations far beyond the LSDA. In particular, fcc Ni has been studied by many experimental [146–152] and theoretical studies [153–155] as a prototype of an itinerant electron FM, since shortcomings of simple one-electron theory are obvious. LSDA calculations for fcc Ni cannot describe various experimental aspects of the electronic structure of Ni. In addition to the feature that valence band PE spectra of Ni [156–158] has a reduced $3d$-band width compared to LSDA results [159] the PE show a satellite feature at a binding energy (BE) of about 6 eV [146, 147, 160–163]. On the other hand, an improved inclusion of correlations for the $3d$ electrons using manybody techniques [153, 154, 164] or in a more recent view applying the LSDA+DMFT method [64, 78]. Within LSDA+DMFT method we can find more or less experimental width of the $3d$-band complex and furthermore is able to predict the 6 eV satellite feature in the valence band.

In Fig. 4.17 we present a comparison of experimental PE results [145] and calculated theoretical data using several theoretical methods [64]. In the upper panel ARPES spectra from (011) surface if Ni along $\overline{\Gamma Y}$ for different emission angles are shown. The dots represent the experimental ARPES data, whereas the red lines are a single-particle spectral functions. Obviously, the LSDA-based data fails to reproduce the experimental data. The energetic positions of the theoretical peaks deviate strongly from the measured ones. In addition, the complex intensity variation that is measured for higher emission angles are not reproduced by the LSDA-based bloch spectral functions. From the second row of this figure, it can be seen that 3BS results provides a significant improvement when compared to the measurements. For the complete range of angles the peak positions agree with the measurments within about 0.1 eV. However, the overall spectral shape of the experimental intensities is different from the calculations due to the neglect of multiple scattering, matrix element and surface effects. In the measurement most peaks seem to be broadened. In addition, the intensity at the normal emission is shifted by 0.1 eV to higher BEs. The intensity shape resulting from the ARPES calculation are presented in the lower panel of Fig. 4.17. Here we obtained very good quantitative agreement between measurment and theory for all angles of emission. In the following we will closely describe spectrum calculated for the emission angle $\Theta = 5°$. The spin-integrated data shows a pronounced double structure with BEs of 0.1 and 0.3 eV. The second peak is reduced in intensity in agreement with the experimental data. Furthermore, the spectral width is also quantitatively reproduced. The calculated BEs are connected to the real part of the self-energy that shifts the energetic positions of peaks due to a dynamical renormalization of the quasi-particle energies. The relative peak intensities, on the other hand, can be traced back to the matrix element effects that are included in the one-step model of PE calculations. This double-peak structure comes from excitation of the exchange splitted d-bands together with a significant amount of surface-state [165].

The second example within this section is devoted to a study of the prominent 6 eV satellite of Ni. As was previously shown by theory [164] and PE experiments [149], the 6 eV satellite has finite spin-polarization. In a recent XPS experimental study the intensity at h$\nu = 150$ eV and the spin ARPES spectra at h$\nu = 66$ eV has been measured. Results for the latter experiment are presented in Figs. 4.18a and c.

The satellite feature is found at a BE of about 6.3 eV in good agreement with all former studies. Furthermore, Fig. 4.18c shows the non-zero spin-polarization, again in agreement with earlier publications [149]. After background correction, the spin-

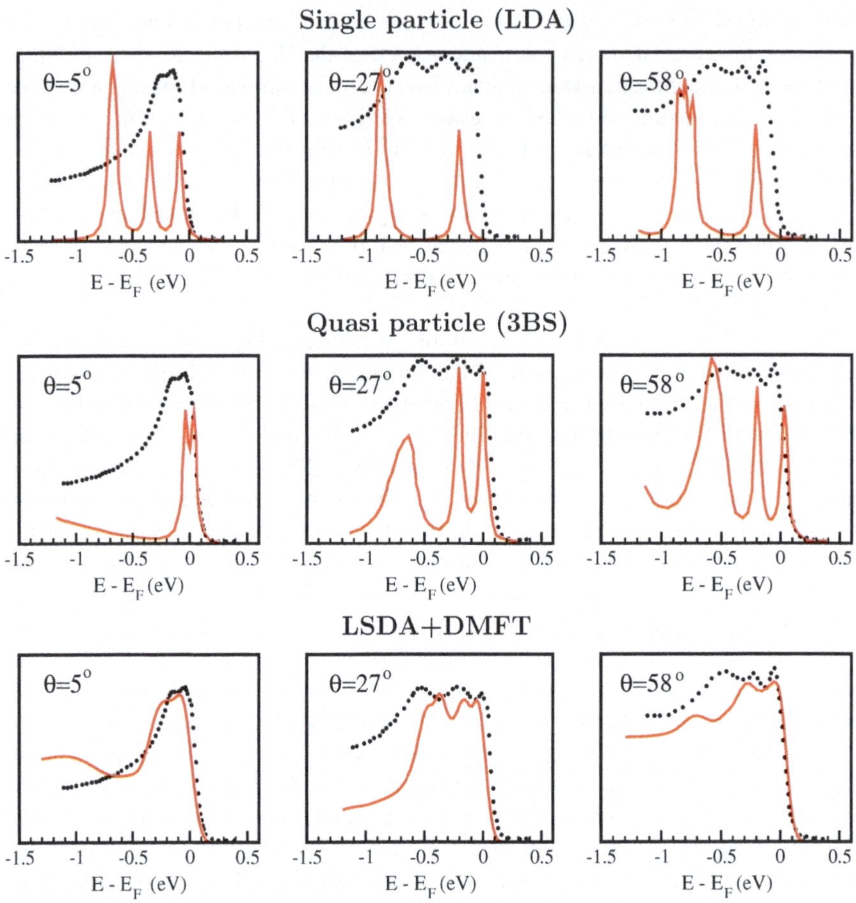

Fig. 4.17 Spin-integrated ARPES spectra from Ni(011) along $\overline{\Gamma Y}$ for three different angles of emission. *Upper row*: comparison between LSDA-based calculation and experiment [145]; *middle row*: comparison between experiment and non-self consistent quasi-particle calculations neglecting matrix element and surface effects [145]; *lower row*: spin-integrated LSDA+DMFT spectra including PE matrix elements (this work). Theory: *solid red line*, experiment: *black dots*. Figure reproduced from [64]

polarization is found to be about 15%. In Fig. 4.18b we compare the measured spectra with a DOS which is calculated based on the LSDA+DMFT method. The values $U = 2.8$ eV and $J = 0.9$ eV are identical with what we used for the previous example. The satellite is found at a BE of ~7.2 eV. This is 1 eV higher in BE compared to the experiments. This is due to the many-body solver used here. The FLEX-solver [86] is based on perturbation theory. As a consequence the energy-dependence of the self-energy is underestimated and this causes this shift of about 1 eV in the BE. The spin- and ARPES calculation is presented in Fig. 4.18d. The predicted spin-polarization of 10% is slightly smaller than the experimental one. Besides these deviations the

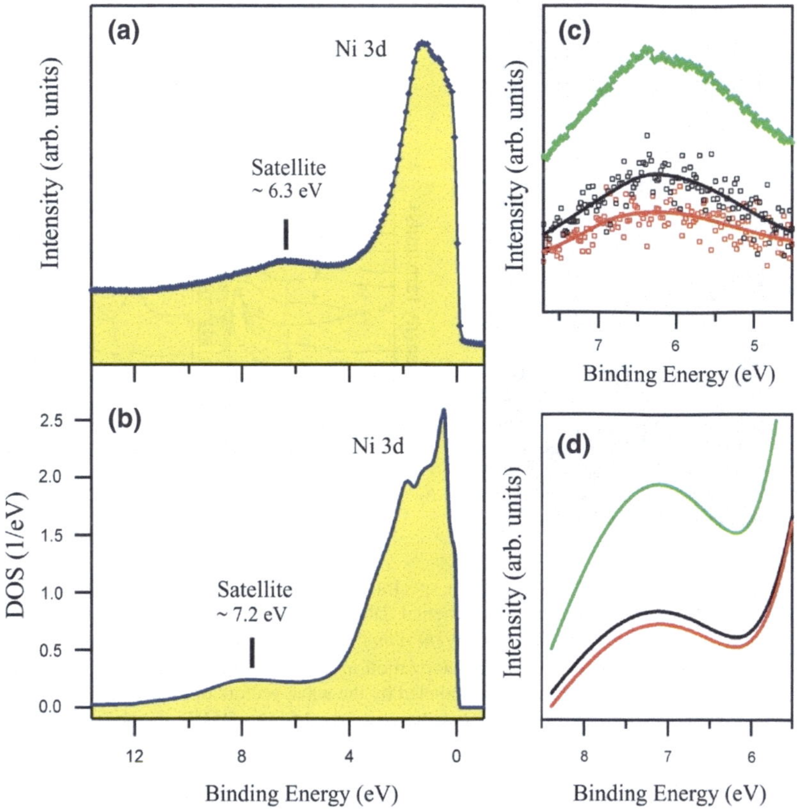

Fig. 4.18 **a** Measured XPS taken at hν = 150 eV. The *Ni 6 eV satellite* structure appears at about 6.3 eV BE. **b** Spin- and angle-resolved PE spectra taken in normal emission at hν = 66 eV with *s*-polarized light. *Open black (dark) squares*: majority spin states, *open red (light) squares*: minority spin states, *solid black (dark)* and *red (light) lines* serve as guides for the eyes. Spin-integrated intensity: *green (gray) thick dotted line*. **c** LSDA+DMFT calculation of the spin-integrated DOS. The satellite feature appears at bout 7.2 eV BE. **d** LSDA+DMFT spin-resolved PE calculation in normal emission at hν = 66 eV for a U value of 3.0 eV: *solid black (dark)* and *red (light) lines* indicate majority and minority spin states, *green (gray) line* shows the spin-integrated intensity. Figure reproduced from [68]

agreement between theory and experiment is very satisfying. Thus we could show the first ARPES calculation in which this spectral feature appears.

The second example in this section is devoted to a spectroscopic investigation of ferromagnetic iron [66]. In the left panel of Fig. 4.19 we show the experimental peak positions together with LSDA+DMFT spectral functions for majority and minority states. In addition to these LSDA+DMFT calculations, correlation effects were also described within the 3BS theoretical approach [166]. Within the 3BS method the self-energy is formulated using a configuration interaction-like description. Three-particle configurations as for example one hole plus one electron-hole pair are taken

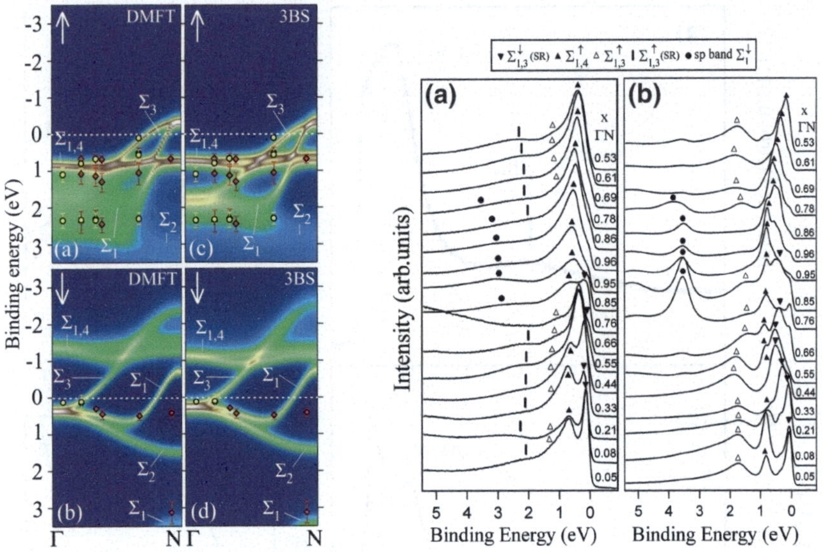

Fig. 4.19 *Left panel*: Spin resolved Bloch spectral functions calculated within LSDA+DMFT and 3BS formalism. Corresponding experimental data points have been deduced from the normal emission spectra along the ΓN direction. *Right panel*: **a** Experimental spin-integrated PE spectra of the Fe(110) surface measured with *p*-polarization in normal emission along the ΓN direction of the bulk Brillouin zone. The curves are labeled by the wave vectors in units of ΓN = 1.55 $Å^{-1}$. **b** Corresponding one-step model calculations based on the LSDA+DMFT method which include correlations, matrix elements and surface effects. Figure reproduced from [66]

into account within the 3BS approach. This result can be directly compared to the ARPES process and allows for a analysis of various contributions to the self-energy, as for example electron-hole lifetime. A more detailed comparison is presented in right panel of Fig. 4.19. Here we show a comparison of experimental ARPES data and theoretical LSDA+DMFT based one-step model calculations of (110) surface of iron along the ΓN of the bulk Brillouin zone with *p*-polarized light. In our LSDA+DMFT studies we use for the on-site Coulomb interaction U a value $U = 1.5$ eV which is deduced from experiment [167] and other theoretical studies [88, 168]. Near the Γ point ($k \sim 0.06$ ΓN), the peak found close to the Fermi level can be ascribed to a $\Sigma_{1,3}^{\downarrow}$ minority surface resonance, as shown on top of Fig. 4.19. In the experiment its $\Sigma_3^{\downarrow}$ bulk component crosses the Fermi energy at $k \sim 0.33$ ΓN. This leads to a reversal of the experimental spin-polarization and to a reduction of the intensity at $k = 0.68$ ΓN in the minority spin channel. The peak at the BE ∼0.7 eV, can be assigned to degenerate $\Sigma_{1,4}^{\uparrow}$ bulk-like majority spin states. A $\Sigma_3^{\uparrow}$ peak at BE ∼1.1 eV dominates the ARPES close to the Γ-point. The broad intensity around 2.2 eV, visible at most of the k-points, but not at the N-point, is connected to a majority surface state denoted as $\Sigma_{1,3}^{\uparrow}$. Around the N-point ($0.76 \leq k \leq 1.0$) and at BE ≥ 3 eV we see a $\Sigma_1^{\downarrow}$ band with strong sp character. The difference between its theoretical and measured intensity can be described by the fact that in the calculations only the Coulomb repulsion between d states is considered, without additional lifetime effects for other sp bands. The comparison of the calculated and measured spectra turned out to be a very detailed check for the U value used in the calculations. This also applies to the self-energy, that was compared to its experimental counterpart derived from the band dispersion and line width.

In summary, spectral function calculations for ferromagnetic iron was dano in such a way that it coherently combine description of electronic correlations, surface emission, multiple-scattering, dipole matrix element related effects that lead to a modification of the relative ARPES intensities. Previously, a similar study was done for hcp Co(0001) [89] and fcc Co(001) [169]. The unified approach allows a detailed and reliable description of high-resolution ARPES spectra of $3d$-ferromagnets. It also allows for a very stringent test of current developments in the field of DMFT and other many-body techniques.

4.2.6.3 Disordered Correlated Alloys: Ni_xPd_{1-x}(001) and Diluted Magnetic Semiconductor $Ga_xMn_{1-x}As$

In this section combination of electronic correlations disorder effects are discussed [62]. The above mentioned combination of KKR band structure method with the DMFT scheme has been generalised to the case of disordered alloys [63, 78]. In this way one can include many-body correlation effects in the electronic structure and PE calculations of this class of disordered materials. Application of this method to the high-energy x-ray PE spectroscopy (HAXPES) spectroscopy is shown in Fig. 4.20 where we HAXPES spectrum for diluted magnetic semiconductor $Ga_{1-x}Mn_xAs$ is shown. An important element for tracing back the nature of the ferromagnetism in dilute magnetic semiconductors is the direct experimental measurement of the electronic states close to the Fermi level. Due to the difficulty in preparation of a good quality surface samples it is very difficult to measure it by UV PE. This problem could be solved by ARPES [55] and angle integrated [56] bulk sensitive HAXPES. In Fig. 4.20 we compare experimental a-c and theoretical data d-f based on a one step model of angle integrated PE. Mn dopant has been described within the CPA approach in combination with DMFT. The theory agrees quantitatively with the experimental results over a broad range of BEs. We found bellow the Fermi level, a feature at 250 meV, which can be seen only by to performing LDA+DMFT instead of LDA calculations. From the theoretical data we conclude that the maximum in the difference spectra has mainly Mn-$3d$ (t_{2g}) angular momentum character. A strong hybridization is present with As $4p$ states which are localized around the impurity.

Figure 4.21 represents a series of ARPES spectra of Ni_xPd_{1-x} as a function of the concentration x. It was calculated and measured for a photon energy $h\nu = 40$ eV with linearly polarized light. The experiments are shown in the left panel and the LSDA+DMFT PE calculations are ploted in the right hand side. Overall agreement between theory and experiments is found for all concentrations. However, starting from $Ni_{0.70}Pd_{0.30}$ it is clearly visible that deviation between theory and experiment

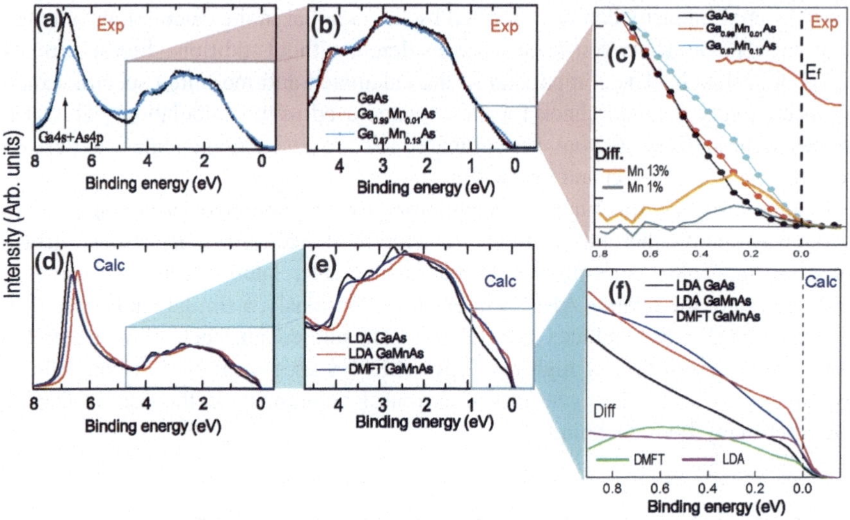

Fig. 4.20 **a–c** HAXPES spectra (5953 eV) for different Mn doping in GaAs. The GaAs spectra have been aligned to the (GaMn)As spectra using the BE position of the As 4*s* core level. No background subtraction was applied. **a** Extended valence band PES ($T = 20$ K), including the As 4*s*, Ga 4*s*, and As 4*p* shallow core levels of GaAs(100) and 13% Mn-doped GaAs. **b** Zoom of the valence band region ($T = 100$ K), showing the spectra from pure GaAs (*black dots*), 1% and 13% Mn-doped GaAs (*red and blue dots*, respectively). **c** High-resolution spectra measured in the vicinity of E_F. Difference spectra, corresponding to the Mn contribution only, are shown in *orange* (13% Mn spectrum (*blue dots*) minus pure GaAs spectrum (*black dots*)) and *grey* (1% Mn spectrum (*red dots*) minus pure GaAs spectrum (*black dots*)]. The reference Fermi level of Au is displayed, offset. **d** Calculated angle-integrated PES (including matrix elements) for photon energy and geometry (*p* polarization) as used in the experiment. **e** Calculated valence-band spectra of GaAs(100) using LDA (*black curve*) and (Ga,Mn)As (13%) using both LDA (*red curve*) and DMFT (*blue curve*). **f** Zoom of the vicinity of E_F with calculated difference spectra, as in **c**, for LDA (*violet curve*) and DMFT (*green curve*). Figure reproduced from [56]

occurs with increasing Pd concentration. This is caused by the Pd *d*-states that are slightly shifted to higher BEs. This is well known from other metals like Cu. It can be described in terms of static correlations in the Pd-states that are not explicitly considered here.

Our study has clearly shown that the electronic properties of the Ni_xPd_{1-x} alloy depend sensitively on the interplay of alloying and electronic correlation. A description within the CPA results in a very good description of the electronic structure of many disordered materials like for example Ni_xPd_{1-x} [62]. This example illustrates that the combination of the CPA with LSDA+DMFT approach provides a powerful tool for electronic structure calculations. In addition its combination with the fully relativistic one-step model of PE takes into account chemical disorder and electronic correlation on same footing and it guarantees a unique analysis of corresponding measured spectroscopic data.

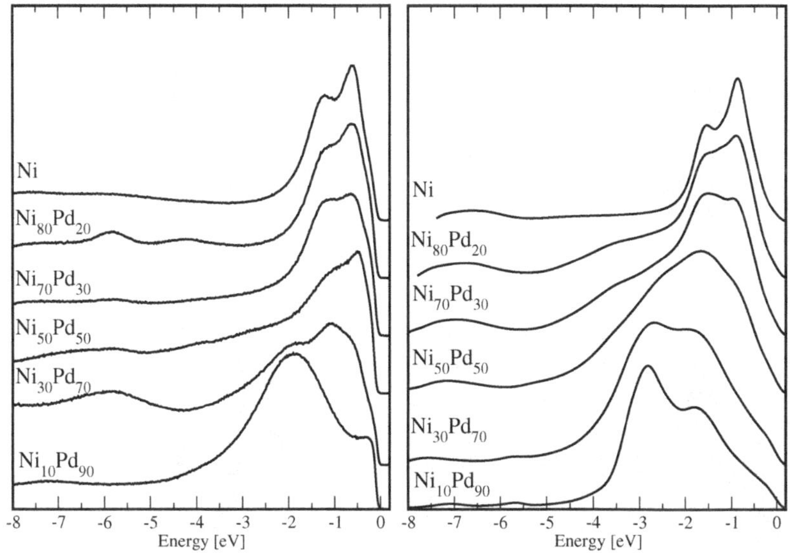

Fig. 4.21 ARPES spectra taken from the $Ni_xPd_{1-x}(001)$ alloy surfaces as a function of the concentration x for a fixed photon energy of $h\nu = 40.0$ eV along ΓX in normal emission. Experimental data shown in the *left panel* calculated spectra shown in the *right panel*. Depending on the concentration x a pronounced shift in spectral weight towards the Fermi level is visible. Figure reproduced from [62]

4.2.6.4 Surface Effects in ARPES

In the following section we picked up an particular example showing power of multiple scattering technique to describe and identify surface states. This example is based on the compound for iron based super conductors, $BaFe_2As_2$. In this material, a long standing discussion and unclear experimental situation concerning possible surface termination has been widely discussed. Using the 1SM of PE one can describe various surface states and can thus understand the origin of these bands. The occurrence of surface-states can be easly predicted within the multiple scattering theory by means of so called determinant criteria [171, 172]. This determinant criteria is formulated from the bulk reflection matrices $\mathbf{R}_b$ and from the scattering properties of surface barrier potential $\mathbf{R}_v$ (which connects the inner potential of the bulk crystal with the vacuum level). The possible appearance of a surface state than given by the following equation:

$$D(E, \boldsymbol{k}) = \det\left(\mathbf{I} - \mathbf{R}_b(E, \boldsymbol{k})\mathbf{R}_v(E, \boldsymbol{k})\right) = 0. \tag{4.15}$$

For better visualization we plot $1/|D(E, \boldsymbol{k})|$ instead of $D(E, \boldsymbol{k})$. If $1/|D(E, \boldsymbol{k})|$ is bigger than 10^3 we identify given spectral feature as a surface states. For values between 10^0 and 10^3 the spectral feature is surface resonance. For values below this value one has bulk states. More details about this criteria can be found in the paper by Braun and Donath [173].

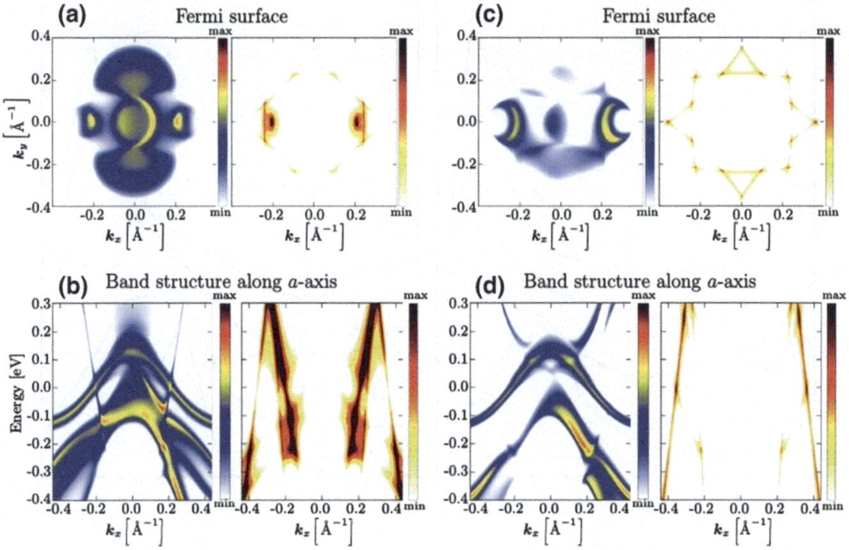

Fig. 4.22 Calculated Fermi surfaces and band structures of $BaFe_2As_2$ for either an As-terminated surface **a** and **b** or a Ba-terminated surface **c** and **d**. The *right side* of each picture shows the corresponding plot of $1/|D(E,\boldsymbol{k})|$, meaning a high intensity indicates a possible surface state if this specific structure can be also identified in the regular electronic structure calculation. Clear surface states can be identified for the As-terminated surface as *bright spots* in the Fermi surface **a** and as corresponding steep bands in the band structure **b**. These surface states are missing for a Ba-terminated surface. Figure reproduced from [170]

This determinant approach is demonstrated in Fig. 4.22. Here we show the band structures and the Fermi surfaces along the a-axis for an As-terminated and a Ba-terminated surface of $BeFe_2As_2$, respectively. The corresponding plot of $1/|D(E,\boldsymbol{k})|$ is shown on the right hand side of each picture. The determinant condition itself without a high intensity in the corresponding ARPES plot is only an indication for a surface state or a surface resonance. Only if a high intensity in the $1/|D(E,\boldsymbol{k})|$ plot agrees with a band in the ARPES one can speak about clear surface character. For example the two high intensity spots in Fig. 4.22a along the a-axis have a surface related origin, more specifically a surface state as the intensity of $1/|D(E,\boldsymbol{k})|$ is in the order of 10^6. This is in agreement with the previous findings for the $\boldsymbol{k}_z$-dispersion in ARPES, which indicates a connection to a surface related phenomenon [170]. Another proof for the surface origin of these states is represented in Fig. 4.22c and d, where the corresponding ARPES Fermi maps and band structures are shown for a Ba-terminated surface. The surface states discussed above for the As-terminated surface have completely disapeared in the Ba-terminated case. The corresponding high values of the $1/|D(E,\boldsymbol{k})|$ are missing in both plots. Overall, the Fermi surface and the dispertion of bands have undergone huge changes for the Ba surface termination. The Ba layer on top of the surface leads to the reduction of the intensity and smearing of the electronic states which are in turn clearly visible in an As-terminated surface. In particular one has to note that the agreement with measured ARPES data is much better for an As-terminated compared to the Ba-terminated surface.

This result suggest the most likely As surface termination of $BaFe_2As_2$. The surface termination in this material is still experimentally not clear and it is under debate [175]. According to first principle calculations only three possible surface termination exists. Is namely a fully As-terminated or a fully Ba-terminated surface and an As surface covered with 50% of Ba atoms [176]. Experimental low-energy electron-diffraction (LEED) and scanning tunneling microscopy (STM) measurements indicates a Ba-terminated surface [177]. However, there are also experimental LEED and STM measurments which clearly favor an As-terminated surface [178]. Above presented ARPES calculations clearly predicts an As-terminated surface. This was shown by agreement of calculations with ARPES measurments.

Acknowledgements We thank the DFG for financial support via Ebe154/32-1, BMBF (05K16WMA) and to the COST action MP1306 EUSpec for travel grants. JM was supported by the CEDAMNF project (CZ.02.1.01/0.0/0.0/15_003/0000358), co-funded by the ERDF as part of the OP RDE program of the Ministry of Education, Youth and Sports (Czech Republic). OS would like to acknowledge support by the GACR (project 17-14840S).

References

1. H. Ebert, D. Ködderitzsch, J. Minár, Rep. Prog. Phys. **74**, 096501 (2011)
2. O.K. Andersen, Phys. Rev. B **12**, 3060 (1975)
3. J. Korringa, Physica **13**, 392 (1947)
4. W. Kohn, N. Rostoker, Phys. Rev. **94**, 1111 (1954)
5. J.S. Faulkner, J. Phys. C Solid State Phys. **10**, 4661 (1977)
6. J.L. Beeby, Proc. R. Soc. Lond. A **302**, 113 (1977)
7. J.S. Faulkner, Phys. Rev. B **19**, 6186 (1979)
8. J.S. Faulkner, G.M. Stocks, Phys. Rev. B **21**, 3222 (1980)
9. B.L. Gyorffy, G.M. Stocks, in *Electrons in Finite and Infinite Structures*, ed. by P. Phariseau, E. Sheire (Plenum, New York, 1976), 144 p
10. Y. Onodera, M. Okazaki, J. Phys. Soc. Jpn. **21**, 1273 (1966)
11. P. Strange, H. Ebert, J.B. Staunton, B.L. Gyorffy, J. Phys. Condens. Matter **1**, 2959 (1989)
12. P. Weinberger, *Electron Scattering Theory for Ordered and Disordered Matter* (Oxford University Press, Oxford, 1990)
13. P. Strange, *Relativistic Quantum Mechanics* (Cambridge University Press, Cambridge, 1998)
14. H. Ebert, in *Electronic Structure and Physical Properties of Solids*, ed. by H. Dreyssé (Springer, Berlin, 2000), 191 p
15. R. Zeller, P.H. Dederichs, Phys. Rev. Lett. **42**, 1713 (1979)
16. I. Turek, V. Drchal, J. Kudrnovský, M. Šob, P. Weinberger, *Electronic Structure of Disordered Alloys, Surfaces and Interfaces* (Kluwer Academic Publishers, Boston, 1997)
17. P. Soven, Phys. Rev. **156**, 809 (1967)
18. P.J. Durham, B.L. Gyorffy, A.J. Pindor, J. Phys. F Met. Phys. **10**, 661 (1980)
19. H. Ebert, Rep. Prog. Phys. **59**, 1665 (1996)
20. M.E. Rose, *Relativistic Electron Theory* (Wiley, New York, 1961)
21. H. Ebert, G.Y. Guo, Solid State Commun. **91**, 85 (1994)
22. W.L. Schaich, Phys. Rev. B **29**, 6513 (1984)

23. H. Ebert, in *Spin-orbit Influenced Spectroscopies of Magnetic Solids*, ed. by H. Ebert, G. Schütz (Springer, Berlin, 1996), 159 p
24. R. Feder, F. Rosicky, B. Ackermann, Z. Physik B **52**, 31 (1983)
25. P. Strange, J. Staunton, B.L. Gyorffy, J. Phys. C Solid State Phys. **17**, 3355 (1984)
26. H. Ebert, J. Phys. Condens. Matter **1**, 9111 (1989)
27. F. Reinert, S. Hüfner, (Springer, Berlin, 2007), pp. 13–53
28. A. Damascelli, Z. Hussain, Z.X. Shen, Rev. Mod. Phys. **75**, 473 (2003)
29. I. Adawi, Phys. Rev. **134**, A788 (1964)
30. G.D. Mahan, Phys. Rev. B **2**, 4334 (1970)
31. W. Schaich, N. Ashcroft, Solid State Commun. **8**, 1959 (1970)
32. C. Caroli, D. Lederer-Rozenblatt, B. Roulet, D. Saint-James, Phys. Rev. B **8**, 4552 (1973)
33. P.J. Feibelman, D.E. Eastman, Phys. Rev. B **10**, 4932 (1974)
34. C.N. Berglund, W.E. Spicer, Phys. Rev. **136**, A1030 (1964)
35. A. Liebsch, Phys. Rev. Lett. **32**, 1203 (1974)
36. D.J. Spanjaard, D.W. Jepsen, P.M. Marcus, Phys. Rev. B **15**, 1728 (1977)
37. J.B. Pendry, Surf. Sci. **57**, 679 (1976)
38. J.F.L. Hopkinson, J.B. Pendry, D.J. Titterington, Comput. Phys. Commun. **19**, 69 (1980)
39. J.B. Pendry, *Low Energy Electron Diffraction* (Academic Press, London, 1974)
40. P. Hohenberg, W. Kohn, Phys. Rev. **136**, B 864 (1964)
41. W. Kohn, L.J. Sham, Phys. Rev. **140**, A 1133 (1965)
42. L.J. Sham, W. Kohn, Phys. Rev. **145**, 561 (1966)
43. J. Braun, Rep. Prog. Phys. **59**, 1267 (1996)
44. B. Ackermann, R. Feder, J. Phys. C Solid State Phys. **18**, 1093 (1985)
45. J. Braun, G. Thrner, G. Borstel, Phys. Stat. Sol. (b) **130**, 643 (1985)
46. B. Ginatempo, P.J. Durham, B.I. Gyorffy, J. Phys. Condens. Matter **1**, 6483 (1989)
47. S.V. Halilov, E. Tamura, D. Meinert, H. Gollisch, R. Feder, J. Phys. Condens. Matter **5**, 3859 (1993)
48. J. Braun, G. Borstel, Phys. Rev. B **48**, 14373 (1993)
49. M. Fluchtmann, M. Grass, J. Braun, G. Borstel, Phys. Rev. B **52**, 9564 (1995)
50. J. Henk, T. Scheunemann, S.V. Halilov, R. Feder, J. Phys. Condens. Matter **8**, 47 (1996)
51. M. Grass, J. Braun, G. Borstel, Prog. Surf. Sci. **46**, 107 (1994)
52. P.J. Durham, J. Phys. F Met. Phys. **11**, 2475 (1981)
53. A.X. Gray, C. Papp, S. Ueda, B. Balke, Y. Yamashita, L. Plucinski, J. Minár, J. Braun, E.R. Ylvisaker, C.M. Schneider, W.E. Pickett, H. Ebert, K. Kobayashi, C.S. Fadley, Nat. Mater. **10**, 759 (2011)
54. C. Papp, L. Plucinski, J. Minár, J. Braun, H. Ebert, C.M. Schneider, C.S. Fadley, Phys. Rev. B **84**, 045433 (2011)
55. A.X. Gray, J. Minár, S. Ueda, P.R. Stone, Y. Yamashita, J. Fujii, J. Braun, L. Plucinski, C.M. Schneider, G. Panaccione, H. Ebert, O.D. Dubon, K. Kobayashi, C.S. Fadley, Nat. Mater. **11**, 957 (2012)
56. J. Fujii, B.R. Salles, M. Sperl, S. Ueda, M. Kobata, K. Kobayashi, Y. Yamashita, P. Torelli, M. Utz, C.S. Fadley, A.X. Gray, J. Braun, H. Ebert, I. Di Marco, O. Eriksson, P. Thunström, G.H. Fecher, H. Stryhanyuk, E. Ikenaga, J. Minár, C.H. Back, G. van der Laan, G. Panaccione, Phys. Rev. Lett. **111**, 097201 (2013)
57. J. Minár, J. Braun, H. Ebert, J. Electron. Spectrosc. Relat. Phenom. **189**, 129 (2013)
58. J. Minár, J. Braun, H. Ebert, J. Electron. Spectrosc. Relat. Phenom. **190** (Part B), 159 (2013)
59. A.X. Gray, J. Minár, L. Plucinski, M. Huijben, A. Bostwick, E. Rotenberg, S. Yang, J. Braun, A. Winkelmann, G. Conti, D. Eiteneer, A. Rattanachata, A.A. Greer, J. Ciston, C. Ophus, G. Rijnders, D.H.A. Blank, D. Doennig, R. Pentcheva, J.B. Kortright, C.M. Schneider, H. Ebert, C.S. Fadley, Europhys. Lett. **104**, 17004 (2013)
60. J. Braun, K. Miyamoto, A. Kimura, T. Okuda, M. Donath, H. Ebert, J. Minár, New J. Phys. **16**, 015005 (2014)
61. J. Braun, J. Minár, S. Mankovsky, V.N. Strocov, N.B. Brookes, L. Plucinski, C.M. Schneider, C.S. Fadley, H. Ebert, Phys. Rev. B **88**, 205409 (2013)

62. J. Braun, J. Minár, F. Matthes, C.M. Schneider, H. Ebert, Phys. Rev. B **82**, 024411 (2010)
63. J. Minár, L. Chioncel, A. Perlov, H. Ebert, M.I. Katsnelson, A.I. Lichtenstein, Phys. Rev. B **72**, 045125 (2005)
64. J. Braun, J. Minár, H. Ebert, M.I. Katsnelson, A.I. Lichtenstein, Phys. Rev. Lett. **97**, 227601 (2006)
65. M. Pickel, A.B. Schmidt, F. Giesen, J. Braun, J. Minár, H. Ebert, M. Donath, M. Weinelt, Phys. Rev. Lett. **101**, 066402 (2008)
66. J. Sánchez-Barriga, J. Fink, V. Boni, I. Di Marco, J. Braun, J. Minár, A. Varykhalov, O. Rader, V. Bellini, F. Manghi, H. Ebert, M.I. Katsnelson, A.I. Lichtenstein, O. Eriksson, W. Eberhardt, H.A. Dürr, Phys. Rev. Lett. **103**, 267203 (2009)
67. J. Braun, J. Minár, H. Ebert, A. Chainani, J. Miyawaki, Y. Takata, M. Taguchi, M. Oura, S. Shin, Phys. Rev. B **85**, 165105 (2012)
68. J. Sánchez-Barriga, J. Braun, J. Minár, I. Di Marco, A. Varykhalov, O. Rader, V. Boni, V. Bellini, F. Manghi, H. Ebert, M.I. Katsnelson, A.I. Lichtenstein, O. Eriksson, W. Eberhardt, H.A. Dürr, J. Fink, Phys. Rev. B **85**, 205109 (2012)
69. K. Kambe, Z. Naturf, **22a**, 322 (1967)
70. J.M. MacLaren, S. Crampin, D.D. Vvedensky, J.B. Pendry, Phys. Rev. B **40**, 12164 (1989)
71. G. Borstel, Appl. Phys. A **38**, 193 (1985)
72. J.S. Faulkner, Prog. Mater. Sci. **27**, 1 (1982)
73. R.O. Jones, O. Gunnarsson, Rev. Mod. Phys. **61**, 689 (1989)
74. S.Y. Savrasov, G. Kotliar, Phys. Rev. B **69**, 245101 (2004)
75. A. Georges, G. Kotliar, W. Krauth, M.J. Rozenberg, Rev. Mod. Phys. **68**, 13 (1996)
76. G. Kotliar, S.Y. Savrasov, K. Haule, V.S. Oudovenko, O. Parcollet, C.A. Marianetti, Rev. Mod. Phys. **78**, 865 (2006)
77. K. Held, Adv. Phys. **56**, 829 (2007)
78. J. Minár, J. Phys. Condens. Matter **23**, 253201 (2011)
79. A.H. MacDonald, S.H. Vosko, J. Phys. C Solid State Phys. **12**, 2977 (1979)
80. R.M. Dreizler, E.K.U. Gross, *Density Functional Theory* (Springer, Heidelberg, 1990)
81. H. Ebert, S. Bornemann, J. Braun, D. Ködderitzsch, S. Lowitzer, S. Mankovskyy, J. Minár, M. Offenberger, S. Polesya, V. Popescu, Recent Developments in KKR Theory, Newsletter of the Ψ_k Network (Scientific highlight of the month) **97**, 79 (2010)
82. A. Gonis, W.H. Butler, *Multiple Scattering in Solids* (Springer, Berlin, 1999)
83. E.N. Economou, *Green Functions in Quantum Physics* (Springer, Berlin, 1990)
84. R. Zeller, J. Deutz, P.H. Dederichs, Solid State Commun. **44**, 993 (1982)
85. E. Tamura, Phys. Rev. B **45**, 3271 (1992)
86. L.V. Pourovskii, M.I. Katsnelson, A.I. Lichtenstein, Phys. Rev. B **72**, 115106 (2005)
87. S. Chadov, *Application of Many-body Perturbation Theory to the Description of Correlated Metals* (Universität München, Doktorarbeit, 2007)
88. S. Chadov, J. Minár, M.I. Katsnelson, H. Ebert, D. Ködderitzsch, A.I. Lichtenstein, Europhys. Lett. **82**, 37001 (2008)
89. J. Sánchez-Barriga, J. Minár, J. Braun, A. Varykhalov, V. Boni, I. Di Marco, O. Rader, V. Bellini, F. Manghi, H. Ebert, M.I. Katsnelson, A.I. Lichtenstein, O. Eriksson, W. Eberhardt, H.A. Dürr, J. Fink, Phys. Rev. B **82**, 104414 (2010)
90. S. Biermann, F. Aryasetiawan, A. Georges, Phys. Rev. Lett. **90**, 086402 (2003)
91. O. Šipr, J. Minár, S. Mankovsky, H. Ebert, Phys. Rev. B **78**, 144403 (2008)
92. V. Drchal, V. Janiš, J. Kudrnovský, Phys. Rev. B **60**, 15664 (1999)
93. O. Šipr, H. Ebert, Phys. Rev. B **72**, 134406 (2005)
94. O. Šipr, M. Košuth, H. Ebert, J. Magn. Magn. Mater. **272–276**, 713 (2004)
95. S. Bouarab, A. Vega, J.A. Alonso, M.P. Iñiguez, Phys. Rev. B **54**, 3003 (1996)
96. O. Šipr, H. Ebert, Czech. J. Phys. **53**, 55 (2003)
97. R. Wienke, G. Schütz, H. Ebert, J. Appl. Phys. **69**, 6147 (1991)
98. G. Schütz, M. Knülle, H. Ebert, Physica Scripta **T49**, 302 (1993)
99. P. Carra, B.T. Thole, M. Altarelli, X. Wang, Phys. Rev. Lett. **70**, 694 (1993)

100. K. Hirsch, V. Zamudio-Bayer, J. Rittmann, A. Langenberg, M. Vogel, T. Möller, B.v Issendorff, J.T. Lau, Phys. Rev. B **86**, 165402 (2012)
101. M. Niemeyer, K. Hirsch, V. Zamudio-Bayer, A. Langenberg, M. Vogel, M. Kossick, C. Ebrecht, K. Egashira, A. Terasaki, T. Möller, B.v Issendorff, J.T. Lau, Phys. Rev. Lett. **108**, 057201 (2012)
102. G.A. Mulhollan, R.L. Fink, J.L. Erskine, Phys. Rev. B **44**, 2393 (1991)
103. C. Liu, S.D. Bader, Phys. Rev. B **44**, 12062 (1991)
104. A.J. Cox, J.G. Louderback, S.E. Apsel, L.A. Bloomfield, Phys. Rev. B **49**, 12295 (1994)
105. J. Honolka, K. Kuhnke, L. Vitali, A. Enders, K. Kern, S. Gardonio, C. Carbone, S.R. Krishnakumar, P. Bencok, S. Stepanow, P. Gambardella, Phys. Rev. B **76**, 144412 (2007)
106. V. Sessi, K. Kuhnke, J. Zhang, J. Honolka, K. Kern, C. Tieg, O. Šipr, J. Minár, H. Ebert, Phys. Rev. B **82**, 184413 (2010)
107. Y. Jinlong, F. Toigo, W. Kelin, Phys. Rev. B **50**, 7915 (1994)
108. K. Lee, Z. Phys. D **40**, 164 (1997)
109. V. Kumar, Y. Kawazoe, Eur. Phys. J. D **24**, 81 (2003)
110. T. Futschek, M. Marsman, J. Hafner, J. Phys. Condens. Matter **17**, 5927 (2005)
111. O. Šipr, H. Ebert, J. Minár, J. Phys. Condens. Matter **27**, 056004 (2015)
112. M. Benfatto, C.R. Natoli, A. Filipponi, Phys. Rev. B **40**, 9626 (1989)
113. K. Hayakawa, T. Fujikawa, K. Nakagawa, I. Shimoyama, R. Tittsworth, P. Schilling, V. Saile, Chem. Phys. **289**, 281 (2003)
114. O. Šipr, J. Vackář, A. Kuzmin, J. Synchr. Rad. **23**, 1433 (2016)
115. G. Calas, L. Cormier, L. Galoisy, P. Jollivet, Comptes Rendus Chimie **5**, 831 (2002)
116. F. Rocca, G. Dalba, P. Fornasini, F. Monti, in *Borate Glasses, Crystals and Melts*, ed. by A.C. Wright, S.A. Feller, A.C. Hannon (The Society of Glass Technology, Sheffield, 1997), 295 p
117. G. Dalba, P. Fornasini, F. Rocca, F. Monti, J. Non-Cryst, Solids **293**, 93 (2001)
118. J. Swenson, L. Börjesson, W. Howells, J. Phys. Condens. Matter **11**, 9275 (1999)
119. O. Šipr, F. Rocca, G. Dalba, J. Synchr. Rad. **6**, 770 (1999)
120. O. Šipr, G. Dalba, F. Rocca, Phys. Rev. B **69**, 134201 (2004)
121. A. Kuzmin, G. Dalba, P. Fornasini, F. Rocca, O. Šipr, Phys. Rev. B **73**, 174110 (2006)
122. A. Kuzmin, Phys. B Condens. Matter **208**, 175 (1995)
123. P.J. Bray, in *Borate Glasses, Crystals and Melts*, ed. by A.C. Wright, S.A. Feller, A.C. Hannon (The Society of Glass Technology, Sheffield, 1992), 1 p
124. O. Šipr, F. Rocca, J. Synchr. Rad. **17**, 367 (2010)
125. O. Šipr, P. Machek, A. Šimůnek, J. Vackář. J. Horák Phys. Rev. B **56**, 13151 (1997)
126. L.A.J. Garvie, A.J. Craven, R. Brydson, Am. Mineral. **80**, 1132 (1995)
127. M. Kasrai, M.E. Fleet, S. Muthupari, D. Li, G.M. Bancroft, Phys. Chem. Miner. **25**, 268 (1998)
128. O. Šipr, A. Šimůnek, J. Vackář, G. Dalba, F. Rocca, Phys. Chem. Glasses Eur. J. Glass Sci. Technol. B **47**, 412 (2006)
129. F. Wilhelm, M. Angelakeris, N. Jaouen, P. Poulopoulos, E.T. Papaioannou, C. Mueller, P. Fumagalli, A. Rogalev, N.K. Flevaris, Phys. Rev. B **69**, 220404(R) (2004)
130. O. Šipr, J. Minár, J. Vackář, H. Ebert, Phys. Rev. B **75**, 134422 (2007)
131. F.J.A. den Broeder, D. Kuiper, A.P. van de Mosselaer, W. Hoving, Phys. Rev. Lett. **60**, 2769 (1988)
132. K. Spörl, D. Weller, J. Magn. Magn. Mater. **101**, 217 (1991)
133. G. Gubbiotti, G. Carlotti, F. Albertini, F. Casoli, E. Bontempi, L.E. Depero, H. Koo, R.D. Gomez, Thin Solid Films **428**, 102 (2003)
134. A. Zunger, S. Lany, H. Raebiger, Physics **3**, 53 (2010)
135. M. Snure, D. Kumar, A. Tiwari, Appl. Phys. Lett. **94**, 012510 (2009)
136. A. Ney, V. Ney, S. Ye, K. Ollefs, T. Kammermeier, T.C. Kaspar, S.A. Chambers, F. Wilhelm, A. Rogalev, Phys. Rev. B **82**, 041202 (2010)
137. G. Ciatto, A. Di Trolio, E. Fonda, P. Alippi, A.M. Testa, A.A. Bonapasta, Phys. Rev. Lett. **107**, 127206 (2011)

138. Q. Ma, J.T. Prater, C. Sudakar, R.A. Rosenberg, J. Narayan, J. Phys. Condens. Matter **24**, 306002 (2012)
139. M. Sawicki, E. Guziewicz, M.I. łukasiewicz, O. Proselkov, I.A. Kowalik, W. Lisowski, P. Dluzewski, A. Wittlin, M. Jaworski, A. Wolska, W. Paszkowicz, R. Jakiela, B.S. Witkowski, L. Wachnicki, M.T. Klepka, F.J. Luque, D. Arvanitis, J.W. Sobczak, M. Krawczyk, A. Jablonski, W. Stefanowicz, D. Sztenkiel, M. Godlewski, T. Dietl, Phys. Rev. B **88**, 085204 (2013)
140. A. Ney, V. Ney, M. Kieschnick, F. Wilhelm, K. Ollefs, A. Rogalev, Appl. Phys. Lett. **115**, 172603 (2014)
141. P. Vachhani, O. Šipr, A. Bhatnagar, R. Ramamoorthy, R. Choudhary, D. Phase, G. Dalba, A. Kuzmin, F. Rocca, J. Alloys Comput. **678**, 304 (2016)
142. A. Ney, K. Ollefs, S. Ye, T. Kammermeier, V. Ney, T. Kasper, S.A. Chambers, F. Wilhelm, A. Rogalev, Phys. Rev. Lett. **100**, 157201 (2008)
143. P.S. Vachhani, G. Dalba, R.K. Ramamoorthy, F. Rocca, O. ipr, A.K. Bhatnagar, J. Phys. Condens. Matter **24**, 506001 (2012)
144. O. Šipr, J. Vackář, P.S. Vachhani, R.K. Ramamoorthy, G. Dalba, A.K. Bhatnagar, F. Rocca, J. Phys. Conf. Ser. **430**, 012128 (2013)
145. F. Manghi, V. Bellini, J. Osterwalder, T.J. Kreutz, P. Aebi, C. Arcangeli, Phys. Rev. B **59**, R10409 (1999)
146. S. Hüfner, G.K. Wertheim, Phys. Lett. A **51A**, 299 (1975)
147. C. Guillot, Y. Ballu, J. Paigné, J. Lecante, K.P. Jain, P. Thiry, R. Pinchaux, Y. Pétroff, L.M. Falicov, Phys. Rev. Lett. **39**, 1632 (1977)
148. A. Kakizaki, K. Ono, K. Tanaka, K. Shimada, T. Sendohda, Phys. Rev. B **55**, 6678 (1997)
149. K.N. Altmann, D.Y. Petrovykh, G.J. Mankey, N. Shannon, N. Gilman, M. Hochstrasser, R.F. Willis, F.J. Himpsel, Phys. Rev. B **61**, 15661 (2000)
150. J. Osterwalder, J. Elec. Spec. Rel. Phenom. **117–118**, 71 (2001)
151. N. Kamakura, Y. Takata, T. Tokushima, Y. Harada, A. Chainani, K. Kobayashi, S. Shin, Phys. Rev. B **74**, 045127 (2006)
152. O. Karis, S. Svensson, J. Rusz, J. Oppeneer, M. Gorgoi, F. Schäfers, W. Braun, W. Eberhardt, M. Mårtensson, Phys. Rev. B **78**, 233105 (2008)
153. A. Liebsch, Phys. Rev. Lett. **43**, 1431 (1979)
154. A. Liebsch, Phys. Rev. B **23**, 5203 (1981)
155. O. Miura, T. Fujiwara, Phys. Rev. B **77**, 195124 (2008)
156. D.E. Eastman, F.J. Himpsel, J.A. Knapp, Phys. Rev. Lett. **40**, 1514 (1978)
157. F.J. Himpsel, J.A. Knapp, D.E. Eastman, Phys. Rev. B **19**, 2919 (1979)
158. W. Eberhardt, E.W. Plummer, Phys. Rev. B **21**, 3245 (1980)
159. C.S. Wand, J. Callaway, Phys. Rev. B **15**, 298 (1977)
160. J. Kanski, P.O. Nilsson, C.G. Larsson, Solid State Commun. **35**, 397 (1980)
161. Y. Sakisaka, T. Kommeda, M. Onchi, H. Kato, S. Masuda, K. Yagi, Phys. Rev. B **36**, 6383 (1987)
162. T. Kinoshita, T. Ikoma, A. Kakizaki, T. Ishii, J. Fujii, H. Fukutani, K. Shimada, A. Fujimori, T. Okane, S. Sato, Phys. Rev. B **47**, 6787 (1993)
163. N. Nakajima, S. Hatta, J. Odagiri, H. Kato, Y. Sakisaka, Phys. Rev. B **70**, 233103 (2004)
164. A.I. Lichtenstein, M.I. Katsnelson, G. Kotliar, Phys. Rev. Lett. **87**, 067205 (2001)
165. W. Eberhardt, E.W. Plummer, K. Horn, J. Erskine, Phys. Rev. Lett. **45**, 273 (1980)
166. C. Calandra, F. Manghi, Phys. Rev. B **50**, 2061 (1994)
167. M.M. Steiner, R.C. Albers, L.J. Sham, Phys. Rev. B **45**, 13272 (1992)
168. M. Cococcioni, S. de Gironcoli, Phys. Rev. B **71**, 035105 (2005)
169. T. Allmers, M. Donath, J. Braun, J. Minár, H. Ebert, Phys. Rev. B **84**, 245426 (2011)
170. G. Derondeau, J. Braun, H. Ebert, J. Minár, Phys. Rev. B **93**, 144513 (2016)
171. P.M. Echenique, J.B. Pendry, J. Phys. C Solid State Phys. **11**, 2065 (1978)
172. E.G. McRae, Rev. Mod. Phys. **51**, 541 (1979)
173. J. Braun, M. Donath, Europhys. Lett. **59**, 592 (2002)
174. M. Yi, D. Lu, J.H. Chu, J.G. Analytis, A.P. Sorini, A.F. Kemper, B. Moritz, S.K. Mo, R.G. Moore, M. Hashimoto, W.S. Lee, Z. Hussain, T.P. Devereaux, I.R. Fisher, Z.X. Shen, Proc. Nat. Ac. Sci. US. **108**, 6878 (2011)

175. J.E. Hoffman, Rep. Prog. Phys. **74**, 124513 (2011)
176. G. Profeta, C. Franchini, K.A.I.L.W. Gamalath, A. Continenza, Phys. Rev. B **82**, 195407 (2010)
177. F. Massee, S. de Jong, Y. Huang, J. Kaas, E. van Heumen, J.B. Goedkoop, M.S. Golden, Phys. Rev. B **80**, 140507 (2009)
178. V.B. Nascimento, A. Li, D.R. Jayasundara, Y. Xuan, J. O'Neal, S. Pan, T.Y. Chien, B. Hu, X.B. He, G. Li, A.S. Sefat, M.A. McGuire, B.C. Sales, D. Mandrus, M.H. Pan, J. Zhang, R. Jin, E.W. Plummer, Phys. Rev. Lett. **103**, 076104 (2009)

Chapter 5
Multichannel Multiple Scattering Theory in *R*-Matrix Formalism

Peter Krüger

Abstract Multichannel multiple scattering theory (MCMS) in R-matrix formulation is introduced for x-ray absorption spectra calculations from spin-orbit split core-levels. A multichannel extension of scattering theory is motivated by the occurrence of strong electron correlation effects of the atomic multiplet type. MCMS is implemented in real-space multiple scattering theory with a correlated particle-hole wave function and the multichannel scattering matrix of the core-level site is computed using the variational R-matrix method. This affords an accurate and numerically efficient treatment of strong particle-hole configuration mixing induced by core-valence Coulomb coupling. Applications of MCMS to $L_{2,3}$-edge spectra of light transition elements are reviewed and shown to give excellent results for metallic and insulating Ca and Ti compounds, where long range band structure effects and particle-hole coupling must be treated on an equal footing.

5.1 Introduction

5.1.1 Single- and Many-Electron Formulation of X-ray Absorption Spectroscopy

In this introductory section, we introduce the theoretical basis of x-ray absorption spectroscopy starting from the exact many-electron formula of the cross-section. We describe its reduction to the often used single-particle theory and discuss the validity and limitations of the latter. According to Fermi's golden rule, the x-ray absorption cross-section can be written as

$$\sigma(\omega_q) \propto \sum_F |\langle F|H_{\text{int}}|G\rangle|^2 \delta(E_F - E_G - \hbar\omega_q) , \tag{5.1}$$

P. Krüger (✉)
Graduate School of Engineering, Chiba University, Chiba 263-8522, Japan
e-mail: pkruger@chiba-u.jp

D. Sébilleau et al. (eds.), *Multiple Scattering Theory for Spectroscopies*,
Springer Proceedings in Physics 204,
https://doi.org/10.1007/978-3-319-73811-6_5

where H_{int} is the interaction between the x-ray field and the matter, $|G\rangle$ is the ground state and $|F\rangle$ are excited (final) eigenstates of the many-electron system, with energies E_G and E_F, respectively. The interaction Hamiltonian is essentially given by $\boldsymbol{A} \cdot \boldsymbol{p}$, where $\boldsymbol{A}$ is the vector potential of the light, $\boldsymbol{p}$ is the electron momentum operator, and a sum over all electrons is understood. Close to an absorption edge from some core-shell c, the dominating final states are those with one hole in that core-shell. However, for nearby core-shells, configuration mixing between states with holes in different shells may occur, as we shall see below.

Theories for XAS can be divided into two big classes depending on whether they are based on the independent particle approximation (IPA) or not. In the IPA, $|G\rangle$ and $|F\rangle$ are assumed to be (antisymmetrized) product functions corresponding to the same effective single-particle Hamiltonian. Then, it is easy to show that (5.1) reduces to

$$\sigma(\omega_q) \propto \sum_f |\langle f|H_{\mathrm{int}}|c\rangle|^2 \delta(\epsilon_f - \epsilon_c - \hbar\omega_q) , \tag{5.2}$$

where $|c\rangle$ and $|f\rangle$ are single-particle eigenstates (spin-orbitals) of, respectively, the core and the unoccupied levels, which energies ϵ_c and ϵ_f. Equation (5.2) is the single particle formula for XAS, and forms the basis of most calculations, especially for K-edge spectra. For the computation of the continuum state $|f\rangle$, or equivalently, the single-particle Green's function, multiple scattering theory is a very convenient technique.

The single-particle formula (5.2) can be derived under somewhat more general conditions than the IPA. But it must be assumed that $|G\rangle$ is the product between a core-state $|c\rangle$ and a single wave function $|\Phi\rangle$ of the other $(N-1)$ electrons, and that $|F\rangle$ is the product of the same function $|\Phi\rangle$ and the excited spin-orbital $|f\rangle$. Physically this means that during the excitation of the core-electron, the remaining $(N-1)$ electrons are mere *spectators* and do not change their state. In reality, the reaction of the other electrons is not at all negligible and may lead to strong spectral changes, so-called core-hole effects. Often the most important effect is the change of the effective one-electron potential due to the creation of the core-hole, that is an additional localized positive charge. As a consequence of the deepened atomic potential, the spectator orbitals contract or *relax*, and thereby screen the core-hole charge. The orbital relaxation can be taken into account in the one-electron theory (5.2) by calculating $|f\rangle$ with a *core-hole potential*, that is an effective single-particle potential obtained by a constrained ground state calculation where one electron is removed from the core shell and added to the valence shell. The prescription to use the core-hole potential instead of the ground state potential is known as the final state rule. For K-edge spectra, the final state rule often yields improved results as compared to the ground state potential. From a many-body point of view, the final state rule raises an orthogonality problem. Indeed, since the ground state potential and the final state potential are different, all the orbitals are different, and so the state of the spectator electrons can no longer be the same. So, $|G\rangle = |\Phi\rangle|c\rangle$ but $|F\rangle = |\tilde{\Phi}\rangle|f\rangle$ with $|\tilde{\Phi}\rangle \neq |\Phi\rangle$. Then $\langle F|H_{\mathrm{int}}|G\rangle = \langle f|H_{\mathrm{int}}|c\rangle \times S$, with $S = \langle\tilde{\Phi}|\Phi\rangle < 1$. The loss of spectral weight expressed by $S < 1$ must be compensated elsewhere in the spectrum. This implies that for the $(N-1)$ electron wave function in the presence of the core-hole, not only the (constraint) ground state $|\tilde{\Phi}_0\rangle$ with energy $\tilde{E}_0$ but also excited states $|\tilde{\Phi}_n\rangle$ with energy $\tilde{E}_n$ will give non-zero contributions to the spectrum. In core-level photoemission, this is directly visible since the energy loss $\tilde{E}_n - \tilde{E}_0$ corresponds to an increase in binding energy, and so the excitations appear as higher energy peaks, such as multiplet, shake-up and shake-off satellites in atoms and charge transfer or plasmon satellites in extended systems. In other words, the core-hole creation is necessarily accompanied by spectral weight transfer to many-body excitations. The success of the one-electron theory for K-edge spectra indicates that in this case, the spectral weight transfer is either small, $S \approx 1$, or much spread out in energy and thus hidden in the background.

In other cases, however, the spectral weight transfer is large and the single-particle theory (5.2) breaks down. This is mostly due to strong configuration mixing in the ground or final states which often occurs in $3d$ and $4f$ electron systems. The effects are strongest when the optical transition goes directly into the open $3d$ or $4f$ shell, i.e. at the transition metal L_{23} and rare earth M_{45} adsorption edges. For these *white lines*, good results are obtained with ligand field multiplet (LFM) theory [1] where (5.1) is evaluated with correlated N-electron wave functions for a single atom whereby intra-atomic configuration mixing (or *multiplet* effects) can be easily taken account for. However, in the case of transition metal L_{23} edges, the final $3d$ states are not fully localized and the photo-excited electron will scatter at the surrounding atoms, which gives rise to ligand field splittings and band broadening of L_{23} edge spectra. The ligand field splittings can, to some extend, be reproduced in the LFM model, but at the cost of empirical parameters which need to be fitted to experiment. Thus the predictive power of the LFM method is very limited. LFM theory essentially performs an exact diagionalization of the many-electron Hamiltonian. Let us note that it is not possible to extend this scheme in a straightforward manner to extended systems (solids and large molecules) because the size of the Hilbert space grows exponentially with electron number. Nonetheless in the last two decades, much progress has been achieved for taking account of multiplet effects in XAS calculations of extended systems on an *ab initio* level using various different approaches. Without going into detail, we mention time-dependent density functional theory [2–5], the Bethe–Salpeter-Equation approach [6, 7], *ab initio* ligand field multiplet theories based on quantum chemistry methods [8, 9] or Wannier-states [10], as well as the multichannel multiple scattering method [11] which is the subject of this chapter.

5.1.2 Multiple Scattering and the Need for a Multichannel Extension

Multichannel multiple scattering theory was first proposed in 1990 in a seminal paper by C.R. Natoli et al. [12]. It is a generalization of multiple scattering theory to many-electron wave functions and was motivated by the many-electron effects observed in x-ray absorption spectra that we have touched upon in the previous section. For single scattering centers, multichannel scattering theory is well known since the 1950s in nuclear physics [13] and since the 1960s in atomic physics [14]. Stated simply, MCMS theory extends multichannel scattering theory from a single atom to a condensed systems by taking account of multiple scattering at the various atomic sites.

In order to introduce our notation and to well understand in what sense MCMS extends standard multiple scattering theory (MST) we shall start by recalling some essential points of standard multiple scattering theory. More details can be found in Chap. 1. Standard MST is a numerical method for solving the single-particle, time-independent Schrödinger equation with fixed energy ϵ for a solid or a finite cluster of atoms. As a band structure method for solids it was first introduced by Korringa [15], Kohn and Rostoker [16] and is known as the KKR method. In MST the potential of the system is written as the sum of local potentials, $V = \sum_i V_i$, where i runs over the atomic sites and each V_i is non-zero only inside the atomic cell No i. The atomic cells must be non-overlapping and are often approximated by spheres. The main characteristic of MST is that it divides the electronic structure problem into a potential part, which can be solved site by site, and a structural part which only depends on the geometry of the system. In the potential part, the scattering response of each local potential is determined and expressed in terms of the phase shift functions $\delta_{i\ell}(\epsilon)$ or equivalently atomic t-matrices $t_{i\ell}(\epsilon)$, where ℓ is the orbital quantum number and we have assumed the local potential to be spherical for simplicity. In the structural part, the propagation of free spherical waves between the scattering sites is determined and expressed in terms of (real-space) KKR structure constants $G_{iL,jL'}$, which represents the transition amplitude between an outgoing free spherical wave centered at site i with angular momentum $L = (\ell, m)$ and a regular spherical wave centered at site j with L'. By letting the waves propagate between all sites, imposing the appropriate boundary conditions for the physical problem at hand, and requiring that the total wave be continuous and differentiable one arrives at consistency equations between the spherical wave amplitudes. They are the multiple scattering equations. A central quantity of MST is the *scattering path operator* τ. Its matrix elements $\tau_{iL,jL'}$ express the transition amplitude between free waves iL and jL' including all possible scatterings. Here $|iL\rangle$ denotes a local solution of the Schrödinger equation at site i with angular momentum L. τ can be calculated by matrix inversion as

$$\tau = \mathbf{M}^{-1} \quad \text{where} \quad M_{iL,jL'} = t_{i\ell}^{-1}\, \delta_{LL'}\delta_{ij} - G_{iL,jL'} \,, \tag{5.3}$$

From the knowledge of τ and the local solutions, it is easy to calculate the total wave function.

In the single-particle theory of XAS, standard MST can be used to calculate the final state wave $|f\rangle$ in (5.2). If the many-electron formula (5.1) is to be used, MST needs to be generalized to a multichannel, i.e. many-particle theory for the calculation of $|F\rangle$.

Multichannel scattering theory has been used since the 1940s to describe collision processes in nuclear physics. In the 1960s it was first applied to electron-atom scattering and photoemission. The need for a multichannel theory is obvious in the case of inelastic scattering phenomena. When an electron with energy ϵ_0 scatters inelastically off an atom, the electron loses a part of its energy to the atom which is thereby excited from its electronic ground state Φ_0 to some excited state Φ_α. Thus each excited state gives rise to a different scattering channel α with a different scattered electron orbital ϕ_α and energy $\epsilon_\alpha = \epsilon_0 + E_0 - E_\alpha$.

So the concept of multichannel scattering arises naturally in the description of inelastic processes. This is relevant for x-ray absorption spectroscopy, because the photoelectron may lose part of its energy in many-body excitations induced by the creation of the core-hole and the inelastic scattering of the photoelectron. These loss processes give rise to satellite structures in photoemission spectra. In the sudden approximation, the different photoemission channels are assumed uncoupled [12]. Then, the x-ray absorption can be calculated as a convolution of the core-level photoemission spectrum (including satellites) and the one-electron x-ray absorption spectrum, as will become apparent in (5.7) below. In contrast, if the photoelectron strongly interacts with the other electrons, the channels are generally coupled and must be solved together. For high photoelectron energies the sudden approximation is good, but in the near threshold region the excitation process may be closer to the adiabatic regime and strong multichannel coupling can occur [12]. From a more technical point of view, one can say that multichannel scattering theory is required whenever there is strong configuration mixing between the excited electron in the continuum state and the other electrons of the system.

5.2 Derivation of Multichannel Multiple Scattering Theory

5.2.1 *Single-Site Multichannel Scattering*

Consider x-ray absorption from a core-shell c of an isolated atom with N electrons. The ground state is $|\Psi_g^N\rangle$ and its energy E_g. Possible final states are eigenstates of the system with one hole in shell c and one electron in the continuum, called the photoelectron in the following. The final state energy is $E = E_g + \hbar\omega_{\boldsymbol{q}}$, where $\hbar\omega_{\boldsymbol{q}}$ is the photon energy. We first consider the eigenstates of the core-excited ionized atom with $(N-1)$-electrons

$$H^{N-1}|\Phi_\alpha^{N-1}\rangle = E_\alpha^{N-1}|\Phi_\alpha^{N-1}\rangle \,. \tag{5.4}$$

Physically, $|\Phi_\alpha^{N-1}\rangle$ describes a possible final state of the ionized atom in the asymptotic limit when the photoelectron is infinitely far away. In this limit the total N-electron final state is a product state $|\Phi_f^N\rangle = |\Phi_\alpha^{N-1}\rangle|\phi_\alpha\rangle$ where $|\phi_\alpha\rangle$ is the free electron wave with energy $\epsilon_\alpha = E^N - E_\alpha^{N-1}$.

Now let us first assume that the total wave is a product state not only in the asymptotic region but everywhere

$$|\Phi_f^N\rangle = |\Phi_\alpha^{N-1}\rangle|\psi_\alpha\rangle\,, \tag{5.5}$$

where now ψ_α is the full (*distorted*) photoelectron wave. Then the optical transition matrix elements have the form

$$\langle\Phi_f^N|H_{\rm int}|\Phi_g^N\rangle = \langle\psi_\alpha|H_{\rm int}|\phi_c\rangle\langle\Phi_\alpha^{N-1}|c_c|\Phi_g^N\rangle\,, \tag{5.6}$$

where the subscript c denotes a core-state and c_c its annihilator. The absorption intensity can then be simplified to

$$\sum_f |\langle\Phi_f^N|H_{\rm int}|\Phi_g^N\rangle|^2\delta(E_f - E_g - \omega_{\boldsymbol q}) = \int {\rm d}\epsilon\, I_{\rm XPS}(\epsilon - \omega_{\boldsymbol q}) I_{\rm XAS}^{\rm IPA}(\omega_{\boldsymbol q})\,, \tag{5.7}$$

where

$$I_{\rm XAS}^{\rm IPA}(\omega_{\boldsymbol q}) \equiv \sum_\nu |\langle\psi_\nu|H_{\rm int}|\phi_c\rangle|^2\delta(\omega_{\boldsymbol q} + \epsilon_c - \epsilon_\nu) \tag{5.8}$$

is the XAS spectrum in the independent particle approximation (5.2) and

$$I_{\rm XPS}(\epsilon) = \sum_\alpha |S_\alpha|^2\delta(\epsilon - E_\alpha + E_g)\,, \quad S_\alpha \equiv \langle\Phi_\alpha^{N-1}|c_c|\Phi_g^N\rangle\,, \tag{5.9}$$

is the many-body core-level XPS spectrum in the sudden approximation. So we see that if the total wave can be factored between the photoelectron and the rest, the XAS spectrum can be expressed as the convolution of the XPS spectrum and the XAS spectrum in the IPA. In this case several excitation channels may exist, but they are uncoupled. This corresponds to the sudden approximation which should be valid in the high energy (EXAFS) region [17]. For low photon energy, however, the photoelectron interacts strongly with the core-hole and the other electrons and substantial channel coupling may occur. In this case, the following full (i.e. coupled) multichannel theory must be used.

In multichannel scattering theory the N-electron final states are expanded over the $(N-1)$ electron eigenstates Φ_α^{N-1} times a single-electron continuum state describing the photoelectron. This is known as the *close coupling expansion*

$$|\Psi^N\rangle = \mathcal{A}\sum_\alpha |\Phi_\alpha^{N-1}\rangle|\phi_\alpha\rangle\,. \tag{5.10}$$

Here, $\mathcal{A}$ is the antisymmetrization operator. Each state Φ_α^{N-1} gives rise to a channel α and in each channel, the photoelectron has a different wave function $\phi_\alpha(\boldsymbol r)$ and a different energy

$$\epsilon_\alpha = E - E_\alpha^{N-1}\,. \tag{5.11}$$

Assuming the states Φ_α to be known, the remaining problem is the calculation of the photoelectron orbitals ϕ_α. We write the total N-electron Hamiltonian as a sum of the Hamiltonian of the ionized atom H^{N-1}, that of the photoelectron h and an interaction term V

$$H^N = H^{N-1} + h + V \ , \tag{5.12}$$

h includes the kinetic energy and the external (nuclear) potential, and V is the Coulomb interaction of the photoelectron (located at $\boldsymbol{r}$) with the other $(N-1)$ electrons.

$$V = \sum_{i=1}^{N-1} \frac{e^2}{|\boldsymbol{r}_i - \boldsymbol{r}|} \ . \tag{5.13}$$

Then, the Schrödinger equation for the N electron system reads

$$0 = (H^N - E)|\Psi^N\rangle = (H^{N-1} + h + V - E)\mathcal{A}\sum_\beta |\Phi_\beta^{N-1}\rangle|\phi_\beta\rangle \ . \tag{5.14}$$

For simplicity we shall disregard the antisymmetrization operator $\mathcal{A}$ in the following. We now project the above equation on a state $\langle\Phi_\alpha|$ and integrate over all electron coordinates except those of the photoelectron ($\boldsymbol{r}$). Using (5.4) and the orthogonality of the eigenstates $|\Phi_\alpha\rangle$ we get

$$0 = \sum_\beta \left[(h + E_\alpha^{N-1} - E)\delta_{\alpha\beta} + \sum_\beta \langle \Phi_\alpha^{N-1}|V|\Phi_\beta^{N-1}\rangle \right] |\phi_\beta\rangle \ . \tag{5.15}$$

This leads to the definition of the interchannel potential [12]

$$V_{\alpha\beta}(\boldsymbol{r}) \equiv \langle \Phi_\alpha^{N-1}|V|\Phi_\beta^{N-1}\rangle \tag{5.16}$$

$$= \int \Phi_\alpha^*(\boldsymbol{r}_1 \dots \boldsymbol{r}_{N-1}) V(\boldsymbol{r}_1 \dots \boldsymbol{r}_{N-1}\boldsymbol{r}) \Phi_\beta(\boldsymbol{r}_1 \dots \boldsymbol{r}_{N-1}) \mathrm{d}\boldsymbol{r}_1 \dots \mathrm{d}\boldsymbol{r}_{N-1} \ . \tag{5.17}$$

Using (5.11), (5.16) we can rewrite (5.15) as

$$(\epsilon_\alpha - h)\phi_\alpha(\boldsymbol{r}) = \sum_\beta V_{\alpha\beta}(\boldsymbol{r})\phi_\beta(\boldsymbol{r}) \ . \tag{5.18}$$

Equations (5.18) are the multichannel equations [12]. They are a system of coupled Schrödinger-like partial differential equations for the various photoelectron orbitals ϕ_α. The orbitals ϕ_α are coupled through the inter-channel potential $V_{\alpha\beta}$ which reflects the interaction between the photoelectron and the other $(N-1)$ electrons. If $V_{\alpha\beta} = 0$ then (5.18) reduces to a set of independent one-electron Schrödinger equations for each channel. If, moreover, only the lowest energy state of the core-ionized atom is kept, then we get a single-particle theory of absorption, essentially equivalent to the final state rule.

The equations (5.18) are complicated because of the presence of the interchannel potential $V_{\alpha\beta}$. In the derivation above, we have, for simplicity, disregarded antisymmetrization of the wave function between the photoelectron and the other electrons. However, this is not justified for low energies, where exchange is strong. If antisymmetrization is taken into account, the interchannel potential is non-local and (5.18) becomes a set of coupled *integro-differential* equations for the wave components ϕ_α. The calculation of the interchannel potential $V_{\alpha\beta}$ with exchange and the integration of the multichannel equations is a difficult problem. It has been successfully applied only to simple cases of electron-atom scattering and valence photoemission from light atoms in the late 1960s [18]. In the early 1970s, this direct method for the solution of the multichannel equations (5.18) has been superseded by the computationally much simpler R-matrix methods [19, 20]. In R-matrix methods, it is assumed that exchange and correlation between the photoelectron and the other electrons is restricted to a finite *reaction* volume. Inside this reaction volume the orbitals ϕ_α are expanded over a set of basis functions. Thereby, the system of integro-differential equations simplifies to a linear algebra problem. More importantly even, the interchannel potential does not need to be computed in coordinate representation but only its matrix elements between the basis functions, which is a great simplification.

5.2.2 Calculation of the Multichannel T-Matrix

We want to calculate the multichannel T-matrix corresponding to a N-electron absorption final state $|\Psi^N\rangle$ as in (5.10). We use time-reversed LEED (low-energy-electron-diffraction) boundary conditions appropriate for photoemission. The total wave function is the sum of a free wave Ψ^0 and a scattered wave Ψ^{sc}.

$$\Psi = \Psi^0 + \Psi^{\text{sc}} . \tag{5.19}$$

The free wave corresponds to the asymptotic region ($r \to \infty$) or the $t \to \infty$ limit in a time-dependent description. For photoemission, the proper choice of the free wave is a plane wave. In photoabsorption, all emission angles of the photoelectron are integrated over and so we may replace the plane waves by regular spherical waves and sum over all angular momenta. Then, the free wave of the photoelectron reads

$$J_{\beta Ls}(\boldsymbol{r}\sigma) \equiv j_\ell(k_\beta r) Y_L(\hat{\boldsymbol{x}}) \chi_s(\sigma) , \tag{5.20}$$

where j_ℓ is a spherical Bessel function, Y_L a spherical harmonic, $\chi_s(\sigma) = \delta_{s\sigma}$ a spin function and $k_\beta^2 \equiv \epsilon_\beta = E - E_\beta$.

The N-electron free wave Ψ^0 is the product between the photoelectron free wave J_{Ls} and an eigenstate of the ionized atom $|\Phi_\beta\rangle$. The total N-electron wave function is

$$|\Psi_{\beta Ls}\rangle = |\Phi_{\beta}\rangle|J_{\beta Ls}\rangle + \sum_{\alpha}|\Phi_{\alpha}\rangle|\phi^{-}_{\alpha,\beta Ls}\rangle \ . \tag{5.21}$$

where antisymmetrization is understood, but the symbol $\mathcal{A}$ has been suppressed for convenience of notation. We are looking for the scattered wave functions $|\phi^{-}_{\alpha,\beta Ls}\rangle$, where the subscripts βLs indicate that they depend on the free wave quantum numbers through the boundary conditions. The superscript "$-$" stands for the time-reversed LEED boundary conditions, that is, ϕ^{-} is a purely incoming spherical waves, which behaves as $\exp(-ikr)/r$ when $r \to \infty$. Consequently ϕ^{-} may be expanded in Hankel functions of the second kind:

$$\phi^{-}_{\alpha,\beta Ls}(\boldsymbol{r}\sigma) = \sum_{L's'} h^{-}_{\ell}(k_{\alpha}r)Y_{L'}(\hat{\boldsymbol{x}})\chi_{s'}(\sigma)A_{\alpha L's',\beta Ls} \ . \tag{5.22}$$

Here the coefficients A are the scattered wave amplitudes. By definition, the multichannel T-matrix $T_{\alpha Ls,\beta L's'}$ is, up to a constant, equal to the amplitude of the scattered wave $|\alpha Ls\rangle$ when the incoming wave is $|\beta L's'\rangle$. So $A = \mathrm{i}kT$. With the definition

$$H^{-}_{\alpha Ls}(\boldsymbol{r}\sigma) \equiv h^{-}_{\ell}(k_{\alpha}r)Y_{L}(\hat{\boldsymbol{x}})\chi_{s}(\sigma) \ , \tag{5.23}$$

equation (5.21) then becomes

$$|\Psi_{\beta Ls}\rangle = |\Phi_{\beta}\rangle|J_{\beta Ls}\rangle + \sum_{\alpha L's'}|\Phi_{\alpha}\rangle|H^{-}_{\alpha L's'}\rangle \mathrm{i}k_{\alpha}T_{\alpha L's',\beta Ls} \ . \tag{5.24}$$

We now combine Φ_{α} with the angular and spin part of the photoelectron wave as

$$\Phi_{\alpha Ls} \equiv \Phi_{\alpha}Y_{L}(\hat{\boldsymbol{x}})\chi_{s}(\sigma) \ . \tag{5.25}$$

Then

$$|\Psi_{\beta Ls}\rangle = \sum_{\alpha L's'}|\Phi_{\alpha L's'}\rangle\frac{1}{r}P_{\alpha L's',\beta Ls}(r) \ , \tag{5.26}$$

where

$$P_{\Gamma'\Gamma} \equiv P_{\alpha L's',\beta Ls}(r) = rj_{\ell'}(k_{\alpha}r)\delta_{\alpha L's',\beta Ls} + \mathrm{i}k_{\alpha}rh^{-}_{\ell'}(k_{\alpha}r)T_{\alpha L's',\beta Ls} \ , \tag{5.27}$$

and the collective index $\Gamma \equiv \alpha Ls$ has been introduced.

The *R*-Matrix

In *R*-matrix methods the interchannel coupling, i.e. the configuration interaction between the continuum state and the rest, is restricted to a finite reaction volume v.

For v we use an (atomic) sphere of radius r_0 around the nucleus. The R-matrix is the generalization of the logarithmic derivative for non-spherical or multichannel potentials. It is defined as

$$\sum_{\Gamma''} R_{\Gamma\Gamma''} \frac{\mathrm{d}}{\mathrm{d}r} P_{\Gamma''\Gamma'} \bigg|_{r=r_0} = P_{\Gamma\Gamma'}(r_0) \,. \tag{5.28}$$

Combining (5.27), (5.28) and the derivative of (5.27) with respect to r, the T matrix can be expressed in terms of the R-matrix (or vice versa). The difference between T and R is that T corresponds to a certain choice of boundary conditions and of the kinetic energy zero (i.e. interstitial potential) needed for the free waves in the outside region (see Chap. 1). In contrast, R only refers to the wave function inside the atomic sphere. It is independent of boundary conditions but nonetheless contains all the information that determines the scattering properties of the atom.

In a channel-diagonal, spin-independent and spherically symmetric potential, it is clear from the definition (5.28) that the R-matrix is diagonal in Γ. The diagonal elements are $P_\ell/[\mathrm{d}P_\ell/\mathrm{d}r]$ for $k = k_\alpha$, i.e. they are the inverse logarithmic derivatives of the partial wave with channel energy ϵ_α. So the R-matrix is the direct multichannel generalization of logarithmic derivatives, from which the phase shifts or the T-matrix can be calculated.

The Variational *R*-Matrix Method

The R matrix can be calculated using various numerical procedures, which lead to different flavors of the R-matrix method [19, 20]. We have used the *variational* also known as *eigenchannel* method [20] which yields the R-matrix directly in diagonal form. The variational R-matrix is based on the remarkable fact that the logarithmic derivative satisfies a variational principle [21]. By choosing as trial functions linear combinations of fixed basis functions, the variational R-matrix method becomes an algebraic eigenvalue problem, as in the common variational methods used for bound state solutions of the Schrödinger equation. The variational R-matrix equations are

$$(E - H - L)|\Psi_k\rangle = Q|\Psi_k\rangle \, b_k \,. \tag{5.29}$$

The solutions are the (inverse) eigenvalues b_k and eigenvectors Ψ_k of the R-matrix. Here E is the total energy, H the Hamiltonian,

$$L \equiv \sum_{i=1}^{N} \delta(r_i - r_0) \frac{1}{r_i} \frac{\partial}{\partial r_i} r_i \,, \tag{5.30}$$

the so-called *Bloch operator* and

$$Q \equiv \sum_{i=1}^{N} \delta(r_i - r_0) , \tag{5.31}$$

is a projection operator onto the surface of the atomic sphere. Because of the singular nature of the operators L and Q, the variational R-matrix method does not lead to a standard eigenvalue problem (as does the Schrödinger equation) but to a generalized eigenvalue (i.e. singular value) problem. We are looking for solutions of (5.29) inside the atomic sphere by expanding the Ψ_k's in a basis

$$|\Psi_k\rangle = \sum_{\Gamma\nu} |\Psi_{\Gamma\nu}\rangle c_{\Gamma\nu,k} , \tag{5.32}$$

with the trial functions

$$|\Psi_{\Gamma\nu}\rangle \equiv \mathcal{A}\left\{|\Phi_\Gamma\rangle \frac{1}{r} P_\nu(r)\right\} . \tag{5.33}$$

In other words, the radial wave functions $P_{\Gamma\Gamma'}(r)$ are expanded over the set of basis orbitals $P_\nu(r)$. The latter must not have all the same logarithmic derivative at $r = r_0$. Ideally, the basis contains functions with very different logarithmic derivatives. In practice, one chooses *open* functions with vanishing logarithmic derivative and *closed* functions with logarithmic derivative equal to $\pm\infty$. The R matrix is then given by

$$R_{\Gamma\Gamma'} = -\sum_k W_{\Gamma k} \frac{1}{b_k} W^{-1}_{k\Gamma'} \quad \text{with} \quad W_{\Gamma k} = \langle \Phi_\Gamma | \Psi_k \rangle|_{r=r_0} . \tag{5.34}$$

It is important to note that the N-electron trial functions (5.33) are states that are totally antisymmetric with respect to exchange of any two electrons, including the photoelectron. Hence exchange interaction is automatically and exactly taken account for in this formalism. As stated earlier, knowledge of the R-matrix is sufficient for the calculation of the multichannel T-matrix of the atom.

5.2.3 From Single to Multiple Scattering

Once the multichannel T-matrices are known for all atoms, multichannel multiple scattering theory can be developed essentially along the same lines as standard multiple scattering theory [12]. The multichannel version of the scattering path operator is given by

$$\tau_{i\alpha Ls,j\alpha' L's'}(E) = \left[\delta_{ij}[\mathbf{T}_i^{-1}]_{\alpha Ls,\alpha' L's'}(E) - \delta_{\alpha\alpha'}\delta_{ss'}k_\alpha G_{iL,jL'}(k_\alpha)\right]^{-1} , \tag{5.35}$$

where E is the total final state energy and $k_\alpha \equiv \sqrt{\epsilon_\alpha - V_0}$ is the wave number of the photoelectron in channel α and V_0 is the interstitial potential.

In the remainder, we assume that the T-matrices of all but the absorber atom ($i = 0$) are diagonal in the channel indices α, i.e.

$$[\mathbf{T}_i]_{\alpha Ls,\alpha' L's'}(E) = \delta_{\alpha\alpha'}[\mathbf{t}_i]_{Ls,L's'}(k_\alpha) \quad \text{for} \quad i \neq 0\,. \tag{5.36}$$

Physically this means that we neglect configuration mixing in electron atom scattering processes except when the scattering occurs at the absorber site. We make this simplification because we are mainly interested in atomic multiplet type correlation effects induced by core-hole. It is clear from (5.35) that this restriction is not at all necessary for the validity or principle feasibility of multichannel multiple scattering theory. It leads, however, to an enormous gain in computation time.

For the absorption process from a core-level at site $i = 0$, only the $i = j = 0$ bloc of the scattering path operator is needed. This block can efficiently be calculated by partitioning the multichannel multiple scattering matrix between the absorber atom ($i = 0$) and its *environment*, consisting of all other atoms ($i \neq 0$). Then it is easy to show [11] that the absorber block of the scattering path operator is given by

$$\tau_{0\alpha Ls,0\alpha' L's'}(E) = \left[[\mathbf{T}_0^{-1}]_{\alpha Ls,\alpha' L's'}(E) - \delta_{\alpha\alpha'}\delta_{ss'}\rho_{LL'}(k_\alpha)\right]^{-1}\,, \tag{5.37}$$

where the reflectivity $\boldsymbol{\rho}$ of the environment has been introduced [22]. Since the T-matrices of the environment atoms are channel diagonal by assumption (5.36), so is the reflectivity $\boldsymbol{\rho}$. Therefore $\rho(k)$ can be calculated separately for each $k = k_\alpha$ in a standard one-electron multiple scattering calculation.

$$\rho_{LL'}(k) = \sum_{ijL_1L_2} G_{0L,iL_1}(k)\tilde{\tau}_{iL_1,jL_2}(k)G_{jL_2,0L'}(k)\,. \tag{5.38}$$

Here $\tilde{\tau}$ is the one-electron scattering path operator of the environment, i.e. the system without absorber atom. The absorption cross section is then given by

$$\sigma(\omega_q) \propto \Im\left[\sum_{\Gamma\Gamma'} M_\Gamma^* \tau_{0\Gamma,0\Gamma'}(E) M_{\Gamma'}\right] \quad \text{with} \quad M_\Gamma \equiv \langle \Psi_\Gamma^{\text{in}}|H_{\text{int}}|\Psi_g\rangle\,. \tag{5.39}$$

Here $|\Psi_\Gamma^{\text{in}}\rangle$ is a solution of the many-electron Schrödinger equation inside the atomic sphere for the final state energy $E = E_g + \hbar\omega_q$. It can be calculated from the eigenvectors $|\Psi_k\rangle$ of the R-matrix [11].

5.3 The MCMS Code

We have implemented the R-matrix multichannel multiple scattering method for x-ray absorption spectra calculations as outlined above, in a computer program named MCMS. Here we give a few computational and technical details of the implementation. The MCMS code is mainly written in the C programming language, but it calls a few Fortran77 routines, namely for the single-electron multiple scattering calculation of the reflectivity ρ, as well as routines from the linear algebra library LAPACK/BLAS [23]. Therefore both a C and a Fortran compiler are required. The MCMS code solves the MCMS equations for a finite cluster of atoms. On input, a structure file and the potentials for all inequivalent atoms need to be provided. The atomic potentials are spherically symmetric, i.e. the muffin-tin or atomic sphere approximation is used. Accepted potential formats are those of the LMTO band structure code [24] or the *potgen* code [25]. The latter is used in *several other methods presented in this collection*.

The MCMS code divides into two main parts: (i) a single-channel cluster calculation and (ii) a multi-channel, single-site calculation.

Part (i). From the potentials, the single-electron atomic t-matrices are computed for all atoms on a fine energy mesh. The t-matrices and structural data are passed to a routine which calculates the reflectivity of the environment ρ of (5.38). This (Fortran) routine has been adapted from the CONTINUUM code by C.R. Natoli [25] and performs a standard real-space multiple scattering calculation. The single electron energy mesh for ρ needs to be quite fine, because in the multichannel calculation, $\rho(\epsilon_\alpha)$ must be evaluated for many energies ϵ_α, which is done by cubic interpolation.

Part (ii). The core-hole potential of the absorber atom is constructed. When the partially screened core-hole potential is used, two potentials are required for the absorber atom: the ground state potential and a statically screened potential corresponding to the final state rule. The latter should be computed self-consistently in a supercell or finite cluster calculation with one (spherically symmetric) core-hole on the absorber site. The radial basis functions for the R-matrix are computed by solving the single-electron Schrödinger equation with the previously constructed absorber atom potential. Solutions for both open ($\mathrm{d}P/\mathrm{d}r = 0$ at $r = r_0$) and closed boundary conditions ($P(r_0) = 0$) are found and then the basis is orthogonalized. Next the N-electron trial functions are set up. The N-electron basis states are all possible Slater determinants in the chosen set of electronic configurations. The matrix elements of the operators L and Q, and of the Hamiltonian, are computed between these Slater determinants using second quantization algebra.

Then, in a loop over the N-electron final state energy, the eigenchannel equations (5.29) which take the form of a generalized eigenvalue problem are solved and the R matrix together with the inner solutions $|\Psi_\Gamma^{\mathrm{in}}\rangle$ are obtained. By matching the photoelectron part of the inner waves to free electron states, the atomic multichannel T-matrix is calculated. The dipole transition matrix elements are computed. The scattering path operator is obtained by matrix inversion of $\mathbf{T}^{-1} - \rho$. Finally the absorption cross section is computed using (5.39).

Computing time. Except for very small systems of ten atoms or less, the standard multiple scattering calculation in part (i), takes much longer than the atomic multichannel in part (ii). For large clusters of a few hundred atoms, a one-electron energy point of part (i) may take a few minutes on a single CPU. As a fine energy mesh with several hundred points is required to get all the fine structure of the $L_{2,3}$-edge spectra, computation of part (i) can easily take tens of hours on a single CPU. On multi-processor machines the energy mesh may be sliced into parts and then computed in parallel. This procedure is however not automatized in the code and requires concatenation of the ρ output files from the different runs.

In the multichannel part (ii), the calculation of a single total energy point usually takes less than a second, and so the whole part (ii) takes only a few minutes on a single CPU.

5.4 $L_{2,3}$ Edge Spectra of Transition Elements

We have applied the MCMS method to the calculation of x-ray absorption spectra at the $L_{2,3}$ edges 3*d* transition metal elements. These spectra are a challenge for theory because they are strongly influenced by local electron correlation effects, especially in the final state caused by core 2*p* spin-orbit interaction and the exchange and Coulomb coupling between the 2*p* core-hole and the 3*d* valence states. These interactions gives rise to a rich multiplet structure in the free atom spectra. The multiplet structures reflect strong configuration mixing of the wave function. Generally, theories based on single determinantal wave functions, such as the Hartree–Fock approximation and density functional theory, badly fail to reproduce these spectra. Instead, full configuration interaction in the *active* space of the open 2*p* and 3*d* shells is required. When the transition metal atom is embedded in a solid or molecule, the valence 3*d* states get partially delocalized and the spectra strongly depend on the ligand field and bonding of the 3*d* orbitals. A reliable description of these extra-atomic effects requires a first-principles method for the electronic structure of the extended system. So for $L_{2,3}$-edge spectra of transition metal ions in molecules and solids a theoretical method is needed that describes both the long range bonding properties of the system and the strong configuration mixing in the 2*p*-3*d* shells of the absorber atom. Traditionally, $L_{2,3}$-edge spectra have been mostly calculated using the ligand-field multiplet model [1] where an empirical ligand field is added to the atomic multiplet Hamiltonian. This approach proved very successful for ionic compounds in cubic symmetry, but it is not satisfactory for low-symmetry and covalent bonding where the number of adjustable parameters increases and the predictive power of the theory becomes very small.

Standard single-particle theories of absorption spectra generally cannot reproduce the rich fine structure observed at the $L_{2,3}$-edges of transition elements. One of the most striking failures of one-electron theory is that it cannot account for the highly non-statistical L_3:L_2 branching ratio of the light 3*d* elements (K-Cr). In one-electron theories the branching (i.e. intensity) ratio of the spin-orbit-split sub-shells

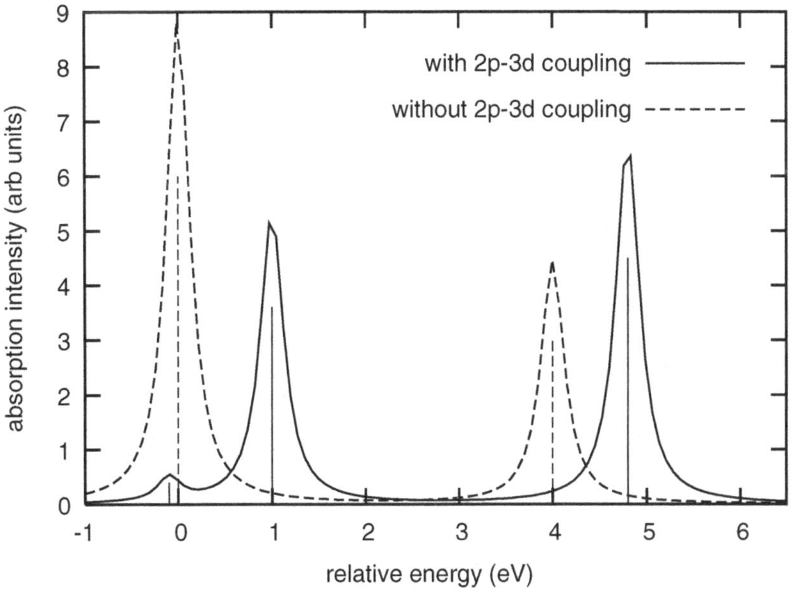

Fig. 5.1 Ca $L_{2,3}$-edge spectrum obtained in atomic multiplet theory, adapted from [26]. *Solid line*: full calculation with $2p$-$3d$ multiplet coupling. *Broken line*: calculation with $2p$-$3d$ coupling switched off

$j_+ = \ell + 1/2$ and $j_- = \ell - 1/2$ is necessarily close to the statistical value $(\ell + 1)/\ell$ coming from the multiplicity of the sub-shells. A close-to-statistical branching ratio is observed at most absorption edges, but not at $L_{2,3}$ edges of $3d$ elements, where the statistical value is 2, but the experimentally observed value varies from about 0.8 to 3 along the series. The main reason for this is the strong $2p$-$3d$ Coulomb and exchange interaction which is of the same order of magnitude as the $2p$ spin-orbit coupling [26]. This leads to final states with large configuration mixing between the $2p_{3/2}$ and $2p_{1/2}$ holes. Therefore this effect can by no means be described in a single-particle framework. In an atomic multiplet calculation, such configuration mixing is naturally included and highly non-statistical branching ratios are found for the early elements in agreement with experiment. Figure 5.1 shows the results of a multiplet calculation for a $(3d^0)$ ground and $(2p^5 3d^1)$ final state configuration, appropriate, e.g. for Ca. It is clear from Fig. 5.1 that the $2p$-$3d$ interaction strongly modifies the spectral shape. In the case of a $(3d^0)$ ground state, it gives rise to a prepeak and a huge reduction of the branching ratios. The angular moment of the hole (j_c) does not have a defined value, but only the total angular moment of the atom $J = j_{2p} + j_{3d}$ is a good quantum number. Due to strong $2p$-$3d$ coupling the angular momentum is shared between hole and electron and so the $2p_{3/2}$ and $2p_{1/2}$ holes are strongly mixed. In any single particle scheme, the $2p$-$3d$ interaction is treated on a mean-field level and the wave function corresponds to a single Slater determinant. As a consequence, strong configuration mixing, i.e. non-separable combinations of several Slater determinants, cannot be described. The hole is described as either purely $2p_{3/2}$ or $2p_{1/2}$, and so the branching ratio is necessarily statistical. It is clear that the $L_{2,3}$-edge spectra of light $3d$ elements are a case of strong channel coupling between the $2p_{1/2}$ and $2p_{3/2}$ final states. It is thus an interesting problem for multichannel multiple scattering theory.

5.4.1 Application of MCMS Theory to Calcium Compounds

Here we summarize the results that we have obtained by applying multichannel multiple scattering theory to the Ca $L_{2,3}$-edges in metallic calcium and for ionic Ca compounds (CaF_2, CaO) [11]. We use a wave function with six electrons describing the Ca $2p$ core-shell of the absorber atom and the photoelectron. In the ground state the $2p$ shell is filled and the electronic configuration is $(2p^6)$ which is a orbital and spin single $S = L = 0$. In the final state, the $(N-1)$-electron functions of the ionized atom, Φ_α in (5.4), are one of the six states of the one-core-hole configuration $(2p^5)$. The eigenstates are (j, m_j), where $j = 1 \pm 1/2$ is the total angular momentum and m_j its z-projection. For the final state, the R matrix is calculated with N-electron trial functions made of all possible states in a $(2p^5, nd^1)$ configuration. The quantum numbers are $\alpha = (j, m_j)$ for the $2p$-hole and $n\ell ms$ for the photoelectron. For simplicity, only d-waves ($\ell = 2$) are included in the basis, because the $p \to d$ transitions are dominating in $L_{2,3}$-edge absorption.

More precisely, the photoelectron orbital components $\phi_\alpha(\boldsymbol{r}, \sigma)$ are developed, inside the Ca atomic sphere, over a set of basis functions $\phi_n(r) Y_{\ell m}(\hat{\boldsymbol{x}}) \chi_s(\sigma)$, where χ is a spin function. Some radial functions $\phi_n(r)$ used in the CaO calculation are shown in Fig. 5.2. Note that n is merely a counter of the radial waves and should not be confused with the principal quantum number of atomic orbitals. There are two major differences. First, the waves ϕ_α are not bound states, but parts of continuum waves. Second, the set of basis orbitals includes several waves with the same number of nodes (inside the atomic sphere) but different logarithmic derivatives on the sphere (e.g. the waves o-0 and c-0 in Fig. 5.2). Open and close functions correspond to boundary condition $\mathrm{d}\phi/\mathrm{d}r(r_0) = 0$ and $\phi(r_0) = 0$, respectively, where r_0 is the sphere (muffin-tin) radius. These basis orbitals are calculated by solving a single-electron, radial Schrödinger equation inside the atomic sphere. The effective potential was obtained in a self-consistent density functional theory calculation in the local-density-approximation (LDA) for the bulk system using the Linear-Muffin-Tin-Orbital (LMTO) method [24]. This mean-field potential used for the construction of the basis functions must not be mistaken for the true photoelectron potential underlying the multichannel method. If the R-matrix basis is sufficiently complete, the potential used for the construction of the basis functions should not matter; any potential should in principle give the same result. Because the N-electron final states Φ^N are linear combinations of the trial functions, the potential *seen* by photoelectron is in fact dynamical because states with different energies (ϵ_α) are mixed. The multichannel potential is moreover non-spherical and spin-dependent, because of mixing of different orbitals ℓm and spins s.

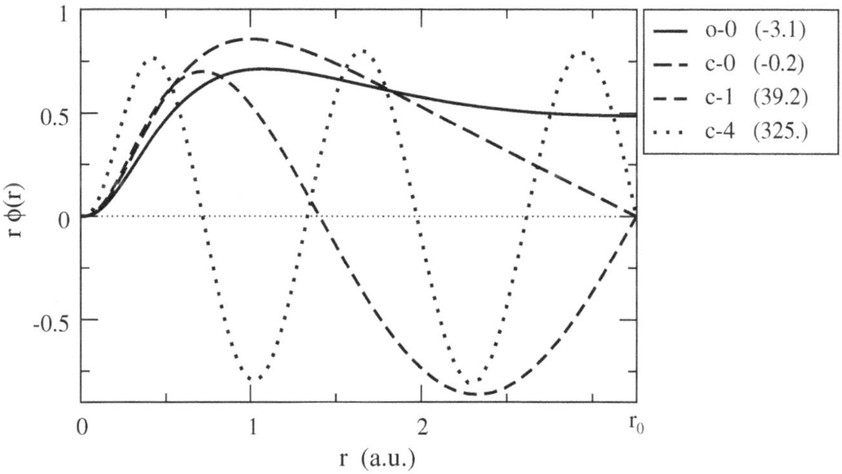

Fig. 5.2 A few radial basis functions $r\phi_n(r)$ used in the expansion of the photoelectron waves $\phi_\alpha(r)$ inside the sphere (of radius r_0) of the absorbing atom (here: Ca in CaO). The functions are labeled as x-i where $x = o, c$ stands for open and closed functions and i is the number of nodes inside the sphere. The orbital energies in Ryd are indicated in parentheses

Since in the R-matrix technique the photoelectron functions ϕ_α are expanded over a basis set, convergence with the number of basis functions must be checked. Figure 5.3 shows the convergence of the CaO spectrum as a function of number of basis functions, labeled as (nc, no) where nc and no are the number of closed and open functions, respectively. Good convergence of the lineshape is found for one open and two closed functions. Including more waves leads to a small shift of the spectrum to lower energy and to some peak intensity reduction, which may be attributed to spectral weight transfer to excitations into the featureless high energy continuum. In many cases, already the minimum set, i.e. one open and one closed function gives a sufficiently converged spectrum.

In the MCMS method, quantum mechanical exchange between electrons at the absorber site is automatically respected because we use antisymmetrized trial functions. From the multiplet analysis in Fig. 5.1 we know that the non-statistical branching ratio is due to the $2p$-$3d$ Coulomb interaction. The Coulomb interaction between two electrons at a distance r_{12} can be expanded in multipoles in the usual way as $1/r_{12} = \sum_{k=0}^{\infty}[r_<^k/r_>^{k+1}]f_k(\hat{\boldsymbol{x}}_1, \hat{\boldsymbol{x}}_2)$, where $\boldsymbol{r}_i = (r_i\,\hat{\boldsymbol{x}}_i)$ and $r_>$ $(r_<)$ is the larger (smaller) of r_1 and r_2. It is well known from atomic physics [27] and impurity model calculations [1, 26] that the monopole part ($1/r_>$ for $k = 0$) is strongly screened by the dielectric response of the other electrons, while the higher order multipoles ($k > 0$) are essentially unscreened. In an atomic multiplet calculation the monopole part of the core-valence Coulomb interaction corresponds to the Slater integral $F^0(2p, 3d)$. By atomic relaxation and solid state screening, the F^0 integral is dramatically reduced and is commonly replaced by a free parameter U_{dc} in Anderson impurity models [1]. The monopole part is spherically symmetric. In a

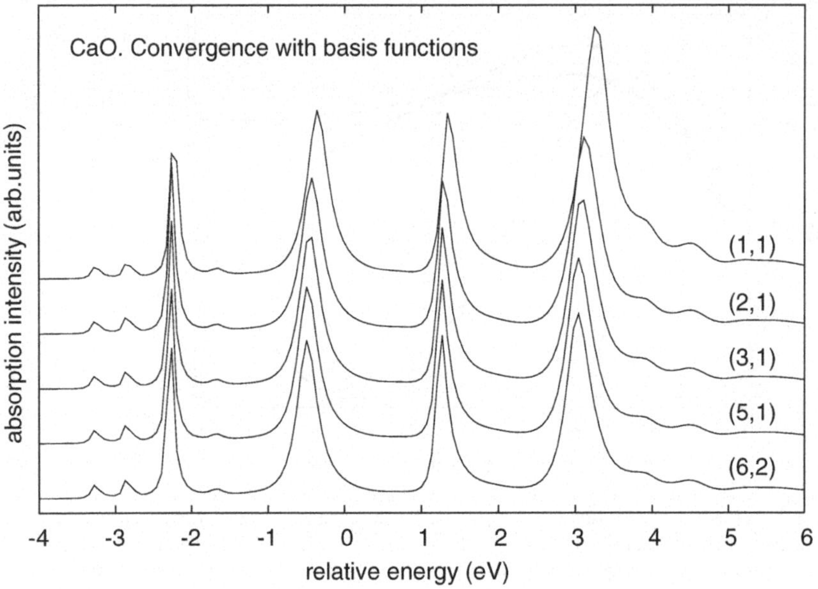

Fig. 5.3 Convergence of the Ca $L_{2,3}$-edge spectrum of CaO as a function of the number of radial basis functions used in the R-matrix method. (nc,no) denote the number of closed and pen functions. The spectra are vertically off-set for clarity

standard single configuration atomic multiplet calculation it is a diagonal operator, i.e. it does not lead to any state mixing but only to a rigid energy shift of the whole spectrum. Therefore, we take the monopole part out of the Coulomb interaction and treat it at the single-electron level by using a screened core-hole potential for the calculation of the final state waves. So for the monopole part of the core-hole interaction, we follow the standard practice in one-electron theories of XAS. The multipole part, however, is treated fully at the many-particle level and leads to configuration mixing. The R-matrix equations are solved for a Hamiltonian of the form $H = \sum_i h(i) + \sum_{i<j} V(i,j)$. Here $h(i)$ is the single particle Hamiltonian operator acting on electron i. It includes the kinetic energy, the effective one-electron potential (with core-hole screening) and the spin-orbit coupling in the case of $2p$ level. The interaction term is the particle-hole Coulomb interaction without the monopole part, $V(i,j) = 1/r_{ij} - 1/r_>$, i.e. the multipole terms are taken unscreened.

For the screening of the monopole term of the electron-hole interaction, we have used a linear mixture between a fully screened core-hole corresponding to the final state rule and an unscreened core-hole. Figure 5.4 shows the spectra of bulk calcium metal, calculated with different choices for the core-hole screening model. Spectrum (gs) corresponds to the LDA ground state potential, without core-hole. For the other spectra a linear mixture was used between the screened ($\alpha = 0$) and unscreened ($\alpha = 1$) core-hole potential. Broken lines correspond to the single-channel theory (i.e. the IPA) obtained by setting V, the multipole part of the particle-hole Coulomb

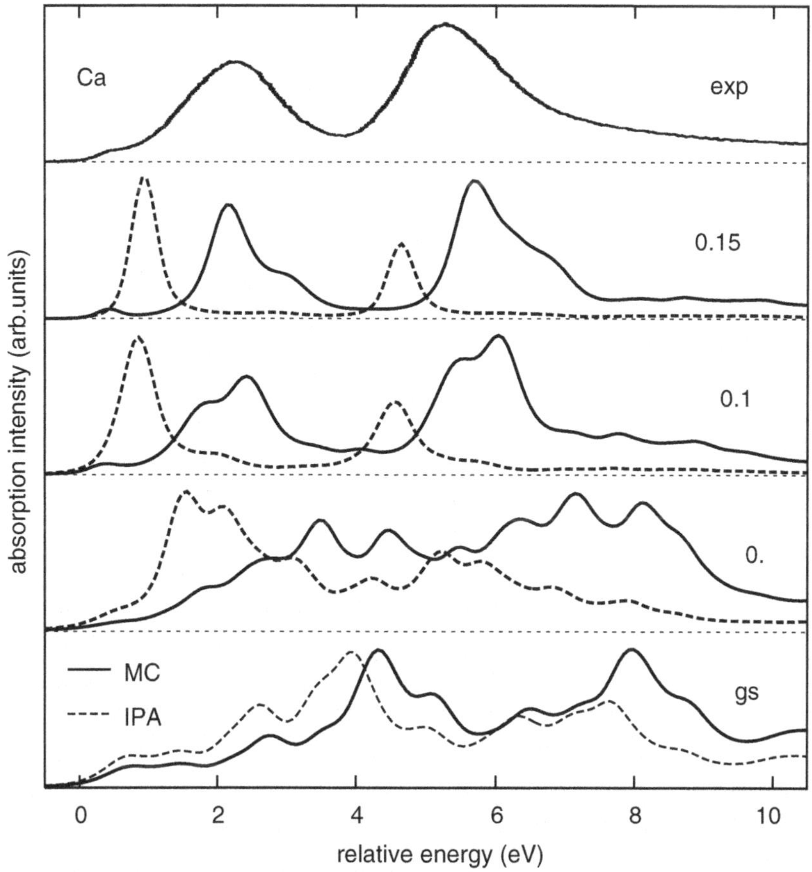

Fig. 5.4 $L_{2,3}$-edge spectra of metallic Ca calculated in various approximation together with the experimental data (exp) from [28]. The multichannel (MC) calculations including the multipole Coulomb interaction V are shown as *solid lines* and the independent particle approximation (IPA, $V = 0$) as *broken lines*. The amount of unscreened core-hole (α-value) is indicated on the right. (gs) corresponds to the ground state potential

interaction, to zero. Solid lines correspond to multichannel theory, with V included. It can be seen that the single-channel calculations yield a L_3:L_2 branching ratio close to the statistical value 2, as expected for the IPA. In contrast, when V is included, the branching ratio is strongly reduced and gets close to the experimental value of about 0.8. These remarks about the branching ratio are independent of the choice of the one-electron core-hole potential. The fine structure of the L_3 and L_2 peaks however, depends considerably on the core-hole potential. If the core-hole is neglected (gs), the lineshape shows the details of the ground state Ca-3d density of states, as we have checked. The L_3 and L_2 peaks are too broad and display too much fine structure, not seen in the experimental spectrum. The details of the Ca-3d band structure is

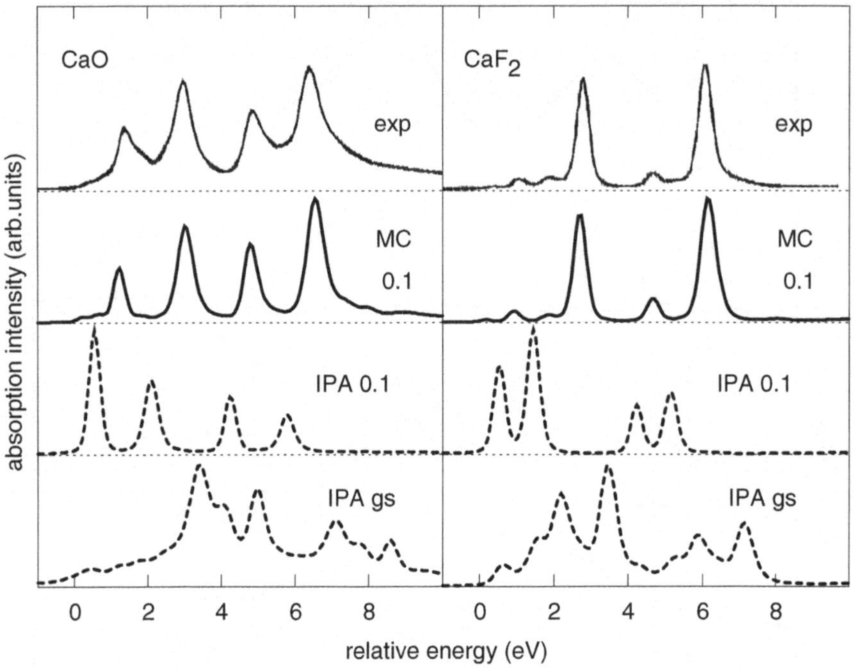

Fig. 5.5 Ca-$L_{2,3}$ edge spectra of CaO and CaF_2. Calculations with multichannel theory (MC, *solid lines*) and in the independent particle approximation (IPA, *broken lines*) using the partially screened core-hole potential with (0.1) unscreened weight or the ground state (gs) potential. (exp) are the experimental spectra

progressively lost when the core-hole strength is increased in the sequence (gs) → (0) → (0.1) → (0.15). This is particularly obvious in the single channel spectra (broken lines) but also valid for the multichannel calculation. It can be seen that the multichannel spectra with $\alpha = 0.1$ or 0.15 agree well with experiment. The only disagreement is that the calculation shows some weak fine structure not seen in the experiment. This hints to some extra broadening mechanism not included in the present theory.

Figure 5.5 shows the spectra for the ionic compounds CaO and CaF_2. Essentially the same conclusions can be drawn as for metallic Ca. Comparing the single-channel spectra without core-hole (IPA gs) and with partially screened hole (IPA 0.1), we see that much of the fine structure in (IPA gs), which reflects the Ca-3*d* ground state density of states, is removed by the core-hole attraction. The single-channel spectrum (IPA 0.1) shows only a doublet splitting of the L_3 and L_2 lines. Analysis of the orbital character of the peaks reveals that the intense peak of the doublet is made of Ca-*d* orbital of t_{2g} symmetry and the weaker peak of e_g symmetry. So the double splitting is a ligand field effect due to the bonding of the Ca-3*d* orbital with the oxygen or fluoride ligand 2*p*-orbitals. By comparing the (IPA gs) and (IPA 0.1) spectra we see that the present real-space multiple scattering method describes well both long range band structure effects and short range ligand field splittings. While the number of peaks and energy splittings of the (IPA 0.1) spectrum roughly agrees with experiment, the intensities are completely wrong. Not only the L_3:L_2 branching ratio is wrong, but also the relative intensity between the two ligand field split lines. When the $2p$-$3d$ interaction V is included in the multichannel calculation (MC 0.1), the peaks shift to higher energy and the intensity ratios change completely. Moreover, weak prepeaks appear at threshold. Clearly, the multichannel spectra (MC 0.1) are in very good agreement with experiment both for CaO and CaF_2.

The relative weight of the unscreened core-hole was introduced as an empirical parameter. However, from the results above it became clear that it is not a free parameter that needs to be adjusted to experiment for each compound. Instead, a value of 0.1 was found suitable for all Ca and Ti compounds that we have studied. This means that the value 0.1 can safely be taken as a default value for the light $3d$ elements.

In conclusion we find that the real-space multiple scattering theory describes well long range band structure and short range ligand field effects both in metallic and ionic Ca compounds. Inclusion of the core-hole dramatically changes the lineshape. The monopole part of the particle-hole Coulomb interaction can be treated at the one-electron level through a partially screened core-hole potential. The multipole part, however, leads to strong channel coupling between the $2p_{3/2}$ and $2p_{1/2}$ core-levels on the one hand and between the different orbital characters (t_{2g} and e_g) of the photoelectron wave on the other hand. Before our multichannel multiple scattering calculation, the CaO and CaF_2 spectra could be reproduced with ligand field multiplet calculations [29]. In these models, the ligand field is described by parameters those values need to be adjusted to experiment, separately for each material. Recently, *ab initio* ligand field calculations have become available basis of quantum chemistry configuration interaction [8]. However, because of the exponential increase of the Hilbert space with system size, these methods are limited to very small clusters. Multichannel multiple scattering theory does not have this limitation. Since we have separated the multiple scattering calculation for the multichannel T-matrix calculation by introducing the reflectively of the environment (see equation above), the limitation in cluster size is the same as in standard one-electron multiple scattering calculations. For the Ca system we have used clusters of about 300 atoms which was sufficient for reproducing the fine structure of the Ca-d density of states of the bulk materials.

Introduction of the Fermi Level

Atomic Ca has a [Ar]$4s^2$ electronic configuration. In Ca compounds, the Ca-$3d$ band is empty in the ground state and the L-edge spectrum reflects the whole empty d-band. In transition metals, the $3d$-band is partially filled in the ground state, so the Fermi energy needs to be introduced which cuts off the occupied part of the spectrum. In the present version of the MCMS method, this is done by replacing (5.37) by the following expression

$$\tau_{0\alpha,0\alpha'} = \left[[\mathbf{T}_0^{-1}]_{\alpha\alpha'} - \delta_{\alpha\alpha'}\mathbf{t}_0^{-1} + \delta_{\alpha\alpha'}(\mathbf{t}_0^{-1} - \rho)/(1-f)\right]^{-1} . \qquad (5.40)$$

Here, $f(\epsilon) \equiv [1 + \exp(-(\epsilon - E_F)/(k_B T)]^{-1}$ is the Fermi function and we have omitted the quantum numbers ℓms of the photoelectron to simplify the notation. The expression $(\mathbf{t}_0^{-1} - \rho)/(1-f)$ is evaluated at ϵ_α, that is the photoelectron energy in channel α. The matrix inverse of this expression coincides with the *scattering path operator* of the absorber atom in the single-electron approximation multiplied by $(1-f)$. This Fermi factor enforces the Pauli principle. In the limit of vanishing correlations we get $[\mathbf{T}_0]_{\alpha\alpha'} \to \delta_{\alpha\alpha'}\mathbf{t}_0$, and (5.40) reduces to the well known single-electron result with Fermi cut-off. Moreover, in the limit $E_F \to -\infty$ we recover the previous result for the empty valence band, (5.37). Let us note that this is a very simple way to introduce the Fermi level. The energy cut-off is applied at the level of the one-particle scattering path operator and so the occupied part of the one-particle spectrum is correctly eliminated. However, the Pauli principle is not fully taken account of in the many-body part of the calculation, because the multichannel T-matrix is calculated as before for a local $(2p^5d^1)$ configuration. Work is under way aiming to enforce the Pauli principle also in the atomic T-matrix calculation.

5.4.2 Application to Titanium Dioxide

Titanium dioxide is a technologically important material with a very rich phase diagram. Titania nanoparticles are used for photovoltaics and photocatalysis. In the search of new titania nanostructures, Ti L-edge absorption spectroscopy is a useful tool for probing the local geometrical and electronic structure of around Ti sites. The overall lineshape of the Ti L-edge spectra can be understood from multiplet theory with an octahedral ligand field [29]. The octahedral field reflects the fact that the Ti ions are in a TiO_6 octahedral coordination which is common for titanite and TiO_2. However, marked differences of the spectra exist between the various phases. It is important to understand the origin of these differences for a correct analysis of new titania nanostructures. We have applied the MCMS method to the Ti $L_{2,3}$-edge spectra of strontium titanite ($SrTiO_3$) and titanium dioxide (TiO_2) in the rutile and anatase phases [30]. From the analysis, we could settle a long-standing debate about the origin of the *fingerprint* spectral fine structure that characterizes the rutile and anatase phases and which is absent in $SrTiO_3$. We further studied more complex phases found in nanostructures, namely lepidocrocite TiO_2 [31] and titanite nanotubes [32] and nanoribbons [33]. Using MCMS theory we could reproduce the Ti L-edge spectra of these low-dimensional systems, relate the spectral features to the electronic structure and understand the pronounced linear dichroism observed in some of these low-symmetry phases [31, 33].

Here we briefly summarize the results obtained for rutile TiO_2 [30]. The rutile crystal is made of both corner and edge sharing TiO_6 octahedra, where the oxygen atoms have a three-fold, planar Ti coordination. The Ti ions sit in the centers of the weakly distorted octahedra and have a point symmetry D_{2h}. The Ti $L_{2,3}$-edge spectrum (Fig. 5.6d) has four prominent peaks, C, D/E, F and G. The four-peak structure is common to $SrTiO_3$ (Fig. 5.6e) and all other (d0) ions in octahedral coordination [29]. It can be understood in a one-particle picture as being due to the spin-orbit splitting of the $2p$-hole state into L_3 and L_2 lines, and the splitting of the Ti-$3d$-like photoelectron states into t_{2g} and e_g-symmetry by the octahedral ligand field. The appearance of the prepeaks A,B is a genuine multiplet effect, and cannot be explained in the one-particle picture as we have already seen in the case of Ca. When comparing the TiO_2 and $SrTiO_3$ spectra (Fig. 5.6d,e), considerable differences in terms of line width and intensity are observed, as well as the fact that in TiO_2 the L_3-e_g peak is split into an asymmetric doublet D-E. In the early 1990s, these differences were attributed to the reduced point symmetry and distortion of the TiO_6 octahedra in TiO_2, on the basis of empirical ligand field multiplet [34] and TiO_6 cluster calculations [35]. However this explanation was challenged by a Crocombette et al. [36]. They showed that multiplet and TiO_6 cluster calculations with realistic parameters computed from the actual atomic structure of rutile TiO_2 does not produce the D-E splitting but a spectrum similar to that of cubic $SrTiO_3$. They thus concluded that the (actually very weak) distortions of the octahedra in rutile TiO_2 cannot explain the characteristic D-E doublet splitting of the L_3-e_g line in TiO_2. Single-particle approaches have also been applied to the Ti L-edge spectra [37, 38] but they lack the prepeak structure and completely fail to account for the main peak intensities even in $SrTiO_3$.

We have computed the Ti L-edge spectra with MCMS method. Because the LDA strongly underestimates the band gap of Ti oxides, we have used a simplified LDA+U approach for the one-electron potential. Namely, we have shifted the LDA potential upward by 2 eV when solving the radial Kohn-Sham equation for the Ti partial waves of d symmetry. This corresponds to an LDA+U correction *from the atomic limit* [39] for an empty d-band. The correction increases the band gap of TiO_2 from $\sim$1.5 eV to $\sim$3.0 eV and brings it to agreement with experiment.

The calculated L-edge spectra for rutile TiO_2 are shown in Fig. 5.6. The multichannel calculation with a (0.1) partially screened core-hole potential (Fig. 5.6c) agrees very well with experiment (Fig. 5.6d) while the IPA with (Fig. 5.6b) or without core-hole (Fig. 5.6a) yields very poor results. We note that all peak positions and intensities of the experimental TiO_2 spectrum are well reproduced in the multichannel calculation, including the prepeaks (A,B) and the L_3-e_g doublet (D,E) which is the long-debated fingerprint of the TiO_2 spectrum. Figure 5.6f shows the multichannel spectrum as a function of cluster size from 1 to 297 atoms. As expected, the single atom spectrum (1) shows the atomic multiplet 3-peaks structure of a (d^0) ion (as in Fig. 5.1). Going to the 7-atom cluster (i.e. a TiO_6 octahedron) multiple scattering with the oxygen neighbors produces the octahedral ligand field splitting and the main four-peak structure is obtained. However, the widths of the 4 peaks are about the same and the D-E splitting is lacking, in disagreement with experiment. It takes cluster sizes of about 60 atoms for these disagreements to disappear. This clearly shows that the peak widths and especially the fingerprint D-E splitting of the TiO_2 spectrum reflect the electronic structure on a length scale of about 1 nm and can thus by no means be explained in an atomic crystal field or TiO_6 cluster model.

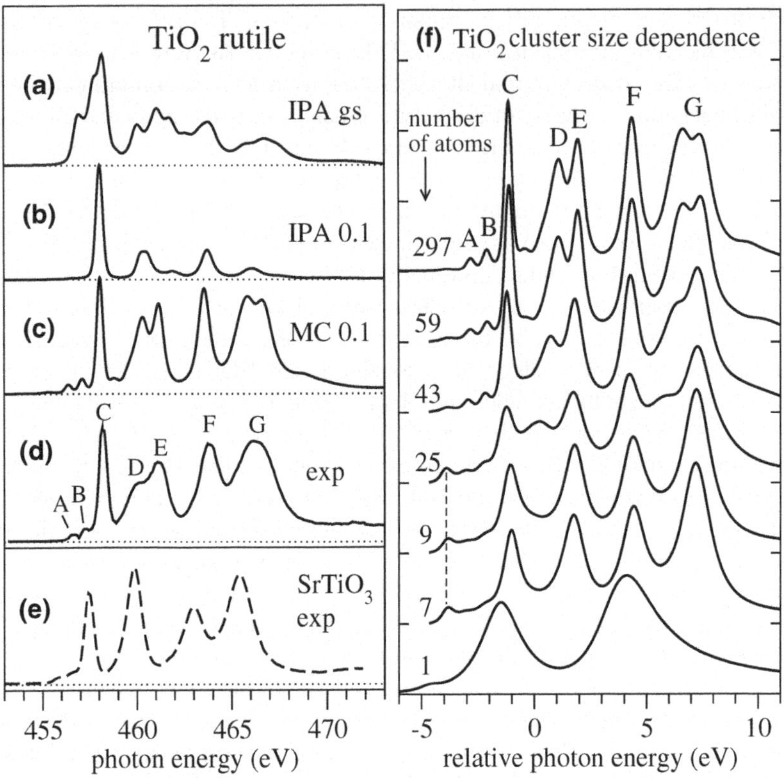

Fig. 5.6 Ti-$L_{2,3}$ edge spectra of rutile TiO_2 **a-d,f** and $SrTiO_3$ **e**. **a-c** TiO_2 spectra calculated in independent particle approximation (IPA) or multichannel theory (MC), with ground state (gs) or partially screened potential with (0.1) unscreened core-hole. **d** Experimental TiO_2 spectrum. **e** Experimental $SrTiO_3$ spectrum. **f** Cluster size dependence of TiO_2 spectrum calculated with MC 0.1

We could also reproduce the anatase spectrum [30]. These findings have settled the 20-year long debate on the *fingerprint* L_3-e_g peak splitting in the Ti *L*-edge spectra of the various TiO_2 phases. It shows the strengths of MCMS which combines (i) an accurate description of the long-range electronic structure of any type of materials by means of multiple scattering theory and (ii) a configuration interaction scheme to account for the strong local electronic correlations in the 2*p*-hole final state.

5.5 Summary and Outlook

In summary, we have presented a multichannel multiple scattering theory with *R*-matrix method for x-ray absorption spectroscopy. The method is implemented with a particle-hole wave function in the MCMS code. It affords a flexible, *ab initio* description of the medium and long-range electronic structure through real-space multiple scattering theory and accounts for local multiplet-type final state effects through a correlated particle-hole wave function. These two aspects catch the main features of the $L_{2,3}$-edge spectra of light $3d$ elements in various materials. Here, only $L_{2,3}$-edge spectra of $3d$ transition elements were shown, but other edges can also be calculated. For example, in actinide $O_{4,5}$-edge spectra, strong channel coupling between the $5d_{3/2}$ and $5d_{5/2}$ sub-shells leads to a large non-statistical branching ratio. We have applied MCMS to the Th $O_{4,5}$-edge spectra and found good results [40]. Compared with other methods that also employ a correlated particle-hole function, in particular the Bethe–Salpeter-Equation approach [7], the MCMS method is numerically extremely light. Indeed, by virtue of a partitioning between the correlated absorber and the uncorrelated rest system, a MCMS calculation is only slightly heavier than a standard, one-electron multiple scattering calculation. Moreover, because it is implemented in real space, it is not restricted to perfect crystals, but can be applied to molecules and nanostructures [32].

In the current implementation of MCMS, the muffin-tin approximation is used for the one-electron potential. The combination of multichannel and full-potential multiple scattering theory is straightforward [12, 41]. The only change needed in the MCMS code is the replacement of the L-diagonal single-electron t-matrices $t_{i\ell}$ in (5.36) by their full potential counterparts $t_{iLL'}$ which may be calculated using the FPMS code (see Chap. 3). Note that because of particle-hole coupling, the multichannel T-matrix is always non-diagonal in the angular momentum L of the photoelectron even in the present version with muffin-tin potentials.

The main limitation of MCMS method in present form is that the many-electron state is a particle-hole wave function. This means that the interaction between the core-hole – photoelectron pair with all other electrons is treated at the mean-field level. This is sufficient for the beginning of a transition series, with a nearly empty d-shell, but becomes questionable for the general case of a partially filled shell in the middle or end the series. We have applied MCMS to bulk Fe [42] and found that the multichannel coupling does not improve the spectra over the independent particle approximation. A version of MCMS for a general $(2p^5 3d^{n+1})$ final state wave function is still under development. Difficulties arise because at threshold, the photoelectron cannot be distinguished from the other electrons, while in the close-coupling and R-matrix formalism it is assumed that only one electron occupies a continuum wave. In order to overcome these problems, a new method has been devised for the solution of the multichannel multiple scattering equations which does not use the R-matrix but goes back to the interchannel potential (5.16). The latter has been be expressed exactly in terms of a multichannel density matrix and the associated multichannel Green's function [41]. This new approach is very general and promising, but its implementation is challenging and has not yet been completed.

Acknowledgements My deepest thanks go to Prof. Calogero "Rino" Natoli, the inventor of multichannel multiple scattering theory. Rino introduced me to this interesting subject and taught me all I know about multiple scattering theory. Without his vision, enthusiasm and great direct contribution, this project could not have been accomplished. Also, I am grateful to many people for advice and fruitful discussions over the years on various aspects of multiple scattering theory and correlation effects in core-level spectroscopy. These people include, in alphabetical order, Dr. Oana Bunau, Dr. Fabiana Da Pieve, Prof. Hubert Ebert, Prof. Frank de Groot, Dr. Keisuke Hatada, Dr. Yves Joly, Prof. Akio Kotani, Dr. Jan Minar, Dr. Didier Sébilleau and Dr. Ondřej Šipr. Finally I would like to thank Dr. Kuniko Hayakawa and Prof. Fabrizio Palumbo for their warm hospitality during my visits at the LNF-INFN in Frascati, where main ideas of this work emerged.

References

1. F. de Groot, A. Kotani, *Core Level Spectroscopy of Solids* (CRC Press, Taylor and Francis, 2008)
2. J. Schwitalla, H. Ebert, Phys. Rev. Lett. **80**, 4586 (1998)
3. A.L. Ankudinov, A.I. Nesvizhskii, J.J. Rehr, Phys. Rev. B **67**, 115120 (2003)
4. G.F.M. Stener, M. de Simone, Chem. Phys. Lett. **373**, 115123 (2003)
5. O. Bunău, Y. Joly, Phys. Rev. B **85**, 155121 (2012)
6. E.L. Shirley, J. Electron Spectrosc. Relat. Phenom. **144**, 1187 (2005)
7. R. Laskowski, P. Blaha, Phys. Rev. B **82**, 205104 (2010)
8. K. Ogasawara, T. Iwata, Y. Koyama, T. Ishii, I. Tanaka, H. Adachi, Phys. Rev. B **64**, 115413 (2001)
9. S.D.D. Maganas, F. Neese, Inorg. Chem. **53**, 6374 (2014)
10. M.W. Haverkort, M. Zwierzycki, O.K. Andersen, Phys. Rev. B **85**, 165113 (2012)
11. P. Krüger, C.R. Natoli, Phys. Rev. B **70**, 245120 (2004)
12. C.R. Natoli, M. Benfatto, C. Brouder, M.F. Ruiz López, D.L. Foulis, Phys. Rev. B **42**, 1944 (1990)
13. A.M. Lane, R.G. Thomas, Rev. Mod. Phys. **30**, 257 (1958)
14. K. Smith, R.J.W. Henry, P.G. Burke, Phys. Rev. **147**, 21 (1966)
15. J. Korringa, Physica **13**, 392 (1947)
16. W. Kohn, N. Rostoker, Phys. Rev. **94**, 1111 (1954)
17. S.H. Chou, J.J. Rehr, E.A. Stern, E.R. Davidson, Phys. Rev. B **35**, 2604 (1987)
18. K. Smith, R.J.W. Henry, P.G. Burke, Phys. Rev. **157**, 51 (1967)
19. P.G. Burke, K.A. Berrington, *Atomic and Molecular Processes: an R-Matrix Approach* (IOP Publishing, Bristol and Philadelphia, 1993)
20. M. Aymar, C.H. Greene, E. Luc-Koenig, Rev. Mod. Phys. **68**, 1015 (1996)
21. W. Kohn, Phys. Rev. **74**, 1763 (1948)
22. D.D. Vvedensky, in *Unoccupied Electronic States*, ed. by J.C. Fuggle and J.E. Inglesfield (Springer, Berlin, 1992), chap. 5, pp. 139–176
23. J. Demmel, J. Dongarra, J. Langou, *Linear Algebra Package*. http://www.netlib.org/lapack/
24. O.K. Andersen, O. Jepsen, Phys. Rev. Lett. **53**, 2571 (1984)
25. C.R. Natoli, D.K. Misemer, S. Doniach, F.W. Kutzler, Phys. Rev. A **22**, 1104 (1980)
26. J. Zaanen, G.A. Sawatzky, J. Fink, W. Speier, J.C. Fuggle, Phys. Rev. B **32**, 4905 (1985)
27. R.D. Cowan, *Theory of Atomic Structure and Spectra* (University of California Press, Berkeley, 1981)
28. F.J. Himpsel, U.O. Karlsson, A.B. McLean, L.J. Terminello, F.M.F. de Groot, M. Abbate, J.C. Fuggle, J.A. Yarmoff, B.T. Thole, G.A. Sawatzky, Phys. Rev. B **43**, 6899 (1991)
29. F.M.F. de Groot, J.C. Fuggle, B.T. Thole, G.A. Sawatzky, Phys. Rev. B **41**, 928 (1990)
30. P. Krüger, Phys. Rev. B **81**, 125121 (2010)
31. L.E. Walle, S. Agnoli, I.H. Svenum, A. Borg, L. Artiglia, P. Krüger, A. Sandell, G. Granozzi, J. Chem. Phys. **135**, 054706 (2011)
32. C. Bittencourt, P. Krüger, M.J. Lagos, X. Ke, G.V. Tendeloo, C. Ewels, P. Umek, P. Guttmann, Beilstein J. Nanotechnol. **3**, 789 (2012). ISSN 2190-4286

33. X. Zhu, A.P. Hitchcock, C. Bittencourt, P. Umek, P. Krüger, J. Phys. Chem. C **119**, 24192 (2015)
34. F.M.F. de Groot, M.O. Figueiredo, M.J. Basto, M. Abbate, H. Petersen, J.C. Fuggle, Phys. Chem. Miner. **19**, 140 (1992). ISSN 1432-2021
35. K. Okada, A. Kotani, J. Electron Spectrosc. Relat. Phenom. **62**, 131 (1993)
36. J.P. Crocombette, F. Jollet, J. Phys. Condens. Matter **6**, 10811 (1994)
37. R. Brydson, H. Sauer, W. Engel, J.M. Thomas, E. Zeitler, N. Kosugi, H. Kuroda, J. Phys. Condens. Matter **1**, 797 (1989)
38. M. Mattesini, J.S. de Almeida, L. Dubrovinsky, L. Dubrovinskaia, B. Johansson, R. Ahuya, Phys. Rev. B **70**, 115101 (2004)
39. M.T. Czyzyk, G.A. Sawatzky, Phys. Rev. B **49**, 14211 (1994)
40. P. Krüger, Unpublished
41. C.R. Natoli, P. Krüger, K. Hatada, K. Hayakawa, D. Sébilleau, O. Sipr, J. Phys. Condens. Matter **24**, 365501 (2012)
42. P. Krüger, J. Phys. Conf. Ser. **190**, 012006 (2009)

Chapter 6
Multiple Scattering in Green's Function Formalism: Single-Channel and Multichannel Versions

Anna Taranukhina, Alexander Novakovich, Calogero R. Natoli and Ondřej Šipr

Abstract In this chapter, we present two versions of the multiple scattering (MS) theory in the real-space electronic Green's function (GF) formalism: single-channel (MS-GF) and multichannel (MCMS-GF). While the first one based on the single-particle picture provides a tool for a precise description of MS processes, the second one allows us to take into account not only MS effects but also electron correlations and spin-orbit coupling on the same footing. Multichannel generalization of the MS-GF method relies on the Dyson integral equation relating the GF of a perturbed system with the GF of the corresponding unperturbed system. The second basic feature of the MCMS-GF approach is the use of the close-coupling method, which via Kohn variational principle for the reactance K-matrix gives rise to a set of the coupled integro-differential equations with the matrix of a potential. An iterative algorithm for solving this system has been developed to evaluate single-site multichannel scattering t-matrices through which the GF of the total many-atom system is expressed. In addition, some numerical aspects concerning the application of both versions are discussed with a focus on x-ray absorption spectroscopy.

A. Taranukhina (✉)
Faculty of Physics, Southern Federal University, Zorge Str. 5,
344090 Rostov-on-Don, Russia
e-mail: aitaranukhina@sfedu.ru

A. Novakovich
Institute of Physics, Southern Federal University, Stachky Ave. 194,
344090 Rostov-on-Don, Russia
e-mail: aanovakovich@sfedu.ru

C. R. Natoli
INFN Laboratori Nazionali di Frascati, Via E. Fermi 40, c.p. 13,
00044 Frascati, Italy
e-mail: calogero.natoli@lnf.infn.it

O. Šipr
Academy of Science of Czech Republic, Cukrovarnicka 10, 162 53 Prague 6,
Czech Republic
e-mail: sipr@fzu.cz

D. Sébilleau et al. (eds.), *Multiple Scattering Theory for Spectroscopies*,
Springer Proceedings in Physics 204,
https://doi.org/10.1007/978-3-319-73811-6_6

6.1 Introduction

The basic equations of the multiple scattering theory (MST) within the *muffin-tin approximation* (MT) to *cluster potentials* have been derived and applied to calculating the electronic structure of polyatomic molecules by Johnson [1, 2]. The theory provides a fast and efficient technique for solving a system of linear differential equations with corresponding boundary conditions on MT-spheres for both continuum and bound states of a physical system. Despite the advantage of the new theory, it was obvious that attempts to use the theory in its original wave function version for calculations of various x-ray spectroscopic response functions would encounter technical and/or first principle difficulties. For example, the theoretical description of the electronic structure of disordered alloys is based on the configuration-averaged single-particle GF [3]; the use of complex potentials also required a new formulation of the theory. The first reformulation of the MST in terms of a *real-space Green's function* (GF) has been given independently in works [4, 5]. The aim of the authors [4] was to develop the theory of x-ray absorption and emission spectra for disordered alloys. A relatively simple method for calculating the GF has been developed by reducing the integral Dyson equation to *a system of linear algebraic equations* for the *GF matrix* in the so-called *angular momentum-site index representation*. Then a procedure of averaging the derived equations over all configurations of the atoms has been devised [6]. The developed MS-GF formalism opened up a wide field of MS theory applications in the description of x-ray spectroscopies, as for instance absorption [7], resonant photoemission [8], elastic [9] and inelastic [10] scattering, resonant diffraction [11], and many others. The MS theory has been developed in a series of the works of Natoli et al. [12] being formulated in terms of both the wave functions and the Green's function not only within MT approximation, but also for general potentials. The GF in this theory is constructed in terms of the *scattering path operator* and has widespread spectroscopy applications. The formulation and applications of the Green's function method on the base of the introduced in [4] GF matrix in *angular momentum-site index representation* is less known. Here we will outline this approach in details.

The MS methods and their applications in the simulation of x-ray spectroscopies rely on one-electron picture, i.e. all electrons are considered moving independently in a common potential field, and a single photon can interact only with a single electron of a physical system under study. Despite limited accuracy of one-electron approximation, in many cases it provides good results. At that time, in many processes, electron-electron interaction essentially modifies the response of a system to the incoming photon considerably altering the probabilities of the processes. The outline of a many-particle description of the x-ray emission and absorption processes was developed by a generalization of MS theory in its wave function approach and the connection between this approach and the Green's function approach was given by Natoli et al. [13]. The multichannel generalization of the MS-GF method (MCMS-GF) on the basis of close-coupling approximation via the Kohn variational principle for the reactance K-matrix was devised and applied to calculating ultraviolet absorption spectra for alkali halides in [14, 15]. The general multichannel theory developed in [13] has been realized in [16] within R-matrix formalism (MCMS) and applied with success to computations of x-ray absorption spectra of various systems (see Chap. 5 of this book). Both methods, MCMS-GF and MCMS, are limited by using MT approximation and by taking into account only electron-hole correlations. However, it is evident that electrons of partly filled d-shells of atoms can strongly couple with electron-hole pair in the final state of an ionization process or between each other in the ground state in compounds. A new general multichannel theory developed in [17] overcomes in full these problems, takes a new higher level of MS theory and indicates the way of a practical implementation of the theory for the description of ground and excited states of correlated systems.

The chapter is organized as follows. Section 6.2 presents the derivation of the equations of the MS-GF method and touches some practical aspects of its application in modeling photoabsorption process. The multichannel reformulation of MS-GF method along with the expressions for computation of x-ray absorption spectra is given in Sect. 6.3.

6.2 MS-GF Formalism: Single-Channel Version

6.2.1 Solution of the Dyson Equation

The starting point of the MS-GF method is the Dyson equation for the single-particle Green's function for a finite system

$$G^{\pm}(\boldsymbol{r},\boldsymbol{r}',\epsilon) = G_0^{\pm}(\boldsymbol{r},\boldsymbol{r}',\epsilon) + \int_{\Omega_n} G_0^{\pm}(\boldsymbol{r},\boldsymbol{r}'',\epsilon)\, V(\boldsymbol{r}'')\, G^{\pm}(\boldsymbol{r}'',\boldsymbol{r}',\epsilon)\, \mathrm{d}\boldsymbol{r}'' \,. \quad (6.1)$$

Here, V is a potential assumed to be local, $G_0^{\pm}$ is the free electron GF given by

$$G_0^{\pm}(\boldsymbol{r},\boldsymbol{r}',\epsilon) = -\frac{1}{4\pi}\,\frac{\mathrm{e}^{\pm\mathrm{i}k|\boldsymbol{r}-\boldsymbol{r}'|}}{|\boldsymbol{r}-\boldsymbol{r}'|}\,, \quad (6.2)$$

where energy ϵ is real, $k = \sqrt{\epsilon}$ is the wavevector of the electron with energy ϵ. Atomic units for lengths and Rydberg units for energies are used throughout this chapter.

It is our purpose to develop a procedure of solving (6.1) in the simplest case of the *muffin-tin approximation* for the potential $V(\boldsymbol{r})$ (see Figs. 1.7 and 1.8 of Chap. 1). In this approximation, the potential of the N-atomic system is spherically symmetrized inside non-overlapping spheres Ω_n of radii b_n and taken to be a constant value in the interstitial region usually defined as the zero level of energies (MT-zero). In addition, we use the so-called *extended continuum* scheme, i.e. without a surrounding sphere enclosing all the atomic spheres. This scheme allows one to treat both continuum and bound states applying the same algorithm. Therefore,

$$
\begin{aligned}
V(\boldsymbol{r}) &= \sum_n v_n(\boldsymbol{r}) \\
v_n(\boldsymbol{r}) &\neq 0, \quad \text{if } \boldsymbol{r} \in \Omega_n \\
v_n(\boldsymbol{r}) &= 0, \quad \text{if } \boldsymbol{r} \notin \Omega_n \,.
\end{aligned}
\tag{6.3}
$$

Accordingly, we introduce local coordinates $\boldsymbol{\rho}_n = \boldsymbol{r} - \boldsymbol{R}_n$ within sphere n with the origin at the center $\boldsymbol{R}_n$ of the atomic site, and the vector $\boldsymbol{R}_{nm} = \boldsymbol{R}_n - \boldsymbol{R}_m$ connecting the origins of sites m and n. Radial variables are denoted by $\rho_n = |\boldsymbol{\rho}_n|$ and angular variables by $\hat{\boldsymbol{\rho}}_n = \boldsymbol{\rho}_n/\rho_n$.

In a first step, we write in the spherical wave representation a number of local functions associated with the truncated atomic potential $v_n(\boldsymbol{\rho}_n) = v_n(\rho_n)$ inside the bounding sphere Ω_n, such as the regular and irregular solutions of the Schrödinger equation (SE) or the Lippmann–Schwinger equation, the t-matrix, and the Green's function. The radial part $R^n_{k\ell}(\rho_n)$ of the regular solution $R^n_{kL}(\boldsymbol{\rho}_n) \equiv R^n_{k\ell}(\rho_n) Y_L(\hat{\boldsymbol{\rho}}_n)$ matches smoothly to the boundary condition at $\rho_n = b_n$

$$
\begin{aligned}
R^n_{k\ell}(b_n) &= \mathrm{e}^{\mathrm{i}\delta^n_\ell} \left[j_\ell(kb_n) \cos\delta^n_\ell - n_\ell(kb_n) \sin\delta^n_\ell \right] \\
&= j_\ell(kb_n) + \mathrm{i} \sin\delta^n_\ell \mathrm{e}^{\mathrm{i}\delta^n_\ell} h^+_\ell(kb_n) \equiv j_\ell(kb_n) - \mathrm{i} h^+_\ell(kb_n)\, t^n_\ell \,,
\end{aligned}
\tag{6.4}
$$

while the radial part $H^n_{k\ell}(\rho_n)$ of the irregular solution $H^n_{kL}(\boldsymbol{\rho}_n) \equiv H^n_{k\ell}(\rho_n) Y_L(\hat{\boldsymbol{\rho}}_n)$ matches smoothly

$$
H^n_{k\ell}(b_n) = -\mathrm{i} h^+_\ell(kb_n) \,, \tag{6.5}
$$

where δ^n_ℓ is the phase shift of the potential $v_n(\rho_n)$ and $t^n_\ell = -\sin\delta^n_\ell \mathrm{e}^{\mathrm{i}\delta^n_\ell}$ is a *partial t-matrix* introduced by definition. Throughout this chapter, we shall use real spherical harmonics $Y_L(\hat{\boldsymbol{\rho}}_n)$ and notations $L \equiv \{\ell, m\}$, $j_L(k\boldsymbol{\rho}) \equiv j_\ell(k\rho) Y_L(\hat{\boldsymbol{\rho}})$, $n_L(k\boldsymbol{\rho}) \equiv n_\ell(k\rho) Y_L(\hat{\boldsymbol{\rho}})$, $h^+_L(k\boldsymbol{\rho}) \equiv h^+_\ell(k\rho) Y_L(\hat{\boldsymbol{\rho}})$, j_ℓ, n_ℓ and h^+_ℓ denote spherical Bessel, Neumann and Hankel functions of the first kind respectively. Writing the different forms of the condition in (6.4), we used known relation $h^+_\ell(z) = j_\ell(z) + \mathrm{i} n_\ell(z)$.

The regular wave function $R^n_{k\ell}(\boldsymbol{\rho}_n)$ satisfies the Lippmann–Schwinger equation

$$
R^n_{kL}(\boldsymbol{\rho}_n) = j_L(k\boldsymbol{\rho}_n) + \int_{\Omega_n} G_0(\boldsymbol{\rho}_n, \boldsymbol{\rho}'_n, k) v_n(\rho'_n) R^n_{kL}(\boldsymbol{\rho}'_n)\, \mathrm{d}\boldsymbol{\rho}'_n \,, \tag{6.6}
$$

which, as one can see, automatically includes the boundary condition (6.4) under the use of the well-known expression for the free GF

$$
G_{0\ell}(\rho, \rho', \epsilon) = -\mathrm{i}\, k\, j_\ell(k\rho_<)\, h^+_\ell(k\rho_>) \,,
$$

and the integral representation of the partial t-matrix introduced in (6.4):

$$\int_{\Omega_n} j_L(k\boldsymbol{\rho}_n) v_n(\rho_n) R^n_{kL'}(\boldsymbol{\rho}_n)\, \mathrm{d}\boldsymbol{\rho}_n \;=\; -\delta_{\ell\ell'}\frac{1}{k}\mathrm{e}^{\mathrm{i}\delta^n_\ell}\sin\delta^n_\ell \;=\; \delta_{\ell\ell'}\frac{1}{k}t^n_\ell\,, \qquad (6.7)$$

where $\delta_{\ell\ell'}$ is the Kronecker symbol.

Now we can express the single- and two-center expansions of the local Green's function in terms of the introduced regular and irregular solutions. Without loss of generality, we shall consider the retarded GF $G^+_n(\boldsymbol{\rho}_n, \boldsymbol{\rho}'_n, \epsilon)$.[1] Henceforth, the sign $+$ on G will be omitted for simplicity.

Let us first consider the *single-center expansion* of the local GF $G_n(\boldsymbol{r}, \boldsymbol{r}', \epsilon)$, i.e. when $\boldsymbol{r}, \boldsymbol{r}' \in \Omega_n$.

This Green's function

$$\begin{aligned} G_n(\boldsymbol{r}, \boldsymbol{r}', \epsilon) &= G_n\left(\left|\boldsymbol{\rho}_n + \boldsymbol{R}_n\right|, \left|\boldsymbol{\rho}'_n + \boldsymbol{R}_n\right|, \epsilon\right) \\ &= \sum_L \frac{1}{\rho_n\rho'_n} G^n_\ell(\rho_n, \rho'_n, \epsilon) Y_L(\hat{\boldsymbol{\rho}}_n) Y_L(\hat{\boldsymbol{\rho}}'_n) \end{aligned} \qquad (6.8)$$

satisfies the differential equations in both variables $\boldsymbol{\rho}$ and $\boldsymbol{\rho}'$:

$$\begin{aligned} \left[\Delta_{\rho_n} - v_n(\rho_n) + \epsilon\right] G_n(\boldsymbol{\rho}_n, \boldsymbol{\rho}'_n, \epsilon) &= \delta(\boldsymbol{\rho}_n - \boldsymbol{\rho}'_n) \\ \left[\Delta_{\rho'_n} - v_n(\rho'_n) + \epsilon\right] G_n(\boldsymbol{\rho}_n, \boldsymbol{\rho}'_n, \epsilon) &= \delta(\boldsymbol{\rho}_n - \boldsymbol{\rho}'_n)\,. \end{aligned} \qquad (6.9)$$

The partial GF $G^n_\ell(\rho, \rho', \epsilon)$ introduced in (6.8) is the solution of the radial equation

$$\left[\frac{\mathrm{d}^2}{\mathrm{d}\rho^2} - \frac{\ell(\ell+1)}{\rho^2} - v_n(\rho) + \epsilon\right] G^n_\ell(\rho, \rho', \epsilon) \;=\; \delta(\rho - \rho')\,. \qquad (6.10)$$

Here and henceforth the index n in the local coordinates $\boldsymbol{\rho}_n$ is dropped, unless necessary.

Evidently, the Wronskian of the functions $k\rho R^n_{k\ell}(\rho)$ and $\rho H^n_{k\ell}(\rho)$ equals unity. Therefore, according to the general method of the Green's function construction [18], one can write

$$G^n_\ell(\rho, \rho', \epsilon) \;=\; k\,\rho_<\,\rho_>\,R^n_{k\ell}(\rho_<)\,H^n_{k\ell}(\rho_>)\,, \qquad (6.11)$$

where $\rho_<$ ($\rho_>$) are the lesser (the greater) of ρ and ρ'.

[1] In a similar way one can derive equations of the MS-GF method making use the Green's function of the required analyticity behavior.

Inserting (6.11) into (6.8), one can write the single-center expansion for the local GF as

$$G_n(\boldsymbol{r},\boldsymbol{r}',\epsilon) = k\sum_{L} R^n_{k\ell}(\rho_<)\, H^n_{k\ell}(\rho_>)\,. \tag{6.12}$$

Now we proceed to the *two-center expansion* of the local GF $G_n(\boldsymbol{r},\boldsymbol{r}',\epsilon)$ around another center m, i.e. when $\boldsymbol{r}\in\Omega_n,\ \boldsymbol{r}'\in\Omega_m,\ n\neq m$. In this case $\rho>\rho'$ and $H^n_{k\ell}(\rho')$ coincides with $-\mathrm{i}h^+_\ell(k\rho')$. Using the well-known re-expansion [19]

$$-\,\mathrm{i}h^+_L(k\boldsymbol{\rho}) = k\sum_{L'} G_0{}^{nm}_{LL'}\; j_{L'}(k\boldsymbol{\rho}_m), \tag{6.13}$$

one can write

$$G_n(\boldsymbol{r},\boldsymbol{r}',\epsilon) = k\sum_{LL'} G_0{}^{nm}_{LL'}\; R^n_{kL}(\boldsymbol{\rho}_n)\; j_{L'}(k\boldsymbol{\rho}'_m), \tag{6.14}$$

where the matrix elements $G_0{}^{nm}_{LL'}$ (the KKR structure factors) are given by

$$G_0{}^{nm}_{LL'} = -4\pi\mathrm{i}\sum_{L''} \mathrm{i}^{\ell-\ell'+\ell''}\, C_{LL'L''}\, h^+_{L''}(k\boldsymbol{R}_{nm})\,, \tag{6.15}$$

with the Gaunt coefficients

$$C_{LL'L''} = \int Y_L(\hat{\boldsymbol{r}})Y_{L'}(\hat{\boldsymbol{r}})Y_{L''}(\hat{\boldsymbol{r}})\,\mathrm{d}\hat{\boldsymbol{r}}\,. \tag{6.16}$$

It is worth noting that for $V_n(\boldsymbol{\rho})=0$ the function $R^n_{kL}(\boldsymbol{\rho}_n)$ coincides with $j_L(k\boldsymbol{\rho}_n)$ so that (6.14) reduces to the well-known two-center re-expansion of the free electron GF [13]

$$G_0(\boldsymbol{r},\boldsymbol{r}',\epsilon) = G_0(\boldsymbol{\rho}+\boldsymbol{R}_n,\boldsymbol{\rho}'+\boldsymbol{R}_m,k) = k\sum_{LL'} G_0{}^{nm}_{LL'}\; j_L(k\boldsymbol{\rho}_n)\; j_{L'}(k\boldsymbol{\rho}'_m)\,. \tag{6.17}$$

In the second step, we develop a method of solving the Dyson equation (6.1) for the full Green's function of the system.

From the *spectral representation* of the full GF already introduced in (2.48) of Chap. 2.

$$G(\boldsymbol{r},\boldsymbol{r}',\epsilon) = \lim_{\eta\to 0+}\sum_j \frac{\psi_j(\boldsymbol{r})\psi^*_j(\boldsymbol{r}')}{\epsilon-\epsilon_j+\mathrm{i}\eta}, \tag{6.18}$$

where $\psi_j(\boldsymbol{r})$ are the eigenstates of the SE corresponding to the energy ϵ_j and the index j runs over discrete and continuum states, it is clear that the GF satisfies the differential equations in its both arguments

$$
\begin{aligned}
(\epsilon - H_{\boldsymbol{r}})\, G(\boldsymbol{r}, \boldsymbol{r}', \epsilon) &= \delta(\boldsymbol{r} - \boldsymbol{r}') \\
(\epsilon - H_{\boldsymbol{r}'})\, G(\boldsymbol{r}, \boldsymbol{r}', \epsilon) &= \delta(\boldsymbol{r} - \boldsymbol{r}'),
\end{aligned} \tag{6.19}
$$

where $H_{\boldsymbol{r}} = -\Delta_{\boldsymbol{r}} + V(\boldsymbol{r})$ is the Hamiltonian of the system ($\hat{H}\psi_j = \epsilon_j \psi_j$).

The general solution of the inhomogeneous equations (6.19) is the sum of a particular solution of the equations (6.19) and the general solutions of the homogeneous equations corresponding to (6.19).

Therefore,

- if $\boldsymbol{r}, \boldsymbol{r}' \in \Omega_n,\ n = 1, \ldots, N$,

$$
G(\boldsymbol{r}, \boldsymbol{r}', \epsilon) = G_n^1(\boldsymbol{r}, \boldsymbol{r}', \epsilon) + k \sum_{LL'} G^{nn}_{LL'} R^n_{kL}(\boldsymbol{\rho}) R^n_{kL'}(\boldsymbol{\rho}'), \tag{6.20}
$$

  where $G_n^1(\boldsymbol{r}, \boldsymbol{r}', \epsilon)$ is a particular solution, the coefficients $G^{nn}_{LL'}$ are to be determined. It is expedient to choose the GF $G_n^1(\boldsymbol{r}, \boldsymbol{r}', \epsilon)$ so that it would satisfy the Dyson equation within the sphere n

$$
G_n^1(\boldsymbol{r}, \boldsymbol{r}', \epsilon) = G_0(\boldsymbol{r}, \boldsymbol{r}', \epsilon) + \int_{\Omega_n} G_0(\boldsymbol{r}, \boldsymbol{r}'', \epsilon)\, v_n(\boldsymbol{r}'')\, G_n^1(\boldsymbol{r}'', \boldsymbol{r}', \epsilon)\, \mathrm{d}\boldsymbol{r}'' ; \tag{6.21}
$$

- if $\boldsymbol{r} \in \Omega_n,\ \boldsymbol{r}' \in \Omega_m,\ n \neq m$,

$$
G(\boldsymbol{r}, \boldsymbol{r}', \epsilon) = k \sum_{LL'} G^{nm}_{LL'}\, R^n_{kL}(\boldsymbol{r})\, R^m_{kL'}(\boldsymbol{r}') , \tag{6.22}
$$

  where coefficients $G^{nm}_{LL'}$ are also to be determined.
  Remembering (6.12) for $\boldsymbol{r}, \boldsymbol{r}' \in \Omega_n,\ n = 1, \ldots, N$, we can express the GF in (6.21) as follows:

$$
G_n^1(\boldsymbol{r}, \boldsymbol{r}', \epsilon) = k \sum_{L} R^n_{kL}(k\rho_<)\, H^n_{kL}(k\rho'_>), \tag{6.23}
$$

  while using (6.14) for $\boldsymbol{r} \in \Omega_n,\ \boldsymbol{r}' \in \Omega_m,\ n = 1..N;\ m = 1..N,\ n \neq m$, we have

$$
G_n^1(\boldsymbol{r}, \boldsymbol{r}', \epsilon) = G_n^1(\boldsymbol{\rho}_n + \boldsymbol{R}_n, \boldsymbol{\rho}_m + \boldsymbol{R}_m, \epsilon) = k \sum_{LL'} G_{0\,LL'}^{\;nm}\, R^n_{kL}(\boldsymbol{\rho}_n) j_{L'}(k\rho'_m). \tag{6.24}
$$

The derivation of the system of equations for the unknown coefficients $G^{nm}_{LL'}$ is based on Dyson equation (6.1) and carried out in the following two cases:

- Case 1: $\boldsymbol{r} \in \Omega_n, \boldsymbol{r}' \in \Omega_m,\ n \neq m$
  Taking into account (6.22), we can rewrite (6.1) as

$$
\begin{aligned}
G(\boldsymbol{r},\boldsymbol{r}',\epsilon) = k\sum_{LL'} G^{nm}_{LL'}\, R^{n}_{kL}(\boldsymbol{r})\, R^{m}_{kL'}(\boldsymbol{r}') &= G_0(\boldsymbol{r},\boldsymbol{r}',\epsilon) \\
&+ \sum_{q\neq n,m} \int_{\Omega_q} G_0(\boldsymbol{r},\boldsymbol{r}'',\epsilon)\, v_q(\boldsymbol{r}'')\, G(\boldsymbol{r}'',\boldsymbol{r}',\epsilon)\, \mathrm{d}\boldsymbol{r}'' \\
&+ \int_{\Omega_n} G_0(\boldsymbol{r},\boldsymbol{r}'',\epsilon)\, v_n(\boldsymbol{r}'')\, G(\boldsymbol{r}'',\boldsymbol{r}',\epsilon)\, \mathrm{d}\boldsymbol{r}'' \\
&+ \int_{\Omega_m} G_0(\boldsymbol{r},\boldsymbol{r}'',\epsilon)\, v_m(\boldsymbol{r}'')\, G(\boldsymbol{r}'',\boldsymbol{r}',\epsilon)\, \mathrm{d}\boldsymbol{r}''.
\end{aligned}
\tag{6.25}
$$

The integrals on the right-hand side (RHS) of (6.25) can be expressed in terms of the introduced coefficients $G^{nm}_{LL'}$ as shown below.

$$
\begin{aligned}
&\int_{\Omega_q} G_0(\boldsymbol{r},\boldsymbol{r}'',\epsilon)\, v_q(\boldsymbol{r}'')\, G(\boldsymbol{r}'',\boldsymbol{r}',\epsilon)\, \mathrm{d}\boldsymbol{r}'' \\
&= k\sum_{LL''L'} G_0{}^{nq}_{LL''}\, t^{q}_{\ell''}\, G^{qm}_{L''L}\, j_L(k\boldsymbol{\rho}_n)\, R^{m}_{kL'}(\boldsymbol{\rho}'_m).
\end{aligned}
\tag{6.26}
$$

The derivation of (6.26) is based on (6.7), (6.17), and (6.22).

$$
\begin{aligned}
&\int_{\Omega_n} G_0(\boldsymbol{r},\boldsymbol{r}'',\epsilon)\, v_n(\boldsymbol{r}'')\, G(\boldsymbol{r}'',\boldsymbol{r}',\epsilon)\, \mathrm{d}\boldsymbol{r}'' \\
&= k\sum_{LL'} G^{nm}_{LL'}\, R^{n}_{kL}(\boldsymbol{\rho}_n) R^{m}_{kL'}(\boldsymbol{\rho}_m) - k\sum_{LL'} G^{nm}_{LL'}\, j_L(k\boldsymbol{\rho}_n) R^{m}_{kL'}(\boldsymbol{\rho}'_m).
\end{aligned}
\tag{6.27}
$$

The derivation of (6.27) is based on (6.6) and (6.22).

$$
\begin{aligned}
&\int_{\Omega_m} G_0(\boldsymbol{r},\boldsymbol{r}'',\epsilon)\, v_m(\boldsymbol{r}'')\, G(\boldsymbol{r}'',\boldsymbol{r}',\epsilon)\, \mathrm{d}\boldsymbol{r}'' \\
&= \int_{\Omega_m} G_0(\boldsymbol{r},\boldsymbol{r}'',\epsilon)\, v_m(\boldsymbol{r}'')\, G^{1}_{m}(\boldsymbol{r}'',\boldsymbol{r}',\epsilon)\, \mathrm{d}\boldsymbol{r}'' \\
&+ k\sum_{L''L'} G^{mm}_{L''L'} \int_{\Omega_m} G_0(\boldsymbol{r},\boldsymbol{r}'',\epsilon) v_m(\boldsymbol{r}'') R^{m}_{kL''}(\boldsymbol{r}'') R^{m}_{kL'}(\boldsymbol{r}')\mathrm{d}\boldsymbol{r}'' \\
&= k\sum_{LL'} G_0{}^{nm}_{LL'}\, j_L(k\boldsymbol{\rho}_n)\, R^{m}_{kL'}(\boldsymbol{\rho}'_m) - G_0(\boldsymbol{r},\boldsymbol{r}',\epsilon) \\
&+ k\sum_{LL'L''} G_0{}^{nm}_{LL''}\, t^{m}_{\ell''}\, G^{mm}_{L''L'}\, j_L(k\boldsymbol{\rho}_n)\, R^{m}_{kL'}(\boldsymbol{\rho}'_m).
\end{aligned}
\tag{6.28}
$$

The derivation of (6.28) is based on (6.6), (6.17), (6.20), (6.21), and (6.24). Inserting (6.26)–(6.28) into (6.25) and canceling several terms, we obtain

$$\sum_{LL'}\left(G^{nm}_{LL'} - G_0{}^{nm}_{LL'} - \sum_{q\neq n}\sum_{L''} G_0{}^{nq}_{LL''}\, t^q_{\ell''}\, G^{qm}_{L''L'}\right) j_L(k\rho_n) R^m_{kL'}(\rho'_m)\,. \qquad (6.29)$$

The functions in (6.29) are independent, therefore

$$G^{nm}_{LL'} = G_0{}^{nm}_{LL'} + \sum_{q\neq n}\sum_{L''} G_0{}^{nq}_{LL''}\, t^q_{\ell''}\, G^{qm}_{L''L'} \quad (n\neq m)\,. \qquad (6.30)$$

- Case 2: $\boldsymbol{r}, \boldsymbol{r}' \in \Omega_n$,
  The derivation of the equations for the Green's function coefficients in this case follows the same lines as in Case 1. One can find

$$G^{nn}_{LL'} = \sum_{q\neq n}\sum_{L''} G_0{}^{nq}_{LL''}\, t^q_{\ell''}\, G^{qn}_{L''L'}\,. \qquad (6.31)$$

Combining (6.30) with (6.31) and defining $G_0{}^{nn}_{LL'} \equiv 0$, we finally obtain

$$G^{nm}_{LL'}(\epsilon) = G_0{}^{nm}_{LL'}(\epsilon)\,(1-\delta_{nm}) + \sum_{qL''} G_0{}^{nq}_{LL''}(\epsilon)\, t^q_{\ell''}(\epsilon)\, G^{qm}_{L''L'}(\epsilon)\,. \qquad (6.32)$$

Here we returned to the coefficients their argument ϵ. Thus, the solution of the Dyson equation is reduced to the simple system of linear algebraic equations (6.32) for the *Green's function matrix in angular momentum-site index representation.*

6.2.2 *Photoabsorption Cross-Section in Green's Function Formalism*

In this section we present an illustrative example of applying the developed MS-GF method to the description of photoabsorption. For this purpose, we will relate the cross-section with the introduced Green's function matrix.

The expression for the total cross-section can be written within the dipole approximation as

$$\sigma(\omega_q) = 4\pi^2\alpha\omega_q \sum_f \left|\int \psi^*_f(\boldsymbol{r})\, \hat{W}(\boldsymbol{r})\, \varphi_i(\boldsymbol{r})\, \mathrm{d}\boldsymbol{r}\right|^2 \delta(\epsilon_i - \epsilon_f + \omega_q)\,, \qquad (6.33)$$

where α is the fine structure constant, ω_q is the photon energy; the initial states and energy are denoted by index i and the final states and energy by f. $\hat{W}(\boldsymbol{r}) = \hat{\boldsymbol{e}}_q \cdot \boldsymbol{r}$ is the dipole transition operator in the so-called *length form*, $\boldsymbol{q}$ and $\hat{\boldsymbol{e}}_{\boldsymbol{q}}$ are the wavevector and polarization of the incoming field respectively. (We shall omit henceforth the unnecessary index $\boldsymbol{q}$ on $\omega_{\boldsymbol{q}}$ and $\hat{\boldsymbol{e}}_{\boldsymbol{q}}$ for simplicity).

Here, we introduce the function $\varrho(\boldsymbol{r}, \boldsymbol{r}', \epsilon) = \sum_j \psi_j(\boldsymbol{r})\psi_j(\boldsymbol{r}')\delta(\epsilon - \epsilon_j)$, where $\psi_j(\boldsymbol{r})$ are the eigenstates of the SE at energy ϵ_j. Note that the completeness of the set of these functions allows us to choose them real for real potentials.

Using the spectral representation of the GF (6.18) and the Sokhotski–Plemelj theorem $\lim\limits_{\eta \to 0+} \frac{1}{x \pm \mathrm{i}\eta} = \mathcal{P}\frac{1}{x} \mp \mathrm{i}\pi\delta(x)$, where $\mathcal{P}$ denotes Cauchy principle part (see Chap. 1), one can write

$$\varrho(\boldsymbol{r}, \boldsymbol{r}', \epsilon) = -\frac{1}{\pi} \Im\left[G(\boldsymbol{r}, \boldsymbol{r}', \epsilon)\right], \tag{6.34}$$

where $\Im$ represents the imaginary part.

Using (6.34) in (6.33) allows one to avoid the summation over the final states:

$$\sigma(\omega) = -4\pi\alpha\omega \left\{ \iint_{\Omega_n} \varphi_i^*(\boldsymbol{r})\hat{W}^*(\boldsymbol{r}) \Im\left[G(\boldsymbol{r}, \boldsymbol{r}', \epsilon)\right] \hat{W}(\boldsymbol{r}')\varphi_i(\boldsymbol{r}')\, \mathrm{d}\boldsymbol{r}\mathrm{d}\boldsymbol{r}' \right\}, \tag{6.35}$$

where $\epsilon = \omega + \epsilon_i$; the integrals are limited to the sphere n of the photoabsorbing atom. This means that one needs to have the GF with $\boldsymbol{r}, \boldsymbol{r}' \in \Omega_n$, i.e. remembering (6.20) and (6.23),

$$G(\boldsymbol{r}, \boldsymbol{r}', \epsilon) = k \sum_L R^n_{kL}(k\rho_<)H^n_{kL}(k\rho_>) + k \sum_{LL'} G^{nn}_{LL'} R^n_{kL}(\rho) R^n_{kL'}(\rho'), \tag{6.36}$$

where the regular and irregular solutions of SE are complex, which makes necessary the calculation of the imaginary part of the irregular solution.

Now we will redefine the regular and irregular solutions and rewrite (6.36) in terms of the *real* solutions. Using the asymptotic forms (6.4) and (6.5) one can show

$$H^n_{kL}(k\rho_>) = N^n_{kL}(\rho_>) - \mathrm{i}\widetilde{R}^n_{kL}(\rho_<), \tag{6.37}$$

where the radial part of the real regular solution $\widetilde{R}^n_{kL}(\rho) = \mathrm{e}^{-\mathrm{i}\delta^n_\ell} R^n_{kL}(\rho)$ matches to the boundary condition

$$\widetilde{R}^n_{kL}(b_n) = j_\ell(kb_n)\cos\delta^n_\ell - n_\ell(kb_n)\sin\delta^n_\ell \tag{6.38}$$

and the radial part of the real irregular solution $N^n_{kL}(\rho)$ matches to the condition

$$N^n_{kL}(b_n) = j_\ell(kb_n)\sin\delta^n_\ell + n_\ell(kb_n)\cos\delta^n_\ell\,. \tag{6.39}$$

The Green's function (6.36) in terms of the new solutions becomes

$$
\begin{aligned}
G(\boldsymbol{r}, \boldsymbol{r}', \epsilon) &= k \sum_{L} \widetilde{R}^n_{kL}(k\rho_<)\, N^n_{kL}(k\rho'_>) \\
&+ k \sum_{LL'} \left[G^{nn}_{LL'} \mathrm{e}^{\mathrm{i}(\delta^n_\ell + \delta^n_{\ell'})} - \mathrm{i}\delta_{LL'} \right] \widetilde{R}^n_{kL}(\rho)\, \widetilde{R}^n_{kL'}(\rho')
\end{aligned}
\tag{6.40}
$$

with the imaginary part

$$
\Im\left[G(\boldsymbol{r}, \boldsymbol{r}', \epsilon)\right] = k \sum_{LL'} \Im\left[G^{nn}_{LL'} \mathrm{e}^{\mathrm{i}(\delta^n_\ell + \delta^n_{\ell'})} - \mathrm{i}\delta_{LL'}\right] \widetilde{R}^n_L(\boldsymbol{r}) \widetilde{R}^n_{L'}(\boldsymbol{r}') \,. \tag{6.41}
$$

The substitution of (6.41) in (6.35) gives

$$
\sigma(\omega) = 4\pi\alpha\omega k \left\{ \sum_{LL'} A^*_{iL}(\epsilon)\, \Im\left[\mathrm{i}\delta_{LL'} - G^{nn}_{LL'} \mathrm{e}^{\mathrm{i}(\delta^n_\ell + \delta^n_{\ell'})}\right] A_{iL'}(\epsilon) \right\} , \tag{6.42}
$$

where

$$
A_{iL}(k) = \int_{\Omega_n} \varphi_i(\boldsymbol{r})\, \hat{W}(\boldsymbol{r})\, R^n_{kL}(\boldsymbol{r})\, \mathrm{d}\boldsymbol{r} \,. \tag{6.43}
$$

6.2.3 Numerical Aspects

In many cases one needs to calculate transitions of the photoelectrons either to the states in continuum rapidly varying with the energy or/and to the states below MT-zero. As will be demonstrated in Chap. 13, modeling the spectra in these cases requires extremely small real energy step, making calculations time consuming. Here we develop an effective numerical procedure of solving this problem.

6.2.3.1 Analytical Continuation of the Green's Function

It is known that there are two alternative ways to account for the finite lifetime of the electron-hole pair. The first one is to convolute a spectrum calculated on the real energy axis with a Lorentzian function of the half width Γ. Another way, to which we will adhere, is a continuation of the GF into the upper semiplane of the complex energy plane [7]. To take into account the Pauli principle, we introduce now the Fermi level and show following [7] that for any photoelectron energy, except for the vicinity of the Fermi energy ϵ_F, the convolution is mathematically equivalent to the use of the GF calculated at the complex energy $\epsilon + \mathrm{i}\Gamma$. The convolution of the cross-section (6.33) is obtained by convoluting the $\varrho(\boldsymbol{r}, \boldsymbol{r}', \epsilon)$ in (6.34):

$$\widetilde{\rho}(\boldsymbol{r},\boldsymbol{r}',\epsilon) = \theta(\epsilon-\epsilon_F)\,\frac{\Gamma}{\pi}\int_{-\infty}^{\infty}\frac{\varrho(\boldsymbol{r},\boldsymbol{r}',\epsilon')}{(\epsilon-\epsilon')^2+\Gamma^2}\mathrm{d}\epsilon'$$

$$= \frac{\Gamma}{\pi}\left[\int_{-\infty}^{\infty}\frac{\varrho(\boldsymbol{r},\boldsymbol{r}',\epsilon'-\epsilon_F)}{(\epsilon-\epsilon')^2+\Gamma^2}\mathrm{d}\epsilon' - \int_{-\infty}^{\epsilon_F}\frac{\varrho(\boldsymbol{r},\boldsymbol{r}',\epsilon'-\epsilon_F)}{(\epsilon-\epsilon')^2+\Gamma^2}\mathrm{d}\epsilon'\right], \tag{6.44}$$

where $\theta(x)$ is the step function. Remembering the relation (6.34), we notice that the first integrand on the RHS is an analytical function of ϵ' if $\Im(\epsilon') > 0$. The second integral in (6.44) decreases rapidly when $\epsilon > \epsilon_F$, and it can be neglected for the transitions to continuum states above ϵ_F. Therefore, one can write

$$\begin{aligned}\widetilde{\rho}(\boldsymbol{r},\boldsymbol{r}',\epsilon) &\cong \frac{\Gamma}{\pi}\int_{-\infty}^{\infty}\frac{\sum_j \psi_j(\boldsymbol{r})\psi_j(\boldsymbol{r}')\delta(\epsilon'-\epsilon_F-\epsilon_j)}{(\epsilon-\epsilon')^2+\Gamma^2}\mathrm{d}\epsilon' \\ &= \frac{\Gamma}{\pi}\sum_j\frac{\psi_j(\boldsymbol{r})\psi_j(\boldsymbol{r}')}{(\epsilon-\epsilon_F-\epsilon_j)^2+\Gamma^2}\,.\end{aligned} \tag{6.45}$$

On the other hand, the spectral representation of the GF (6.18) for the complex energies $\epsilon - \epsilon_F + \mathrm{i}\Gamma$

$$G(\boldsymbol{r},\boldsymbol{r}',\epsilon-\epsilon_F+\mathrm{i}\Gamma) = \sum_j\frac{\psi_j(\boldsymbol{r})\psi_j^*(\boldsymbol{r}')}{\epsilon-\epsilon_F-\epsilon_j+\mathrm{i}\Gamma} \tag{6.46}$$

is the analytical continuation of the GF from the positive real energy semiaxis into the segment $\epsilon > \epsilon_F$ of the upper semiplane of the first sheet of the complex energy plane, where it has no poles. Remembering that we deal with the real eigenfunctions, one finally obtains

$$\begin{aligned}\widetilde{\rho}(\boldsymbol{r},\boldsymbol{r}',\epsilon) &= \frac{\Gamma}{\pi}\sum_j\frac{\psi_j(\boldsymbol{r})\psi_j(\boldsymbol{r}')}{(\epsilon-\epsilon_F-\epsilon_j)^2+\Gamma^2} \\ &= -\frac{1}{\pi}\Im\left[\sum_j\frac{\psi_j(\boldsymbol{r})\psi_j(\boldsymbol{r}')}{\epsilon-\epsilon_F-\epsilon_j+\mathrm{i}\Gamma}\right] = -\frac{1}{\pi}\Im\left[G(\boldsymbol{r},\boldsymbol{r}',\epsilon-\epsilon_F+\mathrm{i}\Gamma)\right].\end{aligned} \tag{6.47}$$

Thus, if the calculations of spectra are carried out for the integration contour shifted by the value of $\Gamma > 0$ into the upper semiplane, we obtain spectra broadened with a Lorentzian function. In addition, this mathematical trick allows one to compute spectra in the range of both continuum and bound electronic states using the same algorithm, since the spectrum is continuous on such a contour.

6.2.3.2 Calculations of the Photoabsorption for Complex Electron Energies

To calculate cross-section (6.35) for the complex energies, it is expedient to use the GF in the form (6.36), with the regular and irregular solutions matching to the boundary conditions (6.4) and (6.5) respectively.

As an example, let us consider K-edge *unpolarized* absorption cross-section, in which case (6.35) is reduced to

$$\sigma(\omega) = -\frac{8\pi\alpha\omega}{3}\left\{\iint_{\Omega_n} \varphi_i(\boldsymbol{r})\boldsymbol{r}\,\Im\left[G(\boldsymbol{r},\boldsymbol{r}',\epsilon-\epsilon_F+\mathrm{i}\Gamma)\right]\boldsymbol{r}'\varphi_i(\boldsymbol{r}')\mathrm{d}\boldsymbol{r}\mathrm{d}\boldsymbol{r}'\right\}$$

$$= -\frac{8\pi\alpha\omega}{3}k\Im\left\{\begin{array}{l}\displaystyle\int_0^{b_n}\int_0^{b_n}\varphi_{1s}(r)\,r^3\,R^n_{k1}(r_<)\,H^n_{k1}(r_>)\,(r')^3\,\varphi_{1s}(r')\,\mathrm{d}r\mathrm{d}r' \\ \displaystyle +\left[\int_0^{b_n}\varphi_{1s}(r)\,r^3\,R^n_{k1}(r)\,\mathrm{d}r\right]^2\frac{1}{3}\sum_{m=-1}^{1}G^{nn}_{1m,1m}(\epsilon-\epsilon_F+\mathrm{i}\Gamma)\end{array}\right\}, \tag{6.48}$$

where spin degeneracy is taken into account by a factor of two. The way to compute the first integral in (6.48) is illustrated in Fig. 6.1.

The integration over the upper triangle (over the variable r) in Fig. 6.1 gives

$$\int_0^{b_n}\int_0^{r'}\varphi_{1s}(r)\,r^3\,R^n_{k1}(r)\,H^n_{k1}(r')\,(r')^3\,\varphi_{1s}(r')\,\mathrm{d}r\mathrm{d}r' \tag{6.49}$$

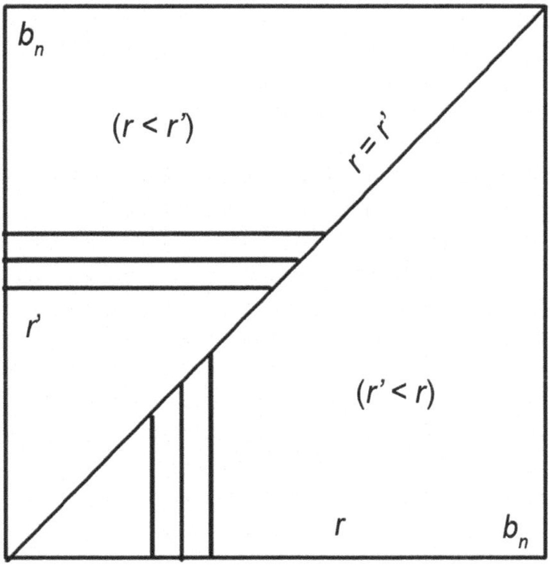

Fig. 6.1 The algorithm of the bivariate integration in (6.48): Horizontal lines show intervals of integration over r from 0 to r in (6.49) and vertical lines over r from 0 to r in (6.50)

and the integration over the lower triangle (over the variable r') gives

$$\int_0^{b_n} \int_0^{r} \varphi_{1s}(r)\, r^3\, R^n_{k1}(r')\, H^n_{k1}(r)\, (r')^3\, \varphi_{1s}(r')\, \mathrm{d}r \mathrm{d}r' \,. \tag{6.50}$$

Therefore, the double integral in (6.48) reduced to

$$2 \int_0^{b_n} \varphi_{1s}(r)\, r^3 H^n_{k1}(r)\, F^n_{k1}(r)\, \mathrm{d}r \,, \tag{6.51}$$

where we put $F^n_{k\ell}(r) = \int_0^r R^n_{k1}(r')(r')^3 \varphi_{1s}(r')\mathrm{d}r'$.

The irregular solution $H^n_{k1}(r)$ matching smoothly to $-\mathrm{i}h^+_1(kr)$ at b_n can be computed by either outward or inward integration of the radial SE.

6.3 MS-GF Formalism: Multichannel Version

As in the one-electron MS-GF method, a derivation of the multichannel equations presented here is based on the Dyson equation for the retarded GF $G(E)$

$$G(E) = {}^0G(E) + {}^0G(E)\, \Delta V\, G(E) \,, \tag{6.52}$$

where E is the full energy of the system under study; ${}^0G(E)$ is the GF of the one-particle approach considered in Sect. 6.2; ΔV is a difference between many-particle and single-particle Hamiltonians.

Reducing the integral Dyson equation (6.1) to the system of the linear algebraic equations (6.32) is based on the expansions (6.12), (6.14) for the local GF and (6.20), (6.22) for the full GF. Similar expansions are not valid for potentials which are non-local in the radial variables. One of the reasons lies in the fact that this potential is an integral operator not allowing the separation of variables $\boldsymbol{r}$ and $\boldsymbol{r}'$. As a consequence, the Wronskian of the radial regular and irregular solutions of the differential equation is not constant. It is worth noting that the introduced potential is non-local not only in radial variables but also in the channel representation. We assume at this point that the non-local potential considered here can be approximated by a sum of separable potentials, keeping however at the same time its nonlocal character in the channel indices. In this case, two-dimensional integrals can be divided into the sum of products of one-dimensional integrals, thus admitting the separation of the radial variables. In this approach, we imply that the Wronskian has a small dependence on the radial variables that can be neglected. So, we will construct the Green's function in terms of the Schrödinger equation solutions calculated with a high accuracy within the variational principle, however, the expansion formulas for the GF in terms of these solutions will be approximate.

It is worth noting that the multichannel generalization of the MS-GF method relies on the MT-approximation to the cluster potential, for this reason the type of correlations which can be treated within this approximation is restricted to the correlations between one electron in a delocalized state and the electrons (or holes) in the states localized inside the MT-spheres.

6.3.1 Single-Site Multichannel Radial Equations

Prior to solving (6.52), we introduce, as before, a number of local functions associated with the truncated atomic potential inside the sphere Ω_n. The many-electron wave function of the N-electron system with one delocalized electron scattered by the rest of the atom with $(N-1)$ electrons can be approximated by a linear combination of the antisymmetrized products, as

$$\Psi^N(\boldsymbol{x}_1, \ldots, \boldsymbol{x}_{N-1}; \boldsymbol{x}) = \frac{1}{\sqrt{N}} \left[\sum_i \hat{A} \varphi_i(\boldsymbol{x})\, \Psi_i^{N-1}(\boldsymbol{x}_1, \ldots, \boldsymbol{x}_{N-1}) \right] \equiv \sum_i \Theta_i \,, \tag{6.53}$$

where $\boldsymbol{x} \equiv (\boldsymbol{r}, s)$ denotes the radial and spin coordinates, $\varphi_i(\boldsymbol{x})$ is the wave function of the scattered electron in a channel i, $\hat{A}$ denotes the operator of permutation of coordinates, which is equal to $+1$ or -1 in the cases of even or odd number of permutation respectively, $\Psi_i^{N-1}(\boldsymbol{x}_1, \ldots, \boldsymbol{x}_{N-1})$ is the Slater determinant for the channel i constructed from the localized orbitals of the $(N-1)$ - electron atom.

Our purpose in this step is to derive the radial Schrödinger equation for the wave function of the scattering electron. We will follow the close-coupling method based on the Kohn variational principle for the reactance K-matrix [20] formulated for a function Ψ^N assumed to be the exact solution of the many-particle problem:

$$\delta\,[\,(\Psi \mid H - E \mid \Psi) - K\,] = 0\,.$$

As the exact solution of the many-particle problem is unknown, it was shown in [21] that the above variation is equivalent to

$$(\delta\Psi \mid H - E \mid \Psi) = 0 \tag{6.54}$$

if Ψ^N is inexact.

Therefore, one first needs to write expressions for the matrix $(\Psi|H - E|\Psi)$ of the Hamiltonian

$$H^N = -\nabla_r^2 + \sum_{j=1}^{N-1} \frac{2}{|\,\boldsymbol{r} - \boldsymbol{r}_j\,|} - \frac{2Z}{|\,\boldsymbol{r} - \boldsymbol{R}_n\,|} + H^{N-1} \tag{6.55}$$

for the set of the trial channel functions Θ_i from (6.53). The matrix elements $(\Theta_i|H - E|\Theta_j)$ are just the same as used in ligand field theory [22]. Their structure depends on the definition of channels. The variation of the radial wave functions of the scattered electron in each Θ_i leads to a *system of coupled integro-differential equations* for these functions. As an illustration, let us consider the simple but interesting example of $2p$ spin-orbit coupling and Coulomb interaction between $2p$ core electrons and s, d electrons in the continuum. In this case H^{N-1} includes the operator of spin-orbit interaction $\sum_{i\in 2p} a(r_i)(\boldsymbol{\ell}_i \cdot \boldsymbol{s}_i)$ (spin-orbit coupling of d electron in the continuum is negligible). The channel can be defined by the set of quantum numbers $i = j_{2p}\mu_j k_i^2 \ell_i m_i \equiv \alpha k_i^2 L_i$, where $j = 1/2, 3/2$; $\mu_j = -j, \ldots, +j$; $L_i \equiv \ell_i m_i$ refers to the angular momentum of the delocalized electron; $k_i^2 = \epsilon_i = E - E_\alpha = E - (E^N - E^{N-1}_{j(2p)})$. Therefore,

$$\varphi_i(\boldsymbol{r}) = P_{\ell_i}(\epsilon_i, r) Y_{L_i}(\hat{\boldsymbol{r}}) \,, \tag{6.56}$$

the Slater determinant $\Psi_i^{N-1}(\boldsymbol{x}_1, \ldots, \boldsymbol{x}_{N-1})$ for the channel i includes corresponding $2p$-orbital

$$\Psi_{i(2p)}(\boldsymbol{x}) = P_{2p}(r) \times \sum_{m+m_s=\mu_j} C^j_{mm_s} Y_{1m}(\hat{\boldsymbol{r}}) \chi^s_{m_s}(\sigma) \,, \tag{6.57}$$

where $P_{2p}(r)$ is a radial wave function of $2p$ electrons calculated in a one-particle mean-field approximation for H^{N-1}; $C^j_{mm_s}$ are the Clebsch–Gordan coefficients, $\chi^s_{m_s}(\sigma)$ is a spin function of $2p$ electrons; $P_{\ell_i}(\epsilon_i, r)$ is a radial wave function of the electron scattered in the channel i, and the Hamiltonian H^N becomes

$$H^N = -\nabla_r^2 + 2V'(\boldsymbol{r}) + \sum_{j\in 2p}^{5} \frac{2}{|\,\boldsymbol{r} - \boldsymbol{r}_j\,|} - \frac{2Z}{|\,\boldsymbol{r} - \boldsymbol{R}_n\,|} + H^{N-1} \,. \tag{6.58}$$

Here $2V'(\boldsymbol{r})$ is a one-particle mean-field potential that includes interaction between the scattered electron and all the localized electrons of the atom except $2p$ electrons. Note that the Slater determinant $\Psi_i^{N-1}(\boldsymbol{x}_1, \ldots, \boldsymbol{x}_{N-1})$ is constructed from the discrete orbitals obtained in the same approximation for H^{N-1} as $2p$-orbitals. By inserting (6.58) into $(\Theta_i|H - E|\Theta_j)$, using the definition of Θ_i given in (6.53), and varying the radial wave functions $P_{\ell_i}(\epsilon_i, r) \equiv P_i(r)$ in each Θ_i, we obtain the system of coupled radial integro-differential equations:

$$\left(\frac{\mathrm{d}^2}{\mathrm{d}r^2} - \frac{\ell_i(\ell_i+1)}{r^2} - V(r) + k_i^2\right) P_i(r) - \sum_j W_{ij}(r, r') P_j(r') + \sum_q \lambda_{iq} P_q(r) \delta_{\ell_i \ell_q} = 0 \,, \tag{6.59}$$

$$W_{ij}(r, r')P_j(r') = 2\sum_q \begin{bmatrix} a_{ij}^q \, y_q\,(P_{2p}P_{2p})\,P_j(r) \\ -c_{ij}^q \, y_q\,(P_{2p}P_i(r))\,P_{2p}(r) \end{bmatrix} , \tag{6.60}$$

$$y_q(fg) = r^{-q-1}\int_0^r f(r')g(r')(r')^q \,\mathrm{d}r' + r^q \int_r^{b_n} f(r')g(r')(r')^{-q-1}\,\mathrm{d}r' , \tag{6.61}$$

where $V(r) = 2V'(r) + 2y_0(P_{2p}P_{2p}) - 2Z/|\boldsymbol{r} - \boldsymbol{R}_n|$ is the one-electron potential within the sphere Ω_n considered in Sect. 6.2. The third term in (6.59) ensures the orthogonality of the radial function of the electron in the continuum to the radial orbitals $P_q(r)$ of the core electrons with $\ell_q = \ell_i$ by means of the Lagrangian multipliers λ_{iq}. Coefficients a_{ij}^q, c_{ij}^q in (6.60) are expressed in terms of the Gaunt coefficients $\int Y_i(\hat{\boldsymbol{r}})Y_{qm_q}(\hat{\boldsymbol{r}})Y_j(\hat{\boldsymbol{r}})\mathrm{d}\hat{\boldsymbol{r}}$ [23]. Note that the first term on the RHS of (6.60) is a local potential, with $q = 2$, whereas the second one is a non-local potential with $q = 1, 3$. Henceforth, as in Sect. 6.2, we will use the local coordinate $\boldsymbol{\rho}$ instead of the global $\boldsymbol{r}$.

Boundary Conditions

The single-site solution of the set of equations (6.59) is a matrix defined by asymptotical conditions. Let us choose them so that the matrix elements $R_{ij}^n(\epsilon_i, \rho) = P_{ij}^n(\epsilon_i, \rho)/(\rho k_j)$ would smoothly match to the conditions:

$$R_{ij}^n(\epsilon_i, \rho) \underset{\rho \to 0}{\longrightarrow} \rho^{\ell_i}\delta_{ij} , \tag{6.62}$$

$$R_{ij}^n(\epsilon_i, b_n) = \sqrt{\frac{k_i}{k_j}}\left[j_{\ell_i}(k_i b_n)\,\delta_{ij} - \mathrm{i}\,h_{\ell_i}^+(k_i b_n)\,t_{ij}^n \right] , \tag{6.63}$$

where t_{ij}^n is the t-matrix in the channel basis introduced by definition.

The coupled set of equations is solved iteratively and independently for each column function which can then be normalized according to the condition (6.63) that also determines the corresponding column of the t-matrix.

Lippmann–Schwinger Equation

The functions $R_{ij}^n(\epsilon_i, \rho)$ satisfy the system of coupled integral equations

$$R_{ij}^n(\epsilon_i, \rho) = j_{\ell_i}(k_i\rho)\delta_{ij} + \sum_q \iint_{\Omega_n} G_{0\ell_i}(\rho, \rho', \epsilon_i)\,\widetilde{V}_{iq}(\rho', \rho'')\,R_{qj}^n(\epsilon_q, \rho'')\,\rho'^2\rho''^2\,\mathrm{d}\rho'\mathrm{d}\rho'' , \tag{6.64}$$

where

$$\widetilde{V}_{iq}(\rho', \rho'') = \left[V(\rho') - \sum_{p} \lambda_{ip} P_p(\rho') \delta_{\ell_i \ell_p} \right] \delta(\rho' - \rho'')\delta_{iq} + W_{iq}(\rho', \rho'') \,, \quad (6.65)$$

and the free GF $G_{0\ell_i}(\rho, \rho', \epsilon_i)$ satisfies the radial equation

$$\left[\frac{1}{\rho^2} \frac{\mathrm{d}}{\mathrm{d}\rho} \left(\rho^2 \frac{\mathrm{d}}{\mathrm{d}\rho} \right) - \frac{\ell_i(\ell_i + 1)}{\rho^2} + \epsilon_i \right] G_{0\ell_i}(\rho, \rho', \epsilon_i) = \frac{\delta(\rho - \rho')}{\rho^2} \quad (6.66)$$

with the known matrix elements in the angular momentum basis

$$G_{0\ell_i}(\rho, \rho', \epsilon_i) = -\mathrm{i} k_i \, j_{\ell_i}(k_i \rho_<) \, h^+_{\ell_i}(k_i \rho_>) \, \delta_{ij} \,. \quad (6.67)$$

Asymptotic Matrices

Writing (6.64) for $\rho = b_n$, substituting the boundary condition (6.63) in the LHS and the boundary value of the GF (6.67) in the RHS, one can obtain the integral representation of the interchannel atomic t-matrix introduced in (6.63):

$$t_{ij}(E) = \sqrt{k_i k_j} \sum_q \iint_{\Omega_n} j_{\ell_i}(k_i \rho) \, \widetilde{V}_{iq}(\rho, \rho') \, R^n_{qj}(\epsilon_q, \rho') \, \rho^2 \rho'^2 \, \mathrm{d}\rho \mathrm{d}\rho' \,. \quad (6.68)$$

It is useful to write the Lippmann–Schwinger set of coupled integral equations in terms of the potential difference as

$$\begin{aligned} R^n_{ij}(\epsilon_i, \rho) &= R^n_{k_i \ell_i}(\rho) \, \delta_{ij} \\ &+ \sum_q \iint_{\Omega_n} G^n_i(\rho, \rho', \epsilon_i) \, \Delta V_{iq}(\rho', \rho'') \, R^n_{qj}(\epsilon_q, \rho'') \, \rho'^2 \rho''^2 \, \mathrm{d}\rho' \mathrm{d}\rho'' \,, \end{aligned} \quad (6.69)$$

where $R^n_{k_i \ell_i}(\rho)$ is the local single-channel regular solution of the Schrödinger equation with the potential $v_n(\rho)$ (see Sect. 6.2); the single-site GF $G^n_i(\rho, \rho', \epsilon_i) \equiv G^n_{\ell_i}(\rho, \rho', \epsilon)/(\rho_n \rho'_n)$ is defined in Sect. 6.2 according to (6.11); $\Delta V_{iq}(\rho, \rho') = \widetilde{V}_{iq}(\rho, \rho') - v_n(\rho')\delta(\rho - \rho')\delta_{iq}$.

Now one can relate the difference between the multichannel and single-channel t-matrices to the difference between the corresponding potentials. Writing (6.69) for $\rho = b_n$, inserting into it the expressions for the single-site atomic GF $G^n_{\ell_i}(b_n, \rho', \epsilon_i) = -\mathrm{i}k_i \, R^n_{k_i \ell_i}(\rho') \, h^+_{\ell_i}(k_i b_n)\delta_{ij}$ and the boundary conditions (6.4), (6.63) for the single-channel and multichannel wave functions respectively, we obtain

$$\Delta t_{ij}(E) \equiv t_{ij} - t_i^n \delta_{ij} =$$
$$\sqrt{k_i k_j} \sum_q \iint_{\Omega_n} R^n_{k_i \ell_i}(\rho)\, \Delta V_{iq}(\rho, \rho')\, R^n_{qj}(\epsilon_q, \rho')\, \rho^2 \rho'^2 \,\mathrm{d}\rho \mathrm{d}\rho' \,. \tag{6.70}$$

In the next section, it is expedient to have the boundary condition (6.63) expressed in terms of the reactance K-matrix. For this, defining the relation between scattering S- and t-matrices similar to that for the single-channel problem:

$$t_{ij} = \frac{\delta_{ij} - S_{ij}}{2\mathrm{i}} \,, \tag{6.71}$$

we first reduce (6.63) to

$$R^n_{ij}(\epsilon_i, b_n) = \frac{1}{2}\sqrt{\frac{k_i}{k_j}} \left[h^-_{\ell_i}(k_i b_n)\, \delta_{ij} + h^+_{\ell_i}(k_i b_n)\, S^n_{ij}\right] .$$

Then, using the well-known relation between unitary S-matrix and Hermitian K-matrix [24]:

$$\mathbf{S} = (\mathbf{I} + \mathrm{i}\mathbf{K})(\mathbf{I} - \mathrm{i}\mathbf{K})^{-1} \,, \tag{6.72}$$

we obtain the desired boundary condition equivalent to (6.63):

$$R^n_{ij}(\epsilon_i, b_n) = \sqrt{\frac{k_i}{k_j}} \sum_q \left[j_{\ell_i}(k_i b_n)\, \delta_{ij} - n_{\ell_i}(k_i b_n)\, K^n_{iq}\right] (\mathbf{I} - \mathrm{i}\mathbf{K})^{-1}_{qj} \,. \tag{6.73}$$

6.3.2 Single-Site Multichannel Green's Function

It is known that the Wronskian for matrices of the regular Φ and irregular F solutions of the differential equation with the symmetrical potential matrix is constant if defined as in [25]:

$$W(F, \Phi) = \widetilde{F}\Phi' - \widetilde{F}'\Phi \,, \tag{6.74}$$

where the tilde over them denotes the matrix transposition and the prime indicates the radial first derivative.

The functions $\Phi_{ij}(\epsilon_i, \rho)$, $F_{ij}(\epsilon_i, \rho)$ satisfy the set of coupled equations

$$\left[\frac{\mathrm{d}^2}{\mathrm{d}\rho^2} - \frac{\ell_i(\ell_i + 1)}{\rho^2} + k_i^2\right] f_{ij}(\rho) - \sum_q \int_{\Omega_n} V_{iq}(\rho, \rho')\, f_{qj}(\rho')\, \mathrm{d}\rho' = 0 \,, \tag{6.75}$$

where $f_{ij}(\rho)$ denotes either $\Phi_{ij}(\epsilon_i, \rho)$ or $F_{ij}(\epsilon_i, \rho)$.

Let us impose for these solutions the following boundary conditions at $\rho = b_n$:

$$\Phi_{ij}(\epsilon_i, b_n) = \frac{1}{\sqrt{k_i}} \left[\bar{j}_{\ell_i}(k_i b_n)\, \delta_{ij} - \bar{n}_{\ell_i}(k_i b_n)\, K_{ij} \right] , \tag{6.76}$$

$$F_{ij}(\epsilon_i, b_n) = \frac{1}{\sqrt{k_i}} \bar{h}^{+}_{\ell_i}(k_i b_n) \sum_p (\mathbf{I} - \mathrm{i}\mathbf{K})^{-1}_{ip} A_{pj} , \tag{6.77}$$

where $\bar{j}_{\ell_i}(k_i\rho) = k_i\rho j_{\ell_i}(k_i\rho)$, $\bar{n}_{\ell_i}(k_i\rho) = k_i\rho n_{\ell_i}(k_i\rho)$, $\bar{h}^{+}_{\ell_i}(k_i\rho) = k_i\rho h^{+}_{\ell_i}(k_i\rho)$; K_{ij} are matrix elements of the K-matrix in the channel basis which is real and symmetrical for the real and symmetrical potential in (6.75). The normalization coefficients A_{pj} should be determined so as the Wronskian between these solutions would be equal to unity allowing us to use them in construction of the single-site multichannel GF.

First, we note that the function

$$\bar{G}(\rho, \rho', E) = \begin{cases} \Phi(\rho)\widetilde{F}(\rho'), & \rho < \rho' \\ F(\rho)\widetilde{\Phi}(\rho'), & \rho > \rho' \end{cases} \tag{6.78}$$

with the matrix elements

$$\bar{G}_{ij}(\rho, \rho', E) = \begin{cases} \sum_q \Phi_{iq}(\epsilon_i, \rho_<) F_{jq}(\epsilon_j, \rho'_>) \\ \sum_q F_{iq}(\epsilon_i, \rho_>) \Phi_{jq}(\epsilon_j, \rho'_<) \end{cases} \tag{6.79}$$

meets requirements to be the single-site multichannel GF, namely:

- satisfies the inhomogeneous system of the coupled integro-differential equations corresponding to (6.75) for both variables ρ and ρ', if $\rho \neq \rho'$

$$\begin{aligned} &\left[\frac{\mathrm{d}^2}{\mathrm{d}\rho^2} - \frac{\ell_i(\ell_i + 1)}{\rho^2} + k_i^2 \right] \bar{G}_{ij}(\rho, \rho', E) \\ &- \sum_q \int_{\Omega_n} \widetilde{V}_{iq}(\rho, \rho'')\, \bar{G}_{qj}(\rho'', \rho', E)\, \mathrm{d}\rho'' = \delta(\rho - \rho')\, \delta_{ij} \end{aligned} \tag{6.80}$$

- satisfies the principle of reversibility

$$\bar{G}_{ij}(\rho, \rho', E) = \bar{G}_{ji}(\rho, \rho', E) \tag{6.81}$$

- satisfies the condition of continuity at $\rho = \rho'$.

Next, assuming that the non-local potential introduced here can be approximated by a potential with separable coordinates and applying the general method of the Green's function construction [18], we determine the normalization coefficient in (6.77). By integrating (6.80) over ρ' in a small interval $[\rho - \Delta, \rho + \Delta]$ and using continuity of $\bar{G}_{ij}(\rho, \rho', E)$ we find

$$\lim_{\Delta \to 0} \frac{\mathrm{d}}{\mathrm{d}\rho'} \bar{G}(\rho, \rho', E)\Big|_{\rho-\Delta}^{\rho+\Delta} = 1 . \tag{6.82}$$

The substitution of (6.78) in (6.82) leads to

$$\Phi(\rho)\widetilde{F}'(\rho) - \Phi'(\rho)\widetilde{F}(\rho) = 1 .$$

Using the boundary conditions (6.76) and (6.77), we obtain

$$\sum_{pq} \left\{ \begin{array}{l} \frac{1}{\sqrt{k_i}}(\bar{j}_i\delta_{ip} - \bar{n}_i K_{ip})\frac{1}{\sqrt{k_j}}k_j\bar{h}_j^{+'}(\mathbf{I} - \mathrm{i}\mathbf{K})^{-1}_{jq}A_{qp} \\ \\ -\frac{1}{\sqrt{k_i}}k_i(\bar{j}'_i\delta_{ip} - \bar{n}'_i K_{ip})\frac{1}{\sqrt{k_j}}\bar{h}_j^{+}(\mathbf{I} - \mathrm{i}\mathbf{K})^{-1}_{jq}A_{qp} \end{array} \right\} = \delta_{ij} ,$$

which gives $A_{qq} = -\mathrm{i}$. Non-diagonal coefficients A_{pq} are put to zero, so as the GF (6.79) would go to the free GF when the potential in (6.75) tends to zero. Finally, $A_{pq} = -\mathrm{i}\delta_{pq}$.

Throughout this chapter we use the GF $G_{ij}(\rho, \rho', E) = \bar{G}_{ij}(\rho, \rho', E)/(\rho\rho')$ for which the expansion (6.79) becomes

$$G_{ij}(\rho, \rho', E) = \begin{cases} \sum_q \bar{\Phi}_{iq}(\epsilon_i, \rho_<)\, \bar{F}_{jq}(\epsilon_j, \rho_>) \\ \\ \sum_q \bar{F}_{iq}(\epsilon_i, \rho_>)\, \bar{\Phi}_{jq}(\epsilon_j, \rho_<) \end{cases} \tag{6.83}$$

being expressed in terms of the functions $\bar{\Phi}_{iq} = \Phi_{iq}/\rho$ and $\bar{F}_{jq} = F_{jq}/\rho$ matching to the conditions at $\rho = b_n$

$$\bar{\Phi}_{ij}(\epsilon_i, b_n) = \sqrt{k_i}\left[j_{\ell_i}(k_i b_n)\,\delta_{ij} - n_{\ell_i}(k_i b_n)\,K_{ij}\right] , \tag{6.84}$$

$$\bar{F}_{ij}(\epsilon_i, b_n) = -\mathrm{i}\sqrt{k_i}\,h^{+}_{\ell_i}(k_i b_n)\,(\mathbf{I} - \mathrm{i}\mathbf{K})^{-1}_{ij} . \tag{6.85}$$

In addition, it is expedient to express the irregular solution $\bar{F}$ in terms of the real regular solution $\bar{\Phi}$. Using the boundary conditions (6.84) and (6.85), we obtain

$$\bar{F}_{iq} = -\mathrm{i}\sum_j \bar{\Phi}_{ij}(\mathbf{I} + \mathbf{K}^2)^{-1}_{jq} + \Delta_{iq} , \tag{6.86}$$

where the irregular solution Δ_{iq} satisfies the condition at $\rho = b_n$

$$\Delta_{iq}(\epsilon_i, b_n) = \sum_j \sqrt{k_i}\left[j_{\ell_i}(k_i b_n)\,K_{ij} + n_{\ell_i}(k_i b_n)\,\delta_{ij}\right](\mathbf{I}+\mathbf{K}^2)^{-1}_{jq}\,.$$

Note that function Δ_{iq} is a real part of the complex irregular solution $\bar{F}_{iq}$.

Similarly, one can write relation between R^n_{ij} matching to the boundary condition (6.73) and $\bar{\Phi}_{ij}$ as

$$R^n_{ij}(\epsilon_i, \rho) = \frac{1}{\sqrt{k_j}} \sum_q \bar{\Phi}_{iq}(\epsilon_i, \rho)\,(\mathbf{I} - \mathrm{i}\mathbf{K})^{-1}_{qj}\,. \tag{6.87}$$

6.3.3 Solution of the Dyson Equation for the Full Green's Function

Here we present a scheme of solving the Dyson equation (6.52) for the full GF in the case when strong many-electron interaction takes place only within one of the atomic spheres in the cluster. This means that ΔV in (6.52) is equal to zero value everywhere with the exception of the sphere n.

The general solution of (6.52) can be written in analogy with the single-channel equation (6.20) (Sect. 6.2) as

$$G_{ij}(\mathbf{r}, \mathbf{r}', E) = G^n_{ij}(\mathbf{r}, \mathbf{r}', E) + \sum_{pq} \sqrt{k_p k_q}\, R^n_{ip}(\mathbf{r})\, G_{pq}(E)\, R^n_{jq}(\mathbf{r}')\,, \tag{6.88}$$

where $\mathbf{r}, \mathbf{r}' \in \Omega_n$; the coefficients $G_{pq}(E)$ are to be determined; $G_{pq}(\mathbf{r}, \mathbf{r}', E) = Y_{L_p}(\hat{\mathbf{r}}) G_{pq}(r, r', E) Y_{L_q}(\hat{\mathbf{r}}')$; $R^n_{ip}(\mathbf{r}) = Y_{L_i}(\hat{\mathbf{r}}) R^n_{ip}(r)$.

Let us rewrite (6.32) for the full single-channel GF in matrix form

$${}^0\mathbf{G} = \mathbf{G}_0 + \mathbf{G}_0\,{}^0\mathbf{t}\,{}^0\mathbf{G}\,, \tag{6.89}$$

which in terms of the inverse GF becomes

$${}^0\mathbf{G}^{-1} = \mathbf{G}_0{}^{-1} - {}^0\mathbf{t}\,, \tag{6.90}$$

where ${}^0\mathbf{t}$ is the single-channel t-matrix for the potential in sphere n.

Similarly, one can write for the full multichannel GF

$$\mathbf{G} = \mathbf{G}_0 + \mathbf{G}_0\,\mathbf{t}\,\mathbf{G}\,, \tag{6.91}$$

$$\mathbf{G}^{-1} = \mathbf{G}_0{}^{-1} - \mathbf{t}\,. \tag{6.92}$$

The subtraction (6.90) from (6.92) gives

$$\mathbf{G}^{-1} = {}^{0}\mathbf{G}^{-1} - \Delta\mathbf{t}\,, \tag{6.93}$$

$$\mathbf{G} = {}^{0}\mathbf{G} + {}^{0}\mathbf{G}\,\Delta\mathbf{t}\,\mathbf{G}\,. \tag{6.94}$$

Writing (6.94) in the channel representation

$$G_{ij} = {}^{0}G_{ij} + \sum_{pq} {}^{0}G_{ip}\Delta t_{pq}\,G_{qj} \tag{6.95}$$

and returning to the notations of Sect. 6.2, i.e. ${}^{0}G_{ij}(E) \equiv G^{nn}_{L_iL_j}(\epsilon_i)\delta_{\epsilon_i\epsilon_j}\delta_{\alpha_i\alpha_j}$, where the coefficients $G^{nn}_{L_iL_j}(\epsilon_i)$ satisfy the coupled algebraic equations (6.32), we find

$$G_{ij}(E) = G^{nn}_{L_iL_j}(\epsilon_i)\delta_{\epsilon_i\epsilon_j}\delta_{\alpha_i\alpha_j} + \sum_{pq}\delta_{\epsilon_i\epsilon_p}\delta_{\alpha_i\alpha_p}G^{nn}_{L_iL_p}(\epsilon_i)\Delta t_{pq}(E)\,G_{qj}(E)\,, \tag{6.96}$$

in which Δt_{pq} is determined by (6.70).

6.3.4 Photoabsorption Cross-Section in the MCMS-GF Method

We consider an application of the developed method for calculating $L_{2,3}$-edge absorption cross-section. The expression (6.33) now becomes

$$\begin{aligned}\sigma(\omega_{\boldsymbol q}) &= 4\pi^2\alpha\omega_{\boldsymbol q}\sum_f \left|\left\langle\Psi_f \mid \hat W \mid \Psi_g\right\rangle\right|^2 \delta(E_g + \omega_{\boldsymbol q} - E_f) \\ &= -4\pi\alpha\omega_{\boldsymbol q}\sum_f \left\langle \Psi_g\hat W \left| \Im\left[\frac{\Psi_f\Psi_f^\dagger}{E - E_f + i\eta}\right]\right| \hat W\Psi_g\right\rangle \\ &= -4\pi\alpha\omega_{\boldsymbol q}\left\langle\Psi_g\hat W \mid \Im\left[G^N(E)\right] \mid \hat W\Psi_g\right\rangle\,,\end{aligned} \tag{6.97}$$

where Ψ_g, Ψ_f and E_g, E_f are respectively the many-electron ground and final eigenvectors and energies of N-electron atom in a cluster; $\hat W = \hat{\boldsymbol e}_{\boldsymbol q}\cdot\sum_{i=1}^N \boldsymbol r_i$ denotes the dipole transition operator; $E = E_g + \omega_{\boldsymbol q}$. Note that in the final state the N-electron absorbing atom includes $2p$ hole and electron in continuum s-, d-states. The many-electron full GF of the cluster $G^N(E)$ can be expended in terms of the complete basis set $\Psi_i^{N-1}(\boldsymbol x_1, \ldots, \boldsymbol x_{N-1})$ introduced in (6.53):

$$G^N(\boldsymbol{x}_1, \ldots, \boldsymbol{x}_{N-1}, \boldsymbol{r}; \boldsymbol{x}'_1, \ldots, \boldsymbol{x}'_{N-1}, \boldsymbol{r}'; E)$$

$$= \hat{A}\hat{A}' \sum_{ij} \left\{ \Psi_i^{N-1}(\boldsymbol{x}_1, \ldots, \boldsymbol{x}_{N-1})\, G_{ij}(\boldsymbol{r}, \boldsymbol{r}', E)\, \Psi_j^{*N-1}(\boldsymbol{x}'_1, \ldots, \boldsymbol{x}'_{N-1}) \right\} . \tag{6.98}$$

Here the operator $\hat{A}$ permutates the coordinates $\{\boldsymbol{x}_1, \ldots, \boldsymbol{x}_{N-1}\}$ with $\boldsymbol{r}$ and $\hat{A}'$ acts similarly on the prime variables.

Writing the expansion (6.83) for $G^n_{ij}(\boldsymbol{r}, \boldsymbol{r}', E)$ as

$$G^n_{ij}(\boldsymbol{r}, \boldsymbol{r}', E) = \begin{cases} \sum_q \bar{\Phi}_{iq}(\epsilon_i, \boldsymbol{r}_<)\, \bar{F}_{jq}(\epsilon_j, \boldsymbol{r}_>) \\ \\ \sum_q \bar{F}_{iq}(\epsilon_i, \boldsymbol{r}_>)\, \bar{\Phi}_{jq}(\epsilon_j, \boldsymbol{r}_<) \end{cases} \tag{6.99}$$

and taking into account (6.86), (6.87), we obtain from (6.88)

$$\begin{aligned} G_{ij}(\boldsymbol{r}, \boldsymbol{r}', E) &= \sum_{pq} \bar{\Phi}_{iq}(\epsilon_i, \boldsymbol{r}) \left[\widetilde{G}_{pq}(E) - \mathrm{i}(\mathbf{I} + \mathbf{K}^2)^{-1}_{pq} \right] \bar{\Phi}_{jp}(\epsilon_j, \boldsymbol{r}') \\ &+ \sum_q \bar{\Phi}_{iq}(\epsilon_i, \boldsymbol{r}_<)\, \Delta_{jq}(\epsilon_j, \boldsymbol{r}'_>) , \end{aligned} \tag{6.100}$$

where $\widetilde{G}_{pq}(E)$ denotes

$$\widetilde{G}_{pq}(E) = \sum_{p'q'} (\mathbf{I} - \mathrm{i}\mathbf{K})^{-1}_{pp'}\, G_{p'q'}(E)\, (\mathbf{I} - \mathrm{i}\mathbf{K})^{-1}_{q'q} . \tag{6.101}$$

Putting together (6.98) and (6.100), dropping again index $\boldsymbol{q}$ on $\omega_{\boldsymbol{q}}$, we reduce (6.97) to

$$\sigma(\omega) = 4\pi\alpha\omega\, \Im \sum_{pq} \left\{ M_q^\dagger \left[\mathrm{i}\,(\mathbf{I} + \mathbf{K}^2)^{-1}_{pq} - \widetilde{G}_{pq}(E) \right] M_p - M_p^\dagger \widetilde{M}_q \delta_{pq} \right\} , \tag{6.102}$$

where dipole transition amplitudes M_q and $\widetilde{M}_q$ are respectively defined as

$$M_q(E) = \sum_i \left\langle \hat{A}\Psi_i^{N-1} \bar{\Phi}_{iq}(\epsilon_i) \mid \hat{W}\Psi_g \right\rangle , \tag{6.103}$$

$$\widetilde{M}_q(E) = \sum_i \left\langle \hat{A}\Psi_i^{N-1} \Delta_{iq}(\epsilon_i) \mid \hat{W}\Psi_g \right\rangle . \tag{6.104}$$

The functions $\Phi_{iq}(\epsilon_i, \boldsymbol{r})$ and $\Delta_{iq}(\epsilon_i, \boldsymbol{r})$ are real, therefore the products of the dipole transition amplitudes in (6.102) are real, thus giving

$$\sigma(\omega) = 4\pi\alpha\omega \sum_{ij} \left\{ M_i \left[(\mathbf{I} + \mathbf{K}^2)^{-1}_{ij} - \Im \widetilde{G}_{ij}(E) \right] M_j \right\} . \tag{6.105}$$

However, if one needs the analytical continuation of the GF into complex energy plane, as described in Sect. 6.2, the calculations of irregular solutions of the coupled integro-differential equations (6.75) are necessary. We can choose either real $\Delta_{iq}(\epsilon_i, \boldsymbol{r})$ or complex $\bar{F}_{iq}(\epsilon_i, \boldsymbol{r})$ as an irregular solution. If we use $\Delta_{iq}(\epsilon_i, \boldsymbol{r})$, the expression (6.102) includes the second term calculated at complex energies. Otherwise, if we use functions $\bar{F}_{iq}(\epsilon_i, \boldsymbol{r})$, it is easy to show following to the same line as in the case of $\Delta_{iq}(\epsilon_i, \boldsymbol{r})$ that (6.102) becomes

$$\sigma(\omega) = 4\pi\alpha\omega \, \Im \sum_{pq} \left\{ -M_q^{\dagger} \left[\widetilde{G}_{pq}(E) \right] M_p - M_p^{\dagger} \widetilde{M}_q \delta_{pq} \right\} , \tag{6.106}$$

$$\widetilde{M}_q(E) = \sum_i \left\langle \hat{A} \Psi_i^{N-1} \bar{F}_{iq}(\epsilon_i) \mid \hat{W} \Psi_g \right\rangle . \tag{6.107}$$

The algorithm of the bivariate integration in $M_i^{\dagger} \widetilde{M}_j$ is similar to that developed in Sect. 6.2.

6.4 Outlook

In this chapter we have presented in details two methods based on the real-space Green's function formalism: single-channel (MS-GF) and multichannel (MCMS-GF). Both approaches reduce the integral Dyson equations to systems of linear algebraic equations for the GF matrices in the angular momentum-site index representation. Since the methods rely on the MT approximation, the MCMS-GF method accounts only for correlations between one delocalized electron and electrons localized inside MT-sphere. One of the ways to go beyond this restriction could be the extension of the MCMS-GF equations to the FPMS (Full Potential Multiple Scattering) theory [26]. The MCMS-GF theory exploits the close-coupling formalism, thus combining the MCMS-GF and FPMS methods is a relatively straightforward task. The second serious assumption that the non-local potential considered here can be approximated by a sum of separable potentials, leading to the same equations of the MCMS-GF method, requires a strict justification.

Acknowledgements A.T. would like to acknowledge financial support from the Ministry of Education and Science of the Russian Federation (project 3.5398.2017/8.9). Parts of this work have been funded by European FP7 MSNano network under Grant Agreement No. PIRSES-GA-2012-317554 and COST Action MP 1306 EUSpec.

References

1. K.H. Johnson, J. Chem. Phys. **45**, 3085 (1966)
2. K.H. Johnson, F.C. Smith Jr., Chem. Phys. Lett. **7**, 541 (1970)
3. A.A. Abrikosov, L.P. Gorkov, I.Ye. Dzyaloshinski, *Quantum Field Theoretical Methods in Statistical Physics*, 2nd edn. (Pergamon Press, Oxford, 1965), p. 365
4. R.V. Vedrinskii, A.A. Novakovich, Phys. Met. Metallogr. **39**(1), 7 (1975)
5. C.A. Ashley, S. Doniach, Phys. Rev. B **11**, 1279 (1975)
6. R.V. Vedrinskii, A.A. Novakovich, Phys. Met. Metallogr. **39**(3), 486 (1975)
7. R.V. Vedrinskii, I.I. Gegusin, V.N. Datsyuk, V.L. Kraizman, A.A. Novakovich, Phys. Status Solidi B **111**, 433 (1982)
8. R.V. Vedrinskii, A.A. Novakovich, A.G. Bermus, M. Élango, A. Ausmees, A. Kikas, E. Nommiste, A. Saar, Phys. Solid State **35**(7) (1993)
9. R.V. Vedrinskii, V.L. Kraizman, A.A. Novakovich, V.Sh. Machavariani, J. Phys.: Condens. Matter **4**, 6155 (1992)
10. R.V. Vedrinskii, V.L. Kraizman, A.A. Novakovich, G.Yu. Machavariani, V.Sh. Machavariani, J. Phys.: Condens. Matter **6**, 11045 (1994)
11. J. Kokubun, K. Ishida, D. Cabaret, F. Mauri, R.V. Vedrinskii, V.L. Kraizman, A.A. Novakovich, E.V. Krivitskii, V.E. Dmitrienko, Phys. Rev. B **69**, 245103 (2004)
12. C.R. Natoli, M. Benfatto, S. Doniach, Phys. Rev. B **34**, 4682 (1986)
13. C.R. Natoli, M. Benfatto, C. Brouder, M.F. Ruitz López, D.L. Foulis, Phys. Rev. B **42**, 1944 (1990)
14. R.V. Vedrinskii, I.I. Gegusin, A.I. Taranukhina, in *Abstract of the 6th International Conference on the X-ray absorption fine structure*, York, 5-11 August 1990
15. A. Taranukhina, R. Vedrinskii, Bull. Am. Phys. Soc. **47**(1), 701 (2002)
16. P. Krüger, C.R. Natoli, Phys. Rev. B **70**, 245120 (2004)
17. C.R. Natoli, P. Krüger, K. Hatada, K. Hayakawa, D. Sébilleau, O. Šipr, J. Phys. Condens. Matter **24**, 365501 (2012)
18. G. Arfken, *Mathematical Methods for Physicist*, 2nd edn. (Academic Press, New York, 1970)
19. C.R. Natoli, M. Benfatto, Phys. Rev. B **34**, 4682 (1986)
20. W. Kohn, Phys. Rev. **74**, 1763 (1948)
21. P. Burke, M. Seaton, in *Methods in Computational Physics*, vol. 10, ed. by B. Alder, S. Fernbach, M. Rotenberg (Academic Press, New York, 1969), p. 9
22. F.M.F. de Groot, J.C. Fuggle, B.T. Thole, G.A. Sawatzky, Phys. Rev. B **41**, 928 (1990)
23. E.U. Condon, G.H. Shortley, *The Theory of Atomic Spectra* (Cambridge University Press, Cambridge, 1991)
24. J.R. Taylor, *Scattering theory* (Wiley, 1972)
25. R.G. Newton, *Scattering Theory of Waves and Particles*, 2nd edn. (Springer, Heidelberg, 2008)
26. K. Hatada, K. Hayakawa, M. Benfatto, C.R. Natoli, Phys. Rev. B **76**, 060102R1-4 (2007)

Chapter 7
MXAN and Molecular Dynamics: A New Way to Look to the XANES (X-ray Absorption Near Edge Structure) Energy Region

Maurizio Benfatto, Elisabetta Pace, Nico Sanna, Cristiano Padrin and Giovanni Chillemi

Abstract X-ray Absorption Near Edge Structure (XANES) spectroscopy is a powerful method to obtain local structural and electronic information around a well-defined absorbing site of matter in many possible different conditions. Few years ago we presented a new method, called MXAN, that allows a complete fit of the XANES energy region in terms of well-defined set of structural parameters. MXAN calculates the photo-absorption cross-section using the full multiple scattering theory and, in this way, the analysis can start from the edge without any limitations in the energy range and polarization conditions. In this paper we present with details the MXAN method and new possibilities, coming from the combination of molecular dynamic simulations and MXAN, in the analysis of structural disordered system. Applications to the solvation spheres analysis of Ni, Cu and Cl ions in water are also presented in details.

7.1 Introduction

X-ray absorption spectroscopy (XAS) is a powerful method for obtaining both electronic and structural information on the absorbing atom site of different types of matter, from biological systems to condensed materials. The low energy part of the XAS spectrum, from the rising edge up to a few hundreds of eV, the so called XANES (x-ray absorption near-edge structure) region, is extremely rich of electronic and structural information. Oxidation state, overall symmetry, distances and angles between atomic species around absorbing site [1] can derived from this part of the XAS spectra. In principle, an almost complete quantitative recovery of the geomet-

M. Benfatto (✉) · E. Pace
LNF INFN, Via Enrico Fermi 44, Frascati, Italy
e-mail: maurizio.benfatto@lnf.infn.it

N. Sanna · C. Padrin · G. Chillemi
CINECA, SuperComputing Applications and Innovation Department,
Via dei Tizii, 6, Roma, Italy

D. Sébilleau et al. (eds.), *Multiple Scattering Theory for Spectroscopies*,
Springer Proceedings in Physics 204,
https://doi.org/10.1007/978-3-319-73811-6_7

rical structure within $6 \div 7$Å from the absorber can be achieved from this part of the experimental spectrum and for long time the fit of XANES data has been an aim of users of this technique, especially in the cases of limited k-range experimental data where a standard EXAFS analysis cannot be performed.

A further advantage of using the XANES data lies in the limited effects of the atomic thermal disorder. This can be easily seen by the consideration that any signal associated with the nth multiple-scattering (MS) event can be written as a sinusoidal function whose argument is given by $kR_{\text{tot}} + F(k, \boldsymbol{R}_i, \ldots, \boldsymbol{R}_n)$ where R_{tot} is the total length of the MS path of order n and the F function depends on the three-dimensional geometry of this path. As a consequence, the associated thermal damping factor contains always a term like $\exp(-k^2\sigma^2)$, coming from the R_{tot} part. This is the dominant term and it is almost equal to 1 in the low-energy part of the spectrum, i.e. for small k values [2].

The possibility to perform quantitative XANES analysis to obtain a structural determination of an unknown compound can be relevant in many scientific fields, like extra-dilute systems, biological systems where the low S/N ratio and the weak scattering power of the light elements limits the k-range of the available experimental data, materials under extreme conditions, and recently the analysis of time-depended data coming from metastable systems living few pico-seconds or even less.

However, the quantitative analysis of the XANES spectra presents some difficulties mainly due to the theoretical approximations needed in the treatment of the potential and the more time-consuming algorithms to calculate the absorbing cross section in the framework of the full multiple-scattering approach [3, 4]. For these reasons, the *XANES analysis* is still considered a *qualitative* technique, used as a help for standard EXAFS studies.

Several years ago Benfatto and Della Longa proposed a fitting procedure, MXAN (Minuit XANES) based on a full MS theory [5], which is able to extract local quantitative structural information around the absorbing atom from experimental XANES data. Since then, the MXAN method has been successfully used for analyses of many known and unknown systems, yielding structural geometries and metrics comparable to x-ray diffraction and/or EXAFS results [6, 7].

In this paper we present a review of the MXAN method describing also the new possibilities available in the latest version of the program. In particular we will show as the link between MXAN and Molecular Dynamic (MD) simulations allows to handle structural disordered and dynamical systems. Some applications will be presented and discussed in details.

7.2 The MXAN Method

The method is based on the comparison between experimental data and many theoretical calculations performed by varying selected structural parameters starting from a well-defined initial geometrical configuration around the absorber. The calculation of XANES spectra are performed within the so-called full MS approach, i.e. the inverse of the scattering path operator is computed exactly, avoiding any a priori selection of the relevant MS paths [3]. The fit procedure is performed in the energy space without the use of any Fourier transforms algorithm; polarized spectra can be easily analyzed because the calculation is performed using the full MS approach [5].

The optimization in the space of the parameters is achieved using the CERN-library MINUIT routines minimizing the square residual function

$$S^2 = m\frac{\sum_{i=1}^{n} w_i\left[(y_i^{\text{th}} - y_i^{\text{exp}})\varepsilon_i^{-1}\right]^2}{\sum_{i=1}^{n} w_i}, \tag{7.1}$$

where m is the number of the independent parameters, n is the number of experimental data points, y_i^{th} and y_i^{exp} are the theoretical and experimental values of the absorption coefficient, ε_i are the individual errors in the experimental data set and w_i is a statistical weight, this last can be chosen using a suitable flag in the MXAN program. For $w_i = 1$, the square residual function S^2 becomes the statistical χ^2 function. A typical fit involves an experimental energy range of about $150 \div 200$ eV from the rising edge, applications to several test cases indicates that the best-fit solution is quite stable and independent from the starting conditions. The achievement of the best-fit condition involves the calculation of many absorption coefficients related to several hundred different geometrical configurations. This is done in a reasonable time with the modern computers.

The MXAN method is based on the standard MS theoretical approach within the muffin-tin (MT) approximation for the shape of the potential and the so-called *extended continuum* scheme to calculate both the continuum and the bound part of the XAS spectrum. It also uses the concept of complex optical potential based on the local density approximation of the self-energy of the excited photoelectron [3]. The total charge density needed to calculate the whole potential is derived by superimposing atomic self-consistent Hartree-Fock charges derived using neutral or non- neutral atoms.

In the MT approximation it is necessary to define the radii of the spheres surrounding all the atoms used in the calculation; they are chosen according to the Norman criterion, with some percentage of overlap between the MT spheres. The potential is recalculated at each step of the minimization procedure keeping fix the overlapping factor. This parameter controls all the MT radii and it can be considered as a free parameter of the theory. At the same time there is the need to define the constant interstitial potential, this can be calculated by averaging the potential over the interstitial volume defined as the difference between the volume of an *outer sphere* (centered on the absorbing atom and encompassing the cluster) and the sum of the volumes of each individual atomic spheres. However in the *extended continuum* scheme the outer sphere is not used and the interstitial potential can also be considered as a free parameter of the theory. Both overlapping factor and the interstitial potential can be optimized during the fit procedure, and in the last version of the program, they are inside the structural loop in order to minimize the computer time and to calculate the statistical correlations with the geometrical parameters. It turns out they are small in most of the cases with a very weak influence in the structural determination. Clearly the overcoming of the MT approximation [8, 9] and the use of a self-consistent field (SCF) potential is the best way to eliminate the arbitrariness that the use of such free parameters introduces in the calculation. However the introduction of the non-MT corrections and the use of SCF potential in a fitting procedure where the geometrical structure changes at each step of computation is complicated and quite time consuming. There is also the risk to stabilize wrong electronic configuration when the geometry is still far from the reality with the possibility to increase the finding of false minima.

On the other hand the optimization of both interstitial potential and MT radii can be a way to mimic the non-MT corrections and the use of a SCF potential for the whole cluster. This is based on the following theoretical considerations: the scattering atomic t-matrices and non-MT corrections depend on the charge density, but they also depend on the MT radii and the interstitial potential via the Wronskians of the theory calculated at the boundary of the MT spheres for the given interstitial potential; it is also possible to write a MS theory without the MT approximation with the same formal structure of the usual MS theory by adding some suitable quantities to scattering matrices and propagators. In fact the total photo-absorption cross section can be written as $\sigma_t \approx \Im(\mathbf{T}+\mathbf{H})^{-1}$ where now $\mathbf{T} = (\mathbf{T}_a)^{-1} + \mathbf{\Delta T}$ and $\mathbf{H} = \mathbf{H}_{\mathrm{MT}} + \mathbf{\Delta H}$. Here $\mathbf{T}_a$ and $\mathbf{H}_{\mathrm{MT}}$ are the t-matrices and the free electron propagators between the atomic sites calculated within the MT approximation. $\mathbf{\Delta T}$ and $\mathbf{\Delta H}$ are the non-MT corrections proportional to the interstitial volume. Indicating with $\mathbf{\Delta} = \mathbf{\Delta T} + \mathbf{\Delta H}$ we can expand the total cross section σ_t in series using $\mathbf{\Delta}$ as parameter:

$$\sigma_t \approx \Im \left\{ \sum_{n=0} (-1)^n \left[(\mathbf{T}_a^{-1} + \mathbf{H}_{\mathrm{MT}})^{-1} \mathbf{\Delta} \right]^n (\mathbf{T}_a^{-1} + \boldsymbol{H}_{\mathrm{MT}})^{-1}) \right\} . \tag{7.2}$$

In this way σ_t becomes $\sigma_t \approx \sigma_{\mathrm{MT}} + \mathrm{corr}(E; V_{\mathrm{int}})$, i.e. the total non-MT photo-absorption cross section can be written as the sum of the total photo-absorption written in the MT approximation and corrections that decrease with the energy and depends on the potential in the interstitial volume [10]. So a judicious choice of MT radii and the interstitial potential can account for the influence of such non-MT corrections and the use of SCF potential or, at least, minimize errors due to the MT approximation, giving a good agreement between theory and experiment and an accurate structural recovery [11].

The self-energy is calculated in the framework of the Hedin-Lundqvist (HL) scheme [12]. The use of the full complex HL potential introduces, in most of the cases, a relevant over damping at low-energies especially in the case of covalent molecular systems. For this reason we have developed in MXAN a phenomenological approach to calculate the inelastic losses based on the convolution of the theoretical spectrum, calculated by using only the real part of the HL potential, with a suitable Lorentzian

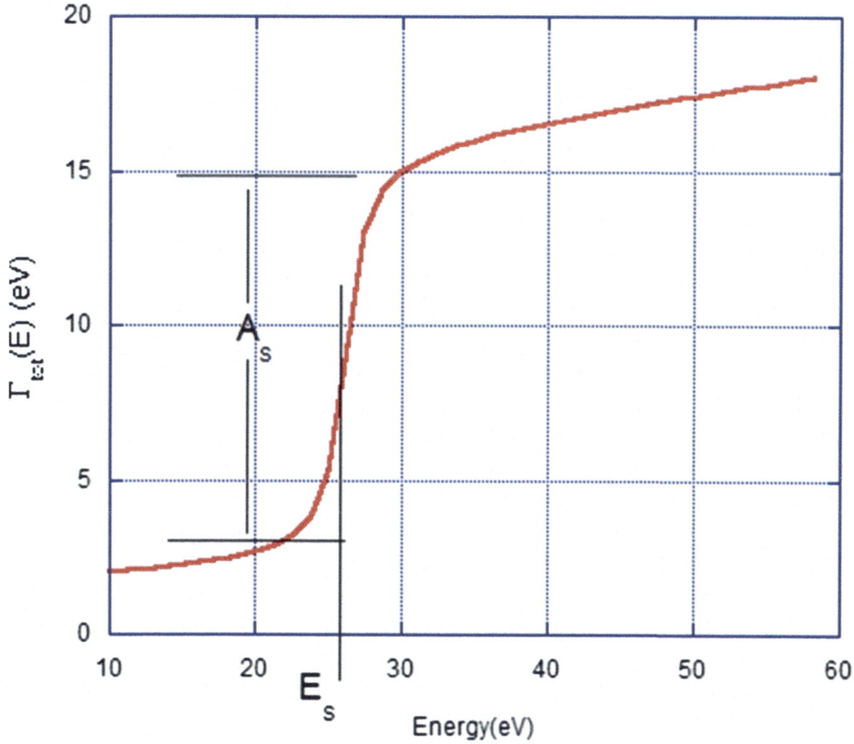

Fig. 7.1 Typical behavior of the total energy dependent width of the Lorentzian function used in the phenomenological damping

function having an energy dependent width of the form $\Gamma_{\text{tot}}(E) = \Gamma_c + \Gamma_{\text{mfp}}(E)$ where Γ_c is the core-hole width while the energy dependent term represents all the intrinsic and extrinsic inelastic processes. All these quantities are expressed in eV. The $\Gamma_{\text{mfp}}(E)$ function is zero below an onset energy E_s, that corresponds in the extended systems to the plasmon excitation energy, and begin to increase from a value A_s following the universal form of the mean free-path in solids [13]. Both the onset energy E_s and the jump A_s are introduced in the $\Gamma_{\text{tot}}(E)$ function via an arctangent functional form to avoid discontinuities and to simulate the electron-hole pair excitations. In Fig. 7.1 we show a typical behavior of the $\Gamma_{\text{tot}}(E)$ function.

Their numerical values are derived at each step of the fit on the basis of a Monte Carlo thermal annealing procedure. In this way the agreement between theory and experimental data is quite good in the whole energy range, starting from the edge.

This phenomenological approach can be justified on the basis of the multi-channel MS theory [14]. In the sudden limit of the theory the net absorption is given by a sum over all the possible excited states of the $(N-1)$-electron system. Assuming that all the electronic channels are near in energy, the total absorption is given by the convolution of the one-particle spectrum, calculated with the full relaxed potential, with a spectral function $A(E)$ that represents the weight of the other electronic channels:

$$\mu_{\text{tot}} = \sum_n \mu_n \underset{\Delta E \to 0}{\to} \int \mu(E - E')A(E')\mathrm{d}E' , \tag{7.3}$$

with the ansatz that the spectral function $A(E)$ is well approximated by the Lorentzian function with the energy dependent width previously defined. μ is the absorption coefficient and E is the energy. It is also possible to demonstrate that in the case of dominance of only one electron channel, the elastic one, the corresponding Green's function G_0 obeys to a Dyson-like with a suitable complex self-energy $\sum^{\text{opt}}(\boldsymbol{r}, \boldsymbol{r}'; E)$. Within the approximations of locality and of homogeneous systems, this is equivalent to a *real* calculation convoluted with a Lorentzian function having an energy depended width. Therefore $\Gamma_{\text{tot}}(E)$ is characterized by parameters which have a clear physical meaning and they are not free to assume any value but are forced within a well-defined interval. To clarify this point we show in Fig. 7.2 the comparison of the mean free path coming from the complex part of the HL potential with the one coming from the use of the phenomenological damping, the core-hole width is included in the calculation. The two curves have the same shape and are almost the same beyond 40 eV from the edge. On the other hand there is a sizeable difference in the in the first ten of eV where the damping coming from the HL complex potential is too strong.

Of course, when contributions from one or more of these excited electronic states are of the same strength of the fully relaxed channel, they must be considered explicitly in the calculation, as in the case of the iron K-edge of Fe^{3+} in water solution [15]. It is also possible to note that the shape of $\Gamma_{\text{mfp}}(E)$ function is similar to the inverse lifetime of the quasi-particle Γ_k derived in GW_0 scheme, without the sharp onset at the plasmon excitation energy to account for the interference between the intrinsic and extrinsic inelastic losses [16].

The experimental resolution is taken into account by a further convolution with an energy independent Gaussian function. In total MXAN has four parameters that completely control the damping procedure. The user can choose how to open them or keep them fixed to some values. Many applications to test cases and unknown systems have shown the reliability of this type of phenomenological approach.

Recently the MXAN code has been modified to allow fitting difference spectrum, i.e. signal coming from the difference of two XANES data. So it is possible to analyze differential transient XAS data, which consist of the difference between the transmission spectra of a unexcited and a laser excited sample. This approach greatly increases the sensitivity of the data to small changes, and at the same time, reduces the influence of systematic errors in the experiment and in the calculation. In this way it is possible to analyze in details transient data coming from samples in states living few picoseconds or even less [17].

The problem of an ab initio quantitative treatment of geometrical disorder in the XANES energy region is still an open question. The only attempt to solve this problem has been made by the use of the so-called *augmented space formalism* [18]. This approach needs to define the distribution functions of the stochastic variables,

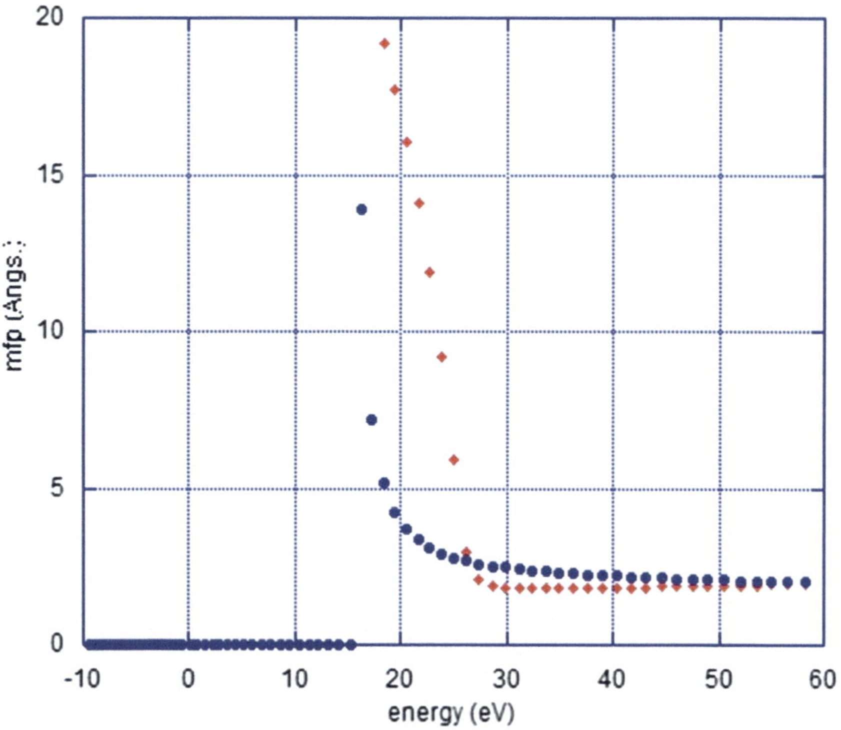

Fig. 7.2 Comparison between the mean free path calculated by the complex part of the HL potential (*blue circles*) and the one coming by the phenomenological approach (*red points*). These values come from the best fit of Ni *K*-edge in water solution

like for example bond distances, but severe difficulties exist to write these functions for structural disorder different from the Gaussian one. A possibility to overcome such difficulties is to combine molecular dynamic (MD) and MXAN simulations [19]. The proper configurational average spectrum is obtained averaging thousands of spectra generated from distinct MD snapshots. Each snapshot is used to generate the XANES signal associated with the corresponding instantaneous geometry, and the averaged theoretical spectrum is obtained by summing all the instantaneous spectra and dividing by the total number of used MD snapshots. Typically each snapshot is taken every 50 fs starting from the time at which the system is supposed to be equilibrated. The important question of what is the right sampling length to have a statistically significant average is solved by imposing that the residual function, defined as the difference between the incremental N and $(N-1)$ averaged spectra is less than 10^{-5}.

This *dynamic MXAN* analysis (hereafter called D-MXAN) is very sensitive to the conformational sampling as demonstrated by its applications to several different systems [20]. In the next paragraph we present a general overview of the classical

MD calculations while in the last we will show the D-MXAN analysis of the XANES data at the K-edges of Ni, Cu and Cl ions dissolved in water.

7.3 Molecular Dynamics Background

Classical Molecular Dynamics (MD) is a well-established method applied to the study of inorganic ad biomolecular molecules in solution [21]. It is based on the numerical integration of $\boldsymbol{f}_i = m_i \boldsymbol{a}_i$ Newton's equation of motion where $\boldsymbol{f}_i$ is the force vector of atom i, m_i is its mass and $\boldsymbol{a}_i$ is the acceleration on atom i. The interacting forces among the atoms in the system are obtained by means of an empirical conservative potential through the standard relation $\boldsymbol{f}_i = -\partial V/\partial \boldsymbol{r}_i$ where $\boldsymbol{r}_i$ is the instantaneous position vector of atom i. A typical form for V, used for biological macromolecules or polymers is

$$
\begin{aligned}
V(\boldsymbol{r}_1, \boldsymbol{r}_2, \ldots, \boldsymbol{r}_n) &= \sum_{\text{bond}} \frac{1}{2} k_b (b_n - b_0)^2 + \sum_{\text{angle}} \frac{1}{2} k_{\theta_0} (\theta_n - \theta_0)^2 + \\
&+ \sum_{\text{improper-dihedral}} \frac{1}{2} k_{\xi_0} (\xi_n - \xi_0)^2 + \sum_{\text{dihedral}} 1 + \cos(m_n \phi_n - \delta_n)^2 + \\
&+ \sum_{\text{nonbonded-pairs}(ij)} \left[\left(\frac{C_{ij}^{(12))}}{r_{ij}^{12}} - \frac{C_{ij}^{(6))}}{r_{ij}^{6}} \right) + \frac{1}{4\pi\varepsilon_0} \frac{q_i q_j}{\varepsilon_r r_{ij}} \right],
\end{aligned}
\tag{7.4}
$$

where the first four equation components describe the *bonded interactions* and the last two the *non-bonded interactions*. The non-bonded component of the potential are the sum of the Lennard-Jones (L–J) potential that describes the van der Waals interactions and the Coulombian potential for the electrostatic interactions. $\boldsymbol{r}_{ij}$ is the distance between atoms i, j; C_6 is the constant in the term describing the dispersion attractive force between atoms; C_{12} is the term that describes interatomic electron cloud repulsion; q_i and q_j are the partial atomic charges for atoms i and j; ε_0 is the vacuum *permittivity*; ε_r is the relative dielectric constant. The interaction between two (stretching), three (bending) and four (planar or tetrahedral force) bonded atoms are described by the first three components of the *bonded potential*, all using an Hookean potential in which b_0, θ_0 and ζ_0 are the equilibrium values while b_n, θ_n and ζ_n are the corresponding instantaneous values at a specific simulation time. The last component of the bonded interaction describe the dihedral angle between four atom in a peptide chain and it is periodic component of the potential, with ϕ the angle, δ the phase and m the number of peaks in a full rotation. The simple two-body potential of equation (7.4) is not accurate enough for the treatment of the Coulombian interaction and corrections based on the Particle-Mesh Ewald method is usually used for the treatment of the long-range electrostatic interactions [22].

From the empirical potential V we can calculate the additive force on atom i due to the interaction with all the other system atoms and then its acceleration. Newton's equations of motion are numerically integrated step by step using Finite Difference methods. These methods use the information available at time t to predict the system's coordinates r and velocities v at a time $t + \delta t$, where δt is a short time interval and through the Taylor expansion of the position at time $t + \delta t$:

$$r(t + \delta t) = r(t) + v(t)\delta t + \frac{1}{2}a(t)\delta t^2 + \cdots . \tag{7.5}$$

Several algorithms can be found in the literature for practical applications [23]. The length of the time step must be small compared to the period of the highest frequency motions being simulated and for the study of ionic solutions, the typical time step is one or two femtosecond.

The equations described for MD simulations are appropriate for the micro-canonical NVE ensemble in which the system loses/gains no net total energy (potential plus kinetic) during the simulation. More realistic simulations, however, are carried out under constant Temperature (T) and/or Pressure (P), i.e. in canonical NVT ensemble or isothermal-isobaric NPT. This is obtained by applying algorithms that model external temperature and pressure baths [24]. A recent and accurate algorithm for the temperature bath is the velocity-rescaling method [25].

The D-MXAN analysis of ionic solutions used three different approaches for the definition of the van der Waals interaction potential energy function describing the interaction between one ion and the solvent molecules:

1. the *pure pair additive potential*, in which the simple L–J potential of equation (7.4) is used;
2. the *effective two body potential*, in which an ad hoc and much more complex function is used for the van der Waals interactions between ion and solvent molecules;
3. the *molecular models*, in which a force field is defined for the bonded $M(H_2O)_n^{n+}$ complex. This approach is necessary when quantum mechanics effects between the ion and the nearest solvent molecules are impossible to be described by any two body potential.

In the following we describe the D-MXAN application for each of these approaches, namely Cl^-, Ni^{2+} and Cu^{2+} ions in waters solution. All the molecular dynamic simulation are done with Gromacs package, version 4.6.5 [26].

7.4 D-MXAN of Chlorine in Water

The solvation sphere studies of ions in water are important by themselves and impact many other fields from chemistry to biology. In particular there is a great interest in the solvation sphere studies of halides, mainly chlorine, iodide and bromide, as many recent reviews indicate. Here we present an application of the D-MXAN method to the study of the solvation sphere of Cl in water solution via the analysis of the Cl K-edge. A more extend study is presented in [27] where experimental and analysis details can be found. This study indicates that chlorine organize the surrounding water molecules into two discrete structurally coherent but axially asymmetric shells.

The D-MXAN method is very sensitive to the conformational sampling. We show in Fig. 7.3a the comparison between the experimental data (blue line) and three calculations obtained with three different L–J parameters, L_E, M_E, and H_E for the SPC/E water model [28], developed by Reif and Hünenberger to reproduce single ion hydration free energies [29].

Only the simulation with the L_E L–J parameters is able to reproduce the spectrum in the whole energy range, few discrepancies remain at the rising edge where the two peaks structure is not well reproduced. This finding is in agreement with the EXAFS analysis of [30]. The MXAN R_{sq} error functions corresponding to the MD simulations using the L_E, M_E and H_E parameters is 16, 23, and 107, respectively. It is worth noting that the C_6 parameter is unchanged in the L_E, M_E and H_E choices. Thus, the only difference between these simulations is in the magnitude of the L–J C_{12} interatomic repulsive parameter that has the value of 126.2, 98.5 and 78.3 (10^{-7} kJ $mol^{-1} nm^{12}$) in the L_E, M_E and H_E parameter, respectively [29]. The D-MXAN analysis can observe definitive differences among the XANES theoretical spectra calculated starting from single MD frames. In panel B of Fig. 7.3 we show the comparison of the theoretical XANES spectrum of Cl obtained from the MD average (red line) and several spectra associated with individual MD configurations (black lines). It is interesting to note the big spread of the calculations associated with the individual MD snapshots indicating a strong variability of the water molecule position around the Cl ion.

Figure 7.4 shows the radial distribution functions and the coordination numbers value for the L_E, M_E and H_E L–J potentials. In Table 7.1 are reported the positions of the maxima of the Cl–O pair correlation function $g_{\text{Cl-O}}$ for the first and second hydration shell.

The first shell maximum of the L_E parameter is in very good agreement with the structural finding of the static determination while the second shell maximum is shifted towards higher values respect with one find in the MXAN structural fit. See Table VI in [27] for more details.

As previously noted there are some problems in the edge region where the first feature lacks of intensity near 3 eV in the energy scale of Fig. 7.3. This is the energy region where previous MXAN analysis indicated the strong influence of the second hydration shell. In order to obtain deeper insight into the second solvation shell geometry, we have performed several MD simulations testing other L–J parameters and MD conditions. In particular we have obtained a better result using the SPC/E water model with L_E parameter and the Encad-shift option for Coulomb potential and ($R_{sq} = 14.3$). A finer refinement has further obtained selecting the MD frames that have the R_{th}, defined as

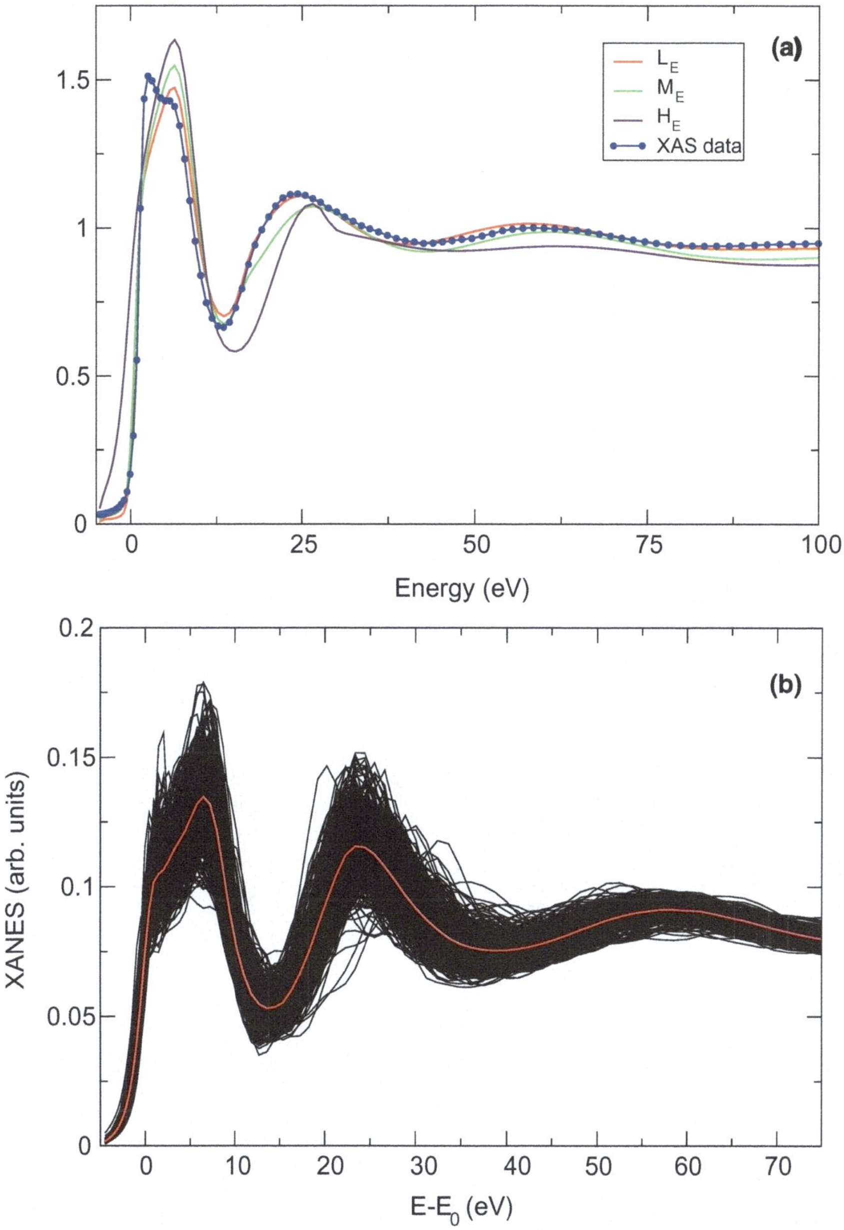

Fig. 7.3 **a** Comparison between the experimental XANES data at Cl *K*-edge and three calculations performed with the D-MXAN method for three different L–J set of parameters, **b** Comparison between the average theoretical XANES data (*red line*) with individual theoretical calculations associated with each configuration

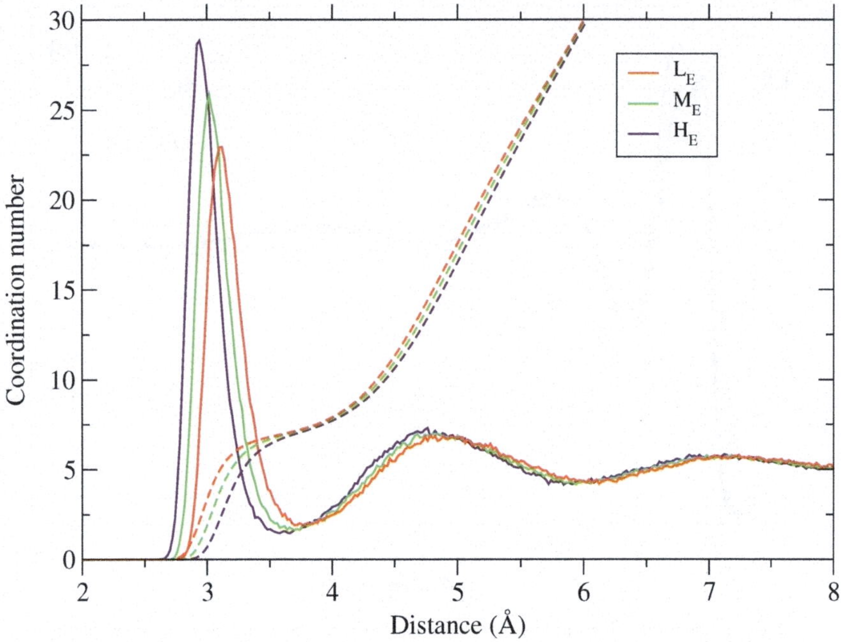

Fig. 7.4 Radial distribution functions and coordination number values for the three potentials used in the D-MXAN analysis

Table 7.1 Positions of the maxima (in Å) of $g_{\text{Cl-O}}$ in the first and second hydration shell

L–J parameters	First shell	Second shell
L_E	3.11	4.93
M_E	3.02	4.82
H_E	2.94	4.76

$$R_{\text{th}} = \frac{\sqrt{\sum_{i=1}^{m} (\sigma_i^{\text{MD}} - \sigma_i^{\text{th}})^2}}{m}, \tag{7.6}$$

less than 7×10^{-4}, where m is the total number of theoretical energy points, $\sigma_i^{\text{MD}}$is the theoretical cross section at the i-energy point for a given MD snapshot and σ_i^{th} is the theoretical cross section at the same energy point obtained using the geometrical structure at the static best fit. This procedure produces a strong increase of the agreement between theory and experiment with the R_{sq} values that goes down to 7.1. The EXAFS parts is still well reproduced and we begin to have the two peaks structure with the right intensity in the very low energy region. See panel a of Fig. 7.5.

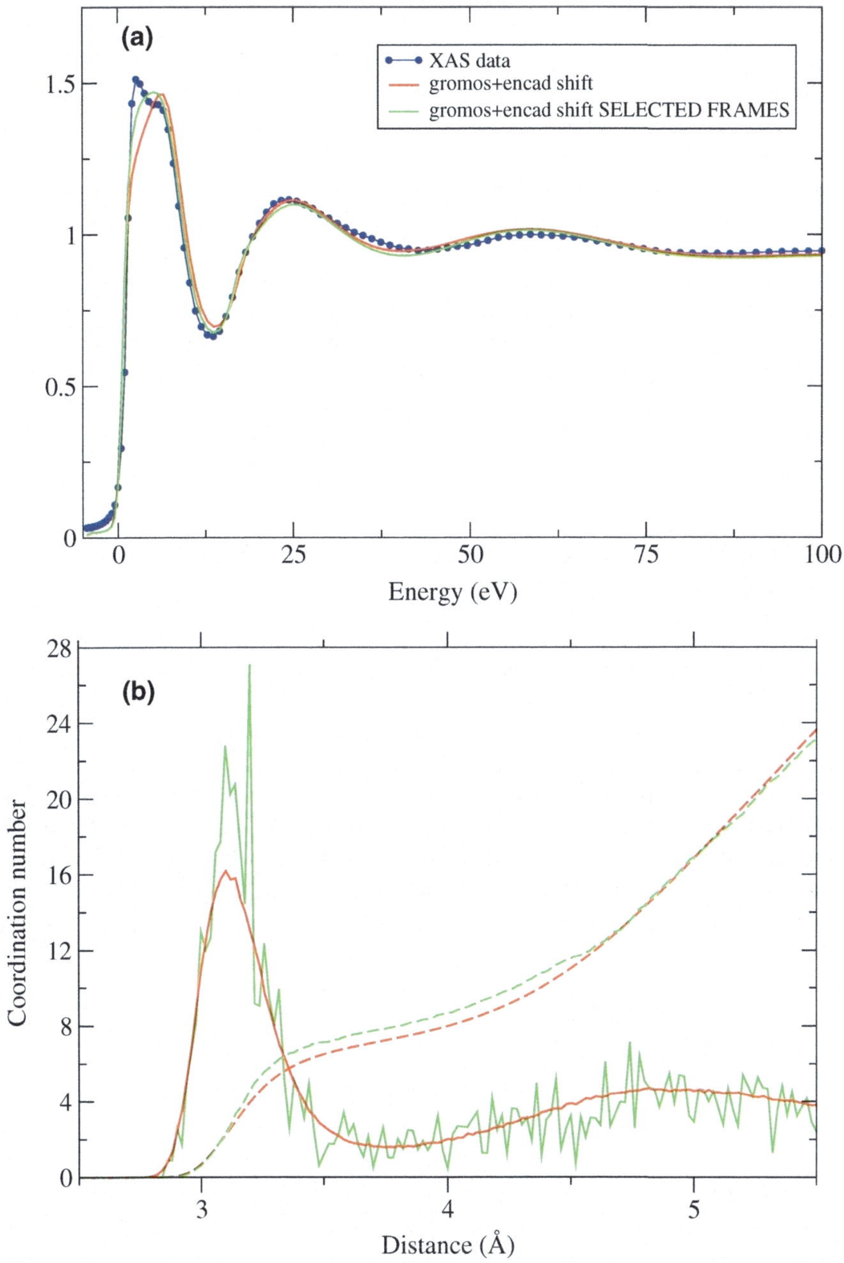

Fig. 7.5 **a** K-edge XAS spectrum of dissolved chloride is reported in *blue points*. The *red line* is the XAS spectrum calculated with LE and Encad-shift options while the *green line* is the one obtained with the selected frames procedure, **b** Comparison between the Cl–O radial distribution function for the first 15 water molecules obtained from the previous MD conditions. The coordination number N is reported for the two MD conditions in *dashed lines*

To better understand the reason of such decrease we report in Fig. 7.5b the comparison between the two $g_{\text{Cl-O}}(r)$ obtained with the two options previously described. This plot clearly shows that the better reproduction of the experimental XANES signal is due to a slightly more compressed hydration structure in the region between 3.1 and 4.6 Å, while the structure is identical before and after this distance range. The discrepancy between the value of the maximum found with the L_E MD and the one found with the static structural fit is also a clear indication of the need of a more compressed structure beyond the first hydration shell. However, it is clear that this slight more compressed structure is not enough to have a reproduction of the XANES data of the same quality obtained with the static fit procedure. A rough comparison with the number obtained in the static fit indicates the need of a further compression of the second hydration shell. This effect is not easy to obtain with a two-body classical potential, since an alteration of the Cl-water or water-water interaction parameters would change the structure of the first hydration shell, as well.

7.5 D-MXAN of Nickel in Water - the Case of the Effective Two Body Potential

A method to include averaged many-body terms in an effective two-body classical potential has been developed by us for the study of Zn(II), Ni(II) and Co(II) in aqueous solution [31] and applied to other aqueous [32, 33] and non-aqueous ionic solutions [34]. The three steps of the computational procedure consist of (i) generation of the ion-solvent ab initio potential energy surface (PES) with the inclusion of the averaged many body ion-water effects by means of the polarizable continuum method [35]; (ii) fitting of the effective two-body potentials from the *ab initio* scans (iii) inclusion of the effective two-body potential in the MD code and system simulation.

A typical form of the effective two-body potential for the interaction of one ion with the aqueous solvent, is:

$$V = \frac{q_{\text{M}} q_{\text{O}}}{\varepsilon_r r_{\text{MO}}} + \frac{A_{\text{O}}}{r_{\text{MO}}^4} + \frac{B_{\text{O}}}{r_{\text{MO}}^6} + \frac{C_{\text{O}}}{r_{\text{MO}}^8} + \frac{D_{\text{O}}}{r_{\text{MO}}^{12}} + E_{\text{O}} e^{\frac{-F_{\text{O}}}{r_{\text{MO}}}} + \sum_{\text{MH}=\text{MH1},\text{MH2}} \frac{q_{\text{M}} q_{\text{H}}}{\varepsilon_r r_{\text{MH}}} + \frac{A_{\text{H}}}{r_{\text{MH}}^4} + \frac{B_{\text{H}}}{r_{\text{MH}}^6} + \frac{C_{\text{H}}}{r_{\text{MH}}^8} + \frac{D_{\text{H}}}{r_{\text{MH}}^{12}} \tag{7.7}$$

Therefore, while the electrostatic interaction is modelled with a Coulombic potential, as in (7.4), six parameters are used to describe the interaction between the ion (M) and the oxygen atoms of the solvent (O) and four parameters for the interaction between M and the hydrogen atoms (H). Fitting of the ab initio PES in the previous equation produce the parameters reported in Table 7.2.

The potential with the parameters of Table 7.2 were included in a modified version of the Gromacs MD package [36] and a system composed by one Ni^{2+} ion and 3270 SPC/E water molecules [28] were simulated for 30 nanoseconds with a time step of

Table 7.2 Ion-water interaction parameters and corresponding standard deviations of the equation in this paragraph

	Parameter	St dev
A_O	$-3.861\ 10^{-1}$	$9.3\ 10^{-2}$
B_O	$6.985\ 10^{-2}$	$9.0\ 10^{-3}$
C_O	$-8.618\ 10^{-4}$	$1.2\ 10^{-4}$
D_O	$3.538\ 10^{-8}$	$5.7\ 10^{-9}$
E_O	$-6.686\ 10^{+4}$	$1.5\ 10^{+4}$
F_O	24.077	$7.1\ 10^{-1}$
A_H	$1.119\ 10^{-1}$	$4.3\ 10^{-2}$
B_H	$-8.688\ 10^{-4}$	$2.3\ 10^{-3}$
C_H	$-1.677\ 10^{-5}$	$3.7\ 10^{-5}$
D_H	$9.748\ 10^{-10}$	$2.8\ 10^{-9}$

1 femtosecond. This method maintains a good description of the many-body interactions, while its computing efficiency allows the investigation of conformational events in the nanosecond time scale, at variance from quantum mechanics (QM) methods.

Panel a of Fig. 7.6 shows the radial distribution for Ni–O and Ni–H (full line, black and red colors, respectively) with their respective integration numbers (dashed lines). The MD simulation perfectly reproduce the Ni(II) strong interaction with the six water molecules constituting the first hydration shell and their strong orientation, as shown by the first peak Ni–H radial distribution function. The experimental value of the residence times of water in the first hydration shell is in the range $10^{-5} - 10^{-7}$ s [37], and we observe no exchange of water molecules between first and second hydration shells during the 30 ns long MD simulation. Panel b of Fig. 7.6 shows the comparison between the experimental XANES data, blue points, and the D-MXAN calculation using both first and second hydration shells. The agreement is quite good in the whole energy range, only small discrepancies are in the 20–30 eV energy range in line with what was found in the static best fit analysis.

7.6 D-MXAN of Copper in Water

Copper is probably the most elusive transition metal ion in water with a lot of different experimental results, highlighting the prevalence of five- or six-fold hydration structure [38, 39]. All studies agree on the stable presence of four equatorial water ligands, which exhibit a uniform Cu–O_w distance of about 1.96 Å, while there are several difficulties in detecting the number and even presence of the axial water ligands. Initial MXAN studies of the K-edge of copper ion dissolved in water were unambiguous in favouring of a axially elongated square pyramid, $[Cu(H_2O)_5]^{2+}$ (Cu–O_{ax} = 2.35Å)

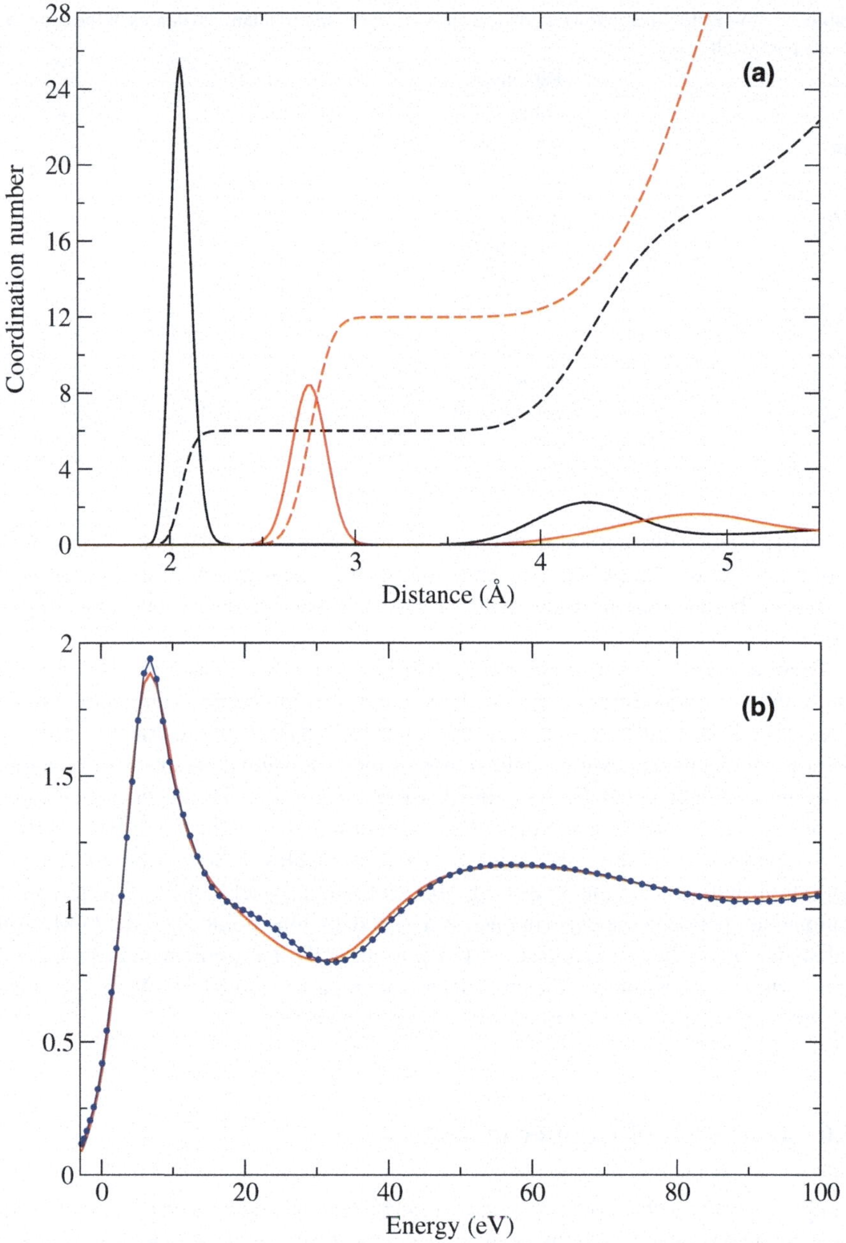

Fig. 7.6 **a** radial distribution function for Ni–O (*black full line*) and Ni–H (*red full line*) ion pair. The respective integration numbers are reported as *dashed lines*, **b** comparison between the experimental XANES data at the Ni *K*-edge (*blue points*) and the D-MXAN calculation (*red full line*)

as the dominant average structure in liquid aqueous solution [40]. More recently a detailed EXAFS and MXAN analysis of a high- resolution Cu K-edge XAS spectrum have been made to further probe the structure of $[Cu(aq)]^{2+}$ in liquid solution [41]. Both analysis converge to a non-centrosymmetric square pyramidal $[Cu(H_2O)_5]^{2+}$ inner coordination sphere with a distinct second shells at ~3.6 Å (EXAFS) or ~3.8 Å (MXAN). However, MXAN analysis revealed that $[Cu(aq)]^{2+}$ could be dimorphic, especially for distances greater than 2.4 Å. A new structure formed by $[Cu(H_2O)_5]^{2+}$ includes an associated but non-bonding axial 2.9 Å water, and produce fits comparable with the pure square pyramidal geometry; including both geometries the quality of the fit further improves. The two structural arrangements are of approximately equal probability in liquid solution. To go deeper in the details of the axial waters dynamic we have performed a D-MXAN analysis of the XAS spectrum [20] starting from the experimental finding that the equatorial geometry of Cu in water solution is formed by the $[Cu(H_2O)_4]^{2+}$ complex with structural parameters coming from the EXAFS and MXAN analysis. A suitable force field for the $[Cu(H_2O)_4]^{2+}$ complex has been built using a standard QM methodology, imposing only the Cu−O distance and the planarity of the $Cu{-}O_4$ atoms with hydrogen atoms free to rotate. This model implicitly presumes that the interaction of the core $[Cu(H_2O)_4]^{2+}$ hydrate with the bulk water molecules is entirely electrostatic [40]. This approach is computationally efficient, and allowed us to carry out a 100-ns long classic MD simulation of this hydrate model in bulk water thus exploring the evolution of the dynamical and structural parameters on a time-scale not accessible to other computational techniques. For example, the present simulation is more than three orders of magnitude longer than that described in Pasquarello (18 ps) [38] or Moin (50 ps) [39]. Details of the experimental solution and the MXAN calculation can be found elsewhere [41]. Two different water models interaction were tested: SPC/E [28] and TIP5P [42] using the same model for equatorial bulk waters. Atomic charges were obtained by means of RESP fitting procedure [43] from the QM calculation of the $[Cu(H_2O)_4]^{2+}$ complex at the Density Functional Theory level (B3LYP functional, 6-311++G** basis set). The atomic charges on the equatorial water obtained by this procedure were employed in the MD simulations accordingly to the model used, i.e. redistributing the charges on the oxygen and hydrogen atoms (SPC/E) or on the hydrogen atoms and the oxygen lone pairs (TIP5P). The obtained non-bonded and bonded parameters are reported in Tables 7.3 and 7.4, respectively.

Table 7.3 Non bonded parameters for the $[Cu(H_2O)_4]^{2+}$ complex. In the TIP5P model the oxygen atom charge is split between the dummy atoms and the oxygen atom charge is put to zero

	Charge (a.u.)	Sigma (nm)	Epsilon (kJ/mol)
Cu	1.3424	1.917748 e^{-01}	5.46273 e^{-01}
O	−0.8128	–	–
H	0.4886	–	–

Table 7.4 Bonded parameters for the $[Cu(H_2O)_4]^{2+}$ complex

Bonds	b_0 (nm)	k_b (kJ mol^{-1})
Cu–O	0.19600	frozen
O–H	0.09572	502416
Angles	q_0 (degree)	k_q (kJ mol^{-1} rad^{-1})
O–Cu–O	90.00	frozen
H–O–H	104.52	628.02
dihedrals	x_0 (degree)	k_x (kJ mol^{-1} rad^{-2})
Cu–O–O–O	0.0	frozen
O–O–O–O	0.0	frozen

The simulated systems are composed by the $[Cu(H_2O)_4]^{2+}$ complex immersed in 1090 bulk water molecules, either SPC/E or TIP5P. After energy minimization, the systems were equilibrated for 5 ns and then two production simulations in the NVT ensemble were carried out, 10 and 100 nanosecond-long for the SPC/E and TIP5P systems, respectively.

In Fig. 7.7a we report the comparison between the experimental data (red points) and the D-MXAN fit obtained by using the TIP5P (black line) and SPC/E (green line) models, sampled along the MD trajectory. The agreement of the TIP5P D-MXAN calculation with the experimental data is quite good in the whole energy range with a square residual function of $R_{sq} = 2.31$. This value is similar to those obtained in the prior *static* MXAN fit ($R_{sq} = 2.12$) [41] and this result is quite remarkable considering that, as already stated, no fit of parameters has been carried out in this case. The SPC/E models produce a quite worse agreement than the previous model with a value of the square residual function of $R_{sq} = 5.35$, more than twice the value of TIP5P. The $g_{\text{Cu-O}}(r)$ radial distribution functions in the distance range 2–4.5 Å with the corresponding oxygen coordination numbers N, for the SPC/E and TIP5P water models are shown in the panel b of Fig. 7.7, and corresponding maximum positions are reported in Table 7.5.

The first hydration shell, formed by the four-equatorial waters which are kept fixed in our model at a distance of 1.94–1.98Å, is completed by the axial water molecules, as highlighted by the presence of the peak at 2.22 Å for both models. The two curves, however, are different in the range 2.7–3.4 Å: i.e. in the SPC/E simulation no water molecules are present (red color in Fig. 7.7b), in marked contrast with the TIP5P model (black color in Fig. 7.7b). This is a key difference, that explains the better reproduction of the XANES curves using the TIP5P data, as reported in the previous paragraph. Note that the coordination number for the TIP5P simulation is five already at 2.3 Å (black dashed line in Fig. 7.7b), indicating that the characteristic broad form of the $g_{\text{Cu-O}}(r)$ peak is due to the behaviour of the sixth water. The relatively large distance range visited by the sixth water molecule is clearly shown when the total $g_{\text{Cu-O}}(r)$ is divided in the partial contribution as obtained by ordering the water oxygen atoms as a function of the Cu–O distance (green curve in Fig. 7.8 panel d).

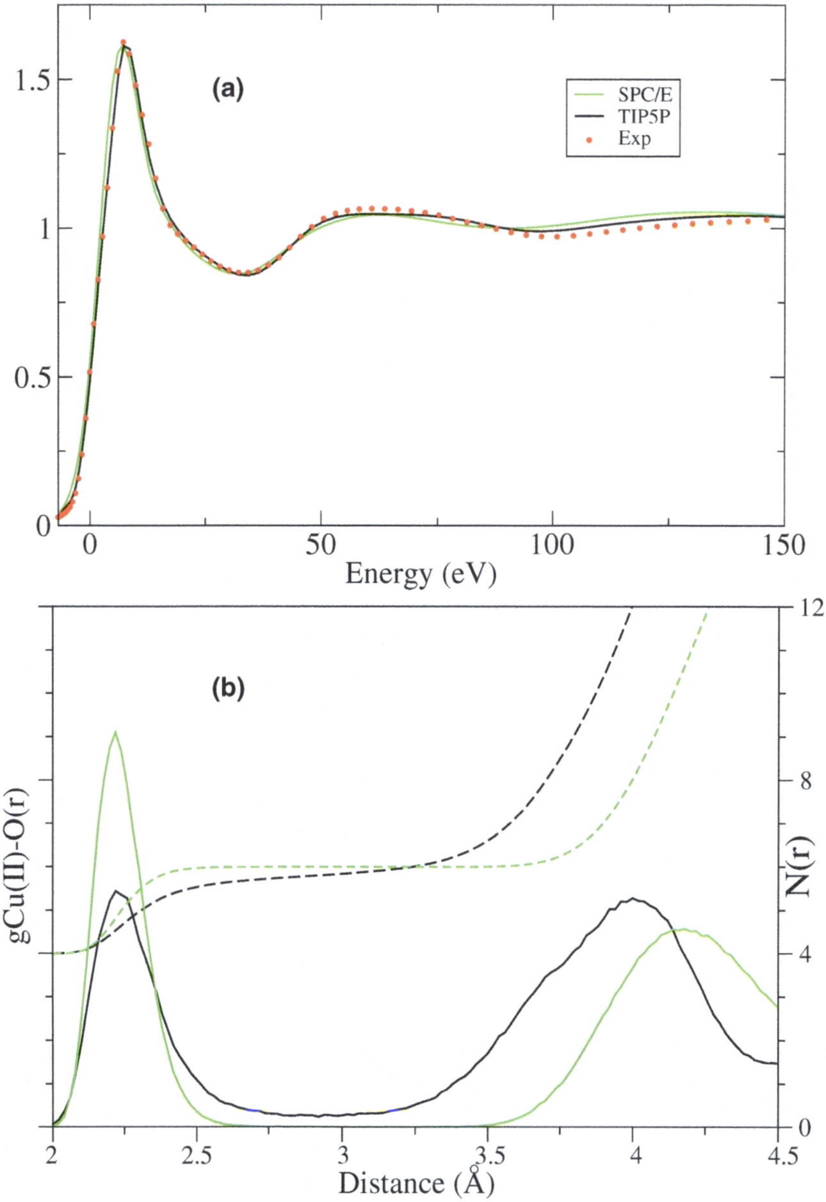

Fig. 7.7 **a** Comparison between the experimental data (*red points*) and the D-MXAN fit obtained by using the TIP5P (*black line*) and SPC/E (*green line*) water models, **b** $g_{\text{Cu-O}}(r)$ radial distribution functions for the TIP5P and SPC/E simulations are shown in *black* and *green lines*, respectively. The corresponding coordination number N is shown in *dashed lines*

Table 7.5 Positions of the Maxima (in Å) of the $g_{\text{Cu-O}}$ and $g_{\text{Cu-H}}$ for the peak formed by the two axial water molecules and for the second hydration shell

Water model	Axial waters $g_{\text{Cu-O}}$	Axial waters $g_{\text{Cu-H}}$	Second shell $g_{\text{Cu-O}}$	Second shell $g_{\text{Cu-H}}$
TIP5P	2.22	2.69	4.00	4.60
SPC/E	2.22	2.64	4.16	4.83

Table 7.6 Structural parameters (in Å) of the $g_{\text{Cu-O}}$ radial distribution function peak formed by the two axial water molecules. Separate contribution of the fifth and sixth oxygen have been calculated by ordering the water oxygen atoms as a function of the Cu–O distance

	Positions at maxima	Median	Interquartile range	Average Cu–O dist	σ
5th	2.20	2.23	0.131	2.24	0.11
6th	2.30	2.47	0.538	2.61	0.38

The structural parameters for the radial distribution peak formed by the two axial water molecules in the TIP5P simulation are reported in Table 7.6. Note that these results are in agreement with the *static* MXAN results both for the obtained axial Cu–O distances, within the statistical errors, and for the contemporary presence of a penta- and hexa-hydrated structure in the first shell of the ion.

The dynamical behaviour of the axial waters is shown in the panel A of Fig. 7.8 where the Cu–O distances as a function of simulated time are reported in different colour for individual water molecules. In some time intervals only an axial water molecule is present in the 2.1–2.5Å Cu–O distance range, while the next closer water molecule is well separated, even farther than 2.9 Å, see for example the 550–950 ps time range. A snapshot of two geometries, the penta- and hexa- hydrated structure, are shown in the insets B and C respectively. At a later time in the simulation, the hexa-coordinated structure is maintained, even though one of the two axial water molecules continuously exchanges with the bulk (see the black and yellow waters in the 1000–1050 ps time range). The two simulations, finally, differ also for the structure of the second hydration shell: i.e. the $g_{\text{Cu-O}}(r)$ peaks are at 4.16 and 4.00 Å in the case of the SPC/E and TIP5P models, respectively, with a corresponding coordination number around 15 and 18 at 4.5 Å. The combined MD/XANES analysis therefore allowed us to appreciate relatively small structural differences in the two simulations that lead to significantly discrepancies with respect to the XANES data reproduction. While the radial distribution functions achieve the convergence after few nanoseconds, to further characterize the dynamic behaviour of the first hydration shell in the TIP5P simulation we carried out a 100 nanosecond long simulation only for this system.

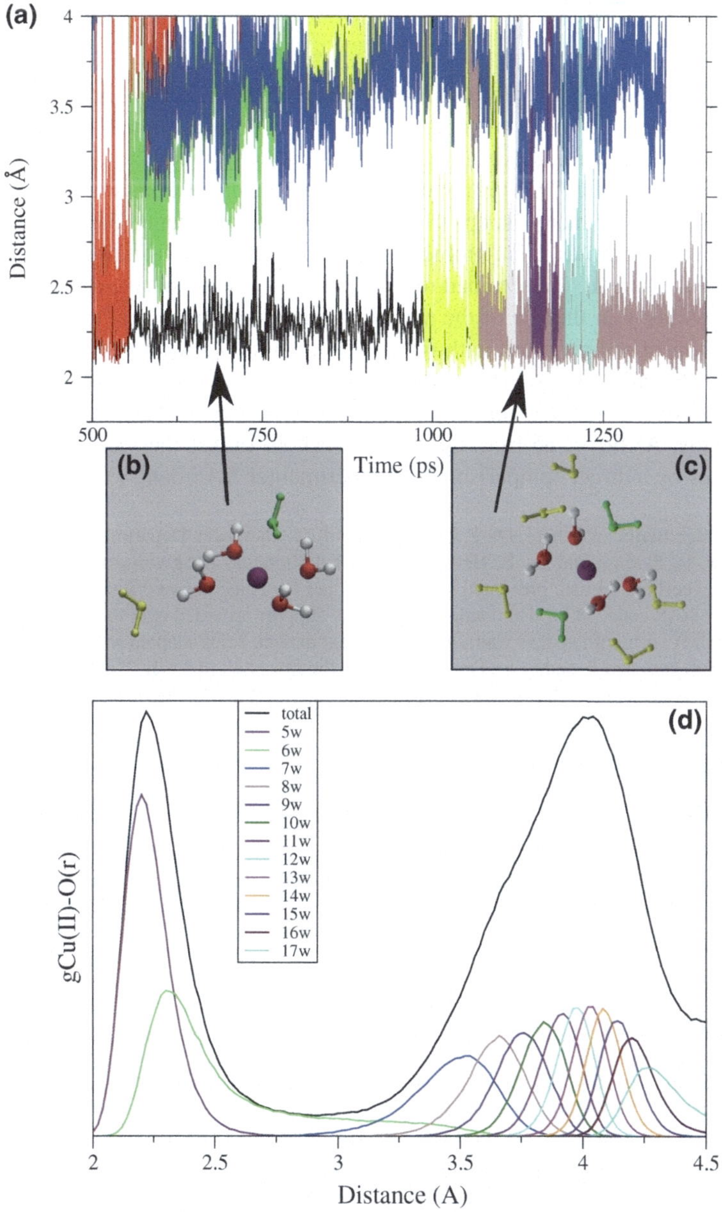

Fig. 7.8 **a** Cu(II)−OW distances as a function of simulation time, in different colors for individual water molecules. In some intervals only an axial water molecule is present in the 2.1–2.5 Å Cu(II)−OW distance range, while the next closer water molecule is well separated, even farther than 2.9 Å (see for example the 550–950 ps time range). In other intervals, two axial water molecules are in the 2.1–2.5 Å range from Cu **b**, **c** Snapshots of the conformational basin *CU5* and *CU6*, respectively. The axial water(s) are in *green color*. Second shell waters in *yellow color*, **d** $g_{Cu\text{-}O}(r)$ for the TIP5P simulation as obtained by ordering the water oxygen atoms as a function of the Cu−O distance. Note the large distance range covered by the sixth water molecule (*green line*)

7.7 Conclusions

In this paper we have presented with details the main ideas that are behind the MXAN package which still remain the only package that allows a quantitative analysis of the XANES energy region within a multiple-scattering theory to derive structural information. MXAN has been presented in the literature in 2001 and in the last 15 years has been applied by many groups to the analysis of experimental data coming from many different samples, from biology to solid state physics. It is almost impossible to summarize here all the obtained results. For this reason we have decided to show a new possibility in the use of the MXAN package, possibility that comes from the combination of the program with molecular dynamic calculations, the D-MXAN analysis. This allow to go beyond the *static* view of the standard fitting procedure and derive some dynamical information that are important to characterize the system under study. At the same time this methods allow to test the validity of the MD procedures by a direct comparison with experimental XAS data.

Acknowledgements We thank Dr. P. Frank, of Stanford Chemistry Department and SSRL Laboratory, Dr. R. Sarangi and Dr. B. Hedman of SSRL Laboratory for giving us the permission of using the experimental data presented in this paper and for constructive criticism, suggestions in the use of MXAN code. M.B. also thanks the financial support provided over the past few years by SLAC for the two-month visit as visiting scientist. Text in Sect. 7.6 is adapted with permission from "Equilibrium between 5- and 6- fold coordination in the first shell of Cu(II)" by G. Chillemi et al., Journal of Physical Chemistry A, **120**, 3958 (2016). Copyright 2016 American Chemical Society.

References

1. C. Meneghini, M. Benfatto, in *Synchrotron Radiation, Basics, Methods and Application*, ed. by S. Mobilio, F. Boscherini, C. Meneghini (Springer, Berlin, 2015), p. 213
2. M. Benfatto, A. Filipponi, C.R. Natoli, Phys. Rev. B **40**, 9626 (1989)
3. T.A. Tyson, K.O. Hodgson, C.R. Natoli, M. Benfatto, Phys. Rev. B **46**, 5997 (1992)
4. J.J. Rehr, R.C. Albers, Rev. Mod. Phys. **72**, 621 (2000)
5. M. Benfatto, S. Della Longa, J. Synchrotron Radiat. **8**, 1087 (2001)
6. S. Della Longa, A. Arcovito, M. Girasole, J.L. Hazemann, M. Benfatto, Phys. Rev. Lett. **87**, 155501 (2001)
7. A. Arcovito, M. Benfatto, M. Cianci, S.S. Hasnain, K. Nienhaus, G.U. Nienhaus, C. Savino, R.W. Strange, B. Vallone, S. Della Longa, PNAS **104**, 6211 (2007)
8. C.R. Natoli, M. Benfatto, S. Doniach, Phys. Rev. A **34**, 4682 (1986)
9. K. Hatada, K. Hayakawa, M. Benfatto, C.R. Natoli, Phys. Rev. B **76**, 060102 (2007)
10. M. Benfatto, S. Della Longa, P. D'Angelo, Phys. Scr. **T115**, 28 (2005)
11. M. Benfatto, S. Della Longa, J. Phys.: Conf. Ser. **190**, 012031 (2009)
12. L. Hedin, S. Lundqvist, *Solid State Physics*, vol. 23 (Academic Press, New York, 1969), p. 1
13. J.E. Müller, O. Jepsen, J.W. Wilkins, Solid State Commun. **42**, 365 (1982)
14. C.R. Natoli, M. Benfatto, C. Brouder, M.F. Ruiz López, D.L. Foulis, Phys. Rev. B **42**, 1944 (1990)
15. M. Benfatto, J.A. Solera, J.G. Ruiz, J. Chaboy, Chem. Phys. **282**, 441 (2002)
16. B. Holm, U. von Barth, Phys. Rev. B **57**, 2108 (1998)
17. W. Gawelda, V.T. Pham, M. Benfatto, Y. Zaushitsyn, M. Kaiser, D. Grolimund, S.L. Johnson, R. Abela, A. Hauser, C. Bressler, M. Chergui, Phys. Rev. Lett. **98**, 057401 (2007)

18. M. Benfatto, Phys. B **208**, 42 (1995)
19. P. D'Angelo, O.M. Roscioni, G. Chillemi, S. Della Longa, M. Benfatto, JACS **128**, 1853 (2006)
20. G. Chillemi, E. Pace, M. D'Abramo, M. Benfatto, J. Phys. Chem. A **120**, 3958 (2016)
21. M.P. Allen, D.J. Tildesley, *Computer Simulation of Liquids* (Oxford Science Publishing, Oxford, 1989)
22. T. Darden, D. York, L. Pedersen, J. Chem. Phys. **98**, 10089 (1993)
23. H.J.C. Berendsen, W.F. van Gunsteren, in *Molecular-Dynamics Simulations of Statistical-Mechanical Systems Enrico Fermi Summer School* (North-Holland, Amsterdam, 1986), p. 43
24. H.J.C. Berendsen, J.P.M. Postma, W.F. van Gunsteren, J.R. Haak, A. Di Nola, J. Chem. Phys. **81**, 3684 (1984)
25. G. Bussi, D. Donadio, M. Parrinello, J. Chem. Phys. **126**, 014101 (2007)
26. S. Pronk, S. Pall, R. Schulz, P. Larsson, P. Bjelkmar, R. Apostolov, M.R. Shirts, J.C. Smith, P.M. Kasson, D. van der Spoel, B. Hess, E. Lindahl, Bioinformatics **29**, 845 (2013)
27. M. Antalek, E. Pace, K.O. Hodgson, G. Chillemi, M. Benfatto, R. Sarangi, P. Frank, J. Chem. Phys **145**, 044318 (2016)
28. H.J.C. Berendsen, J.R. Grigera, T.P. Straatsma, J. Chem. Phys. **91**, 6269 (1987)
29. M.M. Reif, P.H. Hünenberger, J. Chem. Phys. **134**, 144104 (2011)
30. V. Migliorati, F. Sessa, G. Aquilanti, P. D'Angelo, J. Chem. Phys. **141**, 044509 (2014)
31. G. Chillemi, P. D'Angelo, N.V. Pavel, N. Sanna, V. Barone, JACS **124**, 1968 (2002)
32. G. Chillemi, V. Barone, P. D'Angelo, G. Mancini, I. Persson, N. Sanna, J. Phys. Chem. B **109**, 9186 (2005)
33. G. Chillemi, G. Mancini, N. Sanna, V. Barone, S. Della Longa, M. Benfatto, N.V. Pavel, P. D'Angelo, JACS **129**, 5430 (2007)
34. V. Migliorati, A. Zitolo, G. Chillemi, P. D'Angelo, ChemPlusChem **77**, 234 (2012)
35. M. Cossi, N. Rega, G. Scalmani, V. Barone, J. Comput. Chem. **24**, 669 (2003)
36. H.J.C. Berendsen, D. van der Spool, R. van Drunen, Comput. Phys. Commun. **91**, 43 (1995)
37. T. Miyanaga, H. Sakane, I. Watanabe, Bull. Chem. Soc. Jpn. **68**, 819 (1995)
38. A. Pasquarello, I. Petri, P.S. Salmon, O. Parisel, R. Car, E. Toth, D.H. Powell, H.E. Fischer, L. Helm, A. Merbach, Science **291**, 856 (2001)
39. S.T. Moin, T.H. Hofer, A.K.H. Weiss, B.M. Rode, J. Chem. Phys. **139**, 014503 (2013)
40. P. Frank, M. Benfatto, R.K. Szilagyi, P. D'Angelo, S. Della Longa, K.O. Hodgson, Inorg Chem **44**, 1922 (2005)
41. P. Frank, M. Benfatto, M. Qayyam, B. Hedman, K.O. Hodgson, J. Chem. Phys. **142**, 084310 (2015)
42. M.W. Mahoney, W.L. Jorgensen, J. Chem. Phys. **112**, 8910 (2000)
43. C.I. Bayly, P. Cieplak, W. Cornell, P.A. Kollman, J. Phys. Chem. **97**, 10269 (1993)

Chapter 8
GNXAS: Advances in the Suite of Programs for Multiple-Scattering Analysis of X-ray Absorption Data

Fabio Iesari, Keisuke Hatada, Angela Trapananti, Marco Minicucci and Andrea Di Cicco

Abstract This contribution presents some developments of the GNXAS methodology and suite of programs, providing full analysis of raw experimental XAS data through advanced multiple-scattering simulations. The main features of the GNXAS suite of programs, including the basic theoretical background based on an n-body expansion of the x-ray absorption cross section, as well as useful information about the GNXAS flow diagram and practical usage are reviewed. The new GNXAS graphical interface w-GNXAS is specifically illustrated. The recent implementation of Reverse Monte Carlo algorithms allowing for multiatomic structural refinement into the RMC-GNXAS package is also presented in detail with specific examples.

8.1 Introduction

The GNXAS package is an advanced software for x-ray absorption spectroscopy (XAS) data analysis providing multiple-scattering (MS) calculations of the theoretical signals and a rigorous fitting procedure of the raw experimental data.

F. Iesari · M. Minicucci · A. Di Cicco (✉)
Physics Division, School of Science and Technology, University of Camerino, 62032 Camerino (MC), Italy
e-mail: andrea.dicicco@unicam.it

Present Address:
F. Iesari
Faculty of Science, University of Toyama, Gofuku 3190, Toyama 930-8555, Japan

A. Trapananti
Istituto Officina dei Materiali (IOM), Consiglio Nazionale delle Ricerche (CNR), Perugia, Italy

Present Address:
A. Trapananti
Physics Division, School of Science and Technology, University of Camerino, 62032 Camerino (MC), Italy

K. Hatada
Faculty of Science, University of Toyama, Gofuku 3190, Toyama 930-8555, Japan

D. Sébilleau et al. (eds.), *Multiple Scattering Theory for Spectroscopies*, Springer Proceedings in Physics 204,
https://doi.org/10.1007/978-3-319-73811-6_8

X-ray absorption spectroscopy (XAS) is a powerful structural technique for investigating the short-range environment around selected atomic species in ordered and disordered matter. Structural information can be obtained from the analysis of the quantum interference pattern, usually detectable above any inner shell absorption edge, generated by the scattering of the photoelectron on the potential of the surrounding atoms. These oscillations are usually referred to as x-ray absorption fine structure (XAFS) or EXAFS. X-ray absorption spectra can be collected in transmission geometry or using other secondary emission effects, like fluorescence or electron yield. The collection of a spectrum as a function of the x-ray photon energy E allows measurements of the EXAFS structural signal, usually expressed as a function $\chi(k)$ of the photoelectron momentum $k = \sqrt{2m(E - E_e)}/\hbar$ (E_e being the edge energy). XAS experiments are possible using laboratory sources but efficient and accurate experiments are mostly performed, since the 80's, using synchrotron radiation.

The focus of the GNXAS software development project, which started in 1990 and for which a first version working in a VMS Digital environment was ready in 1991 [1], was the possibility to apply EXAFS data analysis to disordered systems and consequently to link the XAS signal to features of the n-atom distribution functions g_n. This is also the origin of the name GNXAS. The GNXAS suite of programs included advanced multiple-scattering (MS) calculations and a rigorous fitting procedure of the raw experimental data [1–3]. GNXAS was the result of over 10 years of research efforts of A. Filipponi, A. Di Cicco and C.R. Natoli, in the field of x-ray absorption spectroscopy and several scientists have contributed to its development including the authors of this contribution (all listed in the website and in the GNXAS handbooks [4, 5]). This software has been widely tested and has been applied to a variety of systems including simple gas-phase molecules, clusters, nanocrystals, molecular solids and liquids, crystalline and liquid metals, solid and molten salts, amorphous solids and glasses, solutions, high-T_c superconductors, and biological matter. A publication list limited to the first 10 years can be found in the first edition of the GNXAS handbook [4].

The present standard GNXAS software distribution includes the main data-analysis programs `crymol`, `phagen`, `gnpeak`, `gnxas`, and `fitheo` as well as several utility programs for visualization (`phaplo`, `readder`). Distribution includes also pre-analysis tools for sample optimization (`xasam`), automatic background subtraction and EXAFS pre-analysis (`jesf`), edge fitting (`fitedg`) and deconvolution (`deconv`). Other auxiliary programs include pair distribution function analysis (`grrec`, `grfit`) and XAS calculation (`grxas`), simulation (`edxrd`) and peak fitting (`peakfit`) of (energy/angular dispersive) XRD patterns.

In the last years, several efforts were devoted to implement alternative model-independent refinement strategies for disordered systems including also x-ray/neutron scattering as a complementary tool (Reverse Monte Carlo), as well as adapting the software to various operating systems and providing a graphical interface. In this contribution, Sect. 8.2 is devoted to a brief presentation of the main characteristics of the GNXAS suite of programs (widely described elsewhere, for which full documentation can be found in [2, 4, 5]). The graphical interface [6] recently developed for GNXAS (w-GNXAS) is illustrated in Sect. 8.3. The last Sect. 8.4 is devoted to the recent implementation of Reverse Monte Carlo algorithms into a new version of the RMC-GNXAS package [7] allowing for multiatomic structural refinement.

8.2 Overview of GNXAS

The GNXAS package has been developed to provide a tool for structural refinement and full XAS data-analysis, using a rigorous fitting procedure of raw experimental data through advanced multiple-scattering simulations, in the framework of the n-body expansion of the x-ray absorption cross-section. The next subsections briefly resume the theoretical background and the practical software implementation of the method. Full details can be found in Refs. [2–5, 8].

8.2.1 Summary of the Theoretical Background for GNXAS

Realistic simulations of the x-ray absorption cross-section are obtained by multiple-scattering (MS) calculations related to a given model structure. In this framework, the polarization averaged XAS cross-section [2] for transitions to a dipole selected final state of angular momentum ℓ_0 can be written as

$$\sigma(\omega) = \sigma_0 \left[\Im \frac{1}{\Im(t_0^{\ell_0})} \frac{1}{2\ell_0 + 1} \sum_{m_0} [\mathbf{T}(\mathbf{I} - \mathbf{G_0 T})^{-1}]_{0,0}^{L_0,L_0} \right] . \tag{8.1}$$

Here σ_0 is the atomic cross-section, $\mathbf{T}$ and $\mathbf{G_0}$ are the atomic scattering (phase-shift) and photoelectron propagator matrices in a local basis, indexed by i, j running over the different atoms, and by a set of angular momenta L, L' (where $L = \{\ell, m\}$).

The $\mathbf{T}$ scattering matrix here is block diagonal ($T_{i,j} = t_i \; \delta_{i,j}$) and, in the MT approximation for the potential, also diagonal on the L indices ($t_i^{L,L'} = t_i^{\ell} \; \delta_{L,L'}$):

$$[\mathbf{T}]_{i,L,j,L'} = t_i \delta_{i,j} \delta_{\ell,\ell'} \delta_{m,m'} \; . \tag{8.2}$$

Matrix elements can be expressed in terms of the ℓth potential phase shift $\delta_i^{\ell}(E)$, as a function of the photoelectron energy E: $t_i^{\ell}(E) = \exp(\mathrm{i}\delta_i^{\ell}(E)) \sin(\delta_i^{\ell}(E))$. Those matrix elements can be calculated by solving the Schrödinger equation for the potential at the corresponding i site.

GNXAS calculations are carried out using the muffin-tin (MT) approximation but it is useful to remark that all MS equations, and in particular the XAS cross-section, remains unaltered if one replaces the MT spheres with space filling cells (no interstitial region), but t^{ℓ} is replaced by $t^{L,L'}$, taking in this way into account the non spherical shape of the cell potential (non MT potentials).

The propagator matrix is composed of null diagonal blocks, (i, i) sites, and non null off-diagonal blocks $G_{i,j}^{L,L'}$ describing the free propagation from site i to site j $(i \neq j)$:

$$G_{i,j}^{L,L'} \neq 0 \quad \text{only if} \quad i \neq j \; . \tag{8.3}$$

The expression for a single propagator block involves $3j$ symbols and is given by:

$$G_{ij}^{LL'} = [4\pi(2\ell+1)(2\ell'+1)]^{1/2} \sum_{\ell_1} (2\ell_1+1)^{1/2} \begin{pmatrix} \ell & \ell' & \ell_1 \\ 0 & 0 & 0 \end{pmatrix}$$
$$\times \begin{pmatrix} \ell & \ell' & \ell_1 \\ m & -m' & m'-m \end{pmatrix} (-1)^{m'} \mathrm{i}^{\ell_1+1} h_{\ell_1}^{+}(kR_{ij}) Y_{\ell,m'-m}(\hat{\boldsymbol{R}}_{ij}) \; . \tag{8.4}$$

Here h_ℓ^+ are Hankel functions, $Y_{\ell,m}$ are the spherical harmonics and $\boldsymbol{R}_{i,j}$ the vector joining site j to site i.

We can see that the two important matrices appearing in the cross-section are related to the chemistry (**T** using only atom indexing) and to the local structure ($\mathbf{G_0}$). There is a strong non-linear relationship between geometry and XAS signal: $\sigma \sim (\mathbf{I} - \mathbf{G_0T})^{-1}$. This non-linearity is the mathematical consequence of the strong coupling of the photoelectron with the surrounding atoms. A first approach to this problem is to use the so-called MS expansion. Where the norm of the $\mathbf{G_0T}$ matrix (maximum modulus of its eigenvalues) satisfies $||\mathbf{G_0T}|| < 1$ then the formal matrix expansion $\mathbf{T}(\mathbf{I} - \mathbf{G_0T})^{-1} = \mathbf{T}(\mathbf{I} + \mathbf{G_0T} + \mathbf{G_0TG_0T} + \mathbf{G_0TG_0TG_0T} + \cdots)$ is convergent and gives rise to the MS series.

The above condition will certainly hold above a given energy since not only the elements of the $\mathbf{G_0}$ matrix decrease like $1/\sqrt{E}$ but also $||\mathbf{T}|| = \max |t_i|$ tends to zero much more rapidly with energy. The convergence threshold is system dependent, typical values range from below the edge to 5–50 eV above the edge.

Writing down the series we obtain:

$$\sigma(\omega) = \sigma_0 \left[1 + \sum_{i\neq 0} \chi_2^{0i0} + \sum_{\substack{i\neq j \\ i\neq 0, j\neq 0}} \chi_3^{0ij0} + \sum_{\substack{i\neq j\neq k \\ i\neq 0, k\neq 0}} \chi_4^{0ijk0} + \cdots \right] , \tag{8.5}$$

where the generic χ_n structural term is:

$$\chi_n^{0i\ldots j0} = \Im \left(\frac{t_0^{\ell_0}}{\Im(t_0^{\ell_0})} \frac{1}{2\ell_0+1} \sum_{m_0} (\Xi^{0i\ldots j0})_{L_0,L_0} \right) , \tag{8.6}$$

and specifically $\Xi^{0i0} = G_{0,i}t_iG_{i,0}t_0$ for χ_2^{0i0}, $\Xi^{0ij0} = G_{0,j}t_jG_{j,i}t_iG_{i,0}t_0$ for χ_3^{0ij0}, and $\Xi^{0ijk0} = G_{0,k}t_kG_{k,j}t_jG_{j,i}t_iG_{i,0}t_0$ for χ_4^{0ijk0}. In this notation it is understood that the internal angular momentum indices have been saturated.

The χ_n signals are oscillating functions of the type:

$$\chi_n(k) = A(k, R)\sin(kR_p + \phi(k, R)) \ , \tag{8.7}$$

where A and ϕ are smooth functions of k and of the geometrical parameters R. The relevant frequency of the signal is the path length R_p.

An important inconvenience of the MS expansion is that it has no simple relationship with the structure (geometry): χ_2 terms probe the relative position of atoms 0 and i, χ_3 terms probe the positions of the atoms 0, i and j. They are sensitive to the two-particle and three-particle distribution respectively. For χ_4 terms the situation becomes more complicated: a generic $0ijk0$ path probes four-particle correlations, but special paths like $0i0k0$ or $0iji0$ and $0i0i0$ probe lower order correlations. In general at order n, in χ_n there are paths involving all particle distributions from 2 to n if n is even or from 3 to n if n is odd.

Of course, in regions of rapid convergence of the series (high-energy limit) a single-scattering approximation may be sufficient and the structural information is limited to the pair distribution (approximation used in the *standard* EXAFS analysis).

A different and potentially more powerful approach to the solution of the XAS structural problem is based on a n-body decomposition of the cross-section [2], which partially avoids the drawbacks related to the MS expansion. The main idea is to reduce the dimension of the problem to the actual local physical quantities of interest, related to n-atom properties where $n \ll N$ (N being the number of atoms of the system).

Within this approach, the total x-ray absorption cross-section for n atoms $\sigma(0, i, j, \ldots, n)$ can be expanded in terms of the irreducible n-body cross sections of lower order:

$$\sigma(0, i, j, \ldots, n) = \sigma_0 + \sum_i \sigma^{(2)}(0, i) + \sum_{(i,j)} \sigma^{(3)}(0, i, j) + \\ + \sum_{(i,j,k)} \sigma^{(4)}(0, i, j, k) + \quad \cdots \quad + \sigma^{(n)}(0, i, j, \ldots, n) \ . \tag{8.8}$$

In (8.8), σ_0 is the atomic absorption (photoabsorber 0), $\sigma^{(2)}(0, i)$ is the pair contribution to the cross-section associated with an additional atom i, $\sigma^{(3)}(0, i, j)$ is the triplet contribution associated with a couple of atoms i, j and so on.

The dimensionless quantities $\gamma^{(n)} = \sigma^{(n)}/\sigma_0$, represent the irreducible n-body contributions to the structural XAS term $\chi(E) = (\sigma(E) - \sigma_0(E))/\sigma_0(E)$. In this way we arrive to an equivalent expansion for the experimentally measurable structural signal $\chi(E)$ that differs substantially from the MS series:

$$\chi(0, i, j, \ldots, n) = \sum_i \gamma^{(2)}(0, i) + \sum_{(i,j)} \gamma^{(3)}(0, i, j) + \\ + \sum_{(i,j,k)} \gamma^{(4)}(0, i, j, k) + \quad \cdots \quad + \gamma^{(n)}(0, i, j, \ldots, n) \ . \tag{8.9}$$

The irreducible n-body $\gamma^{(n)}$ signals [2] are the central quantities in the GNXAS approach, since they are associated with well defined n-body arrangements of the atoms. Using those terms, a simple linear relationship among structure (expressed in terms of $2, 3, 4, \ldots n$-body distributions) and signal is obtained. Due to mean free path effects, the higher order n-body terms are generally smaller than the lower order ones, so that convergence with few terms is expected. Low-order $\gamma^{(n)}$ signals, involving inversions of small matrices, can be easily calculated with different methods (within GNXAS).

As an example, for the two-body term, due to the two-blocks structure of the matrices, only even powers of $\mathbf{TG_0}$ give a contribution:

$$\left\{\begin{pmatrix} t_0 & 0 \\ 0 & t_i \end{pmatrix} \sum_{n=1}^{\infty} \left[\begin{pmatrix} 0 & G_{0,i} \\ G_{i,0} & 0 \end{pmatrix} \begin{pmatrix} t_0 & 0 \\ 0 & t_i \end{pmatrix}\right]^n\right\}^{0,0} =$$
$$t_0 G_{0i} t_i G_{i0} t_0 + t_0 G_{0i} t_i G_{i0} t_0 G_{0i} t_i G_{i0} t_0 + \cdots \tag{8.10}$$

The corresponding MS expansion results:

$$\gamma^{(2)}(0, i) = \chi_2^{0i0} + \chi_4^{0i0i0} + \chi_6^{0i0i0i0} + \chi_8^{0i0i0i0i0} + \cdots \tag{8.11}$$

The number of MS terms contributing to the n-body irreducible γ signals depends on k range, bond distance, and atomic numbers involved. In the EXAFS region, for short bonds terms up to χ_6 are found important whereas for longer bonds χ_4 is usually sufficient. The configurational average is also effective in the damping of these higher order MS signals. A peculiar feature in the MS expansion for $\gamma^{(2)}$ is that the successive terms have leading frequencies multiple of $2R$, R being the distance between atoms 0 and i. As a consequence there is a large frequency difference between the leading term χ_2 and the next order correction χ_4. In general $\gamma^{(n)}$ signals are, like the χ_n, oscillating functions of the photoelectron wavevector modulus k. They show a broader frequency spectrum whose dominant frequency is that associated with the shortest path of the corresponding MS series.

The advantage of incorporating in a few $\gamma^{(n)}$ signals a large (infinite) number of MS terms is evident both from a physical and practical point of views. Indeed different MS terms are not independent if they involve scattering processes on the same sets of atoms, while γ signals are uniquely associated with the given atomic configurations. Convergence properties are also very good, and it can be shown [2] that if the MS series is absolutely convergent also the series of the γ signals will be absolutely convergent.

For calculating a XAS signal associated with a given cluster (model structure) it is necessary to identify the relevant n-body configurations. Identification of configurations can be done assuming a single photoabsorbing site (single photoabsorber or multiple atoms placed in equivalent positions by rotation and translations), and can be generalized for multiple prototypical photoabsorbers. An important property is the hierarchical relationship between different n-body configurations: a n-body configuration contains several $n-m$-body sub-configurations that are not independent. This allows to reduce the number of n-body coordinates just to the independent ones. A typical example is the second shell contribution which usually gives rise to detectable three-body signal. The three-body signal plus the two-body second shell signal is defined as a total 3-body ($\eta^{(3)}$) signal associated with the second shell, using the same set of triplet coordinates.

Of course, even a simple model structure is associated with many different n-body configurations, and a selection procedure is necessary to include those really useful in a XAS data-analysis process. The n-body configurations are naturally ordered according to the corresponding leading oscillation frequency of the corresponding γ signal, which is the length of the shortest path involving all of the atoms in the n-body configuration (roughly corresponding to the position of the Fourier Transform peak).

This is important because any XAS experimental signal shows a natural upper frequency cutoff R_c, limited by experimental resolution, core-hole lifetime and mean free path effects, all affecting the signal intensity. This frequency cutoff selects automatically only a finite number of n-body structures.

Therefore in the interpretation of the signal of a completely unknown structure it is possible to make an a priori selection of the relevant n-body configurations thus limiting considerably the number of unknowns.

We see then that the XAS signal is actually dependent only on a limited number of distribution functions (n-body configurations) which can be easily identified and analyzed.

A realistic simulation of the XAS signal must include a proper configurational average which takes into account thermal and non-thermal disorder in the atomic positions. This is very easy to do using the irreducible $\gamma^{(n)}$ MS signals. In fact, the general expression [2, 9] of the XAS structural term is given in terms of the n-body distribution functions $g_n(r)$ (r is a generic set of n-body coordinates):

$$\begin{aligned}\langle\chi(k)\rangle = & \int_0^\infty dr\, 4\pi r^2 \rho\, g_2(r)\, \gamma^{(2)}(r,k) + \int dr_1 dr_2 d\phi\, 8\pi^2 r_1^2 r_2^2 \sin(\phi) \\ & \times \rho^2 g_3(r_1,r_2,\phi)\gamma^{(3)}(r_1,r_2,\phi,k) + \int dr_1 dr_2 d\phi dr_3 d\Omega\, 8\pi^2 r_1^2 r_2^2 r_3^2 \\ & \times \sin(\phi)\rho^3 g_4(r_1,r_2,\phi,r_3,\Omega)\gamma^{(4)}(r_1,r_2,\phi,r_3,\Omega,k) \cdots \end{aligned} \tag{8.12}$$

The integrals, because of the short range nature of the kernels $\gamma^{(n)}$, are actually limited to a region of linear dimensions of the order of few Å. This is due to the strong electron interaction at moderate kinetic energies limiting the effective electron mean free path.

This equation should be compared with the well known expression for the static structure factor (disordered system) which can be measured by using x-ray or neutron diffraction:

$$S(k) = 1 + \frac{4\pi\rho}{k}\int_0^\infty (g_2(r)-1) r \sin(kr) dr\ . \tag{8.13}$$

The different nature of the kernels makes the structural information on the $g_2(r)$ obtainable in the two cases largely complementary (short vs. medium range information). Moreover, XAS can give information beyond the pair distribution. For crystals, where diffraction gives obviously a richer information about atomic positions, XAS can provide unique data about local correlated vibrations.

The effect of configurational disorder or thermal vibrations can be easily taken into account in the case of small disorder, that is in the presence of a well defined (isolated) first peak, not necessarily Gaussian, in the distribution function. This includes the thermal broadening or moderate structural disorder of molecular or crystalline g_n peaks.

MS signals $\chi_n(k)$, n-body signals $\gamma^{(n)}(k)$ or effective shell signals $\eta^{(n)}(k)$ can be treated in the same way. In all of these cases the signal, indicated generically by $\chi(k)$, can be written in terms of the amplitude $A(k, r)$ and phase $\psi(k, r)$ functions as $\chi(k) = \Im\left[A(k, r)\exp(i\psi(k, r))\right] = A(k, r)sin[\psi(k, r)]$, where r here indicates a set of geometrical coordinates which are sufficient to describe correctly the configuration. Phase $\psi(k, r) = kR_p + \phi(k, r)$ (R_p is the dominant frequency) and amplitude $A(k, r)$ are usually smooth functions of k, r.

For example, considering a specific peak of a distribution function $g_n(r)$ related to a well-defined atomic configuration at equilibrium, we can write [2]:

$$\langle\chi(k)\rangle = \Im\left[\int \mathrm{d}r\, A(k, r)\exp[i\psi(k, r)]P(r)\right], \tag{8.14}$$

where $P(r)$ is the normalized probability density describing a peak of the appropriate n-body distribution function (configuration degeneracy, or coordination number for $n = 2$, being a trivial multiplicative factor). For pair distributions $n = 2$, $P(r) = 4\pi r^2 \rho g_2(r)$ like in the first term of (8.12).

The integral of (8.14) can be easily calculated for selected Gaussian and non-Gaussian functional forms reproducing the peaks of the distribution function, and the result can be expressed in terms of the derivatives of amplitude $A(k, r)$ and phase $\psi(k, r)$ in the space of coordinates [2]. The configurational average results in a damping of the XAS $\chi(k)$ structural signal, in which both amplitude and frequency are affected by the shape of the distribution. Use of simple functional forms, like the Gaussian distribution, considerably simplify the data-analysis because of the limited number of parameters used in the structural refinement process. For example, for pair configurations, a single σ^2 parameter (variance of the interatomic distance) is used to define the broadening of the distance distribution (being the entire distribution defined by 3 parameters: average distance R, variance σ^2, coordination number N). The n-body configurations generally require a larger set of parameters, but very often they are closely interconnected by hierarchical and geometrical constraints. This *peak-fitting* technique, which needs the identification of well-defined peaks of the n-body distributions, is used in the GNXAS method to treat the configurational average of MS signals. It is clearly applicable for molecules and crystals and the same approach can be used to treat the case of moderate structural disorder in glassy covalent systems for which peaks of the g_n distribution are clearly identified. Whenever the peaks of the g_n are not well defined (liquids for example) different average methods should be used. Extensions of the peak-fitting technique to highly disordered systems, using suitable constraints, are discussed in [10, 11]. However, intrinsic limitations of the peak-fitting technique may be overcome by using model-independent methods for which peaks of the distribution functions are not defined a priori, like the Reverse Monte Carlo approach discussed in [7, 12–14] and Sect. 8.4.

MS calculations and proper treatment of the configurational average of n-body signals (8.12) are used for the final step in XAS data-analysis using GNXAS, i.e. the derivation of best-fit structural parameters from the experimentally measured x-ray absorption coefficient $\alpha(E)$ (proportional to the cross-section) above a given deep core level. Structural refinement is obtained by optimizing the model absorption spectrum against the raw x-ray absorption data $\alpha(E_i)$ including noise (not the $\chi(k)$ or filtered data) [3, 8]. The model absorption spectrum $\alpha_m(E_i)$ thus includes directly background modelling:

$$\alpha_m(E_i) = \alpha_{\rm bkg}(E_i) + \alpha_{\rm exc}(E_i) + [1 + \chi_m(E_i)]\,\alpha_0(E_i)\ , \tag{8.15}$$

where $\alpha_0(E_i) = J\sigma_0(E_i)$ is the atomic absorption coefficient (J accounts for the actual absorption jump), $\alpha_{\rm bkg}(E_i)$ is a smooth polynomial background and $\alpha_{\rm exc}(E_i)$ accounts for multi-electron excitation channels. Simultaneous modelling of background and signal ($\chi_m(E_i)$ as obtained by peak distribution modelling of (8.12)) significantly reduces the introduction of systematic errors in the interpretation of the structural signal. The same approach can be used for multi-edge studies (see for example [8, 15]):

$$\alpha_m(E_i) = \alpha_{\rm bkg}(E_i) + \alpha_{\rm exc}(E_i) + \sum_{j=1}^{\rm NXAS} \left[1 + \chi_m^j(E_i)\right] \alpha_0^j(E_i)\ , \tag{8.16}$$

where NXAS is the number of different XAS spectra calculated for a given model structure.

The residual function R which is minimized in the program is the sum over $i = 1, N$ points of the squares of the differences between model and experimental signals in the specified energy interval times a weight function:

$$R = \sum_i [\alpha(E_i) - \alpha_m(E_i, \{\lambda\})]^2 \times W(E_i)\ . \tag{8.17}$$

Here the model signal $\alpha_m(E, \{\lambda\})$ depends on the ensemble of p structural parameters $\{\lambda\}$. The expected value for the residual function R can be calculated for a given normalization of the weight function (for `fitheo` corresponds to the average noise variance usually in the $10^{-6} - 10^{-8}$ range). It should be noted that in many real situations the actual minimum value achieved for the residual function exceeds these limiting values due to the presence of unexplained signal contributions.

Under normal conditions the weight function should mimic the energy dependence of the inverse noise variance $1/\sigma_i^2$, so that the R function becomes a standard χ^2 like statistical function. This can be achieved by a using a simple inverse power law ($\sigma_i^2 \sim k_i^{-w}$ with $w = 1 - 4$) until a satisfactory agreement with the statistical noise is obtained. Moreover, larger w values have the effect of giving an extra-weight to higher energy data, and this choice can be adopted to account for possible inaccuracies of the theoretical model in the near-edge region (the noise variance σ^2 is actually related to the $[\alpha(E_i) - \alpha_m(E_i)]$ random variable).

As it has been defined, the residual function R is expected to follow the χ_N^2 distribution, and under these conditions a full statistical evaluation of the results can be performed as follows:

1. within a given choice for the structural model, the optimal values for the $\{\lambda\}$ structural parameters are the set $\{\lambda_{\min}\}$ such that the residual R is at a minimum.
2. The statistical χ^2 test can be performed to check whether the actual value of R is only due to residual noise or it contains unexplained physical information.
3. A comparison between models containing a different number of structural parameters can be performed, and it can be verified if the reduction of R obtained using more parameters is statistically significant.
4. Statistical errors related to selected confidence intervals can be estimated for the structural parameters, looking at regions in the parameter space for which $R(\{\lambda\}) < R_{\min} + C$, where C depends on the confidence level chosen and $R_{\min}$ corresponds to the expected value (when the residual contains only statistical noise). These regions, in the second-order approximation, are p-dimensional (p is the number of parameters) ellipsoids which provide also an insight onto the correlation among parameters.

8.2.2 GNXAS *Suite of Programs*

The main characteristics of the current standard version of the GNXAS data-analysis suite of programs are:

(1) atomic phase shifts calculations in the muffin-tin approximation are based on atomic self-consistent relativistic calculations. Account for the neighbors is taken.
(2) Inclusion of inelastic losses through complex Hedin-Lundqvist potential.
(3) Calculation of MS signals associated with two, three, and four atom configurations using advanced algorithms and of proper configurational average of MS terms.
(4) Use of an advanced fitting procedure that allows one:
    (a) to fit simultaneously several spectra containing several edges,
    (b) to use directly the raw data without any pre-analysis,
    (c) to account for complex background multi-electron excitation features,
    (d) to use various model peaks for the pair, triplet and quadruplet distribution functions, including non Gaussian models and extremal cases. In all cases absolute values of the structural parameters can be refined.

(e) To treat liquid phase or disordered systems and extract reliable $g(r)$ functions in the short-range limit ($\leq$4 Å).
(f) To perform a rigorous statistical error analysis and plot two-dimensional correlation maps.

(5) To provide a flexible scientific tool for EXAFS data analysis where the user has access to every stage of the calculation. In this sense the GNXAS package is not a black box.
(6) Full modularity that makes easy to interface parts of the GNXAS software with other available software.
(7) Specific software, fully interfaced with GNXAS, that allows to perform calculation and structural refinements of x-ray diffraction data, as well as Reverse Monte Carlo simulations of XAS data.
(8) Software has been developed under VMS and Linux/UNIX architecture but it has been extended to Macintosh and Windows operating systems, also supported by a new graphical interface.

The GNXAS flow diagram is reported in Fig. 8.1, containing also information about the main input and output files and some auxiliary programs that can be used for check, visualization and structural refinement purposes. The reader should refer to the legend and flow diagram reported in Fig. 8.1, and to the detailed GNXAS handbooks [4, 5] for understanding role and meaning of the various input/output files related to a starting "abc" structure-type.

The backbone of the GNXAS package is composed of three main codes identified by the boxes in bold (Fig. 8.1):

(1) `phagen` optical potential and phase-shift generation
(2) `gnxas` n-body XAS signal calculation
(3) `fitheo` structural refinement of the experimental data

Two other main programs are able to prepare automatically the input for `phagen` and `gnxas` starting directly from model molecular position or crystallographic data. These are particularly useful in the case of complex structures where the configuration counting is not trivial. These programs are:

(4) `crymol` allows to treat complex molecular and crystallographic structures providing input information for successive programs; the program identifies prototypical phase-shift atoms and selects a suitable cluster of atoms for successive XAS calculations.
(5) `gnpeak` is based on a general algorithm able to identify inequivalent two-body, three-body and four-body configurations in specified atomic aggregates.

Several other programs were written to facilitate the output readout, in particular:

(6) `phaplo` produces a graphic output for the phase-shift file generated by `phagen` (typically for TopDrawer [17]).
(7) `readder` converts the binary signal files generated by `gnxas` into ASCII files to facilitate visualizing and checking the output (typically for TopDrawer [17]).

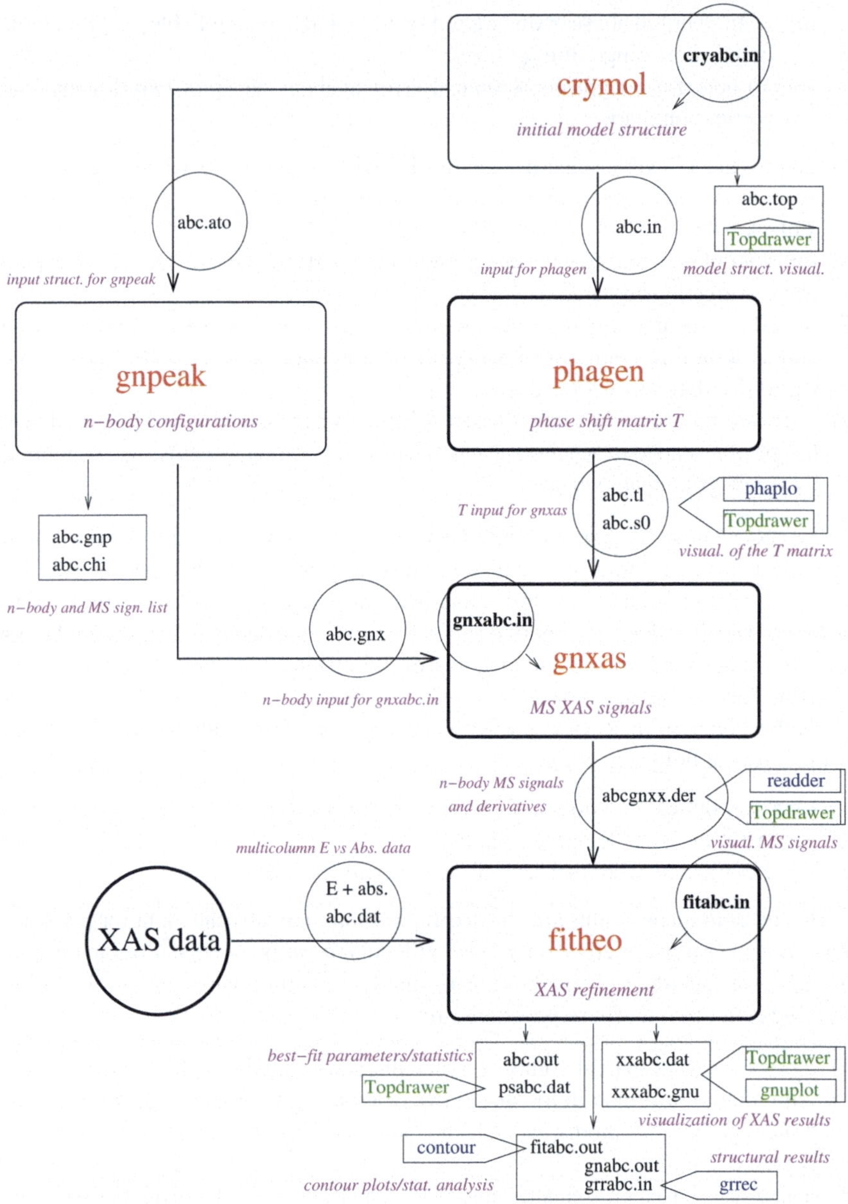

Fig. 8.1 Flow diagram of the GNXAS suite of programs. An input file defining the initial model structure (cryabc.in) may be the starting point of the sequence, composed by the main programs `crymol`, `gnpeak`, `phagen`, `gnxas`, `fitheo` (*red color* in figure). Input files needing specific typing by users are indicated in *bold*. Main output files of the individual programs are *encircled* when needed as an input for successive programs. Other output files containing data for graphics or in tabular form are enclosed in *boxes*. Graphical output is currently designed for specialized plotting programs like `gnuplot` [16] or `Topdrawer` [17]. Auxiliary programs of the GNXAS suite are useful for plotting phase-shifts and MS signals (`phaplo`, `readder`), for statistical error evaluation purpose (`contour`) and for reconstruction of the pair distribution function (`grrec`)

Another important additional program is used for the evaluation of the error bars on best-fit structural parameters:

(8) `contour`, to convert character contour plots in the `fitheo` output and facilitate error analysis.

Three additional programs are also available to facilitate the analysis of disordered systems (including liquids) where the structure is efficiently described in terms of radial distribution functions $g(r)$. These are:

(9) `grfit` is a fitting utility of model radial distribution function data (not shown in Fig. 8.1).
(10) `grxas` is designed to calculate the EXAFS signal associated with a given partial radial distribution function (not shown in Fig. 8.1).
(11) `grrec` allows reconstruction of $g(r)$ after model refinement (also in connection with the `fitheo` output).

Moreover, the GNXAS distribution includes miscellaneous programs (not shown in Fig. 8.1) useful for preparation of the XAS experiments and for pre-analysis:

(12) `xasam` is a program for sample thickness optimization for measurements in transmission geometry.
(13) `jesf` is an automatic routine for EXAFS extraction, Fourier transformation, and noise evaluation. This program is useful for a qualitative data-analysis and it has been routinely used at the several beamlines for on-line data analysis (for this reason it is believed to be used more than 10^6 times).
(14) `fitedg` is a fitting program for edge structures using analytical functions.
(15) `deconv` is a deconvolution program suitable to deconvolve the core-hole lifetime broadening in high quality spectra.

The *extended* GNXAS package includes also several other advanced programs that are available to users upon request (see full description in Ref. [5]):

(16) `edxrd` is a utility program able to calculate energy and angular dispersive x-ray diffraction patterns using an input file for the structure based on `crymol`.
(17) `peakfit` is a second utility program performing multiple peak-fitting refinement with different functions in presence of complex background, including Bragg peaks in x-ray diffraction.

The next sections discuss two important recent advances implemented in the GNXAS package: the design of a graphical user interface (GUI) [6] and the realization of an updated multiatomic RMC refinement package RMC-GNXAS including the `rmcxas` program using input files provided by GNXAS and by a specialized `crymol_rmc` routine that generates the initial atomic configuration (see also [7, 12]). Those programs are also available upon request.

8.3 Overview of the GNXAS Graphical User Interface

In this section we briefly present the graphical interface for the GNXAS suite of programs [6]. This is an ongoing project which aims to help new users to prepare the input files and run all the programs in the GNXAS package.

As previously mentioned, GNXAS was developed in 1990 and at that time text-based input and output files were standard as most programs were running in terminal windows. Since then, both computers and operating systems evolved hugely, and nowadays the standard is a point-and-click scheme. Since the large majority of GNXAS support requests are related to errors and mistyping while preparing the input files, the creation of a graphical interface for using the different programs in the GNXAS package is expected to provide a user-friendly platform solving most of the practical difficulties of the GNXAS new users.

8.3.1 Practical Implementation of the GUI

We have designed a graphical user interface (GUI) called w-GNXAS [6] which controls the text-based operations of GNXAS. This choice is thought to be a good compromise for maintaining the necessary flexibility of accessing each stage of the calculations, within a modern conception of a user-friendly program. The interface, created by using the Python library "wxPython" [18], is platform independent and can be executed on any current operating systems, such as Linux, Windows and Macintosh.

The idea is to provide a simplified window-based panel that facilitates the creation and compilation of the input files for the various programs, allowing to run the GNXAS programs with a simple click and save input set of commands in form of files for successive refinements. Previously existing input files, compatible with those described in the GNXAS handbooks [5], can also be loaded into the w-GNXAS window that will be automatically prepared for visualization and corrections in the specific fields.

Following the scheme of the GNXAS package, each program has its specific tab in the interface as can be seen in the screenshots shown in Fig. 8.2. We have not yet implemented a specific control panel for all the GNXAS programs, but those of the main GNXAS chain are already available: `crymol`, `gnpeak`, `phagen`, `gnxas`, and `fitheo`. A specific panel is also reserved to the program `xasam`, that is used to calculate the optimal sample thickness fro XAS measurements in transmission geometry.

In the next section we will briefly describe some specific details of each program tab.

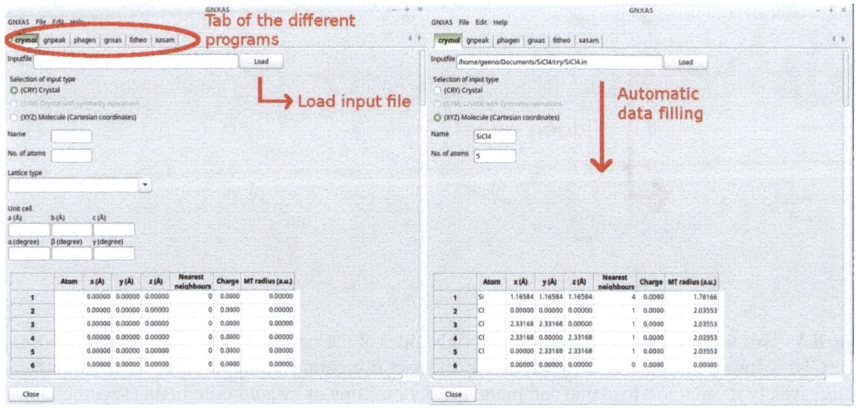

Fig. 8.2 Two screenshots (left and right side) of the control panel of the *w*-GNXAS interface. In the upper part of each screenshot we can see the tab for selecting six different input panels, each one corresponding to a specific GNXAS program. Here, the `crymol` tab is selected and appears highlighted in the main panel. The first input dialog is for loading a previously prepared input file for `crymol`, by clicking the *Load* button, a window for choosing the file will be opened (see *red arrow* in the left screenshot). After selecting the file, the interface will automatically fill the *blank fields* with the input data contained in the loaded command file (hSn_cry.in in this case, see right-side screenshot)

8.3.2 *w*-GNXAS *Control Panels*

As discussed in the preceding section, a single window provides an interface for most GNXAS programs as shown in Fig. 8.2. It is recommended to create a different folder for input and output files of each program in order to prevent confusion between the different files created during the analysis process.

The first tab of the graphical interface, from left to right, controls the input of the `crymol` program. When selected, the first tool appearing in this window is used for loading an existing input file for the `crymol` program. After loading a selected file, the fields corresponding to the input parameters of the `crymol` program are automatically filled (see second screenshot in Fig. 8.2). This is helpful to run or modify a previously saved input file. If no input file is loaded, then the user has to fill the blank fields manually.

The required inputs are of various type, mostly are text-based, but there are also multiple choice options and drop-down lists. In certain cases a specific input may be not required and it will then appear greyed out or hidden. For example, lattice parameters are required only for crystalline systems, therefore selecting the option ‘XYZ’ (molecule) no further unused parameters will be required and the panel related to the’CRY’ (crystal) option will be hidden (see Fig. 8.3, left-side screenshot). When needed, some input parameters are automatically filled in by the GUI software. For example, a cubic lattice type is defined by a single cell parameter a. In this case, all the angles are automatically set to 90° and the b, c parameters are set equal to a (see

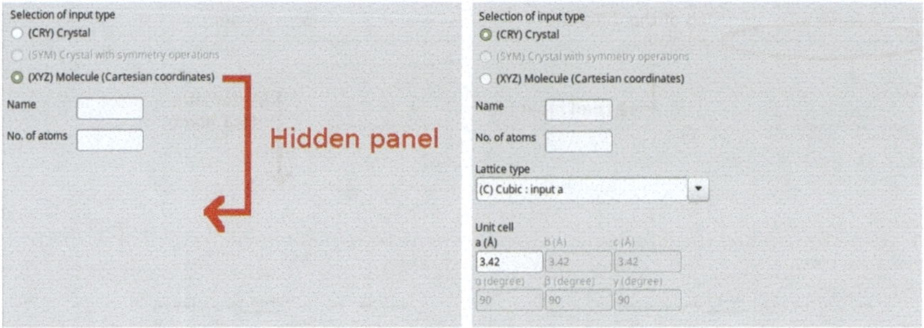

Fig. 8.3 Two screenshot selections (left and right side) of the `crymol` control panel of *w*-GNXAS. The GUI software interface is designed to help the user in creating a suitable set of input parameters (successively saved into a loadable command file) by omitting or by auto-completing specific inputs in certain cases. Two examples are shown here. Left-side screenshot: the window with the lattice parameters is not shown in the molecular XYZ case (hidden panel, *red arrow*). Right-side screenshot: when a cubic lattice is selected (crystal option, CRY), angles will be automatically set to 90° and cell parameters *b* and *c* will be set equal to the input value *a*

Fig. 8.3, right-side screenshot). These are just some examples and the same principle is applied throughout the *w*-GNXAS interface.

After completing all the entries, an input file can be created or updated by using the *Save to a file* button in the lower portion of the window (see Fig. 8.4). The saved file can then be loaded and used to run `crymol` as it is. This is done using the *Load* button in the lower part of the window shown in Fig. 8.4 and then selecting the *Run* button on the right. The log of the `crymol` calculation will appear in the lowest window of the panel, containing information on the run and showing possible errors or warnings occurred.

After the calculation is concluded, in absence of computing problems, the standard output files of each GNXAS program are created in the same directory of the input files and graphical outputs can be plotted using for example TopDrawer [17] and/or gnuplot [16]. In the example reported here, which refers to the molecule $SiCl_4$, one of the data-analysis examples in our website, output files are named using the three characters "abc" = SCl (see Fig. 8.1) therefore the user will find SCl.ato, SCl.in, SCl.top files in the proper directory.

The control panels of the `gnpeak` and `phagen` program are relatively simple, because the inputs are usually automatically generated by the preceding `crymol` program as illustrated also in the GNXAS flow diagram reported in Fig. 8.1.

In the `gnpeak` tab (Fig. 8.5) an input file can not be loaded, neither created, because only four empty fields have to be filled as inputs for operation: (1) the name of .ato file created by `crymol` without the extension; (2) the tolerance distance, which is the minimum distance to consider two paths different; (3) the cut-off distance for the sub-cluster size counting. Both the (2) and (3) parameters can be selected by default (inserting 0.0 as input), being respectively 0.0002 Å and the maximum size of the cluster; (4) the last input is the directory of the file .ato, in which also the outputs (in this example SCl.gnp, SCl.chi, SCl.gnx) will be saved.

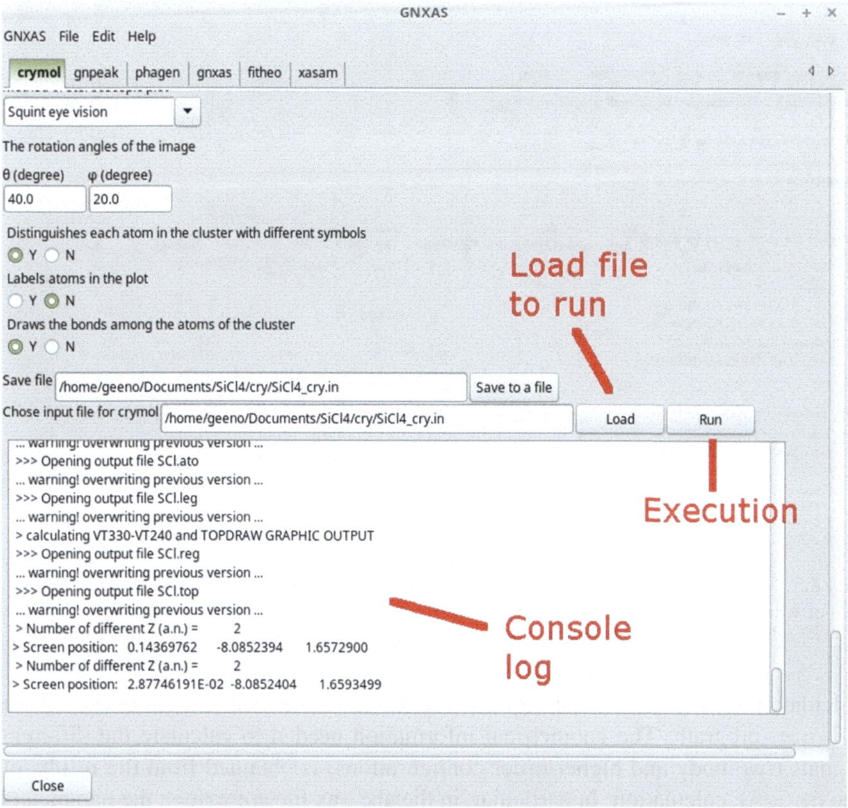

Fig. 8.4 Screenshot of the lowest portion of the `crymol` panel in the *w*-GNXAS interface. The last fields appearing in the panel are used to create an input file for `crymol` using the *Save to a file* button. When selected, all the data inserted by the user will be saved to a file with the appropriate formatting. This file can then be run by loading it and executed with the *Run* button (highlighted in figure). The console dialog (log) is shown in the final window of the panel, showing the output of the calculation

The `phagen` tab has only the load and run part because the input file (in the example considered SCl.in) is created automatically by `crymol` and modifications should only be done by expert users. Again, the important output files related to the phase-shift calculations (SCl.tl, SCl.s0) are created in the same directory of the input file.

The `gnxas` tab (screenshot shown in Fig. 8.6) requires more input parameters by the users and the design of the panel is very similar to that of the `crymol` program. Similarly to the `crymol` window, an existing input file can be loaded with the first entry which will auto-complete all the inputs. Otherwise, the user can write the entire set of input parameters from scratch and then save them to file. A good practice is to load a standard input file for `gnxas` as obtained from the set of data-analysis examples and use it as a template that can be modified by the user for the current

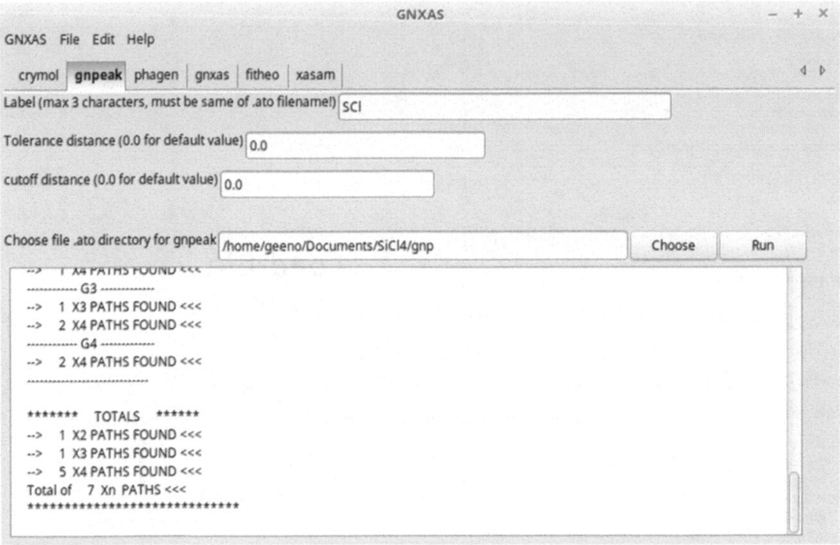

Fig. 8.5 Screenshot of the `gnpeak` window in the program. Only four options can be selected, two of which can be left to the default value (0.0)

calculation. The tl and s0 file requested in the first two fields are produced by the `phagen` program. The geometrical information needed to calculate the different signals (two-body and higher-order configurations) is obtained from the results of the `gnpeak` calculation. In particular, in the abc.gnx file are written the parameters to insert in the g2 and g3 tables (see as an example the text of the SCl.gnx window reported in Fig. 8.6). The `gnxas` window contains the same tools as the `crymol` one, so the input commands can be saved, loaded and run again, checking the results of the calculation.

The `fitheo` window has also a layout analogous to the `crymol` and `gnxas` tab, in which an input file can be loaded. In this case, it is particularly convenient to start from a previously saved input file, that can be taken by one of the data-analysis examples. We refer to the GNXAS handbook [5] for the detailed explanation of all the input needed to run `fitheo` for the structural refinement of XAS data. The table of the fitting parameters is the first and one of the most important entries for this program. All the parameters to be refined are defined within this table. The parameter number (first column labeled as "Number") is used to refer to each specific fitting parameter in the successive input lines (see Fig. 8.7). In this way, if we need to make some adjustment to the value of the parameter or its fitting range, we just change the value in the table without further modification in the input commands.

For simplicity, the current version of the *w*-GNXAS interface does not incorporate all the features of `fitheo`. In particular, with the present version it is possible to perform the structural refinement of only one file containing raw XAS data (single edge) and the number of structural signals that can be inserted for this purpose is

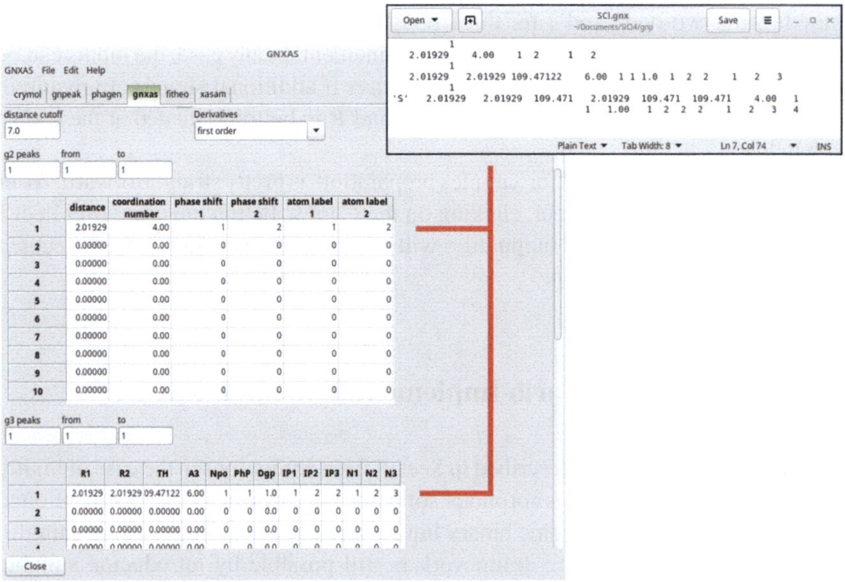

Fig. 8.6 In the gnxas tab the signal corresponding to different geometrical configurations are calculated. These information are contained in the output files of the crymol program. In the example shown, $SiCl_4$, we have one two-body and one three-body signal related to the Si K-edge. The input for the atomic structure can be taken from the output of the gnpeak program (see text of the SCl.gnx window in the upper-right part of the figure)

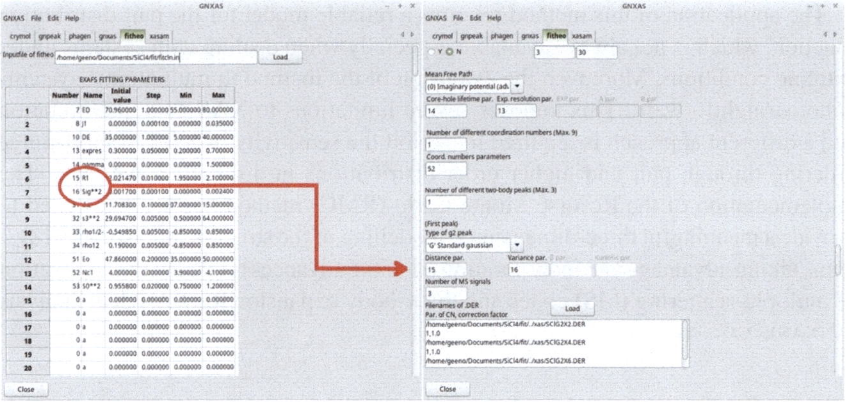

Fig. 8.7 Two screenshots of the fitheo tab in the graphical interface. In the left image the table of fitting parameters is shown: each fitting parameter is defined a number which will be used as a reference throughout the window. In the example shown, the bond distance R1 and variance (Sig**2) for a given atomic configuration are highlighted (see *red arrow*). These parameters are used as structural parameters for the two-body peak of $SiCl_4$ and its various MS signals calculated by gnxas. In order to do so, we insert their respective numbers (15 and 16in the example) in the appropriate boxes (see screenshot in the right-side). Detailed explanations for the fitting parameters and for the fitheo inputs are reported in [5]

limited (3 for two-body and 1 for three-body). In many practical cases this will be sufficient to perform a successful structural refinement. In any case, the inputs can be saved to a file and manually modified by the user if additional signals are required. The input file can then be run with the *Load* and *Run* buttons situated at the end of the window.

The `xasam` window, used for samples preparation, is pretty straightforward. After fulfilling all the requested input, clicking on the *Choose* button the user will select a directory where the input and output files will be saved. The program is then executed by clicking on the *Run* button.

8.4 Reverse Monte Carlo Implementation on GNXAS

The 'peak fitting' approach described in Sect. 8.2 has limitations in the case of highly disordered systems, such as amorphous solids or liquid matter [19]. As mentioned before, for elemental melts, ionic binary liquids and aqueous solutions a meaningful XAS data-analysis within this framework is still possible by introducing suitable physical constraints in the refinement of the short range peak of the pair distribution function provided by diffraction experiments or molecular modelling. The two-body distribution function $g(r)$ is decomposed into a first peak and a long range tail for which the corresponding XAS signals are calculated. The shape of the short range peak is then fitted to the experimental data with a suitable set of constraints. For more details the reader is referred to the original paper [10] (and [11] for binary systems).

The application of this method requires a reliable model for the pair distribution function, which is not always available especially when dealing with systems under extreme conditions. Moreover, the extension of the method to multiatomic systems is not straightforward. This imposes severe limitations to XAS data interpretation and a different approach is required to exploit the sensitivity of XAS to short-range ordering through pair and higher order distributions in disordered materials. The implementation of the Reverse Monte Carlo (RMC) method has been designed to provide a meaningful three-dimensional modelling of the structure of disordered systems, taking advantage and incorporating all of the advances related to the application of multiple-scattering (MS) codes and the n-body expansion for XAS data-analysis (GNXAS).

8.4.1 RMC Background

Reverse Monte Carlo is an *inverse* modelling technique introduced by McGreevy and Pustzai [20] for producing three-dimensional structural models from x-ray and neutron scattering experimental data. The method is a variation of the standard Metropolis Monte Carlo (MMC) algorithm. A series of atomic configurations are iteratively generated by random moves of single atoms until the model gives the best agreement with one or more sets of experimental data within their uncertainties.

Interatomic potentials are not used for RMC and very few assumptions on the structure are required. In principle only number density and chemical composition are needed for RMC modelling, although physical constraints are usually introduced, as discussed in more detail in the original papers (see [21] for a review) and in the following sections. One of the main strengths of the RMC approach is that experimental data from different and complementary techniques are refined and constraint the model simultaneously. Finally, the outcome is a set of three-dimensional structural models for the investigated system on which a full statistical analysis can be performed to derive the pair distribution functions, the distribution of bond angles and to identify specific local atomic arrangements.

Although initially applied to derive structural models by refinement of the structure factor from scattering experiments, the method is very general and finds application with any experimental technique for which the measured signal can be calculated from the atomic coordinates.

The application of the RMC for XAS data-analysis was initially introduced by Gurman and McGreevy [22] and since then a few groups implemented and applied this technique in several works on liquids and glasses as well as crystalline systems (see for example [7, 12–14, 23–25]). As shown in the next section, the RMC-GNXAS software currently represents an advanced tool for XAS structural refinement using the RMC technique.

8.4.2 *Overview of* RMC-GNXAS

The RMC-GNXAS software implemented within the GNXAS package [7] allows Reverse Monte Carlo structural modelling of both molecular and condensed structures. For condensed systems, like liquids, glasses or crystals, the RMC algorithm is applied simultaneously to the XAS signal which probes the local structure around each photo-absorbing species, and to a model pair distribution function $g(r)$. The $g(r)$ function can be obtained by x-ray/neutron diffraction data or computer simulations, and is intended to provide a correct constraint for medium and long range ordering, because XAS is sensitive only to short-range.

The `rmcxas` program, core of the RMC-GNXAS chain, has been now widely optimized and extended to multiatomic systems. Some of the initial features have been therefore upgraded to improve computational efficiency and new capabilities have been added. In particular, multi-edge XAS refinement and proper account for partial pair distributions $g_{\alpha\beta}(r)$ related to the different atomic species (α, β) have been introduced. The underlying algorithm is summarized below and shown in Fig. 8.8. A more detailed description of the calculation steps is given in the following paragraphs.

1. The initial configuration is prepared. For molecular systems, the initial configuration is an ensemble of molecular replicas. For liquids and glasses, the initial

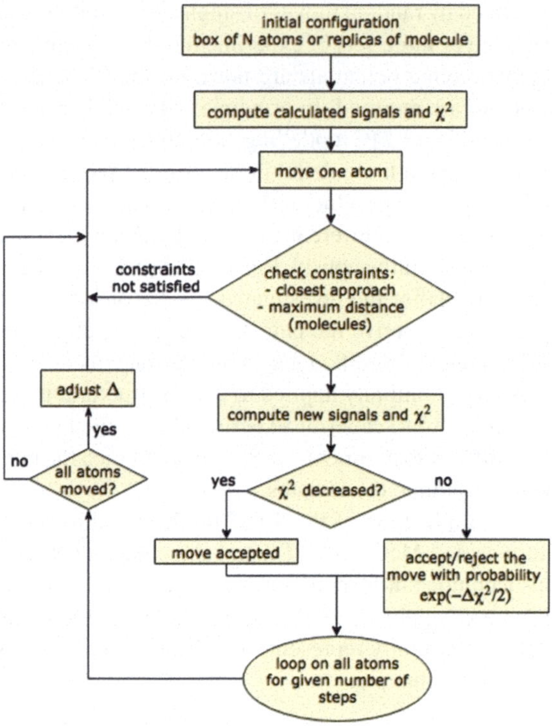

Fig. 8.8 Schematic diagram of the RMC algorithm as implemented in the RMC-GNXAS package

configuration consists in a set of N atoms within a simulation cell of fixed volume V with periodic boundary conditions. Such initial configuration may be generated by randomly choosing atomic positions within the simulation box or derived from prior simulations. It can be also generated from a crystal lattice using the `crymol-rmc` program (see Sect. 8.4.3.1). In any case, it must have the same chemical composition and number density (N/V) of the system under investigation.

2. Calculation of the XAFS signals $\chi^C(k)$ and of the pair distribution function $g^C(r)$ to be compared with the respective experimental quantities, i.e. the $\chi^E(k)$, which is the extracted EXAFS signal, and the model pair distribution function $g^E(r)$ (or the set of partial distribution functions) obtained for example from scattering experiments.
3. The residual χ^2 is calculated using the following equation:

$$\chi^2 = \sum_{n=1}^{N_{\mathrm{EDGE}}} \sum_{i=1}^{N_{\mathrm{XAS}}} \frac{\left[\chi_n^E(k_i) - \chi_n^C(k_i)\right]^2}{\sigma_{n,i}^2} + \sum_{\alpha,\beta} \sum_{j=1}^{N_g} \frac{\left[g_{\alpha\beta}^E(r_j) - g_{\alpha\beta}^C(r_j)\right]^2}{\sigma_{\alpha\beta,j}^2}, \tag{8.18}$$

where the right hand side has two terms. The first corresponds to the XAS part for which the calculated signal is compared with the EXAFS signal determined

by a preliminary data-analysis and $\sigma^2_{n,i}$ are the noise functions associated with the experimental data. The noise functions can be evaluated and stored on specific files for RMC refinement by the auxiliary program `noise`. Since for multiatomic systems simultaneous multiple-edge signal refinement is possible, the first sum indicates a sum over different XAS edges while the second one is a sum over experimental points. The second term in r.h.s., omitted in molecular cases, correspond to the pair distribution functions $g^C_{\alpha\beta}$ calculated from atomic coordinates and compared with model distributions obtained from x-ray/neutron diffraction or computer simulations. For multiatomic systems, it is also possible to include simply the total distribution function $g(r)$ instead of the partials (in some cases partials may be not available or affected by large uncertainty). σ^2_j (or $\sigma^2_{\alpha\beta,j}$ for partials) is the noise associated with the pair distributions.
It is worth noting that both the σ^2 functions and the number of points (N_{XAS}, N_g) control the relative weight of each data set included in the RMC simulation.
4. A new configuration is created by moving one atom at random and the corresponding new signals are calculated. The new residual χ^2_n is evaluated using (8.18). The typical maximum displacements for each atom move is initially set to a value (fraction of atomic distances) which is tuned for reaching acceptance ratios of about 50% during the evolution of the RMC refinement (see below).
5. The atomic move is either accepted or rejected according to the Metropolis-Hastings algorithm. If the new residual is smaller than the older one, $\chi^2_n < \chi^2_o$, the new configuration is accepted. Otherwise, the new configuration is accepted with probability $\exp(-(\chi^2_n - \chi^2_o)/2)$. In this way, the modelling process is able to explore the whole configurational space without being frozen in local minima.
6. If the move is rejected, the atomic positions are maintained. Otherwise the new configuration is saved and the process repeated from step 4 until each atom in the box has been moved.

The RMC loop continues for several cycles until the residual χ^2 reaches an equilibrium value. After convergence, a desired number of *equilibrium* configurations can be saved which are the structural models consistent with the whole set of experimental data and statistically independent.

Before running the RMC simulation, some preliminary calculations and data are required. We explain here different important steps and calculation details involved.

The choice of the initial configuration is one of the main steps of any RMC refinement. As reported in earlier studies [21], the final RMC models should be independent, in principle, of the starting model, although this initial structure may affect the number of steps needed to achieve convergence. Also, for molecular systems, the molecular structure should possibly be pre-defined. Depending on the investigated material, there are several strategies for producing the RMC initial atomic configuration. The simplest approach for disordered systems is to position atoms at random within a simulation box of size chosen to reproduce the experimental density. As a step further, constraints can be applied to prevent atoms to be closer than a cut-off distance (closest approach distance). Another possible approach is to start from a disordered crystal structure with the correct density for the liquid. If a crude model for the interatomic potentials is available, classic Monte Carlo or molecular dynamics simulations could be also used to produce a trial starting configuration for the RMC refinement.

In the current RMC-GNXAS implementation, the initial atomic configuration can be generated by a suitable program (`crymol_rmc`) as a disordered crystalline structure with given Gaussian disorder around each atomic position. Initial data are type and atomic positions within a unit cell, number of atoms and size of the box, calculated in order to obtain the right density of the modelled system. Periodic boundary conditions are applied to overcome finite size effects. For molecular systems, the program generates an ensemble of molecular replicas, with defined number of neighbors and desired Gaussian distribution of intramolecular distances. This program is based on the program `crymol`, part of the GNXAS package.

Before performing the simultaneous RMC refinement of both the XAS signals and the model $g(r)$ (or partials), the initial configuration is usually refined to reproduce the shape of the pair distribution function $g(r)$ only. This is done, for example, by using directly the `rmcxas` program and executing a RMC procedure limited to the $g(r)$. In this way, convergence is achieved faster. This is also important for the successive RMC refinement, because the XAS sensitivity is limited to a few Å around the photoabsorber and the medium and long-range order has to be constrained with information provided by complementary techniques as discussed before.

Density and closest approach distances (i.e. the minimum interatomic distance allowed for each couple of atomic species) are the main constraints usually included in any implementation of the RMC basic algorithm [21]. The choice of closest approach distances is not always obvious and it is generally based on previous experimental determinations of the pair distribution functions or estimated from atomic or ionic radii. In the `rmcxas` program different closest approach distances for different pairs of atoms can be chosen.

An additional constraint can be imposed for molecular cases: a maximum distance which atoms cannot overcome. Different maximum distances can be imposed for different atom type pairs. Since no pair distribution function is present in this case, a maximum distance can prevent atoms to diverge to unphysical distances if no XAS signal for a specific type of atom is present into the simulation.

In (8.18) we compare experimental quantities with the calculated ones from atomic coordinates. For the partial pair distribution functions $g^{C}_{\alpha\beta}(r)$ we start from their definition (see for example [26]) as the ratio of the probability of finding one atom of type β at a distance between r and $r + \mathrm{d}r$ from one atom of type α to the probability of finding one atom of type β at a distance between r and $r + \mathrm{d}r$ of one atom of type α in an ideally homogeneous material. During the RMC refinement, these probabilities are estimated from the counts and binning of distances in calculated histograms. Given a distance interval $[r, r + \Delta r]$, the probability P of finding one interatomic distance in that range is estimated by the ratio of the number n of distances in that interval and all the N distances found in the system. The number of distances available is different if we consider identical or distinct atom types, so we must distinguish the two cases to compute the different partials:

$$P_i^{\alpha\beta} = \frac{n_i^{\alpha\beta}}{N_\alpha N_\beta}; \qquad P_i^{\alpha\alpha} = \frac{2n_i^{\alpha\alpha}}{N_\alpha(N_\alpha - 1)}, \tag{8.19}$$

where N_α and N_β are the number of atoms of type α and β in the actual simulation box, and $n_i^{\alpha\beta}$ ($n_i^{\alpha\alpha}$) are the number of distances in the ith bin of the histogram for different (identical) atom types. The probability for the ideal homogeneous material is given by the ratio of the volume of the spherical shell in the same range $[r, r + \Delta r]$ to the total volume available:

$$\frac{4\pi}{3V}\left[r^3\right]_i^{i+1} . \tag{8.20}$$

So in the end we have:

$$g_{\alpha\beta}(r_i) = \frac{n_i^{\alpha\beta}}{N_\alpha N_\beta} \frac{3V}{4\pi\left[r^3\right]_i^{i+1}} \tag{8.21}$$

$$g_{\alpha\alpha}(r_i) = \frac{2n_i^{\alpha\alpha}}{N_\alpha(N_\alpha - 1)} \frac{3V}{4\pi\left[r^3\right]_i^{i+1}} . \tag{8.22}$$

Since we are dealing with a finite system in the simulations, the calculated $g(r_i)$ values are estimates of the partial pair distribution functions. The uncertainty over the calculated $g^C_{\alpha\beta}(r)$ and $g^C_{\alpha\alpha}(r)$ depends on the number of distances used to compute them, so ultimately on the number of atoms used in the simulation and on the bin size. A too small number of atoms or a too small bin size would yield large statistical uncertainty and 'noisy' calculated $g(r)$. On the other hand, selection of a box with too many atoms increases the calculation time, which scales as N^2, and a too large bin size results in a sparse pair distribution function. A more quantitative analysis of this point can be found in a previous paper [7]. For the total pair distribution function we proceed in an analogous way, but without distinguishing between the different type of atoms.

For the calculation of the total χ^C EXAFS signal, the phase shifts are calculated *ab initio*, by using `phagen`, for a system having structure and chemical composition similar to the liquid or amorphous material modelled by RMC (or for molecules, crystals). Two body $\gamma^{(2)}_{\alpha\beta}(k)$ signals are calculated for a grid of bond distances by using the `gnxas` program. The $\gamma^{(2)}$ signals, given as input to the RMC program, are then interpolated for all the actual distances present in the model structure. In order to take into account damping of the XAFS signal due to experimental resolution and many body effects, the amplitude of the calculated signal is reduced by a factor $\exp(-R_p^2\sigma^2_{\text{exp}}/2k^2)S_0^2$ where R_p is the path length, σ^2_{exp} is the HWHM of a Gaussian energy resolution function and S_0^2 is the amplitude reduction factor. At the present stage, higher order n-body terms related to higher order correlations are not included directly in the computation of χ^C. Signals obtained from a fit of experimental data using `fitheo`, can be taken into account, but these are not calculated directly from the atomic coordinates at each step of the simulation. Direct inclusion of such contributions, essential for extending the applicability to crystalline or complex molecular systems, would increase dramatically computational time, scaling as N^3. Their inclusion need further developments that are presently under evaluation.

At each atom move, N interatomic distances are changed due to the movement of a single atom. Appropriate calculation strategies have been implemented for optimizing the calculation of the variation of the XAS signal and the $g^C_{\alpha\beta}$ associated with the modification of the atomic positions during the RMC loop. The calculation time at each step is therefore considerably reduced and it is possible to increase the size of the model. Utilization of novel hardware devices to further decrease computational time (such as GPUs, for example) is now under development.

The maximum size of the atomic move Δ (usually 0.1–0.3 Å) is also a relevant parameter in RMC modelling, as it determines to which extent the configurations space is explored. In `rmcxas`, the value of Δ is re-adjusted after a selected number of RMC steps (given by the user), in order to have a ratio of accepted RMC moves of ~0.5. This value should be checked at each step because a value too small would be insufficient to change the configuration in a significant way.

8.4.3 A Practical Example: AgBr

In this section we give a short description of the application of RMC-GNXAS for multiatomic systems showing the input files prepared for liquid AgBr [15]. A schematic overview of the work flow with the different steps required to perform a full simulation, is shown in Fig. 8.9.

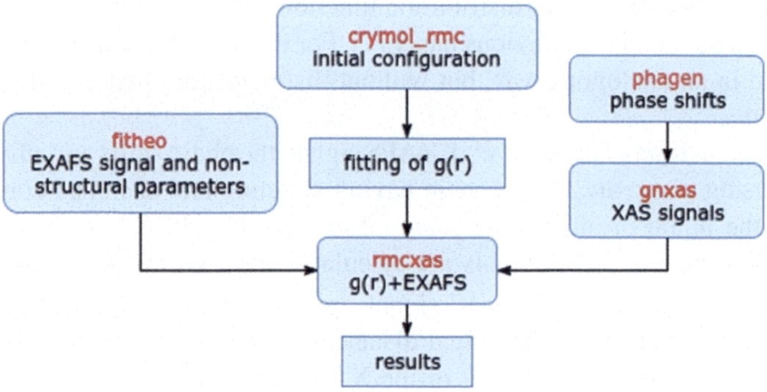

Fig. 8.9 The typical flow diagram of the RMC-GNXAS package is summarized in this figure. A complete RMC refinement of EXAFS data is obtained starting from an initial configuration (prepared by `crymol-rmc`) and the associated γ signals prepared by the standard GNXAS chain (`phagen` and `gnxas`). The initial configuration, usually prepared to reproduce a model pair distribution $g(r)$ (see text), is used by the `rmcxas` program for the RMC refinement of EXAFS data pre-analyzed by the GNXAS program `fitheo`

8.4.3.1 Initial Configuration: crymol_Rmc Program

The very first step of the RMC simulation is to create the initial configuration. As explained in the previous section, this can be done by using the program `crymol_rmc`, which is based on the program `crymol`. The program runs (under Linux) with the command:

crymol_rmc < crymolrmc.in

The input file **crymolrmc.in** (or any other name given by the user) used for the analysis of silver bromide is given below:

```
CRY                                         CARD 1
AgBr                                        CARD 2
C                                           CARD 2.1
6.0911                                      CARD 2.2
8                                           CARD 3
 1 'Ag' 0.5000 0.5000 0.5000 4 0.0 2.31 CARD 4
 2 'Br' 0.0000 0.0000 0.0000 4 0.0 2.31 -
 3 'Ag' 0.5000 0.0000 0.0000 4 0.0 2.31 -
 4 'Br' 0.0000 0.5000 0.5000 4 0.0 2.31 -
 5 'Ag' 0.0000 0.5000 0.0000 4 0.0 2.31 -
 6 'Br' 0.5000 0.0000 0.5000 4 0.0 2.31 -
 7 'Ag' 0.0000 0.0000 0.5000 4 0.0 2.31 -
 8 'Br' 0.5000 0.5000 0.0000 4 0.0 2.31 -
NNY                                         CARD 5
1                                           CARD 6
'Ag' 'Br' 3.0                               CARD 7
7.0                                         CARD 7.1
AgB                                         CARD 8
1,1                                         CARD 9
0.05                                        CARD 10
1                                           CARD 11
1                                           CARD 12
1875.20,0.                                  CARD 13
3,-.2,100.,.05                              CARD 14
5,-1                                        CARD 15
'S',1,40.,20.,'S',' ','S'                   CARD 16
6,6,6                                       CARD 17
0.05                                        CARD 18
0.05                                        -
```

This is the same input file for the `crymol` program, with two additional items at the end (CARD 17 and CARD 18) when the CRY option is selected (CARD 1). It is worth mentioning that when dealing with molecules (XYZ option for CARD 1), the number of molecular replica is needed (an additional CARD 19 is used for the total number of molecules of the calculation).

For the purpose of RMC, not all the cards are actually used. We report here the important ones:

CARD 1: Selects between 'CRY' for simple crystals or 'XYZ' for molecules.

CARD 2: Identifying name of the material.

CARD 2.1: Crystallographic unit cell type. For liquids a cubic unit cell (C) should be built. For molecules this card and the next one are not used.

CARD 2.2: Size of the unit cell given in Å. This is a key parameter, since it defines the density of the system and should be chosen to reproduce the experimental number density of the liquid. In the case of AgBr, the unit cell contains 8 atoms and the resulting number density given by $8/(6.0911)^3 \simeq 0.0354$ atoms/Å^3 in agreement with the literature value [27].

CARD 3: Number of atoms in the unit cell (or in the molecule).

CARD 4: Specific information on each atom of the unit cell (or molecule). The first entry is just a sequential number labeling each atom (up to the number specified by the previous card). The second column is the chemical symbol of the atom (e.g. 'Ag','Br',' C'; note a blank character is required before single-letter symbols). Then there are the x, y and z coordinates: for molecules one must insert Cartesian coordinates in Å, while for CRY, like in the example, the dimensionless fractional coordinates must be specified. The other columns do not affect the configuration and can be ignored.

CARD 5 to CARD 16 are not important for the purpose of RMC initial configuration, apart from CARD 8 which specifies the name for the output file.

CARD 17 n_a, n_b, n_c are the number of cells along each crystallographic direction. $n_a \cdot n_b \cdot n_c$ multiplied by the number of atoms per unit cell (CARD 3) gives the total number of atoms within the simulation box.

CARD 18 Standard deviation on the atomic positions. This card is necessary to produce an initial configuration with Gaussian disorder around average atomic positions. This card has to be repeated for each atomic species.

The `crymol_rmc` program provides as a starting configuration a disordered crystalline or molecular model with well-defined average distances, number and type of atoms and density (or number of molecules). The coordinates of each atom within the simulation box are written in the output file **ABCxyz.rmc** (ABC given at CARD 8), here called **AgBxyz.rmc**. The file contains the coordinates of all the atoms and a header with some useful information:

```
  CRYSTAL    AgBr
 C! Cubic ->  a
   6.0911
   0.03539996   ! Atomic density
     6    6    6   ! Number of cells
    1728   ! Number of atoms
 1    ! Disordered crystal configuration
  2  ! Number of inequivalent atoms
  '   Ag' ! atom
   0.050000 ! Standard deviation
  '   Br' ! atom
   0.050000 ! Standard deviation
 START  ! Initial configuration for RMC run
    -0.416387200  -0.417111397  -0.415618539  '   Ag'
    -0.499927670  -0.497871548   0.498389006  '   Br'
    -0.416463912   0.499605536   0.498553693  '   Ag'
    -0.497876823  -0.415489465  -0.416263878  '   Br'
    ....
```

In particular, the order in which the different atomic species appear in the header is important, since in the input file for the `rmcxas` program they are identified by this number. This is explained in more details below.

As anticipated, for the purpose of XAS refinement of liquids or amorphous systems, the initial configuration given by `crymol_rmc` can be successively refined to reproduce the shape of the pair distribution function $g(r)$ (or the partials for multiatomic systems) obtained by diffraction experiments or computer simulations. This is an important aspect for the successive RMC refinement, because the XAS sensitivity is limited to a few Å and the medium and long-range order has to be constrained using complementary information from different techniques. For this purpose, `rmcxas` can be directly used for any RMC refinement of the model pair distribution of a given model system. The program needs in any case input data files (cards 5.1 and 5.6 as reported below) for the EXAFS signals and the associated noise, and the practical strategy for a simple RMC refinement of the $g(r)$ is to artificially increase the noise level (typical values are 10^{-2}–10^{-4}). When the RMC simulation is converged for a given pair distribution model, the EXAFS signals can be *turned on* improving the accuracy of the structural refinement.

8.4.3.2 RMC Simulation: `rmcxas` program

Once the initial configuration has been built, the Reverse Monte Carlo simulation can be set up by preparing an input file for the `rmcxas` program containing several cards (here called **rmc-gnxas.in**, but an arbitrary name can be chosen). Of course, the full RMC-GNXAS refinement process can proceed only if suitable EXAFS signals are available for the system under consideration (liquid AgBr in the present case). Therefore, as also illustrated in Fig. 8.9, an accurate EXAFS pre-analysis must be carried out for the extraction of the structural $\chi^E(k)$ signals in (8.18). This pre-analysis is usually carried out using the `fitheo` program of the GNXAS chain in order to have also consistent values for the non-structural parameters E_0 and S_0^2 and experimental resolution (see [3, 5] for their definition) to be included in the RMC procedure.

An example is given below:

```
AgBxyz.rmc                                  CARD 1
AgB                                         CARD 2
9.50                                        CARD 3
Y                                           CARD 4
2.00                                        CARD 4.1
1.90                                        -
2.80                                        -
2                                           CARD 5
../data/BrK_AgBr_725K.dat                   CARD 5.1
0                                           CARD 5.2
1.                                          CARD 5.3
2                                           CARD 5.4
13474.                                      CARD 5.5
../data/BrK_AgBr_725K.noise                 CARD 5.6
../data/AgK_AgBr_725K.dat                   CARD 5.1
0                                           CARD 5.2
1.                                          CARD 5.3
1                                           CARD 5.4
25514.                                      CARD 5.5
../data/AgK_AgBr_725K.noise                 CARD 5.6
2                                           CARD 6
3                                           CARD 6.1
1,1                                         CARD 6.2
../data/gr_AgAg_753K_Tas.dat                CARD 6.3
../data/gr_noise_cost.dat                   CARD 6.4
1,2                                         CARD 6.2
../data/gr_AgBr_753K_Tas.dat                CARD 6.3
../data/gr_noise_cost.dat                   CARD 6.4
2,2                                         CARD 6.2
../data/gr_BrBr_753K_Tas.dat                CARD 6.3
../data/gr_noise_cost.dat                   CARD 6.4
1.90,9.50                                   CARD 7
1                                           CARD 8
1.90,.1                                     CARD 8.1
```

```
0.                                          CARD 9
1000                                        CARD 10
1,200                                       CARD 11
800                                         CARD 12
0.05                                        CARD 13
1                                           CARD 14
13478.                                      CARD 15.1
0.725                                       CARD 15.2
1.5                                         CARD 15.3
25518.                                      CARD 15.1
0.825                                       CARD 15.2
1.8                                         CARD 15.3
2                                           CARD 16
1,2                                         CARD 16.1
53                                          CARD 16.2
../xas_Br_K/BABG201.DER                     CARD 16.3
2,2                                         CARD 16.1
44                                          CARD 16.2
../xas_Br_K/BBBG201.DER                     CARD 16.3
2                                           CARD 16
1,1                                         CARD 16.1
52                                          CARD 16.2
../xas_Ag_K/AAAG201.DER                     CARD 16.3
1,2                                         CARD 16.1
53                                          CARD 16.2
../xas_Ag_K/AABG201.DER                     CARD 16.3
5.6,0.4                                     CARD 17
1                                           CARD 18
1                                           -
1                                           -
```

The description of the input lines is as follows:

CARD 1: Initial configuration file name. The program will choose as starting configuration the first one preceded by the line "START". This is useful to select a specific starting configuration when running the simulation several times.

CARD 2: Three characters string used for the output files.

CARD 3: Maximum cutoff distance. If the input distance is greater than the natural cut-off (half of the simulation box), it will be reduced to the natural cut-off.

CARD 4: Use different closest approach distances for each pair of atomic species (Y) or use the same closest approach distance for all the pairs (N).

CARD 4.1:	If the previous card was N, only one closest approach distance has to be inserted. Otherwise a number $b(b+1)/2$ cards are needed, where b is the number of different atomic species present in the material (in the example, $b = 2$).
CARD 5:	Number of experimental EXAFS signals to be used for the RMC refinement (in the example two signals measured at the Ag and Br K-edges). The successive cards (5.1–5.6) have to be repeated for each data set.
CARD 5.1:	EXAFS signal file name.
CARD 5.2:	Scale of the experimental signal: 0 for k (Å^{-1}), 1 for E (eV/keV)
CARD 5.3:	Weight on the input data, if $k^w \chi(k)$ is used.
CARD 5.4:	Number of the photo-absorbing atom. In the example, 'Ag' corresponds to 1 and 'Br' to 2, because they appear in this order in the initial configuration file. The first data set is Br K-edge and hence the photo-absorbing atom entry is 2, while the second is Ag K-edge and the corresponding photo-absorbing atom is 1. In this way there is not a pre-defined order in which the different data set have to be inserted.
CARD 5.5:	Edge energy in eV.
CARD 5.6:	Name of the file containing the noise function for the EXAFS signals ($\sigma_{n,i}^2$ of (8.18)). This can be estimated by using a program called `noise` (included in the package) and written in a specific file, whose filename is given here.
CARD 6:	This card is used for the $g(r)$ refinement. Three options are possible: 0 if no $g(r)$ are used in the simulation, 1 if only the total pair distribution function $g(r)$ is inserted, 2 if partial pair distribution functions are used. If option 0 is selected, the successive card is CARD 8. Instead if option 1 is chosen the successive cards have to be the data file name for the total $g(r)$ and its noise file name.
CARD 6.1:	If option 2 is inserted in the previous card, we have to specify the number n_g of partial $g_{\alpha\beta}(r)$ functions used. The successive cards must be repeated n_g times.
CARD 6.2:	This define the atom types of the corresponding partial pair distribution function, e.g. for Ag-Ag we insert 1,1, for Ag-Br 1,2.
CARD 6.3:	File name of the corresponding partial $g_{\alpha\beta}(r)$.
CARD 6.4:	File name of the noise data for $g_{\alpha\beta}(r)$.
CARD 7:	r_{min} and r_{max} for the $g(r)$ fitting in the RMC. Not needed if CARD 6 is 0.
CARD 8:	Choice of the $g(r)$ histogram binning. Option 0 if the same binning as the input $g(r)$ (total or partial) has to be used. Option 1 if the binning is specified by the user with the additional card 8.1.
CARD 8.1:	This is required only if 1 was chosen in the previous card. It is the r_{min} and the Δr for the $g(r)$ calculated by the program.
CARD 9:	Weight on the EXAFS signal in the output files.
CARD 10:	Number of RMC steps (for each atom).

CARD 11: Respectively, number of RMC steps between successive visualization and storage of configurations.

CARD 12: Number of RMC steps before calculating the average $g(r)$ over different configurations.

CARD 13: Initial maximum size of random displacement (in Å).

CARD 14: Number of RMC steps between updating the maximum size displacement.

CARD 15.1: This and the two following cards 15.2 and 15.3 refer to non-structural parameters (see [3, 5]) determined by a previous analysis (for example by `fitheo`) of the XAFS signals and must be repeated for each signal included. Here the E_0 energy is given.

CARD 15.2: S_0^2 factor.

CARD 15.3: Experimental resolution, HWHM in eV.

CARD 16: Number of different signals inserted to fit the EXAFS data. For each type of signal the successive three cards have to be specified. This must be repeated for each EXAFS data set.

CARD 16.1: Corresponding type of atoms for the signal, like CARD 6.2. For example for Ag-Ag we insert 1,1.

CARD 16.2: Number of .DER files to read.

CARD 16.3: File name of the first .DER file. The program will read the files up to the number specified by CARD 16.2.

CARD 17: Maximum distance cut-off r_c^E and standard deviation σ^E of the Gaussian smoothing used for the integration of the EXAFS signal. The EXAFS signal is calculated from r_c^E up to $r_c^E + 3\sigma^E$ with a Gaussian smoothing in order to avoid truncation problems.

CARD 18: Noise function on the $g(r)$: 0 constant noise, based on average value of the input noise file; 1 accounting for pair statistic; 2 read on the inserted noise files; 3 accounting for pair statistic, but tunable, in this case an additional card is required with the multiplying factor. This card has to be repeated for each $g(r)$ inserted into the fitting procedure.

The program runs with the command

rmcxas < rmc-gnxas.in

The .DER files are created by the `gnxas` program, and specific instructions about the practical use of this program can be found in [4, 5]. In the present example, the Ag-Br Ag K-edge $\gamma^{(2)}$ signals (in form of .DER files) were calculated in a range of distances between 1.90 and 7.00 Å with a step ΔR=0.10 Å, for a total of 52 different .DER files. If the cut-off distance $r_c^E + 3\sigma^E$ defined by CARD 17 results to be greater than the maximum distance R_{der}^M of the corresponding set of .DER files, the program updates the cut-off r_c^E appropriately (using $R_{der}^M - 3\sigma^E$).

While running the `rmcxas` program, many useful files are created to check the input data and for visualization of the simulation. Detailed information is stored

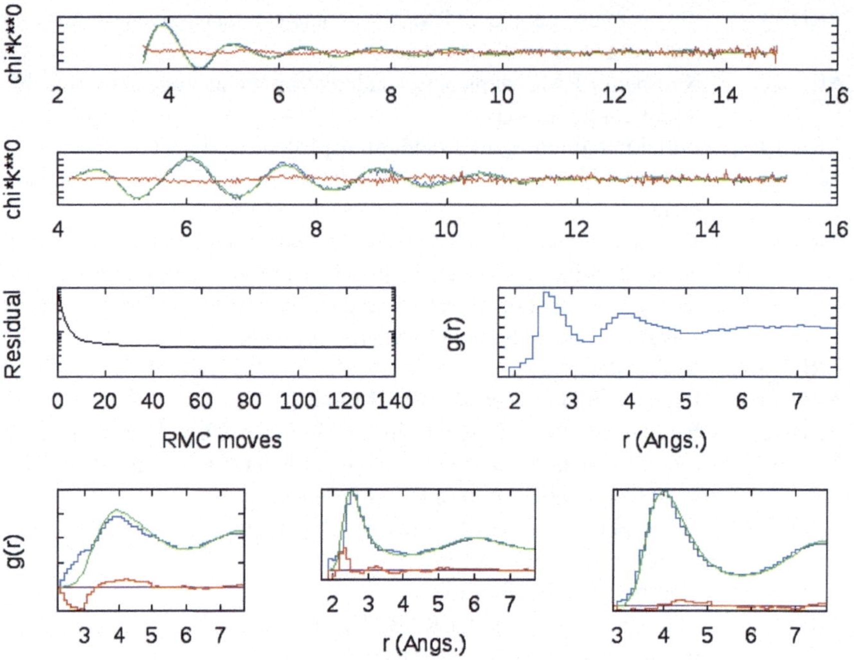

Fig. 8.10 Screenshot of the window created with the command `gnuplot < sh_rmc`, for the RMC simulation of liquid AgBr at 725 K [15]. Starting from the top we have the two Br (upper) and Ag (lower) K-edge experimental and RMC-EXAFS signals, and their difference. The residual (χ^2 of (8.18)) as a function of the RMC steps and the total pair distribution function are shown in the middle of the figure (third row). The lower figures (last row) are the three partial $g(r)$ obtained by the RMC refinement compared with the models obtained by molecular dynamics [27]: from the left, the first is AgAg, the second is AgBr and the last is BrBr

in the folder **chk** and the relevant files can be checked in order to avoid mistakes and misleading results. The **g_rmod** files contain the models of the pair distribution functions used for the fitting and their interpolation if binned on a different mesh, while the **XASinp** and **phainp** files contain respectively the EXAFS data and the phase shifts of the .DER files. Finally the **phacor** files contains the phase shifts corrected for possible 2π jumps. It is advisable to check the smoothness both for distance r and in k (or energy) variables of the phase functions, especially when dealing with large set of .DER files.

Other files are created for visualization and storage of the various data produced by the program which are updated at each loop of the simulation. The files **exXrmc.gnu** contain the total experimental and calculated EXAFS signals and the residual, for each edge under consideration. The files **gam_eX.gnu** contain the individual γ signals contributing to each edge. The information on the partial pair distribution functions at each step is stored in the **radrmcXX.gnu** files, while the information on the total distribution function is in **radrmc-t.gnu**. All these files can be visualized using gnuplot [16]. Finally, a gnuplot macro is created automatically by the program, called **sh_rmc**, which allows monitoring all the previous file in a single window during the RMC simulation. A screenshot of this window for AgBr is shown in Fig. 8.10.

During the simulation, the atomic configurations of the RMC model are appended to the file containing the initial structure AgBxyz.rmc, for selected RMC steps. In this way, an extended set of tridimensional models of the structure is available for successive analysis. After equilibration, those structures are all compatible with the given experimental data and can be used for a detailed structural analysis. Typically, a reasonable statistics is achieved considering 10^2–10^4 configurations obtained after equilibration to a minimum residual value. Specific programs for detailed structural analysis [12, 14], including determination of pair and triplet distributions, spherical-harmonics invariants and common-neighbor analysis are also available from the authors.

References

1. A. Filipponi, A. Di Cicco, T.A. Tyson, C. Natoli, Solid State Commun. **78**, 265 (1991)
2. A. Filipponi, A. Di Cicco, C.R. Natoli, Phys. Rev. B **52**, 15122 (1995)
3. A. Filipponi, A. Di Cicco, Phys. Rev. B **52**, 15135 (1995)
4. A. Filipponi, A. Di Cicco, TASK Quart. **4**, 575 (2000)
5. A. Di Cicco (ed.), *GNXAS. Extended Suite of Programs for Advanced X-ray Absorption Data-analysis: Methodology and Practice* (TASK Publishing, Gdansk, Poland, 2009)
6. K. Hatada, F. Iesari, L. Properzi, M. Minicucci, A. Di Cicco, J. Phys. Conf. Ser. **712**, 012002 (2016)
7. A. Di Cicco, A. Trapananti, J. Phys. Condens. Matter **17**, S135 (2005)
8. A. Di Cicco, Phys. Rev. B **53**, 6174 (1996)
9. A. Filipponi, A. Di Cicco, M. Benfatto, C.R. Natoli, Europhys. Lett. **13**, 319 (1990)
10. A. Filipponi, J. Phys. Condens. Matter **6**, 8415 (1994)
11. A. Trapananti, A. Di Cicco, Phys. Rev. B **70**, 014101 (2004)
12. A. Di Cicco, A. Trapananti, S. Faggioni, A. Filipponi, Phys. Rev. Lett. **91**, 135505 (2003)
13. A. Di Cicco, A. Trapananti, E. Principi, S. De Panfilis, A. Filipponi, Appl. Phys. Lett. **89**, 221912 (2006)
14. A. Di Cicco, F. Iesari, S. De Panfilis, M. Celino, S. Giusepponi, A. Filipponi, Phys. Rev. B **89**, 060102 (2014)
15. A. Di Cicco, M. Taglienti, M. Minicucci, A. Filipponi, Phys. Rev. B **62**, 12001 (2000)
16. T. Williams, C. Kelley, Many others, Gnuplot 5.0: an interactive plotting program, http://gnuplot.sourceforge.net/. Accessed 30 Nov 2016
17. J. Clement, Topdrawer 5.12, http://www.pa.msu.edu/reference/topdrawer-docs/ (1993). Accessed 30 Nov 2016
18. R. Dunn, H. Pasanen, wxPython home page, http://www.wxpython.org/. Accessed 30 Nov 2016
19. A. Filipponi, J. Phys. Condens. Matter **13**(7), R23 (2001)
20. R.L. McGreevy, L. Pusztai, Mol. Simul. **1**, 359 (1988)
21. R.L. McGreevy, J. Phys. Condens. Matter **13**, R877 (2001)
22. S.J. Gurman, R.L. McGreevy, J. Phys. Condens. Matter **2**, 9463 (1990)
23. Y. Wang, K. Lu, C. Li, Phys. Rev. Lett. **79**, 3664 (1997)
24. M. Winterer, J. Appl. Phys. **88**, 5635 (2000)

25. J. Timoshenko, A. Kuzmin, J. Purans, Comput. Phys. Commun. **183**(6), 1237 (2012)
26. G. Evrard, L. Pusztai, J. Phys.: Condens. Matter **17**(5), S1 (2005)
27. C. Tasseven, J. Trulls, O. Alcaraz, M. Silbert, A. Giró, J. Chem. Phys. **106**(17), 7286 (1997)

Part II
Extended knowledge

The chapters of this part correspond to the talks of the joint MSNano-EUSpec conference that was held at the University of Rennes 1 on July 1 and 2, 2016. This conference was organized both as the final conference of the MSNano network and in honor of Rino Natoli at the occasion of his 75th birthday. The talks selected here correspond to those directly related to the multiple scattering descriptions of spectroscopies.

Group photograph of the participants at the joint MSNano-EUSpec conference

Chapter 9
$(e,2e)$ Impact Ionization Processes for Surface Science

Rakesh Choubisa, Didier Sébilleau, Junqing Xu and Calogero R. Natoli

Abstract We present a scattering theoretic approach to the calculation of the cross-section of $(e,2e)$ impact spectroscopy where all the electrons involved are treated within the real space multiple scattering framework. This approach is particularly suited to the reflection geometry at low kinetic energies, with the ejection of a core-level electron. In this case, we expect $(e,2e)$ spectroscopy can be turned into an extremely sensitive surface structure probe.

9.1 Introduction

$(e,2e)$ has a long history, especially in atomic and molecular physics. Originally derived for the $(p, 2p)$ spectroscopy in nuclear physics [1], where p represents a proton, it was proposed in 1966 by Smirnov and coworkers [2] for the investigation of atomic wave functions [3], upon the replacement of the protons by electrons. Since then, it has enjoyed a widespread popularity and is often termed *electron momentum spectroscopy* [4]. Indeed, in the high energy regime (primary energy $\sim$10–50 keV), we can represent the electron by a plane wave. If in addition, the momentum transfer is large ($\boldsymbol{K} = \boldsymbol{k}_{\text{in}} - \boldsymbol{k}_{\text{sc}} \approx \boldsymbol{k}_{\text{ex}}$), the collision between the impinging electron and

R. Choubisa (✉)
Department of Physics, Birla Institute of Technology and Science, Pilani 333031, Rajasthan, India
e-mail: rchoubisa@pilani.bits-pilani.ac.in

D. Sébilleau
Institut de Physique de Rennes (IPR), Univ Rennes, CNRS, UMR 6251, F-35000 Rennes, France
e-mail: didier.sebilleau@univ-rennes1.fr

J. Xu
Department für Geo- und Umweltwissenschaften, Ludwig Maximilians Universität München, Theresienstrasse 41, 80333 München, Deutschland
e-mail: Junqing.Xu@lrz.uni-muenchen.de

C. R. Natoli
Theory Group, INFN-Laboratori Nazionali di Frascati, c. p. 13, 00044 Frascati, Italy
e-mail: calogero.natoli@lnf.infn.it

D. Sébilleau et al. (eds.), *Multiple Scattering Theory for Spectroscopies*, Springer Proceedings in Physics 204,
https://doi.org/10.1007/978-3-319-73811-6_9

the target electron can be described as a *single binary collision* (i.e. many-body interactions are approximated by consecutive two-body interactions) and therefore, the impact approximation is valid [5]. This approximation assumes that the incoming electron interacts only with the ejected electron and neither affects the target nor is affected by it (the *spectator electrons* are frozen). Within this so-called Plane Wave Impulse Approximation (PWIA), the momentum opposite to the recoil ion momentum vector is interpreted as the bound electron momentum. In this case, the cross-section is proportional to the *spectral momentum density* $\varrho(\boldsymbol{q}, \epsilon) = |\phi(\boldsymbol{q}, \epsilon)|^2$ where $\boldsymbol{q}$ is the momentum of the bound electron and ϵ its binding energy. $\phi(\boldsymbol{q}, \epsilon)$ is the Fourier transform of the wave function $\phi(\boldsymbol{r}, \epsilon)$ *before* the collision. $\varrho(\boldsymbol{q}, \epsilon)$ is the probability of finding a bound electron of momentum $\boldsymbol{q}$ and energy ϵ. Hence the name of *wave function mapping* spectroscopy [6, 7] often given to (*e*,2*e*).

Although proposed since the very beginning of (*e*,2*e*), the application of the technique to solids [2] has long been hampered by the low count rate. It is only rather recently that technical developments in the devising of new analyzers, such as the time-of-flight technique [8], have made it possible to use it as a probe of condensed matter.

(*e*,2*e*) in reflection mode was proposed back in 1978 in order to study surfaces by D'Andrea and Del Sole in a theoretical paper [9]. The first experiment was performed in 1992 by Kirschner and coworkers [10] on W(110). But it is not before 1995 that the feasibility of this binding-energy spectroscopy with quasimomentum discrimination was firmly established [11]. However, in this case, as we need to be sensitive to the surface, the experiments have to be performed for primary beams of low energy and grazing incidence in order to minimize the inelastic mean free path and the escape depth of the outgoing electrons. In the case of surfaces, the use of (*e*,2*e*) can give us important information on many different processes. For instance, in their seminal paper, D'Andrea and Del Sole [9] demonstrated that the computed (*e*,2*e*) spectra were very sensitive to the surface reconstruction. It is with this type of applications in mind that we develop here a real space multiple scattering framework to model (*e*,2*e*) experiments. Previous approaches were more interested in the excitation of valence states; we focus here on core states.

9.2 Reduction of the Cross-Section

The cross section of (*e*,2*e*) can be written as

$$\frac{\mathrm{d}^3\sigma}{\mathrm{d}E_1\mathrm{d}\Omega_{\boldsymbol{k}_{\mathrm{sc}}}\mathrm{d}\Omega_{\boldsymbol{k}_{\mathrm{ex}}}} = \frac{1}{8}\,(2\pi)^4\left(\frac{2m}{\hbar^2}\right)^3\frac{k_{\mathrm{sc}}k_{\mathrm{ex}}}{k_{\mathrm{in}}}\sum_{\mathrm{av}}|\langle\Phi^{N+1}_{\boldsymbol{k}_{\mathrm{sc}}\boldsymbol{k}_{\mathrm{ex}}}|T_I|\Phi^{N+1}_{\boldsymbol{k}_{\mathrm{in}}}\rangle|^2\,. \tag{9.1}$$

Here, we have noted $\boldsymbol{k}_{\mathrm{in}}$, $\boldsymbol{k}_{\mathrm{sc}}$ and $\boldsymbol{k}_{\mathrm{ex}}$ respectively the direction of the incoming, scattered and excited (ejected) electron. By convention, the scattered electron is the faster of the two outgoing electrons and the ejected (or excited) electron is the slower.

The sum over the states noted av is a sum over degenerate final states and an average over degenerate initial states.

Here, $|\Phi_{\boldsymbol{k}_{\rm in}}^{N+1}\rangle$ and $|\Phi_{\boldsymbol{k}_{\rm sc}\boldsymbol{k}_{\rm ex}}^{N+1}\rangle$ are eigenstates of the *asymptotic Hamiltonian*, i.e. of the Hamiltonian H_0 *without interaction*. $(N+1)$ is the total number of electrons, the target being assumed to contain N electrons. We can express the initial state as

$$|\Phi_{\boldsymbol{k}_{\rm in}}^{N+1}\rangle \;=\; |\varphi_g^{N-1}\rangle\,|\phi_0\rangle \otimes |\phi_{\boldsymbol{k}_{\rm in}}\rangle \;, \tag{9.2}$$

where $|\phi_0\rangle$ is the initial wave state of the electron that will be excited, $|\phi_{\boldsymbol{k}_{\rm in}}\rangle$ is the wave state of the incoming electron and $|\varphi_g^{N-1}\rangle$ is the ground state of the remaining $(N-1)$ system. Here, $\otimes$ represents the product in the two-electron space.

Likewise, we write the final state as

$$|\Phi_{\boldsymbol{k}_{\rm sc}\boldsymbol{k}_{\rm ex}}^{N+1}\rangle \;=\; |\varphi_{\rm ion}^{N-1}\rangle\,|\phi_{\boldsymbol{k}_{\rm sc}\boldsymbol{k}_{\rm ex}}\rangle \;, \tag{9.3}$$

where $|\varphi_{\rm ion}^{N-1}\rangle$ is the residual ion state and $|\phi_{\boldsymbol{k}_{\rm sc}\boldsymbol{k}_{\rm ex}}\rangle$ is the outgoing two-electron state.

With these notations, the cross-section becomes

$$\begin{aligned}\frac{{\rm d}^3\sigma}{{\rm d}E_1{\rm d}\Omega_{\boldsymbol{k}_{\rm sc}}{\rm d}\Omega_{\boldsymbol{k}_{\rm ex}}} \;&=\; \frac{1}{8}\,(2\pi)^4\left(\frac{2m}{\hbar^2}\right)^3\frac{k_{\rm sc}k_{\rm ex}}{k_{\rm in}}\;\times\\ &\sum_{\rm av}\left|\langle\varphi_{\rm ion}^{N-1}|\,\langle\phi_{\boldsymbol{k}_{\rm sc}\boldsymbol{k}_{\rm ex}}|\,T_I\,|\varphi_g^{N-1}\rangle\,|\phi_0\rangle\otimes|\phi_{\boldsymbol{k}_{\rm in}}\rangle\right|^2 .\end{aligned} \tag{9.4}$$

At this stage, we make a first approximation called *the frozen core approximation* or *the sudden approximation*. This approximation assumes that the ejected electron leaves the atom in a time much less that the time it takes for the remaining electrons of the ion to readjust. In other words, these atoms can be considered as *spectators* of the interaction process. This approximation implies

$$\frac{{\rm d}^3\sigma}{{\rm d}E_1{\rm d}\Omega_{\boldsymbol{k}_{\rm sc}}{\rm d}\Omega_{\boldsymbol{k}_{\rm ex}}} \;=\; \frac{1}{8}\,(2\pi)^4\left(\frac{2m}{\hbar^2}\right)^3\frac{k_{\rm sc}k_{\rm ex}}{k_{\rm in}}\sum_{\rm av}\left|S_0^{N-1}\right|^2\left|T_{fi}\right|^2 , \tag{9.5}$$

where we have noted

$$T_{fi} \;=\; \langle\phi_{\boldsymbol{k}_{\rm sc}\boldsymbol{k}_{\rm ex}}|\,T_I\,|\phi_0\rangle\otimes|\phi_{\boldsymbol{k}_{\rm in}}\rangle \;. \tag{9.6}$$

We are then left with a three-body problem plus the overlap term $\left|S_0^{N-1}\right|^2 = \left|\langle\varphi_{\rm ion}^{N-1}|\varphi_g^{N-1}\rangle\right|^2$ which is independent of the interaction process. $\langle\varphi_{\rm ion}^{N-1}|\varphi_g^{N-1}\rangle$ is usually called the *structure factor* while $\langle\phi_{\boldsymbol{k}_{\rm sc}\boldsymbol{k}_{\rm ex}}|\,T_I\,|\phi_0\rangle\otimes|\phi_{\boldsymbol{k}_{\rm in}}\rangle$ is called the *collision factor*. The separation of the total matrix element into these two contributions is called the *factorization approximation*. For the rest of the discussion, we will consider the structure factor as constant and take it out of the sum. In this approximation, T_I is a two-body transition operator.

9.3 The Two-Electron Differential Cross-Section

We recall that a N-body T-matrix element can be written as [12]

$$\begin{cases} T_{fi} = \langle \Phi_f | V_f | \Psi_i^+ \rangle = \langle \Phi_f | V_f \; \Omega_i^{(+)} | \Phi_i \rangle \quad \textit{post form} \\ \quad\; = \langle \Psi_f^- | V_i | \Phi_i \rangle \;\; = \langle \Phi_f | \; \Omega_f^{(-)\dagger} \; V_i | \Phi_i \rangle \; \textit{prior form} \;, \end{cases} \tag{9.7}$$

where $|\Psi^\pm\rangle$ is an eigenket of H, the full Hamiltonian, $|\Phi_f\rangle$ an eigenstate of $H_f = H - V_f$ and $|\Phi_i\rangle$ an eigenstate of $H_i = H - V_i$. V_i and V_f are the interaction potentials respectively in the *initial channel* and the *final channel*. $\Omega^{(\pm)}$ are Møller wave operators.

Each continuum electron interacts with the cluster potential through

$$V_{\mathrm{xx\ S}} = \sum_{n=1}^{N_{\mathrm{at}}} V_{\mathrm{xx}\ n} \;, \tag{9.8}$$

where N_{at} is the number of atoms in the cluster. Here, xx stands for in, sc or ex, and S for the sample. Within this 3-body approach (incoming electron, bound/ejected electron, cluster), we can define the interactions in the initial and final channel as

$$\begin{cases} V_i \; = V_{\mathrm{in\ S}} + V_{\mathrm{in\ co}} + V_{\mathrm{co\ S}} \\ V_f \; = V_{\mathrm{sc\ S}} + V_{\mathrm{ex\ S}} + V_{\mathrm{sc\ ex}} \;. \end{cases} \tag{9.9}$$

We note that we can suppress the interaction potential $V_{\mathrm{co\ S}}$ between the core state and the sample in the expression of V_i. Indeed, if the calculation of the sample potential V_{S} and of the wave function $\phi_{\mathrm{co}}(\boldsymbol{r})$ have been done properly, this has already been accounted for.

Let us consider now the prior form of (9.7). As the interaction between the incoming electron and the system $V_{\mathrm{in\ S}}$ cannot excite the bound electron, we can reduce V_i to $V_{\mathrm{in\ co}}$ in (9.7) so that we have now $|\Phi_i\rangle$ eigenstate of $H_i = H - V_{\mathrm{in\ co}} = H_0 + V_{\mathrm{in\ S}}$. To keep the standard distorted wave notation, we will write it as $|\chi_i^\pm\rangle \otimes |\phi_{\mathrm{co}}\rangle$. Therefore, (9.7) becomes

$$T_{fi} \; = \; \langle \Phi_f | \; \Omega_f^{(-)\dagger} \; V_{\mathrm{in\ co}} | \chi_i^+ \rangle \otimes |\phi_{\mathrm{co}}\rangle \;, \tag{9.10}$$

where according to (9.9), $|\Phi_f\rangle$ is an eigenstate of H_0. $|\chi_i^+\rangle$ is the result of the interaction of the incoming electron with the sample through the potential $V_{\mathrm{in\ S}}$. It is therefore a *multiple scattering state* which we will rewrite $|\chi_{\mathrm{in}}^\pm\rangle$ to make the difference with the multiple scattering states in the final channel.

As the matrix elements of T involve multiple scattering states $|\chi^\pm\rangle$, let us introduce the multiple scattering Møller wave operators $\omega_{\mathrm{xx}}^{(\pm)}$ associated to the Hamiltonian

$H_{\text{xx}} = H_0 + V_{\text{xx S}}$ by

$$\omega_{\text{xx}}^{(\pm)} \, |\Phi\rangle = (\omega_{\text{xx}}^{(\pm)} \otimes I) \; |\boldsymbol{k}_{\text{xx}}\rangle \otimes |\phi\rangle \; . \tag{9.11}$$

We can now rewrite our result (9.10) in terms of the initial state and final state plane waves solutions of H_0 as

$$T_{fi} = \langle \boldsymbol{k}_{\text{sc}}| \otimes \langle \boldsymbol{k}_{\text{ex}}| \; \Omega_f^{(-)\,\dagger} \; V_{\text{in co}} \; (\omega_{\text{in}}^{(+)} \otimes I) \, |\boldsymbol{k}_{\text{in}}\rangle \otimes |\phi_{\text{co}}\rangle \; . \tag{9.12}$$

This expression is an *exact result* within our 3-body impact approximation.

An important issue in (e,2e) spectroscopy is to find a proper description of the so-called *post-collision interaction* (PCI), i.e. the interaction between the two outgoing electrons. In our formulation, this effect is embedded within the final channel Møller wave operator $\Omega_f^{(-)\,\dagger}$. We know how to compute the individual wave operators $\omega_{\text{xx}}^{(\pm)}$, all multiple scattering codes can do it, but $\Omega_f^{(\pm)}$ is a complicated operator taking into account all the interactions in the final state at the same time. However, we can use the simple first order approximation derived by Briggs [12]

$$\Omega^{(\pm)} \approx \prod_{n=1}^{N} \omega_n^{(\pm)} \; . \tag{9.13}$$

We will call this result *Briggs' first order approximation* (Br1 in the equations). We find then

$$\Omega_f^{(-)\,\dagger}\Big|_{\text{Br1}} = (\omega_{\text{sc S}}^{(-)\,\dagger} \otimes I)(I \otimes \omega_{\text{ex S}}^{(-)\,\dagger}) \, \omega_{\text{sc ex}}^{(-)\,\dagger} = (\omega_{\text{sc S}}^{(-)\,\dagger} \otimes \omega_{\text{ex S}}^{(-)\,\dagger}) \, \omega_{\text{sc ex}}^{(-)\,\dagger} \; , \tag{9.14}$$

where the order of the wave operators is indifferent. $\omega_{\text{sc ex}}^{(-)\,\dagger}$ is the PCI wave operator which describes the (screened Coulomb) interaction between the two outgoing electrons.

Therefore, we can rewrite now the transition matrix element as

$$T_{fi}\Big|_{\text{Br1}} = \underbrace{\langle \boldsymbol{k}_{\text{sc}}| \otimes \langle \boldsymbol{k}_{\text{ex}}| \; \omega_{\text{sc ex}}^{(-)\,\dagger}}_{\text{PCI scattering state}} \; (\omega_{\text{sc S}}^{(-)\,\dagger} \otimes \omega_{\text{ex S}}^{(-)\,\dagger}) \; V_{\text{in co}} \; (\omega_{\text{in}}^{(+)} \otimes I) \, |\boldsymbol{k}_{\text{in}}\rangle \otimes |\phi_{\text{co}}\rangle \; . \tag{9.15}$$

The neglect of PCI amounts to the replacement of the Møller wave operator $\omega_{\text{sc ex}}^{(-)\,\dagger}$ by the identity $I_\otimes$ in the two-electron space.

In order to simplify the notation, let us write as usual

$$\omega_{\text{xx}}^{(\pm)} \; |\boldsymbol{k}_{\text{xx}}\rangle = |\chi_{\boldsymbol{k}_{\text{xx}}}^{\pm}\rangle \; , \tag{9.16}$$

and the Coulomb interaction in the initial channel as V_C as there is no ambiguity (the one in the final channel is contained into the PCI wave operator). Note that V_C is a short-hand notation for the antisymmetrized operator $(\mathcal{A}_{\otimes}^{\dagger} V_C \mathcal{A}_{\otimes})$, where $\mathcal{A}$ is the antisymmetrizer.

We can now rewrite (9.15) as

$$T_{fi}\big|_{\mathrm{Br1}} = \langle \chi^{-}_{\boldsymbol{k}_{\mathrm{sc}}} | \otimes \langle \chi^{-}_{\boldsymbol{k}_{\mathrm{ex}}} | \omega^{(-)\dagger}_{\mathrm{sc\,ex}} \; V_C \, | \chi^{+}_{\boldsymbol{k}_{\mathrm{in}}} \rangle \otimes | \phi_{\mathrm{co}} \rangle \;, \tag{9.17}$$

where the $|\chi^{\pm}\rangle$ are the multiple scattering states computed by the multiple scattering codes.

We can now inject this result into the cross-section

$$\left.\frac{\mathrm{d}^3\sigma}{\mathrm{d}E_1 \mathrm{d}\Omega_{\boldsymbol{k}_{\mathrm{sc}}} \mathrm{d}\Omega_{\boldsymbol{k}_{\mathrm{ex}}}}\right|_{\mathrm{Br1}} = 2\pi^4 \left(\frac{2m}{\hbar^2}\right)^3 \frac{k_{\mathrm{sc}} k_{\mathrm{ex}}}{k_{\mathrm{in}}} \; |S_0^{N-1}|^2 \sum_{\mathrm{av}} \left|T_{fi}\right|^2 \;, \tag{9.18}$$

to obtain a workable formula within the first Briggs approximation.

References

1. V.G. Neudatchin, F.A. Zhivopistsev, Phys. Rev. Lett. **32**, 995 (1974)
2. Yu.F. Smirnov, V.G. Neudatchin, Zh. Eksp. Teor. Fiz., Pis'ma **3**, 298 (1966) (JETP Lett. **3**, 192 (1966)); V.G. Neudatchin, G.A. Novoskol'tseva, Yu.F. Smirnov, Zh. Eksp. Teor. Fiz. **55**, 1039 (1968) (Sov. Phys. JETP **28**, 540 (1969))
3. M. Vos, I.E. McCarthy, Rev. Mod. Phys. **67**, 713 (1995)
4. V.G. Neudatchin, Yu.V Popov, Yu.F Smirnov, Phys. Uspekhi **42**, 1017–1044 (1999)
5. M. Vos, M. Bottema, Phys. Rev. B **54**, 5946 (1996)
6. I.E. McCarthy, E. Weigold, Rep. Prog. Phys. **51**, 299–392 (1988)
7. S. Samarin, R. Herrmann, H. Schwabe, O. Artamonov, J. Electron Spectrosc. Relat. Phenom. **96**, 61–67 (1998)
8. J. Kirschner, G. Kerhervé, C. Winkler, Rev. Sci. Instrum. **79**, 073302 (2008)
9. A. D'Andrea, R. Del Sole, Surf. Sci. **71**, 396–326 (1978)
10. J. Kirschner, O.M. Artamonov, A.N. Terekhov, Phys. Rev. Lett. **69**, 1711 (1992)
11. S. Iacobucci, L. Marassi, R. Camilloni, S. Nannarone, G. Stefani, Phys. Rev. B **51**, 10252 (1995)
12. J.S. Briggs, Phys. Rev. A **41**, 539 (1990)

Chapter 10
Layer-Resolved Photoemission Study of Doped Ag-Supported Ultrathin MgO Films

Thomas Jaouen

Abstract MgO/Ag(001) ultrathin films doped with interfacial Mg atoms are studied with layer-resolved Auger electron diffraction experiments, ultraviolet photoemission measurements, multiple scattering calculations, and density functional theory (DFT) calculations. The Mg atom intercalation at the MgO/Ag(001) interface induces a strong rumpling of the interface layers as well as a lowering of the work function related to interface electronic structure changes. DFT analysis of the metal-oxide interactions responsible for the interface dipole reproduces the experimental observations and reveals that the metal/oxide work function changes essentially originate in an increased electrostatic compression effect.

10.1 Introduction

Metal-supported ultrathin oxide films host unique chemical and physical properties owing to their reduced dimensionality and the enhanced role of the metal/oxide interface [1]. In particular, they have been widely studied both experimentally and theoretically in the field of heterogeneous catalysis due to their pivotal role in controlling charging mechanisms, adsorption properties, and catalytic activation of metal ad-atoms and molecules [2–7]. Defects engineering further allows tuning the work function and the electronic structure at the interface without modifying the oxide overlayer [8–12].

For this purpose, MgO/Ag(001) represents a model system that has been intensively studied both theoretically and experimentally [13]. For example, MgO/Ag(001) interfaces doped with oxygen or magnesium vacancies and impurities have shown to exhibit an enhanced catalytic activity with respect to the dissociation of H_2O [14, 15] due to modifications of the metal/oxide work function [16, 17]. Nevertheless, despite the increasing theoretical understanding of doped metal/oxide interfaces, few

T. Jaouen (✉)
Département de Physique and Fribourg Center for Nanomaterials,
Université de Fribourg, 1700 Fribourg, Switzerland
e-mail: thomas.jaouen@unifr.ch

D. Sébilleau et al. (eds.), *Multiple Scattering Theory for Spectroscopies*,
Springer Proceedings in Physics 204,
https://doi.org/10.1007/978-3-319-73811-6_10

experiments dealing with post-growth interface engineering have been performed due to the practical difficulties inherent to the buried interfaces.

In this chapter, we focus on the electronic and structural properties of experimentally engineered MgO/Ag(001) interfaces. Intercalation of Mg atoms at the interface is achieved by post-growth Mg exposures. Layer-resolved Auger electron diffraction (AED) and ultraviolet photoemission (UPS) experiments are used to demonstrate that Mg intercalation induces a strong rumpling of the interface layers as well as a lowering of the work function. The origin of the induced interface dipole is discussed through DFT calculations and the disentanglement of its charge transfer, rumpling, and electrostatic compression contributions. It is shown that the strong rumpling is a structural response to the interfacial charge transfer and that the metal/oxide work function changes essentially originate in an increased electrostatic compression effect.

10.2 Methods

10.2.1 Experimental

The experiments have been conducted in a multi-chamber ultrahigh vacuum (UHV) system with base pressures below 2×10^{-10} mBar. The MgO monolayers (ML) (1 ML $= 2.10\,\text{Å}$) were epitaxially grown on a freshly cleaned Ag(001) surface by co-evaporation of Mg and O_2 (oxygen pressure $= 5 \times 10^{-7}$ mBar) at 453 K. Mg intercalation have been performed by exposing the MgO films to an Mg flux (2.4×10^{13} atoms/(cm^{2}s)) at a substrate temperature of 513 K for limiting Mg adsorption at the oxide surface.

X-ray and ultraviolet photoelectron spectroscopy (XPS-UPS) have been performed using a hemispherical analyzer (Omicron EA125) with a five-channel detection system, and Al $K\alpha$ and He-I resonance ($h\nu = 21.22$ eV) lines as the x-ray and UV sources, respectively. The total energy resolutions were respectively 0.80 and 0.15 eV for XPS and UPS. The work function of the dielectric system (ϕ_m^*), defined as the energy of the vacuum level (E_{vac}) with respect to the Fermi level of the MgO/Ag(001) system (E_F), is determined from the low-energy cutoff (E_{cut}) of the secondary photoelectron emission: $\phi_m^* = h\nu - (E_F - E_{\text{cut}})$. Auger Electron Diffraction (AED) measurements of the Mg $KL_{23}L_{23}$ Auger transition were done using a two axis manipulator allowing for polar and azimuthal sample rotations with an accuracy better than 0.2° and AED profiles were recorded for polar sample rotations (the polar angle is defined with respect to the surface normal) between −5° and 60° for the (010) and (110) inequivalent emission planes of the cubic structure of the MgO(001) film.

10.2.2 Computational Details

The multiple scattering spherical wave cluster calculations have been performed in the Rehr–Albers framework [18], by using the MsSpec program [19, 20] for clusters containing up to 420 atoms. Details on the calculations are given in [21]. Briefly, the multiple scattering expansion of the photoelectron wave function was carried out up to the fourth order which we checked to be sufficient to achieve convergence for the configurations considered. Following various experimental works [22–24], we assumed pseudomorphic ultrathin MgO films on Ag(001) with interface Mg atoms occupying the substrate hollow sites and an interfacial distance between Ag and O atoms of 2.51 Å. Finally, a broadening of the AED peaks due to the formation of mosaic observed during the growth of the MgO films on Ag(001) [25], was taken into account by averaging the calculations over a cone of 2.5° half angle.

The DFT calculations have been carried out in the generalized gradient approximation (GGA) using the Perdew–Burke–Ernzerhof (PBE) exchange-correlation functional [26]. The calculations used the Projector-Augmented Wave (PAW) formalism [27], implemented in a real-space grid in the GPAW code [28, 29], with a grid spacing of 0.18 Å. The MgO/Ag (001) system was modeled with three layers of MgO on three Ag layers with lattice parameter $a_0 = 4.16$ Å and Ag interface atoms below the oxygen anions. The vacuum region between adjacent slabs was set to $\sim$20 Å. $(\sqrt{2} \times \sqrt{2})a_0$ surface unit cells were used for calculations and Brillouin zone sampling were performed using a Monkhorst–Pack mesh with $4 \times 4 \times 1$ k points [30]. During geometry optimization, only the Ag atoms of the bottom layer have been kept fixed and a tolerance of 0.02 eV/Å was applied.

10.3 Results

10.3.1 Layer-Resolved Auger Transition at the Ultrathin Limit

Figure 10.1a shows normal-emission Mg $KL_{23}L_{23}$ Auger spectra of bulk (18 ML) and ultrathin (2–3 ML) MgO films. For a non-resonant excitation, the photoemission of Auger electrons can be described by a two-step process [31]. A core hole is first created by photon absorption on the 1s core level, and the Auger decay then involves two electrons of the 2p orbital. The Auger electron leaves the atom with a kinetic energy which depends on the binding energies of the 1s and 2p core levels and on U, the on-site Coulomb interaction which takes into account of the Coulomb interaction between the two holes on the L shell in the final state. Since an Auger transition does not obey to the dipolar selection rule, many final states are allowed, explaining the multiplet structure of the Auger transitions. For bulk MgO film [Fig. 10.1a, top spectrum], the Mg $KL_{23}L_{23}$ spectrum exhibits two contributions corresponding to the ^1S and ^1D multiplets of the Mg 2p final state. For ultrathin MgO films grown on

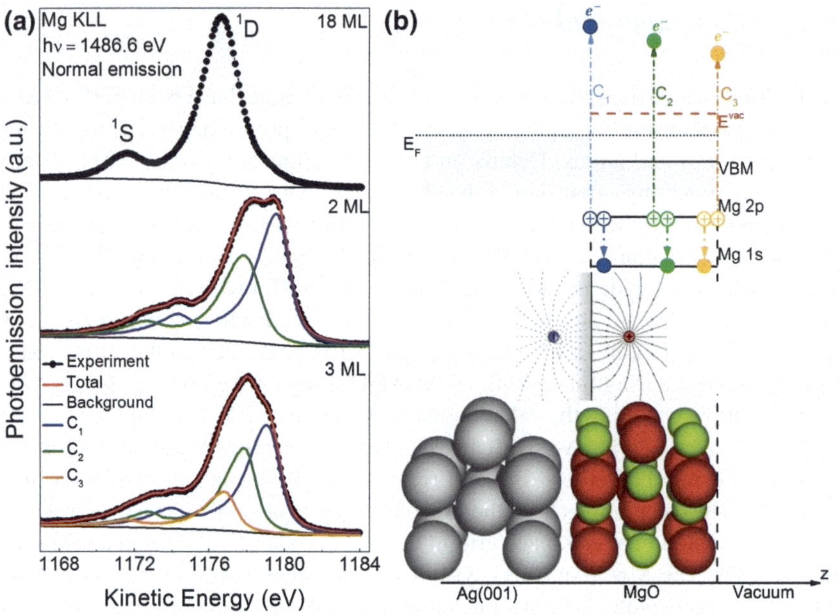

Fig. 10.1 **a** Photoemission spectra of the Mg $KL_{23}L_{23}$ Auger transition of bulk (18 ML, top spectrum) and ultrathin (2–3 ML, middle and bottom spectra, respectively) Ag-supported MgO films. Best fit and layer-by-layer decomposition are also shown. **b** Schematic band diagram illustrating the spectroscopic origin of the Mg $KL_{23}L_{23}$ layer-by-layer resolution for a 3 ML-thick MgO film. The observed increase in the Auger kinetic energy with decreasing distance from the interface has been attributed to the image-potential screening of the localized Mg 2p holes by the metal substrate. The distance-dependent screening of the two-holes in the Auger final state leads to different values of the on-site Coulomb interaction for the Auger transitions involving electrons of the interface, sub-surface and surface layers

Ag (2–3 ML) [Fig. 10.1a middle and bottom spectra], it has been demonstrated that the Mg $KL_{23}L_{23}$ spectra are layer-resolved [32]. As seen Fig. 10.1a, the Auger spectra can be perfectly adjusted using an experimental Mg $KL_{23}L_{23}$ Auger spectrum of a 1 ML MgO/Ag(001) sample and including as many fitting components as there are MgO layers. For 3ML thick MgO film, the Auger spectrum is fitted by three shifted monolayer-Auger components C_1, C_2, and C_3 which correspond to Auger electron emission from the interface, sub-surface and surface oxide layers, respectively.

As previously discussed by Kaindl et al. for Xe multilayers on Pd(001) [33] and Altieri et al. [34] in the case of MgO/Ag(001), the layer-by-layer resolution of the Auger transition reflects the sensitivity of the two holes final states resulting from the Auger decay to extra-atomic relaxation effects. The observed increase in the Auger kinetic energy with decreasing distance from the interface has been attributed to the image-potential screening of the localized Mg 2p holes by the metal substrate that basically leads to different values of the on-site Coulomb interaction for the

Auger transitions involving electrons of the interface, sub-surface and surface layers [Fig. 10.1b].

10.3.2 Mg-Doped MgO/Ag(001) Interfaces

10.3.2.1 Layer-Resolved Atomic Structure Probed by Auger Electron Diffraction

Figure 10.2a shows the variation of the normal-emission Mg $KL_{23}L_{23}$ Auger spectrum upon Mg exposure of a 3 ML MgO sample. The fourth component (labeled C_0) that appears at higher kinetic energy in the Mg $KL_{23}L_{23}$ Auger spectrum is related to electron emission from intercalated Mg atoms in the Ag substrate [32, 35]. As seen in the inset of Fig. 10.2a, its normal-emission intensity gradually increases as a function of the exposition time suggesting that Mg intercalation takes place throughout the Mg exposure likewise.

Figures 10.2b, c respectively show the experimental and simulated layer-resolved AED polar scans of the Mg $KL_{23}L_{23}$ Auger spectra in the (010) emission plane. As the C_1, C_2, and C_3 components of an as-grown Mg $KL_{23}L_{23}$ Auger spectrum correspond respectively to the interface, sub-surface and surface oxide layers, the

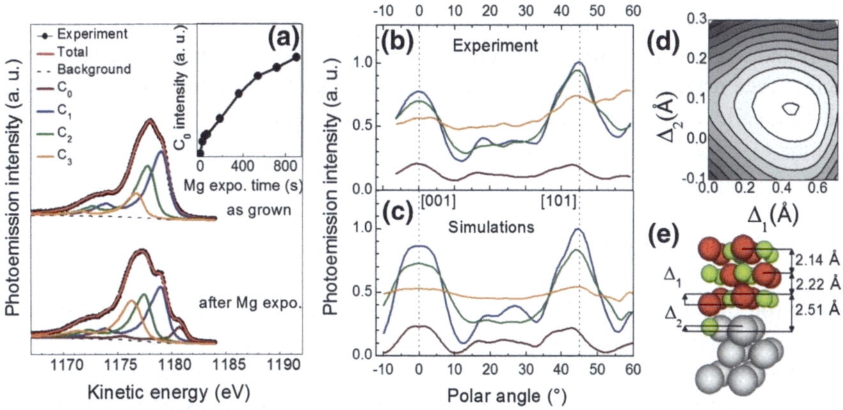

Fig. 10.2 **a** Photoemission spectra of the Mg $KL_{23}L_{23}$ Auger transition of the 3 ML MgO film obtained before and after exposition to an Mg atomic flux. Best fit and layer-by-layer decomposition are also shown. The *inset* shows the evolution of the intensity of the C_0 component as a function of the Mg exposure time. **b** Experimental AED polar scans of the C_0, C_1, C_2, and C_3 Auger components in the (010) emission plane, for a 3 ML-thick MgO film exposed to Mg. **c** Calculated Mg $KL_{23}L_{23}$ AED profiles in the (010) emission plane for a 3 ML MgO system with Mg atoms occupying substitutional sites of the Ag substrate. **d** R_p-factor contour map using Δ_1 and Δ_2 as parameters. These parameters are defined in **e** which shows a sketch of our structural model. The *red*, *green* and *grey* atoms correspond respectively to oxygen, magnesium and silver

intensity distribution of C_3 is nearly isotropic whereas the AED profiles associated with C_1 and C_2 show typical forward scattering peaks along the [001] and [101] directions of the rocksalt structure of the MgO lattice. The C_0 metallic component curve in Fig. 10.1c also shows a well-structured pattern but with forward scattering peaks sharper than those of the C_1 and C_2 components. Such a narrowing is known to originate in defocusing effects related to the multiple scattering events [36, 37]. Therefore it is likely to be observed for electron emission from Mg atoms located *beneath* the oxide layer. The fact that the defocusing effects do not cancel the forward-scattering constructive interferences then indicates that the metallic Mg atoms are located in the MgO/Ag interface region.

Indeed, the simulated AED curves obtained by combined multiple scattering calculations and reliability factor (R_p-factor) analysis [35] (Fig. 10.2d, e) show that the C_0 component is related to electron emission from Mg atoms intercalated in the substitutional sites of the Ag plane just beneath the MgO lattice and that the MgO lattice undergoes a significant distortion at the interface upon Mg intercalation. The R_p factor contour map is shown in Fig. 10.2d (Δ_1 and Δ_2 parameters are defined in the sketch of our structural model Fig. 10.2e). The best agreement between experimental and calculated AED profiles is obtained for an Mg-Ag alloy concentration of about 30% with $d_{\mathrm{Mg-O}} = 2.0$ Å and a R_p-factor of 0.15. The Mg atoms of the interfacial alloy are displaced toward the oxide layer by 0.1 Å and the nearest neighbors O^{2-} ions are displaced downward by 0.4 Å relatively to the Mg^{2+} ions position.

Such a rumpling at the interface upon Mg intercalation has been further predicted by DFT calculations. Bader analysis has shown that the neighboring Mg and Ag atoms of the substrate interface layer are respectively positively (−1.5 electrons per atom) and negatively charged (0.9 electrons per atom). The structural relaxation at the interface originates in the charge transfer between the Mg and Ag atoms of the interfacial alloy that are respectively in attractive and repulsive Coulomb interactions with the O^{2-} ions directly adjacent [35].

10.3.2.2 Induced Electronic Structure Changes

Deposition of ultrathin oxide films on metals induces a metal work function shift and a variation of the tunneling probability for charges at the interface [38–45]. This effect has been understood has coming from the competition between the interfacial charge transfer ($\Delta\phi^{\mathrm{CT}}$), the oxide lattice polarizability and associated rumpling dipole moment ($\Delta\phi^{\mathrm{SR}}$), and the polarization of the metal electrons at the interface induced by the highly ionic oxide layer, the so-called electrostatic compression effect ($\Delta\phi^{\mathrm{comp}}$).

Figure 10.3a shows the low-energy cutoff of the secondary photoelectrons emission for the Ag(001) substrate and for the MgO(3ML)/Ag(001) sample before and after Mg intercalation. The MgO deposition leads to a metal work function shift, $\Delta\phi_m$, of about -1.30 ± 0.05 eV which results in a metal/oxide work function value ϕ_m^* of 3.10 ± 0.05 eV for our reference sample. This decrease is mainly driven by the compression of the metal electron density at the interface induced by the oxide

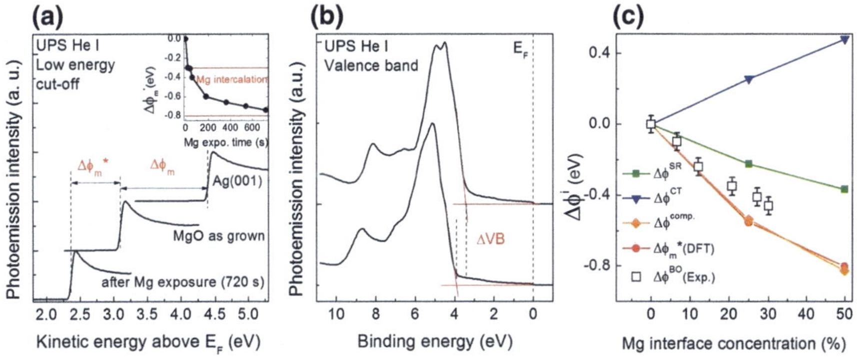

Fig. 10.3 **a** Low-energy cutoff of the secondary photoelectrons emission for the Ag(001) substrate and for the MgO(3ML)/Ag(001) sample before and after exposure to a Mg flux. $\Delta\phi_m$ and $\Delta\phi_m^*$ correspond to the work function variations of the metal substrate and of the metal/oxide system, respectively. The *inset* shows the evolution of $\Delta\phi_m^*$ as a function of the Mg exposure time. **b** He-I UPS spectra showing the valence band region of the MgO reference sample (*top*) and after Mg exposure (*bottom*). The method used for the VBM position determination is also sketched. **c** Evolutions of the calculated $\Delta\phi^{\mathrm{CT}}$, $\Delta\phi^{\mathrm{SR}}$, and $\Delta\phi^{\mathrm{comp}}$ contributions to the calculated MgO(3ML)/Ag(001) work function variation $\Delta\phi_m^*$ (DFT), as a function of the interface Mg atom concentration. Experimental work function changes $\Delta\phi^{BO}$ due to the band-offset (BO) variations at the metal/oxide interface are also shown for comparison

deposition and is particularly strong for wide band gap oxides such as MgO [38, 42–44]. As seen Fig. 10.3a, intercalation of Mg at the metal/oxide interface further modifies the metal/oxide work function. Removing the initial band bending contribution of about 0.30 eV [32], the metal/oxide work function can be diminished over $\Delta\phi^{\mathrm{BO}} = -0.50\,\mathrm{eV}$ depending on the intercalated-Mg concentration [see inset Fig. 10.3a]. Comparison of the valence-band (VB) region of He-I UPS spectra corresponding to the reference MgO/Ag(001) sample (top) and to the Mg-intercalated one (bottom) [Fig. 10.3b] show that this work function diminution is related to a change in the Fermi level pinning position in the MgO band gap.

The physical origin of the interface dipole responsible for the work function changes induced by Mg intercalation at the MgO/Ag(001) interface has been investigated with the help of DFT calculations by disentangling the contributions of charge transfer ($\Delta\phi^{\mathrm{CT}}$), rumpling ($\Delta\phi^{\mathrm{SR}}$), and compression effect ($\Delta\phi^{\mathrm{comp}}$) to the total work function variation of the metal/oxide system ($\Delta\phi_m^*$) [35]. Figure 10.3c shows the comparison between DFT-calculated [$\Delta\phi_m^*$ (DFT)] and experimental work function changes [$\Delta\phi^{\mathrm{BO}}$ (Exp.)], as well as the evolutions of the calculated $\Delta\phi^{\mathrm{CT}}$, $\Delta\phi^{\mathrm{SR}}$, and $\Delta\phi^{\mathrm{comp}}$ contributions as a function of the interface Mg atom concentration. As it can be seen, the experimental $\Delta\phi^{\mathrm{BO}}$ (Exp.) values are in overall good agreement with the DFT-predicted work function changes $\Delta\phi_m^*$ (DFT). In particular, the intercalation of 25% of Mg in the first interface Ag layer results in theoretical work function change of $-0.55\,\mathrm{eV}$ in very good agreement with our experimental findings. The $\Delta\phi^{\mathrm{CT}}$ and $\Delta\phi^{\mathrm{SR}}$ contributions respectively increases and decreases with Mg concentration in such a way that they compensate each other. Indeed, as previously discussed, the local lattice distortion at the interface is a structural response to the interfacial charge transfer. It turns out that the decay of the metal/oxide work function is mainly governed by the progressive increase of the compression effect as demonstrated by the similar evolutions of $\Delta\phi_m^*$ and $\Delta\phi^{\text{comp}}$ on Fig. 10.3c.

It has been shown that the compression effect scales with the inverse of the interface distance between the oxide and the metal [43]. For MgO(3ML)/Ag(001), the interface distance only diminishes by 0.1 Å when the Mg concentration of the interface Ag-Mg alloy varies from 0 to 100% suggesting that the distance reduction effect induced by the Mg intercalation cannot simply explain the enhancement of the compression contribution. A deeper analysis of the interface electronic structure as done for example in BaO/metal systems [46], is therefore required for better enclosing the richness of interaction at the metal/oxide interfaces.

10.4 Conclusion

The mechanisms responsible for the work function changes induced by the Mg atoms intercalation at the MgO/Ag(001) interface have been studied by means of Auger electron diffraction experiments combined with multiple scattering calculation, ultraviolet photoemission spectroscopy, and density functional theory calculations. Intercalated Mg atoms are preferentially intercalated in the substitutional site of the Ag interface plane, lead to a strong work function change related to band-offset variations at the MgO/metal interface and induce a significant rumpling in the interface layers. DFT-based calculations have further shown that the work function reduction mainly reflects the increased electrostatic compression effect, since the coupled charge transfer and rumpling contributions compensate each other. This work gives new insights for controlling the interfacial properties of metal-supported ultrathin oxide films such as work functions and local atomic structure which have a significant impact on their catalytic activities.

Acknowledgements Collaboration and fruitful discussions with P. Aebi, S. Tricot, G. Delhaye, B. Lépine, D. Sébilleau, G. Jézéquel and P. Schieffer are gratefully acknowledged. Skillful technical assistance was provided by A. Le Pottier and Y. Claveau. Parts of this work have been funded by European FP7 MSNano network under Grant agreement n° PIRSES-GA-2012-317554 and by COST Action MP1306 EUSpec. It has also been supported by the Fonds National Suisse pour la Recherche Scientifique through Division II.

References

1. C. Freysoldt, P. Rinke, M. Scheffler, Phys. Rev. Lett. **99**, 086101 (2007)
2. H.-J. Freund, G. Pacchioni, Chem. Soc. Rev. **37**, 2224 (2008)
3. S. Surnev, A. Fortunelli, F.P. Netzer, Chem. Rev. **113**, 4314 (2013)

4. G. Pacchioni, L. Giordano, M. Baistrocchi, Phys. Rev. Lett. **94**, 226104 (2005)
5. M. Sterrer, T. Risse, U. Martinez Pozzoni, L. Giordano, M. Heyde, H.-P. Rust, G. Pacchioni, H.-J. Freund, Phys. Rev. Lett. **98**, 096107 (2007)
6. M. Sterrer, T. Risse, M. Heyde, H.-P. Rust, H.-J. Freund, Phys. Rev. Lett. **98**, 206103 (2007)
7. S. Benedetti, F. Stavale, S. Valeri, C. Noguera, H.-J. Freund, J. Goniakowski, N. Nilius, Adv. Funct. Mater. **23**, 75 (2013)
8. U. Martinez, L. Giordano, G.J. Pacchioni, Chem. Phys. **128**, 164707 (2008)
9. U. Martinez, J.F. Jerratsch, N. Nilius, L. Giordano, G. Pacchioni, H.-J. Freund, Phys. Rev. Lett. **103**, 056801 (2009)
10. J.F. Jerratsch, N. Nilius, H.-J. Freund, U. Martinez, L. Giordano, G. Pacchioni, Phys. Rev. B **80**, 245423 (2009)
11. R. Wlodarczyk, M. Sierka, J. Sauer, D. Löffler, J.J. Uhlrich, X. Yu, B. Yang, I.M.N. Groot, S. Shaikhutdinov, H.-J. Freund, Phys. Rev. B **85**, 085403 (2012)
12. T. Jaouen, G. Jézéquel, G. Delhaye, B. Lépine, P. Turban, P. Schieffer, Appl. Phys. Lett. **100**, 022103 (2012)
13. G. Pacchioni, H. Freund, Chem. Rev. **113**, 4035 (2013)
14. J. Jung, H.-J. Shin, Y. Kim, M. Kawai, J. Am. Chem. Soc. **133**, 6142 (2011)
15. J. Jung, H.-J. Shin, Y. Kim, M. Kawai, J. Am. Chem. Soc. **134**, 10554 (2012)
16. S. Ling, M.B. Watkins, A.L. Shluger, J. Phys. Chem. C **117**, 5075 (2013)
17. S.B. Cho, K.-H. Yun, D.S. Yoo, K. Ahn, Y.-C. Chung, Thin Solid Films **544**, 541 (2013)
18. J.J. Rehr, R.C. Albers, Phys. Rev. B **41**, 8139 (1990)
19. D. Sébilleau, R. Gunnella, Z.-Y. Wu, S. Di Matteo, C.R. Natoli, J. Phys. Condens. Matter **18**, R175 (2006)
20. D. Sébilleau, C. Natoli, G.M. Gavaza, H. Zhao, F. Da Pieve, K. Hatada, Comput. Phys. Commun. **182**, 2567 (2011)
21. D. Agliz, A. Quémerais, D. Sébilleau, Surf. Sci. **343**, 80 (1995)
22. A.M. Flank, R. Delaunay, P. Lagarde, M. Pompa, J. Jupille, Phys. Rev. B **53**, R1737 (1996)
23. O. Robach, G. Renaud, A. Barbier, Phys. Rev. B **60**, 5858 (1999)
24. P. Luches, S. D'Addato, S. Valeri, E. Groppo, C. Prestipino, C. Lamberti, F. Boscherini, Phys. Rev. B **69**, 045412 (2004)
25. J. Wollschläger, D. Erdös, K.-M. Schröder, Surf. Sci. **402–404**, 272 (1998)
26. J.P. Perdew, K. Burke, M. Ernzerhof, Phys. Rev. Lett. **77**, 3865 (1996)
27. P.E. Blöchl, Phys. Rev. B **50**, 17953 (1994)
28. J. Enkovaara et al., J. Phys. Condens. Matter **22**, 253202 (2010)
29. J.J. Mortensen, L.B. Hansen, K.W. Jacobsen, Phys. Rev. B **71**, 035109 (2005)
30. H.J. Monkhorst, J.D. Pack, Phys. Rev. B **13**, 5188 (1976)
31. T. Jaouen, E. Razzoli, C. Didiot, G. Monney, B. Hildebrand, F. Vanini, M. Muntwiler, P. Aebi, Phys. Rev. B **91**, 161410(R) (2015)
32. T. Jaouen, S. Tricot, G. Delhaye, B. Lépine, D. Sébilleau, G. Jézéquel, P. Schieffer, Phys. Rev. Lett. **111**, 027601 (2013)
33. G. Kaindl, T.C. Chiang, D.E. Eastman, F.J. Himpsel, Phys. Rev. Lett. **45**, 1808 (1980)
34. S. Altieri, L.H. Tjeng, F.C. Voogt, T. Hibma, G.A. Sawatzky, Phys. Rev. B **59**, R2517 (1999)
35. T. Jaouen, P. Aebi, S. Tricot, G. Delhaye, B. Lépine, D. Sébilleau, G. Jézéquel, P. Schieffer, Phys. Rev. B **90**, 125433 (2014)
36. M.-L. Xu, J.J. Barton, M.A. Van Hove, Phys. Rev. B **39**, 8275 (1989)
37. H.A. Aebischer, T. Greber, J. Osterwalder, A.P. Kaduwela, D.J. Friedman, G.S. Herman, C.S. Fadley, Surf. Sci. **239**, 261 (1990)
38. T. Jaouen, G. Jézéquel, G. Delhaye, B. Lépine, P. Turban, P. Schieffer, Appl. Phys. Lett. **97**, 232104 (2010)
39. C. Pauly, M. Grob, M. Pezzotta, M. Pratzer, M. Morgenstern, Phys. Rev. B **81**, 125446 (2010)
40. M. Bieletzki, T. Hynninen, T.M. Soini, M. Pivetta, C.R. Henry, A.S. Foster, F. Esch, C. Barth, U. Heiz, Phys. Chem. Chem. Phys. **12** (2010)
41. T. König, G.H. Simon, H.P. Rust, M. Heyde, J. Phys. Chem. C **113**, 11301 (2009)
42. J. Goniakowski, C. Noguera, Interface Sci. **12**, 93 (2004)

43. S. Prada, U. Martinez, G. Pacchioni, Phys. Rev. B **78**, 235423 (2008)
44. L. Giordano, F. Cinquini, G. Pacchioni, Phys. Rev. B **73**, 045414 (2006)
45. G. Butti, M.I. Trioni, H. Ishida, Phys. Rev. B **70**, 195425 (2004)
46. M. Nuñez, M. Buongiorno Nardelli, Phys. Rev. B **73**, 235422 (2006)

Chapter 11
Es2Ms: Interface from Electronic Structure Codes to Multiple Scattering Codes

Naoki Komiya, Fukiko Ota, Junqing Xu and Keisuke Hatada

Abstract We present an interface package, called ES2MS (Xu J et al. Comput Phys Commun, 203:331, 2016, [1]) for passing self-consistent charge density and potential from Electronic Structure (ES) codes to Multiple Scattering (MS) codes. MS theory is based on the partitioning of the space by atomic-size scattering sites, so that the code provides the charge densities and potentials for each scattering site. For pseudopotential codes, the interface solves Poisson equation to construct the all-electron potential on the radial mesh which is used to solve the transition operators (T-matrix) and Green's functions in MS codes. For Gaussian basis set programs, we solve recurrence relations to obtain the Coulomb potential. We describe the algorithms of the interface and an example.

11.1 Algorithm for Peudopotential Codes Output to MS Calculations

The VASP code [2] is a popular plane-wave code for total energy Density Functional Theory (DFT) calculations in condensed matter. Plane-wave codes have the advantage of high accuracy and computational efficiency.

N. Komiya (✉) · F. Ota
Graduate School of Advanced Integration Science, Chiba University, Chiba 263-8522, Japan
e-mail: n_komiya@chiba-u.jp

F. Ota
e-mail: f_ota@chiba-u.jp

J. Xu
Department für Geo- und Umweltwissenschaften, Ludwig Maximilians Universität München, Theresienstrasse 41, 80333 München, Germany
e-mail: Junqing.Xu@lrz.uni-muenchen.de

K. Hatada
Department Chemie, Ludwig-Maximilians-Universität München, Butenandtstr. 5-13, 81377 Munich, Bavaria, Germany
e-mail: keisuke.hatada@cup.uni-muenchen.de

D. Sébilleau et al. (eds.), *Multiple Scattering Theory for Spectroscopies*, Springer Proceedings in Physics 204,
https://doi.org/10.1007/978-3-319-73811-6_11

The Projector-Augmented-Wave (PAW) method [3] is widely used in electronic structure calculations by VASP. In the PAW method, the all-electron Kohn–Sham potential is replaced by a pseudopotential inside the augmentation spheres around the nuclei. This pseudopotential coincides with the all-electron potential only in the space outside the augmentation spheres. However, for MS calculations the all-electron potential is required in all space, especially near the nuclei where the core orbital exhibits a large amplitude. Calculations using the PAW method provide pseudopotential and all-electron charge density. Inside the augmentation spheres, the charge density is interpolated onto the radial mesh of the MS code. The all-electron full-potential is obtained by solving the Poisson equation with proper boundary conditions on the surface of the spheres as given by the pseudopotential. Outside the spheres, the charge density and potential are interpolated from the Cartesian mesh onto the radial mesh.

In order to solve Poisson equation we need to find the regular solution inside an (augmentation) sphere S of radius r_c, with the Dirichlet boundary condition on the sphere, i.e. to solve

$$\nabla^2 V(\boldsymbol{r}) = -4\pi\varrho(\boldsymbol{r}) , \tag{11.1}$$

for $r < r_c$, given the charge density ϱ for all points inside S and the potential V on the surface ($r = r_c$). We introduce the function

$$W(\boldsymbol{r}) \equiv \int_S \frac{\varrho(\boldsymbol{r}')}{|\boldsymbol{r}-\boldsymbol{r}'|}\,\mathrm{d}\boldsymbol{r}' , \tag{11.2}$$

which satisfies Poisson equation but not, in general, the boundary condition. $W(\boldsymbol{r})$ can easily be expanded into a spherical harmonic representation,

$$W(r,\hat{\boldsymbol{r}}) = \sum_L \omega_L(r) Y_L(\hat{\boldsymbol{r}}) , \tag{11.3}$$

$$\omega_L(r) = \int_0^{r_c} \tfrac{r_<^\ell}{r_>^{\ell+1}}\, \varrho_L(r') 4\pi r'^2 \mathrm{d}r', \tag{11.4}$$

$$\varrho_L(r) = \int Y_L^*(\hat{\boldsymbol{r}})\varrho(r,\hat{\boldsymbol{r}})\,\mathrm{d}\hat{\boldsymbol{r}} , \tag{11.5}$$

where $\hat{\boldsymbol{r}} = \boldsymbol{r}/r$. The general solution of the Poisson equation is then the sum of a particular solution, such as W, and the general solution of the corresponding homogeneous (i.e., Laplace) equation. The latter can be expressed as $\sum_L C_L r^\ell Y_L(\hat{\boldsymbol{r}})$ for arbitrary coefficients C_L. By choosing $C_L = [\upsilon_L(r_c) - \omega_L(r_c)]/r_c^\ell$, where

$$\upsilon_L(r) = \int Y_L^*(\hat{\boldsymbol{r}})V(r,\hat{\boldsymbol{r}})\,\mathrm{d}\hat{\boldsymbol{r}} , \tag{11.6}$$

the boundary conditions are satisfied, and therefore the final solution is given by

$$V(r,\hat{\boldsymbol{r}}) = \sum_L \left(\omega_L(r) + [\upsilon_L(r_c) - \omega_L(r_c)]\frac{r^\ell}{r_c^\ell}\right) Y_L(\hat{\boldsymbol{r}}) . \tag{11.7}$$

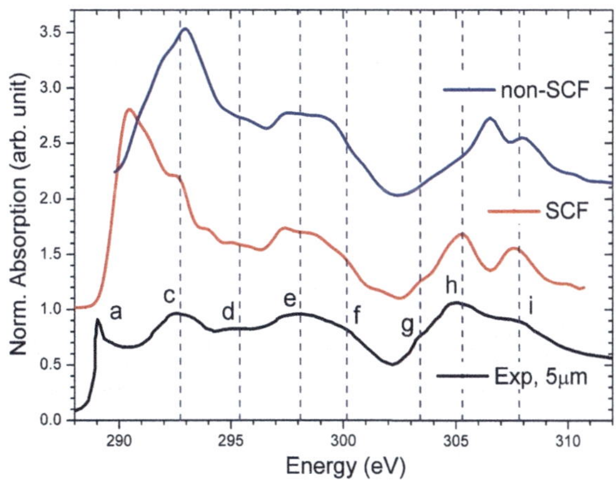

Fig. 11.1 Calculated C K-edge XAS of diamond using non-SCF and SCF potentials [1]

Figure 11.1 shows the calculated carbon K-edge x-ray absorption spectra of a diamond cluster of radius 18 Å by the FPMS code [4, 5] using non-SCF (self-consistent field) and SCF potentials, respectively. The SCF potential is generated by the VASP code and reconstructed by the ES2MS package. It is clear that the SCF result agrees better with the experimental data [6] than the non-SCF result.

11.2 Algorithm for GAUSSIAN Codes Output to MS Calculations

GAUSSIAN 09 [7] is one of the widely used electronic structure calculation codes for molecular systems. The basis sets are constructed by linear combination of Gaussian Type Orbitals (GTOs), which makes it possible to perform a variety of calculations very efficiently.

We describe here the algorithm that uses the GAUSSIAN output for MS theory. In this algorithm, the electron charge density and the Hartree potential are evaluated with and without the Muffin-tin (MT) approximation. In addition, the initial state molecular orbital is reexpanded by a multi-site spherical harmonics. These calculations can be accomplished mostly analytically.

In the MT case, we have already applied this algorithm to compute the photoemission intensity of a pentacene molecule, where the initial state is the highest occupied molecular orbital [8].

11.2.1 Electron Charge Density and Hartree Potential Constructed by Gaussian Basis Sets

As any product of two Gaussian functions is also a Gaussian function, the electron charge density can be expanded by the Hermite Gaussian functions which are defined by

$$\Lambda_{t,u,v}(\boldsymbol{r}_p, \alpha) = \frac{\partial^t}{\partial P_x^t}\frac{\partial^u}{\partial P_y^u}\frac{\partial^v}{\partial P_z^v}\mathrm{e}^{-\alpha r_p^2}\,, \tag{11.8}$$

where $\boldsymbol{r}_p = \boldsymbol{r} - \boldsymbol{P}$. The center $\boldsymbol{P}$ and the exponent α are chosen so that the Gaussian function in (11.8) coincides with the Gaussian function from the product of two GTOs.

The Hartree potential at $\boldsymbol{r}$ is equivalent to the nuclear attraction potential from the *nucleus* at $\boldsymbol{r}$ with the charge of $-e$. Therefore we apply the McMurchie–Davidson scheme [9] which was originally used for the nuclear attraction potential. In this scheme, the following integral is obtained recursively

$$\int \frac{\Lambda_{t,u,v}(\boldsymbol{r}_p, \alpha)}{|\boldsymbol{r} - \boldsymbol{C}|}\,\mathrm{d}\boldsymbol{r} = \frac{2\pi}{\alpha} R^0_{t,u,v}(\alpha,\, \boldsymbol{R}_{pc})\,, \tag{11.9}$$

where $\boldsymbol{R}_{pc} = \boldsymbol{P} - \boldsymbol{C}$ and $\boldsymbol{C}$ is the position of the interacting electron. Furthermore, $R^n_{t,u,v}(\alpha, \boldsymbol{R}_{pc})$ is a partial derivative of the n-th order Boys function $F_n(\alpha R^2_{pc})$ [10]

$$R^n_{t,u,v}(\alpha,\, \boldsymbol{R}_{pc}) = (-2\alpha)^n \frac{\partial^t}{\partial P_x^t}\frac{\partial^u}{\partial P_y^u}\frac{\partial^v}{\partial P_z^v} F_n(\alpha R^2_{pc})\,, \tag{11.10}$$

$$F_n(x) = \int_0^1 \mathrm{e}^{-xt^2} t^{2n}\,\mathrm{d}t\,. \tag{11.11}$$

The partial derivative in (11.10) can be evaluated by the recurrence equation

$$R^n_{t+1,u,v}(\alpha,\, \boldsymbol{R}_{pc}) = t\, R^{n+1}_{t-1,u,v}(\alpha,\, \boldsymbol{R}_{pc}) - X_{pc}\, R^{n+1}_{t,u,v}(\alpha,\, \boldsymbol{R}_{pc})\,. \tag{11.12}$$

where X_{cp} is the x-component of the vector $\boldsymbol{R}_{cp}$. The higher order derivatives of Boys functions are obtained from the lower order derivatives. Similar equations are satisfied for the y- and z-components.

We developed an algorithm to evaluate the spherically averaged electron density in the MT case. The spherical averages of Hermite Gaussian functions in terms of a given center $\boldsymbol{C}$ are obtained by integrating (11.8)

$$\int \Lambda_{t,u,v}(\boldsymbol{r}_p, \alpha)\,\mathrm{d}\hat{\boldsymbol{r}}_c = \frac{2\pi}{\alpha r_c} \sum_{t'u'v'} {}_tC_{t'}\, {}_uC_{u'}\, {}_vC_{v'}\, F_{t-t',u-u',v-v'}\, S_{t',u',v'}\,. \tag{11.13}$$

where the three coefficients such as ${}_tC_{t'}$ are combinations. Here, we introduced the following two terms

$$F_{t,u,v} = \frac{\partial^t}{\partial P_x^t} \frac{\partial^u}{\partial P_y^u} \frac{\partial^v}{\partial P_z^v} \frac{1}{R_{cp}}, \tag{11.14}$$

$$S_{t,u,v} = \frac{\partial^t}{\partial P_x^t} \frac{\partial^u}{\partial P_y^u} \frac{\partial^v}{\partial P_z^v} \mathrm{e}^{-\alpha R_{cp}^2} \sinh(2\alpha r_c R_{cp}), \tag{11.15}$$

where $\boldsymbol{r}_c = \boldsymbol{r} - \boldsymbol{C}$. Both terms can be calculated by recurrence equations [11]. Although they are more complicated than in the full-potential case, it is nevertheless possible to avoid numerically evaluating the angle integration.

The spherically averaged Hartree potential $\bar{V}_c$ is constructed from the spherically averaged density $\bar{\varrho}_c$. The subscript c denotes the center of the spherical average $\boldsymbol{C}$. $\bar{V}_c$ is numerically evaluated using

$$\bar{V}_c(r_c) = \frac{4\pi}{r_c} \int_0^{r_c} r_c'(r_c' - r_c)\, \bar{\varrho}_c(r_c')\, \mathrm{d}r_c' + \int \frac{\varrho(\boldsymbol{r}_c' + \boldsymbol{C})}{r_c'} \mathrm{d}\boldsymbol{r}_c', \tag{11.16}$$

where ϱ is the non-averaged electron density. The latter integral is the solution of Poisson equation for ϱ with the shift of the origin at $\boldsymbol{C}$. It is calculated recursively using (11.12).

11.2.2 Angular Momentum Expansion of Initial State

In general, centers of angular momentum expansion and of a GTO are different. Thus the exponential function in a GTO depends on the direction from the center of the spherical average $\boldsymbol{C}$. With the help of the plane wave expansion formula, this exponential function is expanded into spherical harmonics

$$\mathrm{e}^{-\alpha r_g^2} = 4\pi\, \mathrm{e}^{-\alpha(r_c^2 + R_{cg}^2)} \sum_L (-1)^\ell\, i_\ell(2\alpha r_c R_{cg})\, Y_L(\hat{\boldsymbol{r}}_c)\, Y_L^*(\hat{\boldsymbol{R}}_{cg}), \tag{11.17}$$

where $L = (\ell, m)$, $\boldsymbol{R}_{cg} = \boldsymbol{C} - \boldsymbol{G}$, $\boldsymbol{G}$ is the center of the GTO, and $i_\ell(\rho)$ is a modified spherical Bessel function. The monomial from a GTO can be also expanded easily into spherical harmonics. Then, the angular momentum expansion results into the evaluation of Gaunt coefficients

$$G(L_1 L_2 | L_3) = \int Y_{L_1}(\hat{\boldsymbol{r}}) Y_{L_2}(\hat{\boldsymbol{r}}) Y_{L_3}^*(\hat{\boldsymbol{r}}) \mathrm{d}\hat{\boldsymbol{r}}. \tag{11.18}$$

Table 11.1 List of the radial grids

Code	Radial grid	Formula
SPR- KKR [12]	SPRKKR	$x_i = x_{\min} \exp\{(i-1)d\}$
WIEN2k [13]	WIEN2K	$x_i = x_{\min} \exp\{(i-1)d\}$
VASP [2]	VASP	$x_i = x_{\min} \exp\left\{(i-1)\left(\frac{x_{\max}}{x_{\min}}\right)^{\frac{1}{N_{\mathrm{grid}}-1}}\right\}$
LMTO [14]	LMTO	$x_i = b[\exp\{(i-1)a\}-1]$
Quantum-ESPRESSO [15]	QEPEXP	$x_i = \frac{\exp(x_{\min}+i\,d)}{N_{\mathrm{grid}}}$
	QEPSFT	$x_i = \frac{\exp(x_{\min}+i\,d)-1}{N_{\mathrm{grid}}}$
GPAW [16]	GPAWL1	$x_i = i\,d$
	GPAWL2	$x_i = \frac{i\,a}{N_{\mathrm{grid}}-1}$
	GPAWEX	$x_i = a\{\exp(i\,d)-1\}$
	GPAWIP	$x_i = \frac{i\,a}{1-i\,b}$
	GPAWQR	$x_i = \frac{\left(\frac{i}{N_{\mathrm{grid}}}+a\right)^5}{a} - a^4$
Abinit [17]	ABINIT	$x_i = \frac{\left(\frac{i}{N_{\mathrm{grid}}}+a\right)^5}{a} - a^4 \quad (a = 10^{-2})$
Exciting [18]	EXCCUB	$x_i = x_{\min} + \left(\frac{i-1}{N_{\mathrm{grid}}-1}\right)^3$
	EXCEXP	$x_i = x_{\min}\left(\frac{x_{\max}}{x_{\min}}\right)^{\frac{i-1}{N_{\mathrm{grid}}-1}}$
	EXC3EX	$x_i = 0.5\{\mathrm{erf}(y_i)+1\}\left\{x_{\min} + \left(\frac{i-1}{N_{\mathrm{grid}}-1}\right)^3\right\}$ $+[1-0.5\{\mathrm{erf}(y_i)+1\}]\left\{x_{\min}\left(\frac{x_{\max}}{x_{\min}}\right)^{\frac{i-1}{N_{\mathrm{grid}}-1}}\right\}$ $y_i = 5\,\frac{i-\mathrm{nint}\left(\frac{N_{\mathrm{grid}}}{2}\right)}{N_{\mathrm{grid}}}$ nint(x) rounds its argument to the nearest whole number
GAMESS [19]	GAMLOG	$x_i = -a\ln\left\{1-\left(\frac{2i-1}{2N_{\mathrm{grid}}}\right)^3\right\}$
	GAMEUL	$x_i = a\,\frac{\left(\frac{i}{N_{\mathrm{grid}}+1}\right)^2}{1-\left(\frac{i}{N_{\mathrm{grid}}+1}\right)^2}$
Hermann–Skillman [20]	HERSKI	$x_i = x_{i-1} + h_{i-1}$ $h_0 = 0.0025$ h_i is doubled every 40 points

i: serial number of the points
x_i: position of i-th point
$x_{\min}$: minimum value of x
$x_{\max}$: maximum value of x
N_{grid}: total number of points
a, b, d: parameters

Acknowledgements Parts of this work have been funded by European FP7 MSNano network under Grant Agreement No. PIRSES-GA-2012-317554 and COST Action MP1306 EUSpec and the European FP7 MS-BEEM (Grant Agreement No. PIEF-GA-2013-625388).

Appendix

There are many electronic structure codes in material science. Each of them uses a different kind of radial mesh, optimized for the computation of the potentials and of the electronic charge densities. Our interface module performs also transformations between the different radial grids whenever necessary.

When a set of original data points is given, this interface interpolates these points by the cubic spline method, calculates the data points of the new radial grid chosen by the user, and returns the set of these new data points. The forms of the radial grids which can be used in this interface are as shown in Table 11.1. The parameters in these formulas can be set up by the user.

References

1. J. Xu, C.R. Natoli, P. Krüger, K. Hayakawa, D. Sébilleau, L. Song, K. Hatada, Comput. Phys. Commun. **203**, 331 (2016)
2. G. Kresse, D. Joubert, Phys. Rev. B **59**, 1758 (1999)
3. P. Blöchl, Phys. Rev. B **82**, 17953 (1994)
4. K. Hatada, K. Hayakawa, M. Benfatto, C.R. Natoli, Phys. Rev. B **76**, 060102(R) (2007)
5. K. Hatada, K. Hayakawa, M. Benfatto, C.R. Natoli, J. Phys. Condens. Matter **22**, 185501 (2010)
6. Y.K. Chang, H.H. Hsieh, W.F. Pong, M.-H. Tsai, F.Z. Chien, P.K. Tseng, L.C. Chen, T.Y. Wang, K.H. Chen, D.M. Bhusari, J.R. Yang, S.T. Lin, Phys. Rev. Lett. **82**, 5377 (1999)
7. M.J. Frisch, G.W. Trucks, H.B. Schlegel, G.E. Scuseria, M.A. Robb, J.R. Cheeseman, G. Scalmani, V. Barone, B. Mennucci, G.A. Petersson, H. Nakatsuji, M. Caricato, X. Li, H.P. Hratchian, A.F. Izmaylov, J. Bloino, G. Zheng, J.L. Sonnenberg, M. Hada, M. Ehara, K. Toyota, R. Fukuda, J. Hasegawa, M. Ishida, T. Nakajima, Y. Honda, O. Kitao, H. Nakai, T. Vreven, J.A. Montgomery Jr., J.E. Peralta, F. Ogliaro, M. Bearpark, J.J. Heyd, E. Brothers, K.N. Kudin, V.N. Staroverov, R. Kobayashi, J. Normand, K. Raghavachari, A. Rendell, J.C. Burant, S.S. Iyengar, J. Tomasi, M. Cossi, N. Rega, J.M. Millam, M. Klene, J.E. Knox, J.B. Cross, V. Bakken, C. Adamo, J. Jaramillo, R. Gomperts, R.E. Stratmann, O. Yazyev, A.J. Austin, R. Cammi, C. Pomelli, J.W. Ochterski, R.L. Martin, K. Morokuma, V.G. Zakrzewski, G.A. Voth, P. Salvador, J.J. Dannenberg, S. Dapprich, A.D. Daniels, Ö. Farkas, J.B. Foresman, J.V. Ortiz, J. Cioslowski, D.J. Fox, Gaussian09 Revision E.01, Gaussian Inc. Wallingford CT 2009
8. N. Komiya, K. Hatada, F. Ota, P. Krüger, T. Fujikawa, K. Niki, J. Electron Spectrosc. Relat, Phenom. (2017) (submitted)
9. T. Helgaker, P. Jorgensen, J. Olsen, *Molecular Electronic-Structure Theory* (Wiley, New York, 2000), pp. 45–56
10. S.F. Boys, Proc. R. Soc. A **200**, 542 (1950)
11. N. Komiya, Master thesis, Chiba University, Chiba, Japan, 2016
12. H. Ebert, D. Ködderitzsch, J. Minár, Rep. Prog. Phys. **74**, 096501 (2011)
13. P. Blaha, K. Schwarz, P. Sorantin, S.B. Trickey, Comput. Phys. Commun. **59**, 399 (1990)

14. O.K. Andersen, O. Jepsen, Phys. Rev. Lett. **53**, 2571 (1984)
15. P. Giannozzi, S. Baroni, N. Bonini, M. Calandra, R. Car, C. Cavazzoni, D. Ceresoli, G.L. Chiarotti, M. Cococcioni, I. Dabo, A. Dal Corso, S. Fabris, G. Fratesi, S. de Gironcoli, R. Gebauer, U. Gerstmann, C. Gougoussis, A. Kokalj, M. Lazzeri, L. Martin-Samos, N. Marzari, F. Mauri, R. Mazzarello, S. Paolini, A. Pasquarello, L. Paulatto, C. Sbraccia, S. Scandolo, G. Sclauzero, A.P. Seitsonen, A. Smogunov, P. Umari, R.M. Wentzcovitch, J. Phys. Condens. Matt. **21**, 395502 (2009)
16. J.J. Mortensen, L.B. Hansen, K.W. Jacobsen, Phys. Rev. B **71**, 035109 (2005)
17. X. Gonze, J.-M. Beuken, R. Caracas, F. Detraux, M. Fuchs, G.-M. Rignanese, L. Sindic, M. Verstraete, G. Zerah, F. Jollet, M. Torrent, A. Roy, M. Mikami, Ph Ghosez, J.-Y. Raty, D.C. Allan, Comput. Mater. Sci. **25**, 478–492 (2002)
18. A. Gulans, S. Kontur, C. Meisenbichler, D. Nabok, P. Pavone, S. Rigamonti, S. Sagmeister, U. Werner, C. Draxl, J. Phys. Condens. Matt. **26**, 363202 (2014)
19. M.W. Schmidt, K.K. Baldridge, J.A. Boatz, S.T. Elbert, M.S. Gordon, J.H. Jensen, S. Koseki, N. Matsunaga, K.A. Nguyen, S. Su, T.L. Windus, M. Dupuis, J.A. Montgomery Jr, J. Comput. Chem. **14**, 1347–1363 (1993)
20. F. Hermann, S. Skillman, *Atomic Structure Calculations* (Prentice-Hall, Englewood Cliffs, 1963)

Chapter 12
L-Edge Absorption and Dichroism in Low Symmetry $3d^0$ Compounds

Peter Krüger and Calogero R. Natoli

Abstract Inorganic compounds containing light transition metal ions in a formal ($3d^0$) electronic configuration often have a complex atomic structure and several polymorphs may co-exist, especially in nanomaterials. X-ray absorption spectra at the metal L-edge probes the symmetry, electronic state and bonding of the transition metal ion. In low symmetry crystals, linear dichroism is common and reflects a complicated ligand field. We show that the metal $L_{2,3}$-edge spectra, which are dominated by strong particle-hole multiplet coupling, are well described with multichannel multiple scattering theory. The ligand field and band effects are included *ab initio* by the anisotropic scattering of the excited electron, whereby complex dichroism can be accounted for. Titanite nanostructures and calcium carbonate are taken as examples.

12.1 Introduction

Many minerals contain a light $3d$ element (K, Ca, Ti, V), oxygen and other first row element ligands. For each compound, several polymorphs with different, complex crystal structures may coexist. The metal ion often occupies a octahedral or tetrahedral site, but strong distortion is common and the point symmetry can be as low as C_1. $L_{2,3}$-edge absorption spectroscopy is a powerful means for probing the electronic state of the metal atom, its local atomic environment and bonding. If the system has cubic or tetrahedral symmetry, the spectra can often be well described using the ligand field multiplet model [1]. In these high symmetry cases, the metal $3d$ states are split into two levels, t_2 and e, and so the effect can be reproduced by introducing one empirical parameter (10Dq). However, the number of ligand field parameters rises to three in tetragonal symmetry and to even larger numbers for lower symmetry. When so many quantities are taken as free fitting parameters, it not clear what can be

P. Krüger (✉)
Graduate School of Engineering, Chiba University, Chiba 263-8522, Japan
e-mail: pkruger@chiba-u.jp

C. Natoli
INFN-LNF Frascati, Frascati, Italy
e-mail: calogero.natoli@lnf.infn.it

D. Sébilleau et al. (eds.), *Multiple Scattering Theory for Spectroscopies*,
Springer Proceedings in Physics 204,
https://doi.org/10.1007/978-3-319-73811-6_12

learned from the theoretical analysis. Therefore, in low symmetry compounds, the usefulness of the ligand field model becomes very questionable. Predictive computational methods are needed to extract the structural and electronic information from the data. The standard first principles electronic structure methods, such the Hartree-Fock and density functional theory, rely on the independent particle approximation. This approximation breaks down for transition metal $L_{2,3}$-edge spectra because of strong intra-atomic correlation effects. For the light $3d$-elements the crucial effect is the Coulomb and exchange coupling between the excited (photo-) electron and the core-hole. This effect requires the use of a correlated particle-hole wave function, as employed in the Bethe-Salpeter-equation (BSE) approach [2]. However, BSE is a numerically very heavy method and has so far only be applied to high-symmetry crystal structures [3]. Alternatively, the multichannel multiple scattering (MCMS) method can be used [4] which is presented in detail in Chap. 5 of this book. The MCMS method is numerically light and formulated in a real-space multiple scattering framework. Therefore low symmetry and non-periodic (nano-) systems present no difficulty. In this short chapter, we report MCMS calculations of polarization dependent L-edge absorption spectra of titanite nanocrystals and calcium carbonate.

12.2 Titanite Nanocrystals

Titanium dioxide is an important material for photocatalysis and photovoltaic applications. One-dimensional systems such as nanotubes and nanoribbons are promising for enhanced device performance because of their large surface-to-bulk ratio and strongly anisotropic carrier transport. Using polarization dependent x-ray absorption spectra, we have recently investigated the electronic structure of individual sodium-hydrogen titanite nanoribbons [5], which can be easily transformed to TiO_2 nanoribbons by ion exchange and calcination. The titanite nanoribbons are made of Ti_3O_7 layers which are separated by sodium and hydrogen cations, hydroxyl groups and water. They have a monoclinic crystal structure and the most likely composition is $NaTi_3O_6(OH){\cdot}2H_2O$ [6], which contains three inequivalent Ti sites at the centres of distorted TiO_6 octahedra, see Fig. 12.1. We have calculated the Ti $L_{2,3}$-spectra using MCMS. The one-electron potential was generated with the linear muffin-tin orbital (LMTO) code [7] in the local density approximation. The atomic sphere approximations was used and empty spheres were included for approximate space filling. The multiple scattering calculations were done with clusters containing about 150 atoms and 100 empty spheres depending on the absorber site. The calculation of core-level energies is a difficult problem beyond the scope of this paper. Here we have aligned the calculated spectra of each Ti site to the experimental threshold. The other computational details are the same as in our previous works on Ti oxides [8]. We consider linear polarized light along the monoclinic crystal axes $a = x$ and $b = y$ and along z. The polarization dependent spectra, averaged over all Ti absorber sites (not shown) agree well with experiment [5]. Here we discuss the spectra of three individual Ti sites, shown in Fig. 12.2.

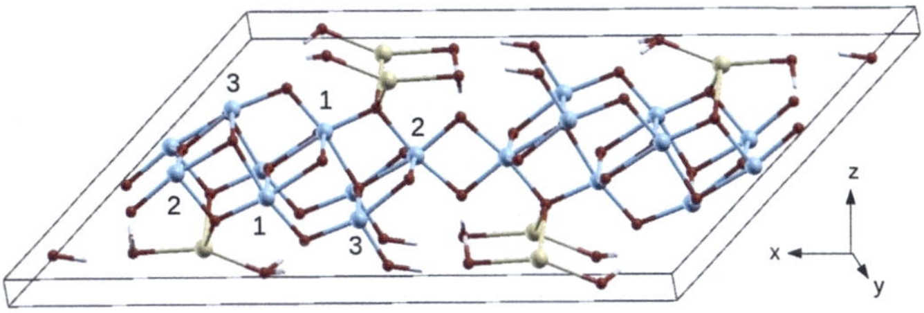

Fig. 12.1 Unit cell of $NaTi_3O_6(OH){\cdot}2H_2O$ crystal structure. Na in *yellow*, Ti in *light blue*, O in *red* and H in *white*. The three inequivalent Ti atoms are numbered 1–3

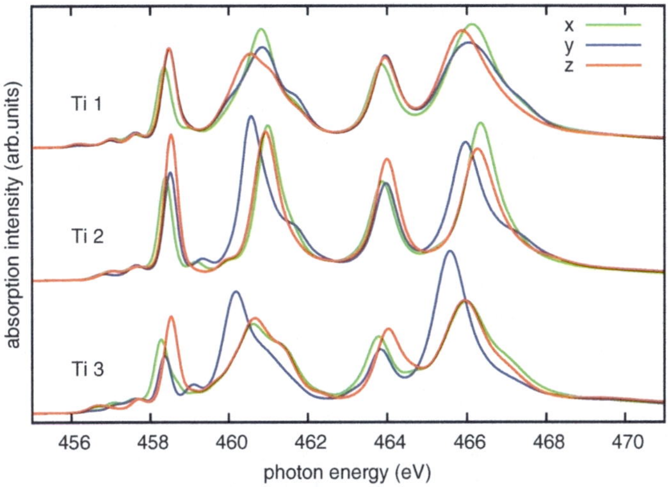

Fig. 12.2 Calculated Ti $L_{2,3}$-edge spectra of $NaTi_3O_6(OH){\cdot}2H_2O$ for the three inequivalent Ti sites and linear polarized light along cartesian axes x, y or z

The spectra of the three sites vary considerably, reflecting their very different local structure. A strong polarization dependence is observed, the largest difference is seen between the y-polarization and the x, z polarizations for sites Ti2 and Ti3. Also in the O K-edge spectra, the strongest difference was found between y-polarization and x or z polarization [5]. For the O K-edge, this linear dichroism can be understood from the fact that the Ti-O bonds of the oxygen atoms on the *surface* of the Ti_3O_7 sheets, all lie in the (a, c) plane, resulting in an approximately uniaxial anisotropy of the Ti-O electronic states along b. The Ti-O bond orientation is probably also an important reason for the observed dichroism at the Ti L-edge. However, the explanation is much more complicated at the L-edge, because there is no one-to-one correspondence between the three polarization directions and the $2p \rightarrow 3d$ transitions, and the transitions mix through $2p$ spin-orbit and multiplet interactions. It is seen from Fig. 12.2 that the dichroic effect increases in the sequence from Ti1 over Ti2 to Ti3. This reflects the increasingly anisotropic environment. The optimized Ti-O bond lengths are in the ranges 1.83–2.13, 1.88–2.10, 1.76–2.23 Å, for Ti1, Ti2,

Ti3, respectively. The range is by far largest for Ti3, reflecting a strongly distorted TiO_6 octahedron, which leads to a large dichroism. However, this kind of reasoning based on Ti–O bond lengths, is not enough since it would imply a slightly weaker dichroism for Ti2 than for Ti1, in contrast to observation. Previously, we have shown that in rutile and anatase TiO_2 [8], the spectral fine structure, especially the L_3-e_g feature (at 459–462 eV), reflects the geometrical and electronic structure of the system on a much longer length scale than the first nearest neighbor (oxygen) shell. In $NaTi_3O_6(OH)\cdot 2H_2O$, the three Ti sites strongly differ in the second neighbor shell which is made of Ti cations at a distance 2.9–3.2 Å and Na cations at 3.4–3.5 Å. Site Ti1 is surrounded by six Ti and two Na ions, Ti2 has four Ti and two Na, and Ti3 has five Ti and zero Na neighbors. Inspection of Fig. 12.1 reveals that this second neighbor cation distribution is rather isotropic for Ti1 but highly anisotropic for Ti2 and Ti3. Thus the ligand field of Ti2 and Ti3 is more anisotropic than that of Ti1, resulting in an enhanced dichroism. From this discussion we conclude that linear dichroism at the metal L-edge is a sensitive probe of the anisotropy of the local electronic structure in low symmetry compounds, but its comprehension requires theoretical methods beyond crystal field or first ligand-shell cluster models.

12.3 Calcium Carbonate

As a second example we discuss two common polymorphs of calcium carbonate ($CaCO_3$) namely calcite and aragonite. The local atomic structures are shown in Fig. 12.3. The point symmetry at the Ca site is low in both cases: point group S_6 in calcite and C_s in aragonite. Looking only at the first coordination shell, calcite has six oxygen ligands in a very nearly octahedral coordination with equal Ca-O bond lengths (2.36 Å) and almost right O-Ca-O angles (87.4°). Aragonite has nine oxygen ligands at bond lengths 2.41–2.54 Å. In the independent particle picture, the Ca L_3 and L_2-edge spectra are both proportional to the unoccupied Ca-d density of states (DOS). The ground state Ca-d DOS as calculated in the local density approximation with the LMTO method are shown in Fig. 12.4 a. In both compounds the $3d$-band is 7–9 eV wide, but the mean width is clearly larger in calcite, whose DOS splits

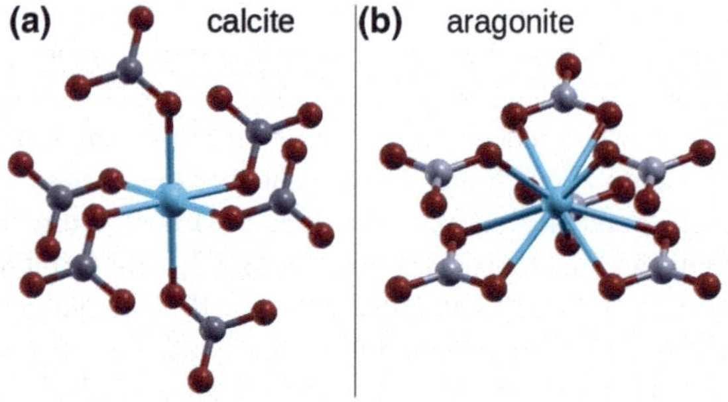

Fig. 12.3 Atomic structure of $CaCO_3$ in **a** calcite and **b** aragonite phases. Ca in *light blue*, C in *dark grey*, O in *red*. One Ca site and its nearest neighbor CO_3 units are shown

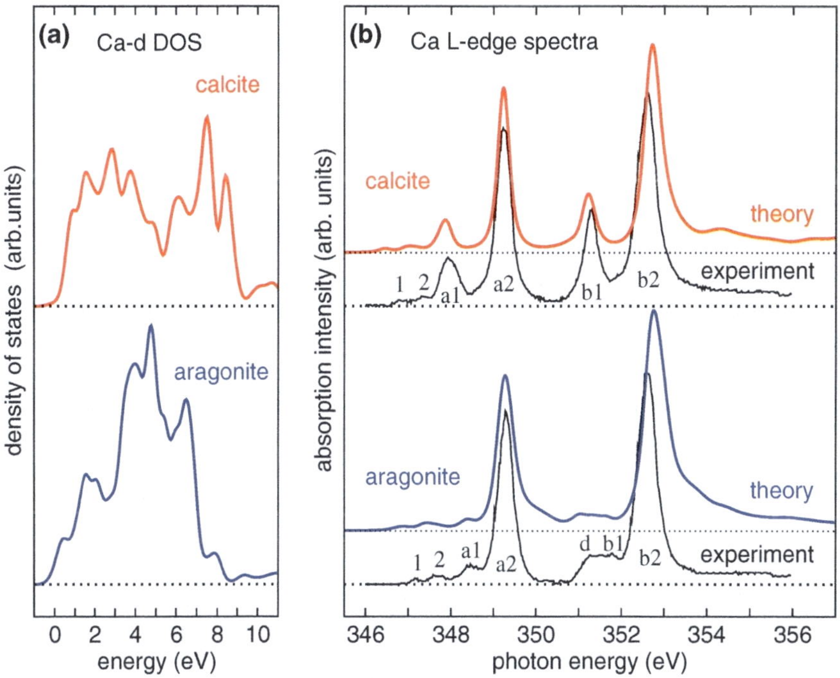

Fig. 12.4 Calcium carbonate ($CaCO_3$) in calcite (*top*) and aragonite phases (*bottom*). **a** Ground state Ca-d partial density of states. **b** Ca L-edge spectra calculated in multichannel multiple scattering theory [9] along with experimental data taken from [10]

into two parts (t_{2g} and e_g) typical for an octahedral ligand field. The aragonite DOS shows some fine structure, too, but the main weight is located in the center of the band, indicating a weaker ligand field. Figure 12.4 b shows the Ca L-edge spectra [9], calculated in MCMS theory with clusters of 177 (176) atoms and 286 (68) empty spheres for calcite (aragonite). As for the Ti spectra before, a partially screened core-hole potential with 10% unscreened hole was used. The spectra were broadened with a Lorentzian and a Gaussian of 0.2 eV FWHM each, and energy-shifted so as to align the (a2) peak with the experimental data taken from [10]. Apart from a small overestimation of some peak splittings, the calculated spectra agree very well with the data. All peaks are reproduced with the correct intensities. In the calcite spectra the a1 and b1 peaks have appreciable intensity. The simple four-peak structure (a1, a2, b1, b2) is similar to CaO [4] and characteristic for metal-oxides with a formal ($3d^0$) ground state and octahedral symmetry [1]. The energy splitting and intensity ratio of the a1:a2 (and b1:b2) doublet is, however, smaller in calcite than in CaO [4] which shows that the ligand field is weaker in calcite. This fact cannot be understood on the basis of a CaO_6 model, because the octahedra are almost identical in the two systems. The Ca-O bond length is even by 2% smaller in calcite than in CaO (2.40 Å) which suggests a larger ligand field, which in turn should lead to an increase in a1-a2 splitting and a1: a2 intensity ratio in contrast to observation. Clearly, second and further coordination shells must be taken into account for a proper description of the L-edge spectra. In the aragonite spectrum, the a1, b1 peaks are very weak and the b1 peak splits into a broad doublet (d-b1). This reflects the weak but complicated ligand field due to the very low symmetry coordination of the Ca ion. In recent polarization dependent measurements [11], substantial linear dichroism was observed at peak d, which could be reproduced with our MCMS calculations [9].

In summary, we have shown that $L_{2,3}$-edge spectra of low symmetry ($3d^0$) compounds are very sensitive to the local atomic and electronic structure around the metal ion. Linear dichroism is weaker than at K-edges, but nonetheless observable. It may be useful for identifying the crystal orientation in nanostructures. The L-edge spectra, which feature strong final state correlation effects, are generally very well reproduced with MCMS calculations. This is because MCMS theory combines a proper treatment of the strong particle-hole multiplet coupling with a precise description of the long-range electronic structure of the system. Owing to the flexibility of the real-space multiple scattering method and the efficient implementation of the MCMS equations in the R-matrix scheme, arbitrarily complex structures can be handled with a very modest computational cost.

Acknowledgements P. K. is grateful to Dr Carla Bittencourt, Prof. Adam Hitchcock and Dr Xiaofeng Zhu for many fruitful discussions.

References

1. F.M.F. de Groot, J.C. Fuggle, B.T. Thole, G.A. Sawatzky, Phys. Rev. B **41**, 928 (1990)
2. E.L. Shirley, J. Electron Spectrosc. Relat. Phenom. **144**, 1187 (2005)
3. R. Laskowski, P. Blaha, Phys. Rev. B **82**, 205104 (2010)
4. P. Krüger, C.R. Natoli, Phys. Rev. B **70**, 245120 (2004)
5. X. Zhu, A.P. Hitchcock, C. Bittencourt, P. Umek, P. Krüger, J. Phys. Chem. C **119**, 24192 (2015)
6. I. Andrusenko, E. Mugnaioli, T.E. Gorelik, D. Koll, M. Panthofer, W. Tremel, U. Kolb, Acta Crystallogr., Sect. B: Struct. Sci. **67**, 218 (2011)
7. O.K. Andersen, O. Jepsen, Phys. Rev. Lett. **53**, 2571 (1984)
8. P. Krüger, Phys. Rev. B **81**, 125121 (2010)
9. P. Krüger, C.R. Natoli, J. Phys. Conf. Ser. **712**, 012007 (2016)
10. P. Rez, A. Blackwell, J. Phys. Chem. B **115**, 11193 (2011)
11. R.A. Metzler, P. Rez, J. Phys. Chem. B **118**, 6758 (2014)

Chapter 13
Multichannel L-Absorption Calculations by Analytical Continuation of Green's Function into Complex Energy Plane

Anna Taranukhina, Alexander Novakovich and Vladislav Kochetov

Abstract X-ray $L_{2,3}$ absorption spectra are dominated by transitions to d states which are either continuum states rapidly varying with the energy or collapsed bound states formed in the field of the $2p$ hole in the absorbing atom. Computing such spectra requires extremely small step in the real energy scale making the calculations time-consuming even in a single-particle approach. In this paper we present an efficient procedure of solving this problem within the multichannel multiple scattering real–space Green's function (MCMS-GF) method which takes into account the strong impact of electron-hole correlations on $L_{2,3}$-edge spectra. A new numerical scheme is based on the analytical continuation of the Green's function to complex energies. The applicability of this approach is demonstrated by computing the K $L_{2,3}$-edge spectrum for potassium fluoride.

13.1 Introduction

It has been shown in works [1, 2] that within the Green's function (GF) formalism the convolution of the XANES spectra with a Lorentzian function of a half width Γ is mathematically equivalent to calculating the GF for the integration contour shifted by the value of $\Gamma > 0$ into the upper semiplane for any photoelectron energy except the vicinity of the Fermi energy ϵ_F. The numerical aspects of this issue are considered in the framework of the multiple scattering Green's function (MS-GF)

A. Taranukhina (✉) · V. Kochetov
Faculty of Physics, Southern Federal University, Zorge Str. 5,
344090 Rostov-on-Don, Russia
e-mail: aitaranukhina@sfedu.ru

A. Novakovich
Institute of Physics, Southern Federal University, Stachky Ave. 194,
344090 Rostov-on-Don, Russia
e-mail: aanovakovich@sfedu.ru

V. Kochetov
Université de Rennes-1, 35042 Rennes Cedex, France
e-mail: vkochetov@sfedu.ru; vladislav.kochetov@etudiant.univ-rennes1.fr

D. Sébilleau et al. (eds.), *Multiple Scattering Theory for Spectroscopies*,
Springer Proceedings in Physics 204,
https://doi.org/10.1007/978-3-319-73811-6_13

method in Chap. 6. As an illustrative example, it is expedient to apply the developed methods to the description of photoabsorption for the systems which are very sensitive to both one-particle mean-field potential and electron-hole correlations. For this, we consider K $L_{2,3}$-edge spectrum of KF. The results presented in Sect. 13.2 for K L-edge obtained with MS-GF method demonstrate the advantage of computing photoabsorption spectra by the analytical continuation of the GF to complex energies. Section 13.3 gives the extension of this numerical procedure to the MCMS-GF method and illustrates the applicability of the MCMS-GF formalism in modeling photoabsorption process.

13.2 K L-Edge Spectrum of KF Within MS-GF Method

Here we show the advantage of calculating photoabsorption spectra by the analytical continuation of the GF to complex energies.

Throughout this chapter, atomic units for lengths and Rydberg units for energies are used. Following the derivation of the K-edge cross-section in Sect. 6.2, one can write the expression for L-edge *unpolarized* absorption cross-section in the dipole approximation as

$$\begin{aligned}\sigma(\omega) &= -\frac{8\pi\alpha\omega}{3}\left\{\iint_{\Omega_n}\varphi_i(\mathbf{r})\,\mathbf{r}\,\Im\left[G(\mathbf{r},\mathbf{r}',\epsilon-\epsilon_F+\mathrm{i}\Gamma)\right]\mathbf{r}'\varphi_i(\mathbf{r}')\,\mathrm{d}\mathbf{r}\,\mathrm{d}\mathbf{r}'\right\}\\ &= -\frac{8\pi\alpha\omega}{3}k\,\Im\left\{\begin{aligned}&2\int_0^{b_n}\int_0^{b_n}\varphi_{2p}(r)\,r^3R^n_{k2}(r_<)H^n_{k2}(r_>)r'^3\varphi_{2p}(r')\,\mathrm{d}r\mathrm{d}r'\\ &+\left[\int_0^{b_n}\varphi_{2p}(r)\,r^3R^n_{k2}(r)\,\mathrm{d}r\right]^2\frac{2}{5}\sum_{m=-2}^{2}G^{nn}_{2m,2m}(\epsilon-\epsilon_F+\mathrm{i}\Gamma)\end{aligned}\right\},\end{aligned} \tag{13.1}$$

where the absorbing atom is denoted by index n; the Green's function matrix element $G^{nn}_{2m,2m}$ satisfies the coupled algebraic equations (see Chap. 6, Sect. 6.2)

$$G^{nm}_{LL'}(\epsilon) = G_0{}^{nm}_{LL'}(\epsilon)(1-\delta_{nm}) + \sum_{qL''} G_0{}^{nq}_{LL''}(\epsilon)t^q_{\ell''}(\epsilon)G^{qm}_{L''L'}(\epsilon)\,, \tag{13.2}$$

where the matrix elements of the free GF $G_0{}^{nm}_{LL'}(\epsilon)$ are the well-known KKR structure constants

$$\begin{aligned}G_0{}^{nm}_{LL'} &= -4\pi\mathrm{i}\sum_{L''}\mathrm{i}^{\ell-\ell'+\ell''}C_{LL'L''}\,h^+_{L''}(k\mathbf{R}_{nm})\\ C_{LL'L''} &= \int Y_L(\hat{\mathbf{r}})Y_{L'}(\hat{\mathbf{r}})Y_{L''}(\hat{\mathbf{r}})\mathrm{d}\hat{\mathbf{r}}\,.\end{aligned} \tag{13.3}$$

The functions $R^n_{k2}(r)$ and $H^n_{k2}(r)$ in (13.1) are respectively the regular and irregular solutions of the Schrödinger equation for a cluster potential inside the sphere n associated with the absorbing atom

$$\left[\frac{1}{r^2}\frac{\mathrm{d}}{\mathrm{d}r}\left(r^2\frac{\mathrm{d}}{\mathrm{d}r}\right) - \frac{\ell(\ell+1)}{r^2} - v_n(r) + (\epsilon + \mathrm{i}\Gamma)\right] f(r, \epsilon + \mathrm{i}\Gamma) = 0\,, \quad (13.4)$$

where $f(r, \epsilon + \mathrm{i}\Gamma)$ denotes either $R^n_{k2}(r)$ or $H^n_{k2}(r)$ which match to the boundary conditions

$$\begin{aligned} R^n_{k\ell}(b_n) &= \mathrm{e}^{\mathrm{i}\delta^n_\ell}\left[j_\ell(kb_n)\,\cos\delta^n_\ell - n_\ell(kb_n)\,\sin\delta^n_\ell\right] \\ H^n_{k\ell}(b_n) &= -\mathrm{i}h^+_\ell(kb_n)\,, \end{aligned} \quad (13.5)$$

where $k = \sqrt{\epsilon + \mathrm{i}\Gamma}$. It is worth noting that we excluded the Fermi energy from the consideration because d-bands in the ground states of alkali halides are empty and well-separated from the bound states by a wide band gap.

The calculation is carried out for the cluster containing 251 atoms. The non-self-consistent muffin-tin cluster potential is constructed according to the following scheme. First, the electron self-consistent potentials of the free atoms (neutral for all atoms except the absorber which includes a hole) are computed by the Herman–Skillman procedure [3] using *$X\alpha$ approximation* for a local spin-independent exchange potential with parameter α chosen according to Schwarz [4]. In addition, we take into account that the $X\alpha$ potential for occupied electronic states has a proper exchange part and a term which almost cancels *self-interaction*. Since there is no need to correct unoccupied states for the self-interaction, the $X\alpha$ potential for these states is weakened by reducing Schwarz's α. Second, we use superimposing Herman-Skillman potentials for construction of non-self-consistent cluster potential: inside MT-spheres it is computed as a sum of atomic potentials plus arbitrary constant shifts. These shifts simulate the corrections caused by some redistribution of electron charge in the system. The proposed model can be justified in part if taking into account that calculated spectra are often determined by the correct relative positions of the atomic levels and/or scattering resonances rather than by the fine details of potential behavior inside their spheres in a cluster. Therefore, changing the potentials inside the spheres by relatively small constant shifts, one can simulate some self-consistent potential. The K L_3 -edge spectra of KF calculated for real and complex energies using the numerical procedure described in Sect. 6.2.3.2. are shown in Figs. 13.1 and 13.2.

Figure 13.1 displays the spectra computed for complex energies with the imaginary part Γ and for real energies with the subsequent convolution by a Lorentzian function with the same Γ. Both spectra are obtained with the energy step of 0.01 Ry in 124 points covering 0.1–1.33 Ry. In the second case, one can observe the effect of a *loss* of the first t_{2g} peak because the energy step of 0.01 Ry is too large to *catch* extremely sharp resonance state in the continuum.

Figure 13.2 shows that the calculation for real energies leads to the same result as for complex energies if only the energy step equals to 0.00005 Ry. To cover the energy region 0.4–1.0 Ry, one needs to calculate spectrum in 12,000 points, as opposed to 124 points covering 0.1–1.33 Ry in the case of complex energies.

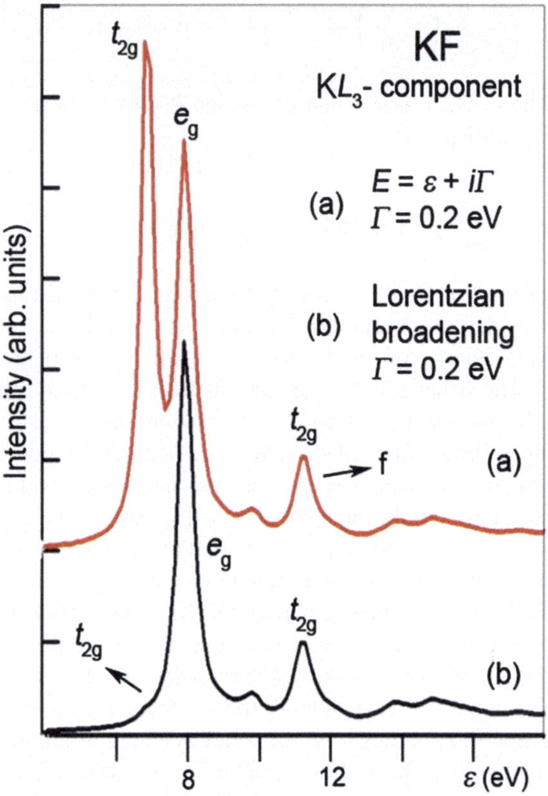

Fig. 13.1 K L_3-edge unpolarized absorption cross-section for KF calculated with the single-channel MS-GF method showing the comparison between the calculations for real and complex energies

13.3 K $L_{2,3}$-Edge Spectrum of KF Within MCMS-GF Method

In this section we present an example of applying the MCMS-GF method to the calculation of the K $L_{2,3}$-edge cross-section for KF by the analytical continuation of the GF to complex energies. We use the expression for absorption cross-section (6.106) derived within the MCMS-GF in Sect. 6.3.4 of this book:

$$\sigma(\omega) = 4\pi\alpha\omega\Im\sum_{pq}\left\{-M_q^\dagger\left[\tilde{G}_{pq}(E+\mathrm{i}\Gamma)\right]M_p - M_p^\dagger\tilde{M}_q\,\delta_{pq}\right\}. \quad (13.6)$$

The formulas for all functions included in (13.6) can be found in Sect. 6.3. One needs only to remember that now the coupled algebraic equations for the multichannel GF and the coupled set of integro-differential equations for the regular and irregular solutions should be solved for complex energies with the imaginary part equal to Γ.

The results of calculations using the single-channel and multichannel methods for complex energies with $\Gamma = 0.2$ Ry are presented in Fig. 13.3 in comparison with the

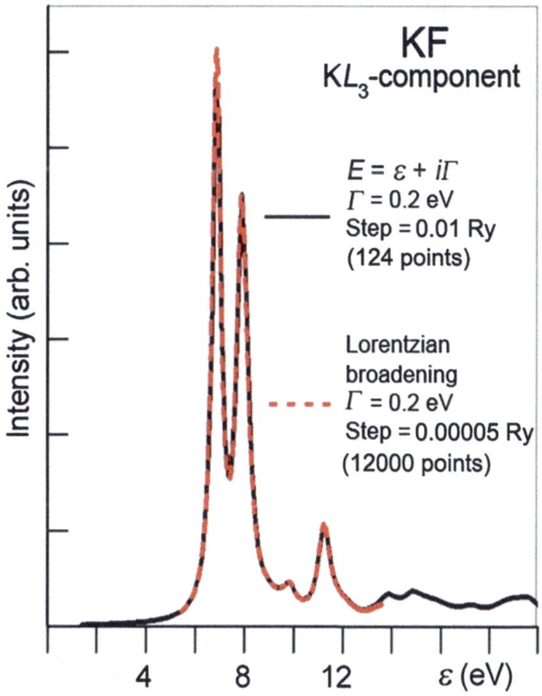

Fig. 13.2 K L_3-edge unpolarized absorption cross-section for KF calculated with the single-channel MS-GF method for real and complex energies with the energy steps differing by two orders

experimental data [5, 6]. The theoretical spectra are shifted in energy to match to the position of the first peak on the experimental curve (exp2). The multichannel calculation takes into account four coupled channels in the final state: $2p^{-1}\left(j=\frac{1}{2},\ \frac{3}{2}\right)t_{2g}$ and $2p^{-1}\ \left(j=\frac{1}{2},\ \frac{3}{2}\right)e_g$. One can see a good agreement between the spectrum calculated with the MCMS-GF method and the experiment over a wide range of photon energies. The reason of the only disagreement in the vicinity of the weak peak marked by "a" on the experimental curve which is absent in the theoretical spectrum has to be investigated by additional testing an accuracy of numerical procedures near the MT-zero.

The analysis of calculation for L_3 component presented in Fig. 13.1 shows that the peak "f" has the symmetry t_{2g}. This peak is displayed as a shoulder of the peak e_g in the L_2 component at $L_{2,3}$-edge (Fig. 13.3). In turn, peak "g" is the same peak "f" which belongs to the L_2 component shifted by the value of the spin-orbit splitting. Thus, both "g" and "f" peaks are due to the multiple scattering effects. It is worth noting that in the multichannel spectrum the symmetry of peaks "f" and "g" is not pure t_{2g} because of the strong multichannel coupling.

Acknowledgements The authors especially thank Calogero Renzo Natoli for collaboration, fruitful discussions, comments and suggestions on this line of researches. We are grateful to Didier Sébilleau for collaboration; without his enthusiasm and leadership the project could not be successful. We also thank Peter Krüger for stimulating discussion on the topic of research. A.T. would like to

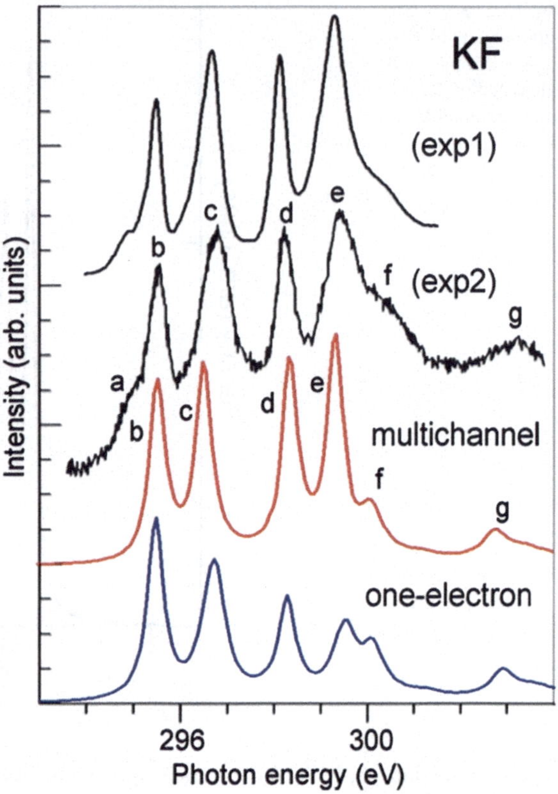

Fig. 13.3 K $L_{2,3}$-edge unpolarized absorption cross-section for KF calculated with the single-channel (MS-GF) and multichannel (MCMS-GF) methods for complex energies. Curves **(exp1)** and **(exp2)** are the experimental data taken from [5, 6] respectively

acknowledge financial support from the Ministry of Education and Science of the Russian Federation (project 3.5398.2017/8.9). Parts of this work have been funded by European FP7 MSNano network under Grant Agreement No. PIRSES-GA-2012-317554 and COST Action MP 1306 EUSpec.

References

1. R.V. Vedrinskii, I.I. Gegusin, V.N. Datsyuk, V.L. Kraizman, A.A. Novakovich, Phys. Status Solidi B **111**, 433 (1982)
2. Ch. Brouder, M. Alouani, K.H. Bennemann, Phys. Rev. B **54**, 7334 (1996)
3. F. Hermann, S. Skillman, *Atomic Structure Calculations* (Prentice-Hall, Englewood Cliffs, 1963)
4. K. Schwarz, Phys. Rev. B. **5**, 2466 (1972)
5. F. Sette, B. Sinkovic, Y.J. Ma, C.T. Chen, Phys. Rev. B. **39**, 11125 (1989-II)
6. M. Yanagihara, H. Maezawa, T. Sasaki, Y. Iguchi, J. Phys. Soc. Jpn. **54**, 3628 (1985)

Chapter 14
Ballistic Electron Emission Microscope by Real Space Multiple Scattering Theory

Keisuke Hatada and Didier Sébilleau

Abstract Ballistic Electron Emission Microscope (BEEM) is a microscope to investigate Schottky barrier based on Scanning Tunneling Microscope (STM) setup. The theoretical scheme widely used for STM is mostly focusing on an electric current from the tip tunneling through the vacuum to the sample surface. However, this model is not applicable for BEEM, since in the BEEM case, electrons tunneling through the vacuum are transported in the material over a very long range. We propose a theoretical model based on the real space full potential multiple scattering theory in order to describe this transport phenomena within the one electron picture. It is analogous to the theoretical model of angle resolved photoemission, except that the electron is emitted from the tip. This framework describes the tunneling effect and the multiple scattering in the tip and the sample and between them. Moreover this theory can be applied for non-Hermitian Hamiltonian, so that the loss of electrons at the Schottky barrier can be mimicked by introducing an imaginary part in the optical potential.

14.1 Introduction

BEEM was proposed by William Kaiser and L. Douglas Bell [1, 2] on 1998 in order to characterize the structural and electronic properties of the interfaces involved in the multilayers, and study the transport through all the layers. It is based on STM and consists in the injection of electrons from the nanometer sized STM tip across a tunneling gap into a thin metal layer that forms with a semiconductor substrate a Schottky barrier (see Fig. 14.1). A small part of these electrons will travel ballistically

K. Hatada (✉)
Department Chemie, Ludwig-Maximilians-Universität München, Butenandtstr. 5-13, 81377 Munich, Bavaria, Germany
e-mail: keisuke.hatada@cup.uni-muenchen.de

D. Sébilleau
Institut de Physique de Rennes (IPR), Univ Rennes, CNRS, UMR 6251, F-35000 Rennes, France
e-mail: didier.sebilleau@univ-rennes1.fr

D. Sébilleau et al. (eds.), *Multiple Scattering Theory for Spectroscopies*, Springer Proceedings in Physics 204,
https://doi.org/10.1007/978-3-319-73811-6_14

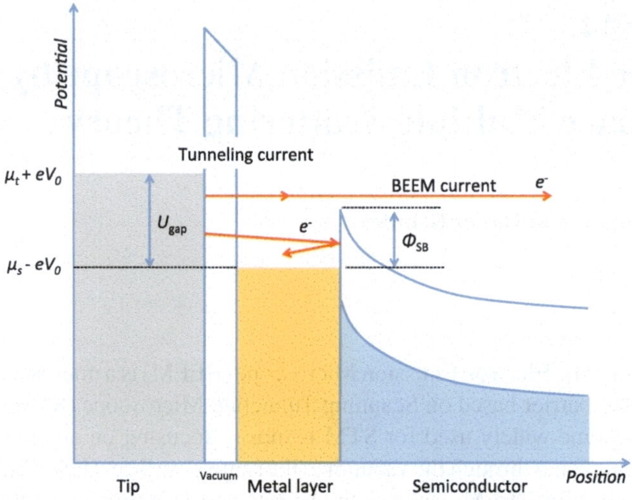

Fig. 14.1 A brief sketch of BEEM current based on STM setup with potential energy. μ_t and μ_s are chemical potentials for the tip and sample, respectively. Φ_{SB} is Schottky barrier. eV_0 is an applied field

(i.e. encountering only elastic scattering by the atoms of the multilayer). At the interface metal/semiconductor, the Schottky barrier will only allow a fraction of them to pass (those with higher energy) and be detected as the BEEM current [3]. Being of excellent spatial resolution (~1 nm), it therefore allows to study the spatial dependence of transport in buried interfaces such as those used to study spin injection and has rapidly become a key tool for these studies.

Prior theoretical models for BEEM have been implemented by means of ***k***-space band calculations [4] and using a tight-binding method [5, 6]. The ***k***-space band model is very convenient to interpret the phenomena in a qualitative way, however the diffusion process of the hot electron in real space is not considered. Likewise, while the tight binding method can treat the transport problem in the real space, as it relies on localized bases, it is not suitable to describe the long-range nature of the travelling electrons.

The purpose of this study is to devise a new description of BEEM using the flexible Multiple Scattering (MS) framework, in order to better understand the transport properties of the spin carried by electrons in heterostructures associating magnetic electrodes and semiconductors that are at the core of the research for spin injection. MS theory gives a good representation for a large-scale system within real space formalism [7, 8]. We apply here the full potential Multiple Scattering theory [9, 10] to BEEM (MSBEEM) to describe the transport phenomena accurately.

14.2 Theory

The theoretical model which we employ here is that a probe electron in the tip travels to the sample by tunneling then goes to the contact at the bottom of the sample in real space (see Fig. 14.2). The main part of the theory is based on the real space full potential multiple scattering (FPMS) theory. Since there is a loss of electrons stopped at the Schottky barrier, we add an imaginary part of the potential to mimic the phenomena. One of the advantages of multiple scattering theory is that the theory is able to treat non-Hermitian Hamiltonians. Our multiple scattering model of MSBEEM describes the interactions of the probe electron from the tip traveling through the sample to the collector attached at the bottom of the sample. For the collector, indium metal is used in many cases, so that there is no potential barrier between the semiconductor and the collector and therefore we can well approximate the probe electron by a plane wave at the collector. In any case, the electron which participates to the current goes to the direction of the collector, so that we just integrate the plane wave-like electron over the hemisphere in the collector. Thus we use a time reverse picture of the process [11] from the collector to the tip via the sample. This picture is very similar to the angle resolved photoelectron diffraction from a molecular orbital of a molecule [12]. In order to recover the STM result, we just need to perform the integration of the escaped electrons from the tip to the sample over the whole angle instead of the hemisphere. For the MSBEEM formalism, we start from Lippmann–Schwinger equation,

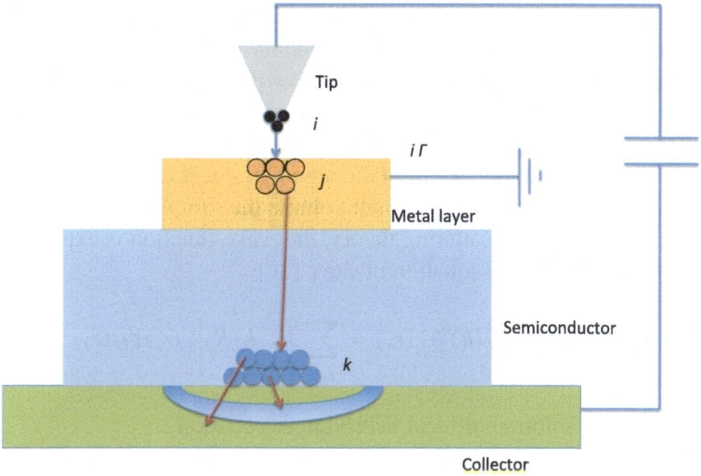

Fig. 14.2 A schematic model for the BEEM transport. Electrons in atomic sites (indicated as i) in the tip are transmitted to the atomic sites j on the surface of the metal. The electrons having higher energy than the Schottky barrier go through to the semiconductor. These electrons are collected in the collector at the bottom from the atomic sites k. The electrons which are blocked by the Schottky barrier are directed to the ground

$$\psi(\boldsymbol{r};\boldsymbol{k}) = \phi_0(\boldsymbol{r};\boldsymbol{k}) + \int \mathrm{d}\boldsymbol{r}' G_0(\boldsymbol{r}-\boldsymbol{r}';\boldsymbol{k}) V(\boldsymbol{r}')\psi(\boldsymbol{r}';\boldsymbol{k}) \,, \tag{14.1}$$

where $\psi(\boldsymbol{r};\boldsymbol{k})$ is the wave function for the probe electron as a function of a wavevector $\boldsymbol{k}$ and the position vector $\boldsymbol{r}$, the first term in the right hand is the free solution, $\phi_0(\boldsymbol{r};\boldsymbol{k}) = \mathrm{e}^{\mathrm{i}\boldsymbol{k}\cdot\boldsymbol{r}}$, G_0 is the free Green's function and V is the potential of the system. In this chapter the energy unit is chosen as Rydberg unit. The detail of the concept of FPMS can be found in Chap. 3 and in [10]. We assume that the optical potential is local and complex, especially in the metal layer, which is attached to the ground, in order to describe the loss.

The flux of electron $\mathrm{d}\phi$ from the tip-sample system to the cable at the bottom of the sample is related with the differential cross section,

$$\mathrm{d}\phi_{\mathrm{sc}} R^2 = \mathrm{d}\phi_{\mathrm{in}} \frac{\mathrm{d}\sigma}{\mathrm{d}\Omega} \,, \tag{14.2}$$

where the subscript sc and in stand for scattered and incoming electron, respectively. R is the radius of the sphere, which covers the tip and the sample, with its origin at the bottom of the sample and $\mathrm{d}\sigma/\mathrm{d}\Omega$ is the differential cross section. The probability density of the current for the scattered electron on a sphere which involves the tip and the sample is,

$$\oint \boldsymbol{j}\cdot\mathrm{d}\boldsymbol{S} = \int \mathrm{d}\phi_{\mathrm{sc}} R^2 \mathrm{d}\Omega = \int \mathrm{d}\phi_{\mathrm{in}} \frac{\mathrm{d}\sigma}{\mathrm{d}\Omega}\mathrm{d}\Omega = \frac{\hbar k}{m_e}\int \mathrm{d}\hat{\boldsymbol{k}}_{\mathrm{in}} \int_{\mathrm{tip}} \mathrm{d}\boldsymbol{r} |\Theta\psi(\boldsymbol{r};\boldsymbol{k})|^2 \,, \tag{14.3}$$

where m_e is the mass of the electron and Θ is the time-reversal operator. We have used the time-reversed process, since the boundary condition is that the collected electron behaves like a plane wave in the cable without interacting again with the system. It also implies that there was no electron current in the distant past. The volume integral $\int_{\mathrm{tip}} \mathrm{d}\boldsymbol{r}$ is performed in the region of the source of electrons. In the case of tip to sample tunneling, we limit the region of integration to the tip, while for opposite bias, this region will over a small volume the sample.

In terms of the multiple scattering theory, the wave function is expanded by the local solution of Schrödinger equation in site i [10],

$$\psi(\boldsymbol{r}_i;\boldsymbol{k}) = \sum_L B^i_L(\boldsymbol{k}) \underline{\Phi}_L(\boldsymbol{r}_i) = \sum_{LL'} B^i_L(\boldsymbol{k}) \underline{R}_{L'L}(r_i) Y_{L'}(\hat{\boldsymbol{r}}_i) \,, \tag{14.4}$$

where $\underline{\Phi}_L(\boldsymbol{r}_i)$ is the solution of local Schrödinger equation,

$$[\nabla_i^2 + k^2 - V(\boldsymbol{r}_i)]\Phi_L(\boldsymbol{r}_i) = 0 \,, \tag{14.5}$$

and normalized as $\underline{\Phi}_L(\boldsymbol{r}_i) = \sum_{L'} S^{-1}_{LL'} \Phi_{L'}(\boldsymbol{r}_i)$ where matrix S is the Wronskian matrix with the spherical Bessel function [10]. The subscript L is the notation for the angular momentum, $L \equiv (\ell, m)$.

The coefficient B^i_L is estimated from the following equality, $\mathrm{d}\hat{\boldsymbol{r}} \equiv \mathrm{d}\sigma$

$$\sum_{j=1}^{N} \int_{S_j} \left[G_0^+(\boldsymbol{r}' - \boldsymbol{r}; \boldsymbol{\kappa})\nabla\psi(\boldsymbol{r}; \boldsymbol{k}) - \psi(\boldsymbol{r}; \boldsymbol{k})\nabla G_0^+(\boldsymbol{r}' - \boldsymbol{r}; \boldsymbol{\kappa})\right] \cdot \boldsymbol{n}_j \, \mathrm{d}\sigma_j$$
$$= \int_{S_o} \left[G_0^+(\boldsymbol{r}' - \boldsymbol{r}; \boldsymbol{\kappa})\nabla\psi(\boldsymbol{r}; \boldsymbol{k}) - \psi(\boldsymbol{r}; \boldsymbol{k})\nabla G_0^+(\boldsymbol{r}' - \boldsymbol{r}; \boldsymbol{\kappa})\right] \cdot \boldsymbol{n}_o \, \mathrm{d}\sigma_o \ . \quad (14.6)$$

Here $\Omega_o = \sum_j \Omega_j$, with surface S_o, centered at the origin o and $G_0^+(\boldsymbol{r}' - \boldsymbol{r}; \check{}\,)$ is the free Green's function with outgoing wave boundary conditions satisfying the equation $(\nabla^2 + \kappa^2)\, G_0^+(\boldsymbol{r}' - \boldsymbol{r}; \boldsymbol{\kappa}) = \delta(\boldsymbol{r}' - \boldsymbol{r})$, where $\kappa^2 = E - V_0$ and V_0 is an arbitrary constant equal to the assumed value of the cell potential outside the cell domain. This identity is valid for all $\boldsymbol{r}'$ lying in the neighborhood of the origin of each cell, since in this case the integrands are continuous with their first derivatives. In this context we shall use two distinct $\boldsymbol{k}$-vectors, defined respectively with $k = \sqrt{E}$ and $\kappa = \sqrt{E - V_0}$. The latter will appear in the expansion of the Green's function $G_0^+(\boldsymbol{r}' - \boldsymbol{r}; \boldsymbol{\kappa})$ by spherical functions [13]. Obviously $k = \kappa$ for $V_0 = 0$.

From the above equality we obtain an explicit form of the coefficient B^i_L

$$B^i_L\,(\boldsymbol{k}) \;=\; \sum_{j\,L'} \tau^{i\,j}_{L\,L'}\; A^j_{L'}(\boldsymbol{k})\;, \quad (14.7)$$

where the amplitude $A^j_{L'}$ is

$$A^j_L\,(\boldsymbol{k}) \;=\; -\mathrm{i}^{\ell}\; Y_L\,(\hat{\boldsymbol{k}}\,)\,\mathrm{e}^{\mathrm{i}\,\boldsymbol{k}\cdot \boldsymbol{R}_{jo}}\;, \quad (14.8)$$

which imposes the condition of plane wave. It satisfies the following relation,

$$\int \mathrm{d}\hat{\boldsymbol{k}}\; A^i_L(\boldsymbol{k}) \left[A^j_{L'}(\boldsymbol{k})\right]^* = J^{ij}_{LL'}\;. \quad (14.9)$$

The matrix $\boldsymbol{\tau}^{-1}$ is so called *multiple scattering matrix*, while $\boldsymbol{\tau}$ is the scattering path operator matrix,

$$\boldsymbol{\tau} = (\mathbf{T}^{-1} - \mathbf{G}_0)^{-1}\;, \quad (14.10)$$

and when real spherical harmonics are chosen, the matrix is symmetric, $\tilde{\boldsymbol{\tau}} = \boldsymbol{\tau}$.

Combining these results with (14.3), the electric current is

$$J = 2e \int \mathrm{d}k\; k^2\; F(f(k^2)) \int_{\Omega} \mathrm{d}\hat{\boldsymbol{k}}_{\mathrm{in}} \int_{\mathrm{tip}} \mathrm{d}\boldsymbol{r}\; \frac{\hbar k}{m_e} |\langle A(\boldsymbol{k})|\boldsymbol{\tau}|\underline{\Phi}(\boldsymbol{r})\rangle|^2$$
$$= \frac{2e}{\hbar} \int \mathrm{d}E\; F(f(E)) \int_{\Omega} \mathrm{d}\hat{\boldsymbol{k}}_{\mathrm{in}} \int_{\mathrm{tip}} \mathrm{d}\boldsymbol{r}\; |\langle A(\boldsymbol{k})|k\boldsymbol{\tau}|\underline{\Phi}(\boldsymbol{r})\rangle|^2\;, \quad (14.11)$$

where the function F describes the window of energy by the Fermi distribution function f. This range corresponds to the U_{gap}, that is to say from $\mu_s - eV_0$ to $\mu_t - eV_0$. For the change of variable, we have used the relation $E = \hbar^2 k^2/2m$ for the plane wave. By contrast, for STM, which corresponds to the absorption case, integration must be performed over the whole solid angle.

Acknowledgements Parts of this work have been funded by European FP7 MSNano network under Grant Agreement No. PIRSES-GA-2012-317554 and COST Action MP1306 EUSpec and the European FP7 MS-BEEM (Grant Agreement No. PIEF-GA-2013-625388).

References

1. W. Kaiser, L. Bell, Phys. Rev. Lett. **60**, 1406–1409 (1988)
2. L.D. Bell, W.J. Kaiser, Phys. Rev. Lett. **61**, 2368 (1988)
3. W. Yi, A.J. Stollenenwerk, V. Narayanamurti, Surf. Sci. Rep. **64**, 169–190 (2009)
4. M. Hervé, S. Tricot, Y. Claveau, G. Delhaye, S. Di Matteo, P. Schieffer, P. Turban, Appl. Phys. Lett. **103**, 202408 (2013)
5. P.L. de Andres, F.J. Garcia-Vidal, K. Reuter, F. Flores, Prog. Surf. Sci. **66**, 3 (2001)
6. Y. Claveau, Ph.D. thesis, University of Rennes 1, France (2014). arXiv:1501.06458 [cond-mat.mes-hall]
7. J. Xu, K. Hatada, D. Sébilleau, L. Song. arXiv:1604.04846 [quant-ph]
8. J. Xu, P. Krüger, C.R. Natoli, K. Hayakawa, Z. Wu, K. Hatada, Phys. Rev. B **92**, 125408 (2015)
9. K. Hatada, K. Hayakawa, M. Benfatto, C.R. Natoli, Phys. Rev. B **76**, 060102(R) (2007)
10. K. Hatada, K. Hayakawa, M. Benfatto, C.R. Natoli, J. Phys. Condens. Matter **22**, 185501 (2010)
11. G. Breit, H.A. Bethe, Phys. Rev. **93**, 888–890 (1954)
12. N. Komiya, K. Hatada, F. Ota, P. Krüger, T. Fujikawa, K. Niki, J. Electron Spectrosc. Relat. Phenom. (2017). (submitted)
13. C.R. Natoli, M. Benfatto, C. Brouder, M.F. Ruiz-López, D.L. Foulis, Phys. Rev. B **42**, 1944–1968 (1990)

Chapter 15
Fully Relativistic Multiple Scattering Calculations for General Potentials

Hubert Ebert, Jürgen Braun and Sergiy Mankovsky

Abstract The formal basis for fully relativistic Korringa–Kohn–Rostoker (KKR) or multiple scattering calculations for a general potential including a non-local part is reviewed. In particular the need to distinguish between right and left hand side solutions to the Dirac equation and the explicit expression for the electronic Green's function are discussed.

15.1 Introduction

Recently there is strong interest in the impact of spin-orbit coupling on the electronic structure of solids and surfaces. In this context one may mention spin transport phenomena [1], magneto-dichroic phenomena in spectroscopy [2] or the occurrence of the topological surface states in topological insulators [3]. The multiple scattering or KKR (Korringa–Kohn–Rostoker) formalism supplies a perfect platform to deal with spin-orbit driven effects as it gives direct access to the electronic Green's function (GF). Indeed a corresponding relativistic version of the KKR-GF technique is available [4, 5]. Also the treatment of a non-spherical shape of the local potential (full potential - FP) has been incorporated [6–8]. Here we deal with technical aspects of FP calculations and the treatment of a non-local self-energy.

H. Ebert (✉) · J. Braun · S. Mankovsky
Department Chemie, Ludwig-Maximilians-Universität München, Butenandtstr. 5-13, 81377 Munich, Bavaria, Germany
e-mail: Hubert.Ebert@cup.uni-muenchen.de

J. Braun
e-mail: Juergen.Braun@cup.uni-muenchen.de

S. Mankovsky
e-mail: Sergiy.Mankovskyy@cup.uni-muenchen.de

D. Sébilleau et al. (eds.), *Multiple Scattering Theory for Spectroscopies*, Springer Proceedings in Physics 204,
https://doi.org/10.1007/978-3-319-73811-6_15

15.2 Hamiltonian and Radial Equations

The starting point of our considerations is the following Hamiltonian:

$$\hat{\mathcal{H}}(z) = \hat{\mathcal{H}}^0 + \hat{\mathcal{V}} + \hat{\Sigma}(z) = \hat{\mathcal{H}}^1 + \hat{\Sigma}(z) , \tag{15.1}$$

that consists of the free electron Hamiltonian ($\hat{\mathcal{H}}^0$), a Hermitian local potential ($\hat{\mathcal{V}}$) and an energy-dependent, complex and non-local self-energy ($\hat{\Sigma}(z)$). For later use $\hat{\mathcal{H}}^0$ and $\hat{\mathcal{V}}$ are combined to the Hermitian local Hamiltonian $\hat{\mathcal{H}}^1$. The aim is to derive an expression for the real space representation of the associated Green's function operator $\hat{G}(z)$ defined by

$$\left(z - \hat{\mathcal{H}}(z)\right) \hat{G}(z) = 1 , \tag{15.2}$$

i.e. $\hat{G}(z)$ is the right-inverse of $\left(z - \hat{\mathcal{H}}(z)\right)$.

Within the four-component Dirac formalism [9] the real space representation of $\hat{\mathcal{H}}^1$ may be written as [10]:

$$\hat{\mathcal{H}}^1(\boldsymbol{r}) = -\mathrm{i}c\boldsymbol{\alpha} \cdot \nabla + \frac{1}{2}c^2(\beta - 1) + \bar{V}(\boldsymbol{r}) + \beta\boldsymbol{\sigma} \cdot \boldsymbol{B}(\boldsymbol{r}) + e\boldsymbol{\alpha} \cdot \boldsymbol{A}(\boldsymbol{r}) , \tag{15.3}$$

where $c\boldsymbol{\alpha}$ is the electronic velocity operator and the local potential involves a spin-independent part $\bar{V}(\boldsymbol{r})$, an effective magnetic field ($\boldsymbol{B}(\boldsymbol{r})$) coupling only to the spin and vector potential ($\boldsymbol{A}(\boldsymbol{r})$) coupling to the electronic current.

Restricting the self-energy $\Sigma(\boldsymbol{r}, \boldsymbol{r}', z)$ to have only a spin-independent and -dependent part one may write analogously:

$$\Sigma(\boldsymbol{r}, \boldsymbol{r}', z) = \Sigma^V(\boldsymbol{r}, \boldsymbol{r}', z) + \beta\boldsymbol{\sigma} \cdot \boldsymbol{\Sigma}^B(\boldsymbol{r}, \boldsymbol{r}', z) \tag{15.4}$$

$$= \sum_{\Lambda\Lambda'} \phi_\Lambda(\boldsymbol{r})\, \Sigma_{\Lambda\Lambda'}(z)\, \phi^\dagger_{\Lambda'}(\boldsymbol{r}') , \tag{15.5}$$

where we assumed that $\Sigma(\boldsymbol{r}, \boldsymbol{r}', z)$ can be represented sufficiently accurate by an expansion into a product of suitable basis functions $\phi_\Lambda(\boldsymbol{r})$.

Using spherical coordinates the electronic wave functions associated with (15.1), (15.3) and (15.4) consist of their large and minor components with radial wave functions $g_\Lambda(r, z)$ and $f_\Lambda(r, z)$ and spin angular functions (spinors) $\chi_\Lambda(\hat{\boldsymbol{r}})$ and $\chi_{-\Lambda}(\hat{\boldsymbol{r}})$, respectively. Here $\Lambda = (\kappa, \mu)$ and $-\Lambda = (-\kappa, \mu)$ combine the conventional spin-orbit and magnetic quantum numbers κ and μ, respectively [9]. Using a superposition of bi-spinors with individual character Λ when dealing with the Hamiltonian in (15.2) for an isolated atom (single-site problem) one is led to the following set of radial Dirac equations:

	$s_{1/2}$	$p_{1/2}$	$p_{3/2}$	$d_{3/2}$	$d_{5/2}$
$s_{1/2}$ $\kappa = -1$	A B			A B	A B
$p_{1/2}$ $\kappa = +1$		C D	C D		C D
$p_{3/2}$ $\kappa = -2$		C D	E C D F	E F	E C D F
$d_{3/2}$ $\kappa = +2$	A B		F E	F A B E	F A B E
$d_{5/2}$ $\kappa = -3$	A B	D C	D F E C	F A B E	D F A B E C

Fig. 15.1 Wave function coupling scheme in the Λ-representation for a magnetic atom at the origin of a hcp lattice with $\Sigma(\boldsymbol{r}, \boldsymbol{r}', z) = 0$, the magnetization along $\hat{z}$ and the xz-plane being a mirror plane. Within each sub-block the magnetic quantum number μ runs from $-j$ to $+j$ with $j = |\kappa| - 1/2$

$$
\begin{aligned}
&\begin{pmatrix} z & c\left(\frac{\partial}{\partial r} + \frac{-\kappa+1}{r}\right) \\ c\left(\frac{\partial}{\partial r} + \frac{\kappa+1}{r}\right) & -(z + c^2) \end{pmatrix} \begin{pmatrix} g_\Lambda(r, z) \\ f_\Lambda(r, z) \end{pmatrix} \\
&\quad - \sum_{\Lambda'} \begin{pmatrix} V^+_{\Lambda\Lambda'}(r) & -\mathrm{i}\, U_{\Lambda -\Lambda'}(r) \\ \mathrm{i}\, U_{-\Lambda\Lambda'}(r) & -V^-_{-\Lambda -\Lambda'}(r) \end{pmatrix} \begin{pmatrix} g_{\Lambda'}(r, z) \\ f_{\Lambda'}(r, z) \end{pmatrix} \\
&\quad - \sum_{\Lambda'} \int r'^2\, \mathrm{d}r' \begin{pmatrix} \Sigma^+_{\Lambda\Lambda'}(r, r', z) & g_{\Lambda'}(r', z) \\ -\Sigma^-_{-\Lambda -\Lambda'}(r, r', z) & f_{\Lambda'}(r', z) \end{pmatrix} = 0\,, \qquad (15.6)
\end{aligned}
$$

with the matrix element functions $V^\pm_{\Lambda\Lambda'}(r)$ for example given by:

$$
V^\pm_{\Lambda\Lambda'}(r) = \int \mathrm{d}\hat{\boldsymbol{r}}\; \chi^\dagger_\Lambda(\hat{\boldsymbol{r}}) \left(\bar{V}(\boldsymbol{r}) \pm \boldsymbol{\sigma} \cdot \boldsymbol{B}(\boldsymbol{r})\right) \chi_{\Lambda'}(\hat{\boldsymbol{r}})\,. \qquad (15.7)
$$

The symmetry of a considered system will decide which terms occur within an angular momentum expansion of the various potential functions (picking rules) in (15.3) and (15.4) [11]. This will lead to a restricted set of non-vanishing potential matrix element functions (see e.g. (15.7)) occurring within the coupled set of radial equations (15.6). Accordingly, any solution $\phi_\Lambda(\boldsymbol{r}, z) = \sum_{\Lambda'} \phi_{\Lambda'\Lambda}(\boldsymbol{r}, z)$ to the Dirac equation will have only a restricted number of coupled terms in the sum over Λ'. An example for the resulting coupling scheme is given in Fig. 15.1, where the letters within a column indicate the non-vanishing terms $\phi_{\Lambda'\Lambda}(\boldsymbol{r}, z)$.

The fact that $\hat{G}(z)$ has to be the right- as well as left-inverse to $\left(z - \hat{\mathcal{H}}(z)\right)$ leads to the demand to consider not only the standard right-hand side (RHS) solutions to the Dirac equation but also the left-hand side (LHS) (indicated by "$\times$") to the corresponding adjoined Dirac equation. This leads to a second set of coupled radial equations for the wave functions $g^\times_\Lambda(r, z)$ and $f^\times_\Lambda(r, z)$ that differ in the matrix element functions. Fortunately, the symmetry of the single site problem leads often to identical matrix element functions for the RHS and LHS solutions implying that one does not have to distinguish these at least concerning their radial part. This favourable situation that sometimes can been forced by the choice of the coordinate system is indicated by a symmetric single site t-matrix [10].

15.3 Single Site t-Matrix and Green's Function

The t-matrix operator $\hat{t}(z)$ for the single-site problem may be introduced when dealing with the Dyson equation for the Green's function operator:

$$\hat{G}(z) = \hat{G}^0(z) + \hat{G}^0(z)\,\hat{t}(z)\,\hat{G}^0(z)\;, \tag{15.8}$$

where $\hat{G}^0(z)$ represents the reference system that normally is the free electron system described by the Hamiltonian $\hat{\mathcal{H}}^0$. Using the real space representation of $\hat{G}^0(z)$ in terms of the relativistic form of the spherical Bessel $j_\Lambda(\boldsymbol{r}, z)$ and Hankel functions $h^+_\Lambda(\boldsymbol{r}, z)$ [4, 6, 7] expressions can be given that allow to determine $t_{\Lambda\Lambda'}(z)$ from a complete set of solutions to (15.6) [10].

To get a corresponding expression for the real space representation of $\hat{G}(z)$ it is convenient to introduce RHS solutions $R_\Lambda(\boldsymbol{r}, z)$ and $H_\Lambda(\boldsymbol{r}, z)$ to the Dirac equation imposing the following behavior (specified by the label Λ) outside the single-site potential regime ($r > r_{\text{crit}}$):

$$R_\Lambda(\boldsymbol{r}, z) = \sum_{\Lambda'} j_\Lambda(\boldsymbol{r}, z)\,\delta_{\Lambda\Lambda'} - \mathrm{i}\bar{p}\,h^+_{\Lambda'}(\boldsymbol{r}, z)\,t_{\Lambda'\Lambda}(z) \tag{15.9}$$

$$H_\Lambda(\boldsymbol{r}, z) = h^+_\Lambda(\boldsymbol{r}, z)\;, \tag{15.10}$$

where $\bar{p}$ represents the relativistic electron momentum; together with LHS solutions $R^\times_\Lambda(\boldsymbol{r}, z)$ and $H^\times_\Lambda(\boldsymbol{r}, z)$ defined analogously [10]. This leads for the single site Green's function $G^n(\boldsymbol{r}, \boldsymbol{r}', z)$ for the atomic site n to [6, 10]:

$$G^n(\boldsymbol{r}, \boldsymbol{r}', z) = -\mathrm{i}\bar{p}\sum_{\Lambda} R_\Lambda(\boldsymbol{r}, z)\,H^\times_\Lambda(\boldsymbol{r}', z)\,\theta(r' - r) \\ + H_\Lambda(\boldsymbol{r}, z)\,R^\times_\Lambda(\boldsymbol{r}', z)\,\theta(r - r')\;. \tag{15.11}$$

However, inspection shows that this expression satisfies the inhomogeneous Dirac-like equation corresponding to (15.2) for any r and $r' < r_{\text{crit}}$ only in the case of a local potential; i.e. for $\Sigma(\boldsymbol{r}, \boldsymbol{r}', z) = 0$. To arrive nevertheless at a useful expression for the Green's function $G^n(\boldsymbol{r}, \boldsymbol{r}', z)$ in case of a non-local potential, one can set up in a first step the real space representation of the Green's function operator $G^1(z)$ connected with the local Hamiltonian $\mathcal{H}^1$ using (15.11) with the corresponding solutions $R^1_\Lambda(\boldsymbol{r}, z)$ and $H^1_\Lambda(\boldsymbol{r}, z)$.

In a next step $G^n(\boldsymbol{r}, \boldsymbol{r}', z)$ is determined on the basis of the Dyson equation

$$\hat{G}^n(z) = \hat{G}^1(z) + \hat{G}^1(z)\,\hat{\Sigma}(z)\,\hat{G}^n(z)\;, \tag{15.12}$$

where the self-energy $\Sigma(z)$ is treated as a perturbation. Equation (15.12) leads for a general self-energy $\Sigma(\boldsymbol{r}, \boldsymbol{r}', z)$ to an expression for $G^n(\boldsymbol{r}, \boldsymbol{r}', z)$ that has the spatial arguments $\boldsymbol{r}$ and $\boldsymbol{r}'$ coupled; i.e. a sum of products of terms that depend only on $\boldsymbol{r}$ or $\boldsymbol{r}'$, respectively. In case that $\Sigma(\boldsymbol{r}, \boldsymbol{r}', z)$ can be represented in terms of a basis as given by (15.5), on the other hand, such a factorization gets possible leading to [10]:

$$G^n(\boldsymbol{r}, \boldsymbol{r}', z) = G^1(\boldsymbol{r}, \boldsymbol{r}', z) + \sum_{\Lambda\Lambda'} \rho^1_\Lambda(\boldsymbol{r}, z)\, \Gamma_{\Lambda\Lambda'}(z)\, \rho^{1\times}_{\Lambda'}(\boldsymbol{r}', z)\ , \quad (15.13)$$

with the renormalized self-energy matrix

$$\underline{\Gamma}(z) = \left(\underline{1} + \underline{\Sigma}(z)\, \underline{\tilde{G}}^1(z)\right)^{-1} \underline{\Sigma}(z)\ , \quad (15.14)$$

and the auxiliary functions

$$\rho^1_\Lambda(\boldsymbol{r}, z) = \int \mathrm{d}^3r'\, G^1(\boldsymbol{r}, \boldsymbol{r}', z)\, \phi_\Lambda(\boldsymbol{r}') \quad (15.15)$$

$$\tilde{G}^1_{\Lambda\Lambda'}(z) = \int \mathrm{d}^3r \int \mathrm{d}^3r' \phi^\dagger_\Lambda(\boldsymbol{r})\, G^1(\boldsymbol{r}, \boldsymbol{r}', z)\, \phi_{\Lambda'}(\boldsymbol{r}')\ , \quad (15.16)$$

that are connected with the local Hamiltonian $\mathcal{H}^1$.

Finally, the Green's function $G(\boldsymbol{r}, \boldsymbol{r}', z)$ for an extended solid is given by adding the back scattering term to $G^n(\boldsymbol{r}, \boldsymbol{r}', z)$:

$$G(\boldsymbol{r}, \boldsymbol{r}', z) = G^n(\boldsymbol{r}, \boldsymbol{r}', z) + \sum_{\Lambda\Lambda'} R_\Lambda(\boldsymbol{r}, z)\, G^{nn}_{\Lambda\Lambda'}(z) R^\times_{\Lambda'}(\boldsymbol{r}', z)\ . \quad (15.17)$$

The later term involves the regular solutions $R_\Lambda(\boldsymbol{r}, z)$ associated to the full Hamiltonian $\hat{\mathcal{H}}(z) = \hat{\mathcal{H}}^1 + \hat{\Sigma}(z)$. The multiple-scattering nature of the term is expressed here in terms of the structural Green's function matrix $\underline{G}^{nm}(z)$ [4]. This step completes the derivation of an expression for the Green's function in case of a non-local but site-diagonal self-energy.

References

1. K. Tauber, M. Gradhand, D.V. Fedorov, I. Mertig, Phys. Rev. Lett. **109**, 026601 (2012)
2. H. Ebert, Rep. Prog. Phys. **59**, 1665 (1996)
3. M.Z. Hasan, C.L. Kane, Rev. Mod. Phys. **82**, 3045 (2010)
4. P. Weinberger, *Electron Scattering Theory for Ordered and Disordered Matter* (Oxford University Press, Oxford, 1990)
5. R. Feder, F. Rosicky, B. Ackermann, Z. Physik B **52**, 31 (1983)
6. E. Tamura, Phys. Rev. B **45**, 3271 (1992)
7. X. Wang, X.-G. Zhang, W.H. Butler, G.M. Stocks, B.N. Harmon, Phys. Rev. B **46**, 9352 (1992)

8. T. Huhne, C. Zecha, H. Ebert, P.H. Dederichs, R. Zeller, Phys. Rev. B **58**, 10236 (1998)
9. M.E. Rose, *Relativistic Electron Theory* (Wiley, New York, 1961)
10. H. Ebert, J. Braun, D. Ködderitzsch, S. Mankovsky, Phys. Rev. B **93**, 075145 (2016)
11. K. Kurki-Suonio, Isr. J. Chem. **16**, 115 (1977)

Chapter 16
Relativistic Effects and Gauge Invariance in Photon Absorption and Scattering

Nadejda Bouldi and Christian Brouder

Abstract The interaction of light with matter is often described within the semi-classical framework where light is represented by an external time-dependent vector potential. We show that the transition probabilities so obtained suffer from two drawbacks: their semi-relativistic limit is delicate and they are not gauge invariant. We describe how a gauge-invariant transition probability can be obtained, for which the semi-relativistic limit is well defined. In this way, we obtain a new spin-position term which is not negligible in x-ray magnetic circular dichroism at the K-edge of transition metals.

16.1 Introduction

The description of relativistic effects in x-ray spectroscopy has become necessary due to the experimental developments that permit more and more precise measurements. The use of a semi-relativistic description is physically appealing because it allows to distinguish the effects related to the relativistic description of matter from the familiar non-relativistic effects.

We show here, however, that taking existing semi-relativistic Hamiltonians as a starting point to determine the absorption or scattering cross-sections might be delicate because perturbation theory mixes the semi-relativistic states coming from two different Hamiltonians. We propose an alternative approach in which we start from the fully-relativistic cross-sections and we apply a Foldy-Wouthuysen transformation to the wave functions.

N. Bouldi (✉) · Ch. Brouder
Institut de Minéralogie, de Physique des Matériaux et de Cosmochimie (IMPMC), UPMC Univ Paris 06, CNRS, UMR 7590, IRD, MNHN, case 115, 4 place Jussieu, 75252 Paris cedex 05, France
e-mail: nadejda.bouldi@impmc.upmc.fr

Ch. Brouder
e-mail: Christian.Brouder@impmc.upmc.fr

D. Sébilleau et al. (eds.), *Multiple Scattering Theory for Spectroscopies*, Springer Proceedings in Physics 204,
https://doi.org/10.1007/978-3-319-73811-6_16

We also discuss the question of gauge invariance of transition probabilities. We conclude that, to obtain gauge invariant results, we must work in a framework where the states describe both the electronic system and the photons. We show that Quantum Electrodynamics (QED) provides such a framework in which gauge invariance of transition probabilities has been established for large classes of gauges.

Finally, we give the semi-relativistic expansion of the absorption and scattering cross sections. We find that a new relativistic term (called spin-position term) must be including in both cross sections. More details will be a given a forthcoming publication [1].

16.2 Time-Dependent Perturbation Theory

To fix notation, we give a fast description of time-dependent perturbation theory. Assume that a time-independent Hamiltonian H_0 is perturbed by a time-dependent term $H_1(t)$ starting at t_0. Setting $\hbar = 1$, the state of the perturbed system at time $t \geq t_0$ is given by $|\psi(t)\rangle = V(t, t_0)|\phi_g\rangle$, where $|\phi_g\rangle$ is the ground state of H_0 and $V(t, t_0) = \mathrm{e}^{-\mathrm{i}H_0 t} V_I(t, t_0)\mathrm{e}^{\mathrm{i}H_0 t_0}$ with

$$V_I(t, t_0) = T\mathrm{e}^{-\mathrm{i}\int_{t_0}^{t} H_I(\tau)d\tau} .$$

The interaction Hamiltonian is $H_I(\tau) = \mathrm{e}^{\mathrm{i}H_0\tau} H_1(\tau)\mathrm{e}^{-\mathrm{i}H_0\tau}$ and T is the time-ordering operator (we omit the adiabatic switching factor for simplicity). Note that $|\psi(t_0)\rangle = |\phi_g\rangle$. The transition probability to the eigenstate $|\phi_n\rangle$ of H_0 at time t is

$$P_{ng}(t) = |\langle\phi_n|\psi(t)\rangle|^2 = |\langle\phi_n|V(t, t_0)|\phi_g\rangle|^2 = |\langle\phi_n|V_I(t, t_0)|\phi_g\rangle|^2 . \quad (16.1)$$

For $n \neq g$ the perturbative expansion gives us, up to second order,

$$P_{ng}(t) = \Big|\langle\phi_n| \int_{t_0}^{t} H_I(\tau)\mathrm{d}\tau - \mathrm{i} \int_{t_0}^{t} \mathrm{d}\tau H_I(\tau) \int_{t_0}^{\tau} \mathrm{d}\tau' H_I(\tau')|\phi_g\rangle\Big|^2 + \ldots \quad (16.2)$$

16.3 Gauge Invariance

The principle of gauge invariance has become a cornerstone of particle physics: observable quantities must be independent of the gauge chosen to describe electric and magnetic fields.

In quantum mechanics, gauge transformations consist in both a change of the potentials $\boldsymbol{A}' = \boldsymbol{A} + \nabla \Lambda$, $\phi' = \phi - \partial_t \Lambda$ and a change of the phase of the wavefunctions $\psi'(\boldsymbol{r}, t) = M_\Lambda \psi(\boldsymbol{r}, t)$ where, $M_\Lambda = \mathrm{e}^{\mathrm{i}e\Lambda(\boldsymbol{r},t)}$. The time-dependent Dirac and Schrödinger equations are invariant under gauge transformation [2] but the time independent ones are not. Indeed, a Hamiltonian $H(\boldsymbol{A}, \phi) = f(\boldsymbol{p} - e\boldsymbol{A}) + e\phi$ where f is some function, becomes $H(\boldsymbol{A}', \phi') = M_\Lambda H(\boldsymbol{A}, \phi) M_\Lambda^\dagger - e\partial_t \Lambda$ (the term $\partial_t \Lambda$ arises because of the electric potential ϕ) while a *gauge invariant operator* $O(\mathbf{A}, \phi)$ should satisfy $O(\mathbf{A}', \phi') = M_\Lambda O(\mathbf{A}, \phi) M_\Lambda^\dagger$ for every function $\Lambda(\boldsymbol{r}, t)$ [2].

If the gauge is modified by a time-dependent term $\Lambda(\mathbf{r}, t)$, the perturbation $H_1 = H(\mathbf{A}, \phi) - H_0$ is modified into $H_1' = H(\mathbf{A}', \phi') - H_0$. We first consider the lack of gauge-invariance for the matrix elements of the perturbation. We compute the difference between these matrix elements in two different gauges:

$$\langle \Psi^m | H_1' - H_1 | \Psi^n \rangle = \langle \Psi^m | M_\Lambda H M_\Lambda^\dagger - H - e \frac{\partial \Lambda}{\partial t} | \Psi^n \rangle \,. \tag{16.3}$$

Using the Baker–Campbell–Hausdorff formula at first order $M_\Lambda H M_\Lambda^\dagger \approx H + [\mathrm{i}e\Lambda, H_0]$, we obtain:

$$\langle \Psi^m | H_1' - H_1 | \Psi^n \rangle \approx \mathrm{i}e(E_i - E_f) \langle \Psi^m | \Lambda | \Psi^i \rangle - e \langle \Psi^m | \frac{\partial \Lambda}{\partial t} | \Psi^n \rangle \,, \tag{16.4}$$

that is non-zero in the general case. Note that, with the Dirac Hamiltonian, this first order calculation is exact because $H_1' - H_1 = c\boldsymbol{\alpha} \cdot \nabla \Lambda + \partial \Lambda / \partial t$ and $c\, \boldsymbol{\alpha} \cdot \nabla \Lambda = \mathrm{i}[H_0, \Lambda]$.

More generally, the gauge-transformed wavefunction obtained in perturbation is $|\psi'(t)\rangle = V'(t, t_0)|\phi_g\rangle$, where $V'(t, t_0) = \mathrm{e}^{\mathrm{i}e\Lambda(t)} V(t, t_0) \mathrm{e}^{-\mathrm{i}e\Lambda(t_0)}$ [3]. We have lost the usual gauge transformation $|\psi\rangle \mapsto e^{ie\Lambda}|\psi\rangle$ because perturbation theory imposes $\psi'(t_0) = |\phi_g\rangle$. The transition probability becomes

$$P_{ng}^\Lambda(t) = |\langle \phi_n | \psi'(t) \rangle|^2 = \Big| \int \mathrm{d}\boldsymbol{r} \mathrm{d}\boldsymbol{r}' \phi_n^*(\boldsymbol{r}) \mathrm{e}^{\mathrm{i}e\Lambda(\boldsymbol{r},t)} V(\boldsymbol{r}, t; \boldsymbol{r}', t_0) \mathrm{e}^{-\mathrm{i}e\Lambda(\boldsymbol{r}',t_0)} \phi_g(\boldsymbol{r}') \Big|^2 .$$

Since $\Lambda(\boldsymbol{r}, t)$ is arbitrary, generally $P_{ng}^\Lambda(t) \neq P_{ng}^0(t) = P_{ng}(t)$ defined in (16.1) and the transition probability is not gauge invariant. Moreover, the transition rate is expressed in terms of the derivative of the transition probability with respect to time. This adds a factor $e\, \partial_t \Lambda(\boldsymbol{r}, t)$ to the integral which can be as large as we want, spoiling the gauge invariance of the transition rate. In other words, when the light beam of an absorption or scattering experiment is represented by a time-varying external potential, then the cross section is not gauge invariant. As reviewed in [1], several solutions of this gauge-dependence problem were proposed, but the only one which is fully satisfactory is to use quantum electrodynamics, where the photon is not represented by a classical electromagnetic field but by a state in Fock space.

We now discuss the semi-relativistic expansion of transition probabilities.

16.4 Foldy–Wouthuysen Transformation and Transition Probabilities

The semi-relativistic expansion is widely used in quantum physics. We highlight why the straightforward way to obtain semi-relativistic cross section that would consist in considering that the system is described by a FW Hamiltonian and computing the transition probabilities between eigenstates of this Hamiltonian is, in fact, not valid.

In the Dirac theory, the state of the particles is described by four-components wave functions $\Psi_D = \begin{pmatrix} \chi_1 \\ \chi_2 \end{pmatrix}$. The two spinors that compose these wavefunctions, χ_1 and χ_2, are called upper and lower components.

For positive energy states, the upper component is called the large component and the lower component is called the small component. In the *non relativistic limit*, the small (i.e. lower) component vanishes. For negative energy states, on the other hand, it is the upper component that vanishes.

The Dirac Hamiltonian H has the form $H = \begin{pmatrix} H_{11} & H_{12} \\ H_{21} & H_{22} \end{pmatrix}$, where each H_{ij} is a 2×2 matrix. If H is independent of time, the idea of the Foldy–Wouthuysen (FW) transformation is to apply a unitary operator U_H such that, even for finite light velocity, the upper and lower components of $\psi_{\text{FW}} = U_H \Psi_D$ are decoupled. In other words, the transformed Hamiltonian $H^{\text{FW}} = U_H H U_H^\dagger$ must be block diagonal.

If the Dirac Hamiltonian H is explicitly time dependent, the Foldy–Wouthuysen state $|\psi_{\text{FW}}\rangle$ is a solution of the time-dependent Schrödinger equation for the Foldy–Wouthuysen Hamiltonian [4]:

$$H^{\text{FW}} = U_H H U_H^\dagger - \mathrm{i} U_H \frac{\partial U_H^\dagger}{\partial t}, \tag{16.5}$$

which must be block diagonal. However, H^{FW} must not be used to calculate matrix elements of H because $\langle \Phi_D | H | \Psi_D \rangle = \langle \phi_{\text{FW}} | U_H H U_H^\dagger | \psi_{\text{FW}} \rangle \neq \langle \phi_{\text{FW}} | H^{\text{FW}} | \psi_{\text{FW}} \rangle$ [5].

Using the dynamics given by H^{FW} is also dangerous to calculate transition probabilities. Indeed, even if we know the exact Foldy–Wouthusen transformation U_H, we obtain $|\Psi_{\text{FW}}(t)\rangle = U_{H(t)} |\Psi_D(t)\rangle$. On the other hand, $|\Phi_D^g\rangle$ being an eigenstate of H_0, its Foldy–Wouthuysen transformation is $|\Phi_{\text{FW}}^g\rangle = U_{H_0} |\Phi_D^g\rangle$. Thus, the Foldy–Wouthuysen transition probability of (16.1) becomes

$$|\langle \Psi_{\text{FW}}(t) | \Phi_{\text{FW}}^g \rangle|^2 = |\langle \Psi_D(t) | U_{H(t)}^\dagger U_{H_0} | \Phi_D^g \rangle|^2 \neq P_{ng}(t),$$

because $U_{H(t)} \neq U_{H_0}$.

On the other hand, we cannot use $U_{H(t)}$ to transform $|\Phi_D^g\rangle$, because the state $|\Phi_{\text{FW}}^g\rangle$ would become time-dependent. The solution is to express the cross sections in terms of Dirac wavefunctions $\langle \Psi_D | T | \Phi_D \rangle$, where T is a transition operator and $|\Phi_D\rangle$ and $|\Psi_D\rangle$ are solutions of a time-independent Dirac equation H_0, and then write

$|\Phi_D\rangle = U^{\dagger}_{H_0}|\Phi_{\rm FW}\rangle$. This transforms T into $T_{\rm FW} = U_{H_0} T U^{\dagger}_{H_0}$, ensures the conservation of the transition probability and gives a semi-relativistic meaning to the cross sections [1].

16.5 Semi-relativistic Cross-Sections

We apply the approach of the previous paragraph to the absorption and scattering cross-sections. From this point, we will work in a *monoelectronic* framework. The FW transformation in a many-body framework is developed elsewhere [1].

16.5.1 Absorption Cross Section

The absorption cross section is calculated by assuming that initially the electron is in the relativistic state $|I\rangle$ that can be transformed into the Foldy–Wouthuysen eigenstate $|i\rangle$, with energy E_i, and that a photon $\boldsymbol{q}, \boldsymbol{e_q}$ is present. In the final state there is no photon and the electron is in state $|F\rangle$ ($|f\rangle$ after transformation).

We showed that the absorption cross section can be written [1]:

$$\sigma = 4\pi^2\, \alpha_0\, \hbar\omega_q \sum_f |\langle f|T_{\rm FW}|i\rangle|^2 \delta(E_f - E_i - \hbar\omega_q)\,,$$

where $T_{\rm FW}$ is:

$$\begin{aligned} T_{\rm FW} &= \boldsymbol{e_q}\cdot\boldsymbol{r} + \frac{\mathrm{i}}{2}(\boldsymbol{e_q}\cdot\boldsymbol{r})\,(\boldsymbol{q}\cdot\boldsymbol{r}) - \frac{\hbar}{4m^2c^2}\boldsymbol{\pi}\cdot(\boldsymbol{e_q}\times\boldsymbol{\Sigma}) \\ &\quad - \frac{\beta}{2m\omega_q}(\boldsymbol{e_q}\times\boldsymbol{q})\cdot(\hbar\boldsymbol{\Sigma}+\boldsymbol{\Lambda})\,, \end{aligned}$$

with $\hbar\omega_q = E_f - E_i$, $\boldsymbol{\pi} = \boldsymbol{p} - e\boldsymbol{A}_0$ and $\boldsymbol{\Lambda} = \boldsymbol{L} - e\,\boldsymbol{r}\times\boldsymbol{A}_0$.

It corresponds to the usual formula for the cross section [6] with $\boldsymbol{\Lambda}$ instead of $\boldsymbol{L}$ and one extra relativistic term that we call spin-position, because it can be rewritten using $\boldsymbol{\pi} = (m/\mathrm{i}\hbar)[\boldsymbol{r}, H_0^{\rm FW}] + O(c^{-2})$, where $H_0^{\rm FW}$ is the time-independent Foldy–Wouthuysen Hamiltonian, to get:

$$-\frac{\hbar}{4m^2c^2}\langle f|\boldsymbol{\pi}\cdot(\boldsymbol{e_q}\times\boldsymbol{\Sigma})|i\rangle = \frac{\mathrm{i}\hbar\omega_q}{4mc^2}\langle f|(\boldsymbol{e_q}\times\boldsymbol{r})\cdot\boldsymbol{\Sigma}|i\rangle\,.$$

We call *spin-position operator* the operator $(\boldsymbol{e_q}\times\boldsymbol{r})\cdot\boldsymbol{\Sigma}$. Its evaluation at the K-edge of materials shows that it gives a significant contribution to XMCD [7].

16.5.2 Scattering Cross Section

The scattering cross section is calculated by assuming that initially the electron is in the relativistic state $|I\rangle$ with a photon $\boldsymbol{q}_i, \boldsymbol{e}_i$ and that in the final state the electron is in state $|F\rangle$ with a scattered photon $\boldsymbol{q}_f, \boldsymbol{e}_f$. We do not consider the special case when $\boldsymbol{q}_i, \boldsymbol{e}_i = \boldsymbol{q}_f, \boldsymbol{e}_f$. The scattering cross-section can be resonant only if the intermediate state $|l\rangle$ involved in the transition is a positive energy state so that the condition $E_l = E_i + \hbar\omega_i$ can be satisfied. The semi-relativistic expansion of the resonant scattering term therefore writes as a sum over intermediate states of positive energy:

$$\frac{\mathrm{d}^2\sigma}{\mathrm{d}\Omega \mathrm{d}\omega_f} = \left(\frac{r_e m}{\hbar^2}\right)^2 \frac{\omega_f}{\omega_i} \sum_f \delta_{if} \left| \sum_l \Delta E^{li} \Delta E^{fl} \frac{\langle f|T'^{fl}_{\mathrm{FW}}(\boldsymbol{e}_f)|l\rangle\langle l|T^{li}_{\mathrm{FW}}(\boldsymbol{e}_i)|i\rangle}{E_i - E_l + \hbar\omega_i + \mathrm{i}\gamma} \right|^2,$$

with $\delta_{if} = \delta(E_f + \hbar\omega_f - E_i - \hbar\omega_i)$, $\Delta E^{ij} = E_i - E_j$,

$$T^{ij}_{\mathrm{FW}}(\boldsymbol{e}_i) = \boldsymbol{e}_i \cdot \boldsymbol{r} + \frac{\mathrm{i}}{2}(\boldsymbol{e}_i \cdot \boldsymbol{r})(\boldsymbol{q}_i \cdot \boldsymbol{r}) - \frac{\hbar}{4m^2c^2}\boldsymbol{\pi} \cdot (\boldsymbol{e}_i \times \boldsymbol{\Sigma}) - \frac{(\boldsymbol{e}_i \times \boldsymbol{q}_i) \cdot (\hbar\boldsymbol{\Sigma} + \boldsymbol{\Lambda})}{2m\Delta E^{ij}},$$

and

$$T'^{ij}_{\mathrm{FW}}(\boldsymbol{e}_f) = \boldsymbol{e}^*_f \cdot \boldsymbol{r} - \frac{\mathrm{i}}{2}(\boldsymbol{e}^*_f \cdot \boldsymbol{r})(\boldsymbol{q}_f \cdot \boldsymbol{r}) - \frac{\hbar}{4m^2c^2}\boldsymbol{\pi} \cdot (\boldsymbol{e}^*_f \times \boldsymbol{\Sigma}) + \frac{(\boldsymbol{e}^*_f \times \boldsymbol{q}_f) \cdot (\hbar\boldsymbol{\Sigma} + \boldsymbol{\Lambda})}{2m\Delta E^{ij}}.$$

We recover the usual monoelectronic resonant part of the scattering cross section [8] with the additional spin-position operator that also appeared in the absorption cross section. The semi-relativistic expansion of this cross section in the non resonant case is more subtle because it involves negative energy intermediate states. This question will be addressed in a future publication.

References

1. N. Bouldi, Ch. Brouder, Eur. Phys. J. **90**, 246 (2017)
2. C. Cohen-Tannoudji, B. Diu, F. Laloë, *Mécanique Quantique* (Hermann, Paris, 1973)
3. D.H. Kobe, K.-H. Yang, Phys. Rev. A **32**, 952 (1985)
4. L.L. Foldy, S.A. Wouthuysen, Phys. Rev. **78**, 29 (1950)
5. M.M. Nieto, Phys. Rev. Lett. **38**, 1042 (1977)
6. Ch. Brouder, M. Alouani, K.H. Bennemann, Phys. Rev. B **54**, 7334 (1996)
7. N. Bouldi, N. J. Vollmers, C.G. Delpy-Laplanche, Y. Joly, A. Juhin, Ph. Sainctavit, Ch. Brouder, M. Calandra, L. Paulatto, F. Mauri, U. Gerstmann, Phys. Rev. B **96**, 085123 (2017)
8. M. Blume, J. Appl. Phys. **51**, 3615 (1985)

Chapter 17
Towards Accurate and Large-Scale Density-Functional Calculations with the Korringa–Kohn–Rostoker Method

Rudolf Zeller

Abstract Development of advanced 21st century applications profits increasingly from a basic quantum-mechanical understanding of material properties. Often, density-functional theory is used to reduce the work to the solution of simple effective one-particle equations. Nevertheless, for all but the smallest systems, considerable computer resources are required and accurate calculations for large systems are difficult. One attempt to overcome this problem is KKRnano, a computer code recently developed in Jülich, which is based on the multiple-scattering Korringa–Kohn–Rostoker (KKR) Green-function method. In the present contribution it will be described how this code enables to treat systems with many thousand atoms and how the use of non-local angular projection potentials provides new insight for obtaining accurate forces and total energies.

17.1 Introduction

In density-functional calculations, usually the Schrödinger equation is solved by expanding the single-particle wavefunctions into a set of basis functions with expansion coefficients found by minimizing the total energy. An alternative method was suggested by Wigner and Seitz [1] in their pioneering work *On the Constitution of Metallic Sodium*. They divide the crystal into non-overlapping cells centred at the atoms and determine single-cell solutions which are joined by wavefunction matching at the cell surfaces. However, the matching is numerically very demanding so that the cellular method of Wigner and Seitz outlived its usefulness when the multiple-scattering KKR method appeared. In this method the outgoing scattered wave from the potential in one cell is analytically transformed into incoming waves at the other cells by assuming free propagation between different scattering events. This assumption was historically guaranteed by confining the potential to non-overlapping muffin-tin spheres. For space-filling potentials, however, free prop-

R. Zeller (✉)
Institute for Advanced Simulation, Forschungszentrum Jülich GmbH and JARA, 52425 Jülich, Germany
e-mail: ru.zeller@fz-juelich.de

D. Sébilleau et al. (eds.), *Multiple Scattering Theory for Spectroscopies*, Springer Proceedings in Physics 204,
https://doi.org/10.1007/978-3-319-73811-6_17

agation is missing which prevents that a multiple-scattering picture can be used in a full-potential treatment. Nevertheless, it was conjectured and backed up in a number of publications that the KKR equations remain valid with minor or no modification. On that account, a full-potential Green-function (GF) method [2], particularly for impurity calculations, was developed in Jülich since 1990. This method is the basis of KKRnano which was designed from the outset to treat large systems on massively parallel supercomputers [3, 4]. Below the main features of KKRnano will be described. It will also be demonstrated that a full-potential treatment can be validated rigorously by using the concept of non-local angular projection potentials and that this concept helps to identify and remove problems for accurate force and total-energy calculations.

17.2 Large Systems

Due to the advance of supercomputing power, calculations for systems with more than one thousand atoms are possible today by standard density-functional codes. For larger systems, however, the applicability of these codes is limited because the computational effort scales with the third power of the number of atoms. KKRnano overcomes this limitation by a linear-scaling implementation which is achieved as described below. The strategy is to determine the density $n^n(\boldsymbol{r})$ from the diagonal part of the density matrix

$$\varrho^{nn'}(\boldsymbol{r},\boldsymbol{r}',T) = -\frac{1}{\pi}\Im\left[\int_{-\infty}^{\infty} f(\epsilon-\mu,T)G^{nn'}(\boldsymbol{r},\boldsymbol{r}';\epsilon+\mathrm{i}0^+)\mathrm{d}\epsilon\right], \tag{17.1}$$

where $\boldsymbol{r}$ and $\boldsymbol{r}'$ are vectors originating at the centers of cell n and n'. This integral, which contains the Fermi–Dirac function for temperature T and chemical potential μ, is conveniently evaluated by contour integration [5]. Dropping the parametric dependence on ϵ, the Green's function is given by

$$G^{nn'}(\boldsymbol{r},\boldsymbol{r}') = \delta_{nn'}G_s^n(\boldsymbol{r},\boldsymbol{r}') + \sum_{\ell\ell'}^{\infty}\sum_{mm'} R_{\ell m}^n(\boldsymbol{r})G_{\ell m\ell' m'}^{nn'}R_{\ell' m'}^{n'}(\boldsymbol{r}'), \tag{17.2}$$

where $R_{\ell m}^n(\boldsymbol{r})$ and $G_s^n(\boldsymbol{r},\boldsymbol{r}')$ are functions that depend only on the potential confined to cell n. The Green's function matrix elements are obtained by

$$G_{\ell m\ell' m'}^{nn'} = G_{\ell m\ell' m'}^{\mathrm{r},nn'} + \sum_{n''}\sum_{\ell''\ell'''}^{\infty}\sum_{m''m'''} G_{\ell m\ell'' m''}^{\mathrm{r},nn''}\Delta t_{\ell'' m''\ell''' m'''}^{n''}G_{\ell''' m'''\ell' m'}^{n''n'}, \tag{17.3}$$

from the difference of the single-cell t matrices of the system and a reference system and the Green-function matrix elements of the reference system. In KKRnano the matrix equation (17.3) is solved iteratively without any loss of accuracy [3] by the

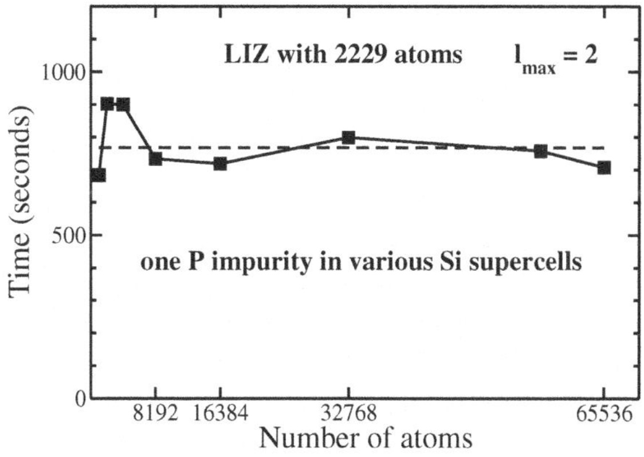

Fig. 17.1 Wall-clock time on the Jülich supercomputer JUQUEEN for one self-consistency step for 2048–131072 cells half occupied by atoms. One MPI task and four OpenMP threads are used for each cell. The *broken line* indicates the average as a guide for eye. The time remains almost constant when the number of processors is increased with the number of atoms

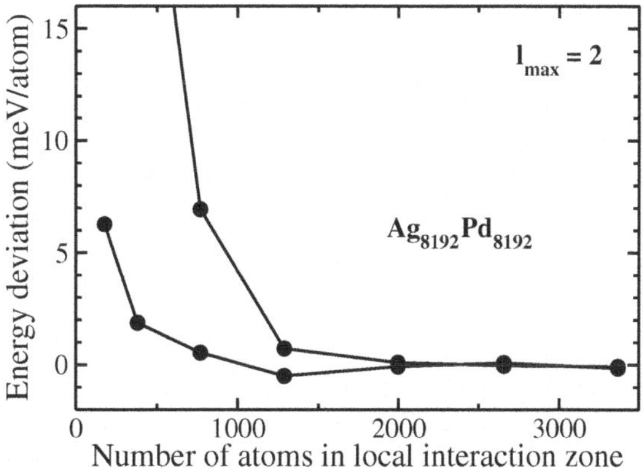

Fig. 17.2 Total energies for a disordered Ag-Pd alloy with 16384 atoms in the supercell using local interaction zones (LIZ) of different size. The reference energy (with zero deviation) is for a LIZ containing all 16384 atoms. Upper and lower curves are for (i) neglecting all GF matrix elements outside the LIZ or (ii) neglecting only elements with $\ell > 0$

transpose-free quasi-minimal residual method [6]. This has a number of advantages: (i) easy parallelization because different values n' do not mix, (ii) quadratically scaling computational effort achieved by not operating on zero matrix elements, which occur if the repulsive reference system of the screened KKR method [7] is used, and (iii) linearly scaling effort achieved by utilizing the decay of the density matrix through neglect of small Green's function matrix elements outside local interaction zones (LIZ) around each atom. These features lead to excellent computational performance as displayed in Fig. 17.1, where the nearly constant wall-clock times indicate that both linear scaling and parallel efficiency are almost perfect. Since linear scaling reduces accuracy, it is important to assess how large the deviations are. As Fig. 17.2 shows, the deviations are controllable and below millielectron-volts per atom if the LIZ are large enough.

17.3 Accurate Forces and Total Energies

For accurate calculations of forces and total energies a full-potential approach is unavoidable [2]. An approach, which does not rely on the Schrödinger or Lippmann–Schwinger equation and matching boundary conditions, is the direct determination of the Green's function from an integral Dyson equation [8]. Then, the only mathematical difficulty are conditionally convergent double angular-momentum sums. Recently, this problem was resolved by using a mathematical model [9] with finite sums restricted to $\ell \leq \ell_{\mathcal{P}}$ and showing that limit $\ell_{\mathcal{P}} \to \infty$ is clearly defined [10]. In this work the cell potentials were understood as non-local angular projection potentials

$$V^n(r,\hat{\boldsymbol{r}},\hat{\boldsymbol{r}}') = \sum_{\ell\ell'}^{\ell_{\max}} \sum_{mm'} Y_{\ell m}(\hat{\boldsymbol{r}}) Y_{\ell' m'}(\hat{\boldsymbol{r}}') V^n_{\ell m \ell' m'}(r) \,. \tag{17.4}$$

An important consequence of (17.4) is that only $(\ell_{\max}+1)^4$ functions $\mathcal{R}^n_{\ell' m' \ell m}(r)$ in the angular-momentum expansion for $R_{\ell m}(\boldsymbol{r})$ must be determined numerically. The other terms are Bessel function because the integral

$$\int_0^\infty r'^2 G^0_{\ell'}(r,r') \sum_{\ell''}^{\ell_{\max}} \sum_{m''} V^n_{\ell' m' \ell'' m''}(r') \mathcal{R}^n_{\ell'' m'' \ell m}(r') \mathrm{d}r' \,, \tag{17.5}$$

which appears in the coupled radial integral equations for $\mathcal{R}^n_{\ell' m' \ell m}(r)$, vanishes for $\ell' > \ell_{\max}$ according to (17.4), but also for $\ell > \ell_{\max}$ as shown in [9]. This result can be used to express the density exactly by a finite number of terms with analytically given dependence on the angular variables [10] and to conclude that the full-potential treatment in the KKR-GF method is not responsible for systematic force and total-energy errors.

In the past, it has been speculated that the standard expression

$$V^n_{\ell m}(r) = \frac{4\pi r^\ell}{2\ell+1} \sum_{n'} \int_{n'} \mathrm{d}\boldsymbol{r}' n^{n'}(\boldsymbol{r}') \frac{Y_{\ell m}(\boldsymbol{r}' + \boldsymbol{R}^{n'} - \boldsymbol{R}^n)}{|\boldsymbol{r}' + \boldsymbol{R}^{n'} - \boldsymbol{R}^n|^{\ell+1}} \,, \tag{17.6}$$

used in the spherical-harmonics expansion of $V^n(\boldsymbol{r})$, is responsible because it is not valid for all points in the cells. These near-field errors, however, are easily avoided if (17.6) is used for a fixed value $r_\mathrm{c} \ll |\boldsymbol{r}' + \boldsymbol{R}^{n'} - \boldsymbol{R}^n|$ and simple one-center electrostatic problems [11] are used to obtain $V^n_{\ell m}(r)$ for $r \neq r_\mathrm{c}$ from the density between r_c and r. The real problem for forces is an expansion for $r' < |\boldsymbol{R}^{n'} - \boldsymbol{R}^n|$, which often is used to evaluate (17.6) with the help of density moments. Because too many moments are needed for arbitrary cells, it is better to use an isoparametric integration method [12] which is presently implemented in KKRnano to obtain rather accurate forces.

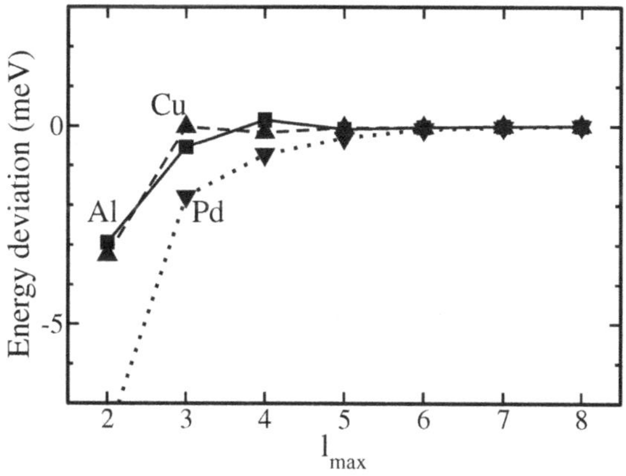

Fig. 17.3 Total energy results using $\ell_{\mathcal{P}} = 8$ and angular projection potentials with different values of $\ell_{\max}$ during the self-consistency steps, but $\ell_{\max} = 8$ for the evaluation of the total-energy functional after self-consistency is reached. The *solid line* is for Al, the *broken line* for Cu and the *dotted line* for Pd

The situation for total energies is more complicated. Since the potential obtained from the Poisson equation is approximated by (17.4) for calculating the single-particle energies, these are determined with a different potential than the double-counting terms. This inconsistency is mitigated if the total-energy functional is evaluated after the self-consistency steps with a higher value for $\ell_{\max}$ [9]. Then much improved total energies as in Fig. 17.3 are obtained.

Acknowledgements The author gratefully thanks P.F. Baumeister, S. Blügel, M. Bolten, M. Bornemann, P.H. Dederichs, T. Hater, T. Fukushima, R. Kováčik, M. Ogura, D. Pleiter, E. Rabel, A. Thiess, I. Yafneh, who have contributed to the development of KKRnano, and acknowledges computing time granted by the JARA-HPC Vergabegremium and provided on the JARA-HPC Partition part of the supercomputer JUQUEEN at Forschungszentrum Jülich.

References

1. E. Wigner, F. Seitz, Phys. Rev. **43**, 804–810 (1933)
2. N. Papanikolaou, R. Zeller, P.H. Dederichs, J. Phys. Condens. Matter **14**, 2799–2823 (2002)
3. R. Zeller, J. Phys. Condens. Matter **20**, 294215 (2008)
4. A. Thiess, R. Zeller, M. Bolten, P.H. Dederichs, S. Blügel, Phys. Rev. B **85**, 235103 (2012)
5. K. Wildberger, P. Lang, R. Zeller, P.H. Dederichs, Phys. Rev. B **52**, 11502 (1995)
6. R.W. Freund, SIAM, J. Sci. Comput. **14**, 470–482 (1993)
7. R. Zeller, P.H. Dederichs, B. Újfalussy, L. Szunyogh, P. Weinberger, Phys. Rev. B **52**, 8807–8812 (1995)
8. R. Zeller, J. Phys. C **20**, 2347–2360 (1987)
9. R. Zeller, J. Phys. Condens. Matter **25**, 105505 (2013)
10. R. Zeller, J. Phys. Condens. Matter **27**, 306301 (2015)
11. D.M.C. Nicholson, W.A. Shelton, J. Phys. Condens. Matter **14**, 5601–5608 (2002)
12. A. Alam, S.N. Khan, B.G. Wilson, D.D. Johnson, Phys. Rev. B **85**, 045105 (2011)

Chapter 18
3D Atomic Structure Analysis Around Local Active Atoms by Two-Dimensional Photoelectron Diffraction and Holography

Hiroshi Daimon

Abstract The atomic structure around specific atom in solid, such as dopant in semiconductor, has become possible to be analyzed directly and precisely recently by new methods of stereophotography and photoelectron holography with the aid of new analysis code and two kinds of two-dimensional display-type analyzers. Based on the development of these new technologies a project of 3D active-site science has been launched to develop a new science of local atomic arrangement which is a key of functional materials.

18.1 Introduction

The atomic structure analysis around local specific atom, such as a dopant in semiconductor, is a key of the functional materials but has been impossible by a standard structure analysis method such as an x-ray diffraction because this kind of active site has no translational symmetry. EXAFS opened a new way to study this kind of local structure, but the obtained information is only one dimensional. Two methods to analyze the 3D atomic structure around this kind of specific atoms with no translational symmetry are *stereography of atomic arrangement* [1] and *photoelectron holography* [2]. Recently their accuracy improved dramatically by a development of a new analysis code [3] and effective two-dimensional display-type electron spectrometers, such as DIANA (Display-type spherical mirror Analyzer) [4] or DELMA (Display-type Ellipsoidal Mesh Analyzer) [5–7]. These techniques as well as *fluorescent x-ray holography* [8] received renewed attentions recently. Hence a project of *3D active-site science* [9] supported by JSPS Grant-in-Aid for Scientific Research on Innovative Areas, Grant Number 26105001 has been launched.

H. Daimon (✉)
Nara Institute of Science and Technology (NAIST), Takayama, Ikoma, Nara 630-0192, Japan
e-mail: daimon@ms.naist.jp

D. Sébilleau et al. (eds.), *Multiple Scattering Theory for Spectroscopies*, Springer Proceedings in Physics 204,
https://doi.org/10.1007/978-3-319-73811-6_18

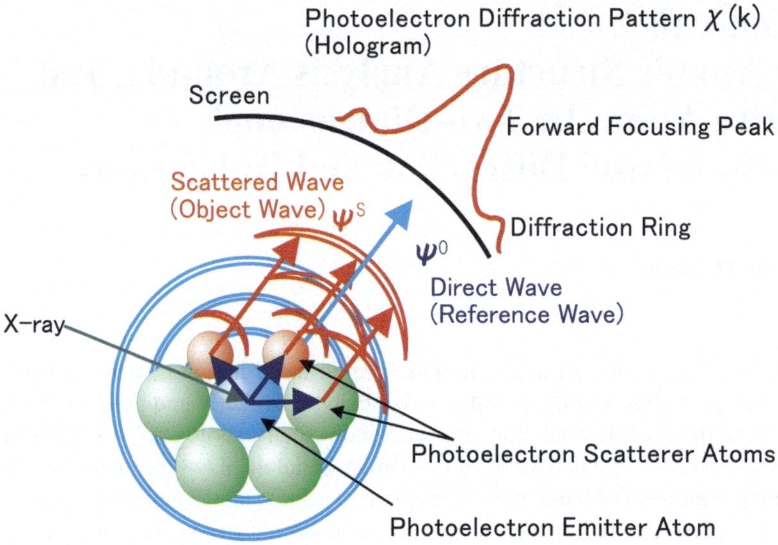

Fig. 18.1 Photoelectron diffraction and holography

18.2 3D Atomic Structure Analysis Around Specific Atoms with No translational Symmetry

18.2.1 Stereography of Atomic Arrangement

Atomic stereography [1] is a new method to realize a dream to see three dimensional atomic arrangement directly by our eyes. This technique is based on a phenomenon of photoelectron diffraction, which is shown in Fig. 18.1.

When an x-ray excites an inner core electron of an atom, a photoelectron is emitted from this emitter atom. The photoelectron is scattered by surrounding atoms and produces scattered waves. The interference pattern at a distance is called photoelectron diffraction pattern. Strong forward focusing peaks appear along the direction connecting the emitter atom and the scatterer atoms as shown in Fig. 18.1. Hence a photoelectron diffraction pattern is a projected atomic image around the photoelectron emitter atom. If the incident x-ray is a circularly polarized light, the forward focusing peak was found to rotate around the circularly polarized light axis [10] as shown in Fig. 18.2. The rotation angle Δ was expressed as

$$\Delta = \frac{m}{kR\sin^2\theta}, \tag{18.1}$$

where k is the wave number of photoelectron directing toward the forward focusing peak, R is the interatomic distance between the emitter and the scatterer atoms, θ is

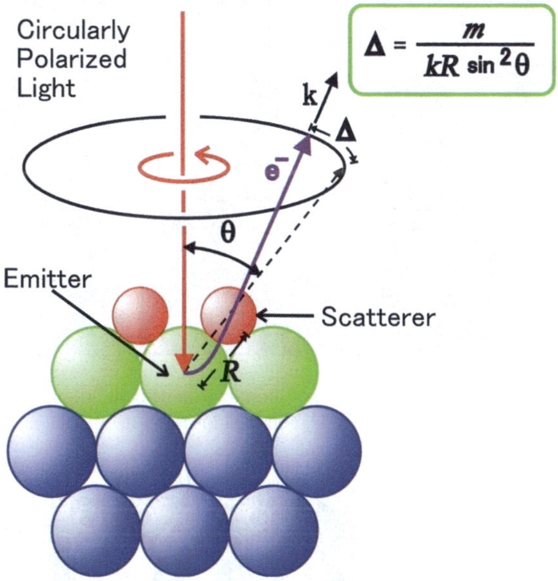

Fig. 18.2 Circularly polarized light photoelectron diffraction

the angle between $\boldsymbol{k}$ and the circularly polarized light propagating vector, and m is the effective magnetic quantum number of the photoelectron around the light axis. An important feature is that it is inversely proportional to the interatomic distance. This feature is used to realize the stereography of atomic arrangement because it is exactly the same as parallax in our stereoview [1].

Stereophotograph of atomic arrangement consists of two pictures of 3D atomic arrangement including parallax viewed by two eyes at right and left side of the photoelectron emitter atom. The viewpoint atom can be selected by selecting the photoelectron kinetic energy. An example of the stereophotograph [11] is shown in Fig. 18.3. This stereophotograph was taken using In 4*d* photoelectrons from the In atoms of InP(001) surface. The direction of the view is [111], and the closest atom A is the nearest neighbour P atom. When you see the left picture with your left eye and the right picture with your right eye you can image the 3D atomic arrangement around In atom directly in your brain.

18.2.2 Photoelectron Holography

Because the photoelectron pattern in Fig. 18.1 is the interference pattern between the direct photoelectron spherical wave and the wave scattered from surrounding objects, it was suggested by Szöke [2] in 1986 that the pattern can be considered as a hologram. Since then many efforts have been made to reconstruct the atomic arrangement using holographic reconstruction algorism, which are based on Fourier

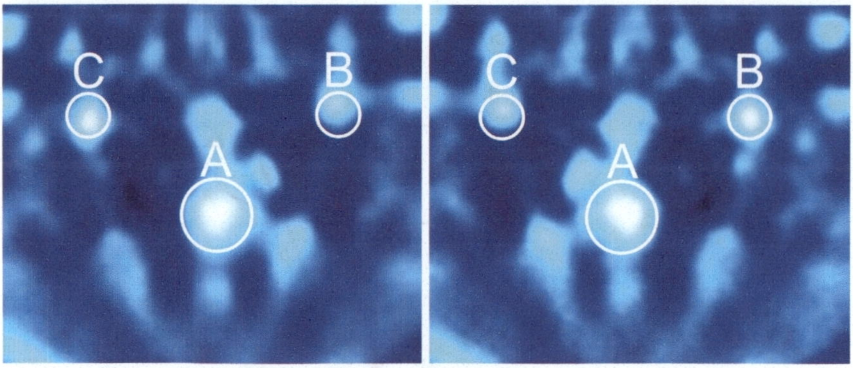

Fig. 18.3 Stereophotograph of atomic arrangement around In atom in InP(001) surface [11]. The direction of the view is [111], and the closest atom A is the nearest neighbour P atom

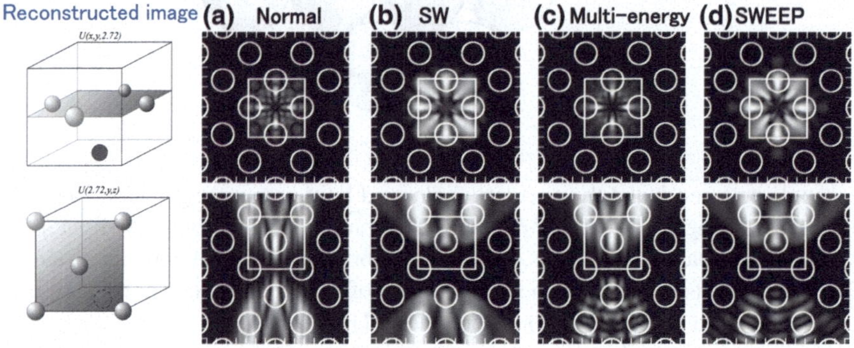

Fig. 18.4 Holographically reconstructed image of Si crystal using various methods. **a** Normal, **b** SW, **c** Multi-energy, and **d** SWEEP [12]

transformation [12]. However the reconstructed image is limited to several atoms around the emitter atom, and the accuracy of the position was not accurate enough to use as an analysis tool (Fig. 18.4).

In 2005 Matsushita developed a new analysis code called SPEA- MEM [3], and succeeded to reconstruct more than hundred atoms from only one hologram. Recent example is the analysis of 3D imaging of intercalant atom in superconducting graphite [13]. The local atomic structure around an intercalant atom in curious superconductor has been precisely imaged.

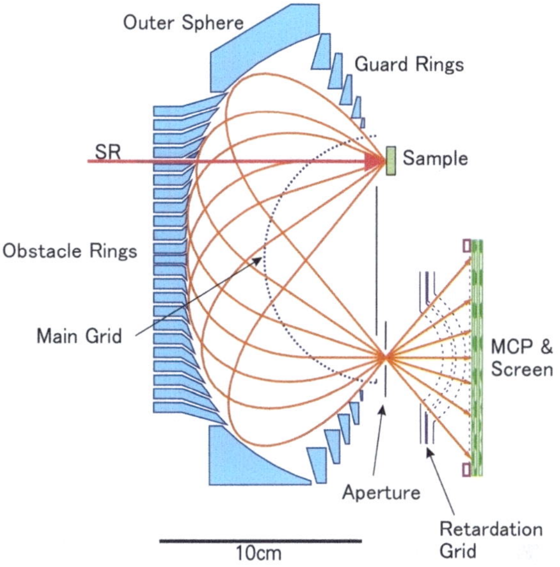

Fig. 18.5 Display-Type spherical mirroranalyzer

18.3 Two-Dimensional Display-Type Electron Spectrometers

18.3.1 DIANA (Display-Type Spherical Mirror Analyzer)

Two-dimensional photoelectron diffraction pattern, which is necessary for the analysis of stereophotography and also for photoelectron holography, is effectively measured by display-type two-dimensional photoelectron analyzer. Display-type spherical mirror Analyzer (DIANA) [4] has been used for a long time to measure photoelectron diffraction patterns as well as three-dimensional valence band mapping. A schematic diagram of DIANA is shown in Fig. 18.5. It consists of a hemi-spherical main grid, many outer electrodes, an exit aperture, retardation grids, MCP and screen. The main grid is grounded, and the electric field between the main grid and the outer electrodes is spherically symmetric. Photoelectrons emitted from the sample make linear trajectory until the main grid. Inside the electric field between the main grid and the outer electrodes, the trajectory is an ellipse obeying Kepler's law. All electrons of the pass energy in different directions converge exactly to the exit aperture and diverge again and detected on the screen. The obstacle rings play a role of low-pass filter, and the retardation grids play as a high-pass filter. Hence only one kinetic-energy electrons are detected. The acceptance angle can be up to ±90°, but the size of MCP limits the detection angle to ±60°.

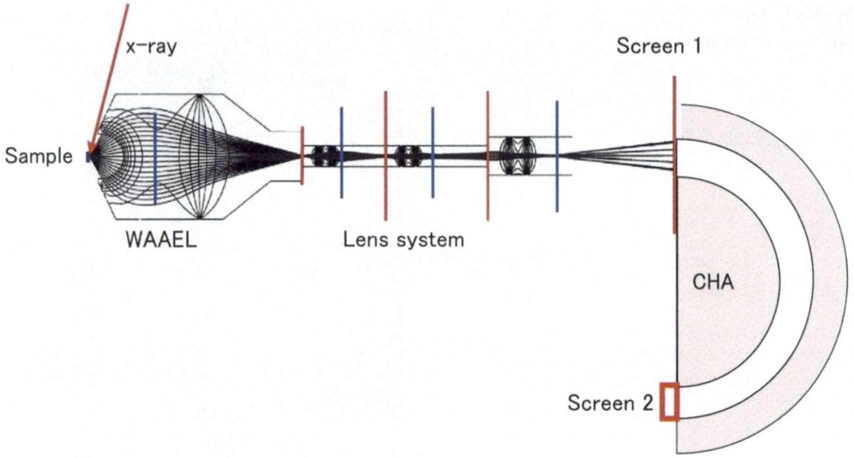

Fig. 18.6 Display-type Ellipsoidal Mesh Analyzer (DELMA)

18.3.2 DELMA (Display-Type Ellipsoidal Mesh Analyzer)

Recently Display-type Ellipsoidal Mesh Analyzer (DELMA) [6, 7] has been developed to realize high-energy resolution and microscopic measurement. A schematic diagram of DELMA is shown in Fig. 18.6. It consists of a Wide Acceptance Angle Electrostatic Lens (WAAEL), lens system, screen 1, CHA (concentric hemispherical analyzer) and screen 2. We have developed a wide acceptance angle electrostatic lens (WAAEL) using an ellipsoidal mesh lens [5], which can converge up to $\pm 60°$ electrons to one point by eliminating the spherical aberration. Because the divergence angle is reduced to 1/5, usual einzel lens can be used afterward, and magnified image of the sample or an angular distribution of the emitted electrons can be projected on the screen 1. Although WAAEL has an energy resolution of about 1%, CHA is used when higher energy resolution is needed.

18.4 3D Active-Site Science Project

Two methods *stereography of atomic arrangement* and *photoelectron holography* to analyze the 3D atomic structure around specific atoms with no translational symmetry have been realized, and *fluorescent x-ray holography* [8] has also been developed recently, a project of *3D active-site science* [9] supported by JSPS Grant-in-Aid for Scientific Research on Innovative Areas, Grant Number 26105001 has been launched in 2014. This project aims to realize a new field of science of functions created by local atomic arrangement with no translational symmetry. It reveals local functional structure for the first time and will make revolution in wide range of fields in physics, chemistry, electronic devices, and bio-materials.

18.5 Summary

Two methods, stereophotography and photoelectron holography, to analyze atomic structure around specific atom in solid, have been reviewed. The development of two-kinds of display-type analyzers and a new analysis code realized fast and precise analysis of local non-periodic structure for the first time. Based on the development of these new technologies a project of 3D active-site science has been launched to develop a new science of local atomic arrangement which is a key of functional materials.

References

1. H. Daimon, Phys. Rev. Lett. **86**, 2034 (2001)
2. A. Szöke, A.I.P. Conf. Proc. **147**, 361 (1986)
3. T. Matsushita, A. Yoshigoe, A. Agui, Europhys. Lett. **71**, 597 (2005)
4. H. Daimon, Rev. Sci. Instrum **59**, 545 (1988)
5. H. Matsuda, H. Daimon, M. Kato, M. Kudo, Phys. Rev. E **71**, 066503 (2005)
6. H. Daimon, H. Matsuda, L. Toth, F. Matsui, Surf. Sci. **601**, 4748 (2007)
7. L. Toth, H. Matsuda, T. Shimizu, F. Matsui, H. Daimon, J. Vac. Soc. Jpn. **51**, 135 (2008)
8. K. Hayashi, N. Happo, S. Hosokawa, W. Hu, T. Matsushita, J. Phys. Condens. Matter **24**, 093201 (2012)
9. http://www.en.3d-activesite.jp/
10. H. Daimon, T. Nakatani, S. Imada, S. Suga, Y. Kagoshima, T. Miyahara, Jpn. J. Appl. Phys. **32**, L1480 (1993)
11. T. Matsumoto, F. Matsui, T. Matsushita, K. Goto, Y. Kato, H. Daimon, e-J. Surf. Sci. Nanotechnol. **7**,181 (2009)
12. T. Nakatani, H. Nishimoto, H. Daimon, S. Suga, H. Namba, T. Ohta, Y. Kagoshima, T. Miyahara, J. Synchrotron Rad. **3**, 239 (1996)
13. F. Matsui, R. Eguchi, S. Nishiyama, M. Izumi, E. Uesugi, H. Goto, T. Matsushita, K. Sugita, H. Daimon, Y. Hamamoto, I. Hamada, Y. Morikawa, Y. Kubozono, Sci. Rep. 36258 (2016)

Chapter 19
Plasmon Losses in Core Photoemission Spectra

Takashi Fujikawa and Kaori Niki

Abstract A short review is presented on theoretical approach to plasmon losses in core-level photoemission from simple metals. Full multiple elastic scatterings of photoelectrons are taken into account before and after the plasmon losses within the quantum cumulant (Landau) formula that can describe both intrinsic and extrinsic losses, and overall features of core level photoemission. Far from normal emission and in low energy excitation, we cannot apply these simple exponential formulas.

19.1 Photoemission Intensities

Main XPS (no loss) bands excited from a core ϕ_c measuring photoelectrons with momentum $\boldsymbol{p}$ and energy ϵ_p are described in terms of the damping photoelectron wave function $f_{\boldsymbol{p}}^-$ under the influence of the optical potential. A basic theoretical formula to study photoemission processes should be built on a sound many-body theory. Here we apply Keldysh Green's function approach. Our discussion is focused on photoexcitation from a core orbital ϕ_c. No-loss band measuring photoelectrons with momentum $\boldsymbol{p}$ and energy ε_p is given by

$$I^0(\boldsymbol{p};\omega) = 2\pi|\langle f_{\boldsymbol{p}}^-|\Delta|\phi_c\rangle S_0|^2\delta(E_0+\omega-E_0^*-\epsilon_p)\ . \tag{19.1}$$

Main ingredients are photoelectron wave function (Dyson orbital) $f_{\boldsymbol{p}}^-$ for which non-Hermitian optical potential also work in addition to Hermitian Hartree potential. The incident photon energy is ω and Δ is the one-electron photon-electron interaction operator. The intrinsic no-loss amplitude S_0 should be close to 1. The ground state energies without and with core hole are E_0 and E_0^*. The amplitude $\langle f_{\boldsymbol{p}}^-|\Delta|\phi_c\rangle$ is given by a closed renormalization full multiple scattering formula [1]. The formula is quite

T. Fujikawa (✉) · K. Niki
Graduate School of Advanced Integration Science, Chiba University, Yayoi-cho 1-33, Inage, Chiba, Japan
e-mail: tfujikawa@faculty.chiba-u.jp

K. Niki
e-mail: niki@chiba-u.jp

D. Sébilleau et al. (eds.), *Multiple Scattering Theory for Spectroscopies*, Springer Proceedings in Physics 204,
https://doi.org/10.1007/978-3-319-73811-6_19

useful to calculate the x-ray photoelectron diffraction (XPD) patterns. Finite-order photoelectron diffraction (XPD) formula has been applied to calculate the depth distribution function. In contrast to the conventional XPD calculations, finite-order multiple scattering calculations give no satisfactory convergence for the depth profile: Full multiple scattering calculations provide us with reasonable results [1].

19.2 Loss Bands and Quantum Landau Formula

In contrast to the no-loss band, the interference between the intrinsic and extrinsic losses plays an important role. Both of the many-body scattering [2] and Keldysh Green's function [3] approaches give a very similar formula. We can use a formula for the single-loss band exciting the mth mode whose excitation energy is ω_m

$$I^1(\boldsymbol{p};\omega) = 2\pi \sum_m |\langle f_{\boldsymbol{p}}^- | \Delta | \phi_c \rangle S_m + \langle f_{\boldsymbol{p}}^- | v_m g(\epsilon_p + \omega_m) \Delta | \phi_c \rangle S_0|^2 \\ \times \delta(E_0 + \omega - E_0^* - \omega_m - \epsilon_p) , \tag{19.2}$$

where v_m is the fluctuation potential responsible for the intrinsic (see (19.3)) and the extrinsic excitation of the mth mode. The causal Green's function g in the second term describes the photoelectron propagation suffering elastic scatterings before the extrinsic loss influenced by v_m. We should note that the intrinsic and the extrinsic losses can interfere.

A simple perturbation theory gives a useful formula for the intrinsic loss amplitude S_m,

$$S_m = -\frac{\langle c|v_m|c\rangle}{\omega_m} \exp(-a/2), \; a = \sum_n \frac{|\langle c|v_n|c\rangle|^2}{\omega_n^2} . \tag{19.3}$$

Well above the threshold, the causal Green's function $g(\varepsilon)$ is approximated by the retarded Green's function $g^r(\epsilon)$. Again we apply the site T-matrix expansion to the photoelectron wave function $f_{\boldsymbol{p}}^-$ and $g^r(\epsilon_p + \omega_m)$. Rather high-energy ($\epsilon_p > 100\,\text{eV}$) excitation and near normal emission allow us to use quite a simple expression for the extrinsic loss amplitude [4]

$$\langle f_{\boldsymbol{p}}^- | v_m g(\epsilon_p + \omega_m) \Delta | \phi_c \rangle \approx \langle f_{\boldsymbol{p}}^- | \Delta | \phi_c \rangle \tau_m^{ex}(\boldsymbol{p}) , \\ \tau_m^{ex}(\boldsymbol{p}) = (2\pi)^{3/2} \int \mathrm{d}\boldsymbol{r}\, f_{\boldsymbol{p}}^{-*}(\boldsymbol{r}) v_m(\boldsymbol{r}) g_0(\boldsymbol{r} - \boldsymbol{R}_A : p') , \tag{19.4}$$

where g_0 is the damping propagator with p'; $p'^2/2 = p^2/2 + \omega_m$.

To recover the lowest sum $I^0(\boldsymbol{p};\omega) + I^1(\boldsymbol{p};\omega)$ and also satisfy the normalization condition, the overall photoemission profile is now written by the exponential form with aid of the approximation (19.4)

$$
\begin{aligned}
I^{\infty}(\boldsymbol{p};\omega) &= I^{0}(\boldsymbol{p};\omega) + I^{1}(\boldsymbol{p};\omega) + \ldots \\
&= |\langle f_{\boldsymbol{p}}^{-}|\Delta|\phi_c\rangle|^2 \int_{-\infty}^{\infty} \mathrm{d}t \exp[\mathrm{i}(\omega + E_0 - E_0^* - \epsilon_p)t] \\
&\times \exp\left[\int_0^{\infty} \mathrm{d}\epsilon\, \beta(\epsilon)(\mathrm{e}^{-\mathrm{i}t\epsilon} - 1)\right] ,
\end{aligned} \tag{19.5}
$$

and

$$
\beta(\epsilon) = \sum_m |\tau_m^{\mathrm{ex}}(\boldsymbol{p}) + S_m/S_0|^2 \delta(\epsilon - \omega_m) . \tag{19.6}
$$

This exponential form (19.5) is known as Landau formula which was derived on the basis of classical transport theory. Similar quantum formula is derived by Hedin [2], where simple time-reversed LEED function is used instead of renormalized damping photoelectron wave function $f_{\boldsymbol{p}}^{-}$. This generalization is crucial to discuss angular and energy dependence of plasmon losses, except for the grazing photoemission.

We obtain an alternative expression for $\beta(\epsilon)$ in terms of causal screened Coulomb interaction W

$$
\beta(\epsilon) = -\frac{1}{\pi} \int \mathrm{d}\boldsymbol{r}\mathrm{d}\boldsymbol{r}' f_A^*(\boldsymbol{r}') f_A(\boldsymbol{r}) \Im\left[W(\boldsymbol{r},\boldsymbol{r}';\epsilon)\right] , \tag{19.7}
$$

where f_A has both of the intrinsic (first term of (19.8)) and the extrinsic loss effects (second term)

$$
\begin{aligned}
f_A(\boldsymbol{r}) &= -\frac{|\phi_c(\boldsymbol{r})|^2}{E_c} + (2\pi)^{3/2} f_{\boldsymbol{p}}^{-*}(\boldsymbol{r}) g_0(\boldsymbol{r} - \boldsymbol{R}_A; p') , \\
E_c &= \omega + E_0 - E_0^* - \epsilon_p .
\end{aligned} \tag{19.8}
$$

It is often reasonable to split $\beta(\epsilon)$ in a low-energy part $\beta_1(\epsilon)$ and a high-energy part $\beta_2(\epsilon)$ by use of a small critical energy ϵ_0,

$$
\beta(\epsilon) = \beta_1(\epsilon) + \beta_2(\epsilon), \beta_1(\epsilon) = \beta_0\theta(\epsilon_0 - \epsilon) .
$$

The low-energy part describes x-ray singularity for metallic systems: Only the intrinsic term is important because of the small energy denominator of the first term in (19.8). The high-energy part describes the plasmon losses taking both intrinsic and extrinsic ones into account. We thus have from (19.5)

$$
\begin{aligned}
I^{\infty}(\boldsymbol{p};\omega) &= 2\pi|\langle f_{\boldsymbol{p}}^{-}|\Delta|\phi_c\rangle|^2 \int \mathrm{d}\epsilon\, D_1(E_c - \epsilon) D_2(\epsilon) \\
&\times \exp\left[-\int_0^{\infty} \mathrm{d}\epsilon \beta(\epsilon)\right] ,
\end{aligned} \tag{19.9}
$$

where low-(D_1) and high-energy (D_2) spectral functions are defined by

$$D_j(t) = \exp\left[\int_0^\infty d\epsilon\ \beta_j(\epsilon)e^{-i\epsilon t}\right], (j = 1, 2) .$$

The high-energy spectral function can be expanded as

$$D_2(\omega) = \delta(\omega) + \beta_2(\omega) + \frac{1}{2}(\beta_2 * \beta_2)(\omega) + \frac{1}{6}(\beta_2 * \beta_2 * \beta_2)(\omega) + \ldots \quad (19.10)$$

where $(A * B)(\omega) = \int d\epsilon\ A(\omega - \epsilon)B(\epsilon)$ is the convolution of A and B. The first term contributes to the main (no loss) band, the second to the single plasmon, and the third to the double plasmon losses, and so on. Substitution of (19.10) in (19.9) yields the overall photoemission feature excited from the core function ϕ_c on site $\boldsymbol{R}_A$

$$I^\infty(\boldsymbol{p};\omega) = 2\pi|\langle f_{\boldsymbol{p}}^-|\Delta|\phi_c\rangle|^2 \exp\left[-\int_0^\infty d\epsilon\ \beta(\epsilon)\right] \times\left[D_1(E_c) + (D_1 * \beta_2)(E_c) + \frac{1}{2}(D_1 * \beta_2 * \beta_2)(E_c) + \ldots\right] . \quad (19.11)$$

As pointed out before $\beta(\epsilon)$ depends on $\boldsymbol{p}$, so that plasmon loss bands show the angular dependence. At low energy region ($\epsilon_p < 200\,\text{eV}$), prominent angular dependence of the plasmon loss structure is observed, whereas weak dependence is observed in the high energy region [5].

19.3 Discussion

The quantum Landau formula originally derived by Hedin et al. can explain overall plasmon loss features accompanied by core-level photoemission; they neglect elastic scatterings before and after the losses [2]. Ohori et al. have studied the applicability of the quantum Landau formula over wide energy range [6]. Their results show that the high-energy approximation used in the cumulant approximation works so well in the range, $\epsilon_p > 60\,\text{eV}$.

Kazama et al. have applied the theoretical approach presented above to the analyses of the angular and energy dependence of the plasmon losses from Al and Na 2s levels [7]. The depth profiles of single-loss spectra calculated with full multiple scatterings decay faster than without elastic scatterings. The single elastic scattering model gives rise to unexpected large loss intensities from deep emitters, which clearly demonstrates the importance of full multiple elastic scatterings to analyze loss bands.

Niki et al. have also calculated the surface and bulk plasmon satellites associated with Li 1s photoemission at $\omega = 170\,\text{eV}$ from a Li metal taking full multiple elastic scatterings [8]. Although Li is quite a weak scatterer, they find rather prominent effects from elastic scatterings on depth profile and loss band structure.

In these practical calculations they have used simple approximations used in electron gas systems for the fluctuation potential and the screened Coulomb interaction W. In some cases, however, band effects, size effects are to be considered. The former should be important for semiconductors, semimetals as graphene, and transition metals. The latter should be inevitable for the analyses of photoemission spectra from nano systems.

References

1. H. Shinotsuka, H. Arai, T. Fujikawa, Phys. Rev. B **77**, 085404 (2008)
2. L. Hedin, J. Michiels, J. Inglesfield, Phys. Rev. B **58**, 15565 (1998)
3. T. Fujikawa, H. Arai, J. Electron Spectrosc. Relat. Phenom. **123**, 19 (2002)
4. T. Fujikawa, M. Kazama, H. Shinotsuka, e-J. Surf. Sci. Nanotechnol. **6**, 263 (2008); 045110 (2014)
5. T. Uwatoko, H. Tanaka, K. Nakayama, S. Nagamatsu, K. Hatada, T. Konishi, T. Fujikawa, T. Kinoshita, A. Harasawa, A. Kakizaki, J. Surf. Sci. Jpn. **22**, 497 (2001)
6. Y. Ohori, H. Shinotsuka, M. Kazama, T. Fujikawa, e-J. Surf. Sci. Nanotechnol. **10**, 145 (2012)
7. M. Kazama, H. Shinotsuka, Y. Ohori, K. Niki, T. Fujikawa, L. Köver, Phys. Rev. B **89**, 045110 (2014)
8. K. Niki, N. Yamamura, Y. Ohori, M. Kazama, T. Fujikawa, L. Köver, Surf. Interface Anal. **46**, 924 (2014)

Chapter 20
Theory of Pump-Probe Ultrafast Photoemission Spectra

Takashi Fujikawa and Kaori Niki

Abstract Keldysh Green's function approach is used in order to derive a practical formula to study pump-probe ultrafast photoemission spectra. The pump pulse is strong whereas the probe pulse can be treated by use of a perturbation theory: We expand the full Green's function in terms of electron-probe pulse interaction and renormalized Green's function including pump pulse excitation. Random Phase Approximation (RPA)-boson approach is introduced to estimate some important factors. Well above the edge, ultrafast photoelectron diffraction provides us with information on the transient structural change after the pump pulse excitation. In addition to these slow processes, the rapid oscillation plays an important role related to electronic excitation by pump pulse.

20.1 Introduction

A pump-probe photoemission technique is now a promising tool to investigate nonequilibrium dynamics of excited molecules and solids on a femto-second time scale. So far some theoretical methods have been proposed to study the nonequilibrium dynamics observed in ultrafast photoemission spectra excited by pump laser pulse [1, 2]. They have provided us with quite interesting information on electron dynamics in strongly correlated systems. These approaches are based on the intrinsic approximation, so that extrinsic losses and resonant processes cannot be discussed. Braun et al. have proposed an interesting one-step theory of pump-probe photoemission based on Keldysh Green's function theory in a fully relativistic four-component formalism [3]. The extrinsic and further complicated many-body effects are not considered there. A much simpler approach based on a model Hamiltonian is

T. Fujikawa (✉) · K. Niki
Graduate School of Advanced Integration Science, Chiba University,
Yayoi-cho 1-33, Inage, Chiba, Japan
e-mail: tfujikawa@faculty.chiba-u.jp

K. Niki
e-mail: niki@chiba-u.jp

D. Sébilleau et al. (eds.), *Multiple Scattering Theory for Spectroscopies*,
Springer Proceedings in Physics 204,
https://doi.org/10.1007/978-3-319-73811-6_20

developed by Lee, and applied to attosecond resonant photoemission of copper dichloride [4].

Rolles et al. have presented time-resolved femtosecond x-ray photoelectron diffraction (XPD) measurements on laser-aligned 1,4-dibromobenzene molecules which were compared with density functional theory calculations [5]. They try to extract useful information about time-dependent structural changes after the pump pulse irradiation. They, however, observe no clear time dependence: They use rather low kinetic energy ($\epsilon_p = 20$ and 35 eV). It is, however, not exactly XPD from molecules fixed in space. First successful photoelectron diffraction measurement is reported by Nakajima et al. from laser aligned I_2 molecule using X-ray free-electron laser (XFEL) pulses [6]. The XPD patterns of the I_2 molecules aligned to the polarization vector of the XFEL are well explained by multiple scattering calculations.

Ulter et al. have also studied the same problems using Kadanoff–Baym equation [5]. These papers are focused on electron dynamics after the pump pulse irradiation.

Kazama et al. have studied the sensitivity of the XPD angular patterns to the structural changes at rather high energy region ($\epsilon_p > 100$ eV) [7, 8]. They show that the XPD patterns are sensitive to the structural changes, and the time-dependent XPD can be a promising tool to study nuclear dynamics after the laser pump excitation. We still have a question whether we can directly observe the time-dependent XPD spectra from the pump-probe XPS. Kuleff and Cederbaum have reviewed their recent theoretical results and some important applications after molecules are exposed to ultrashort laser pulses [9]. They focus on electron and nuclear dynamics after the excitation, in particular on charge migration. Detailed discussion on pump-probe XPS spectra has not been given there. Here our recent related work is briefly reviewed which incorporates all effects in ultrafast XPS spectra [10].

20.2 Ultrafast XPS Theory

In the pump-probe photoemission experiments, we use two photon pulses; pump and probe XFEL pulses. The photoemission current $\boldsymbol{j}$ at the detection site $\boldsymbol{r}$ ($r \to \infty$) and time t is given within the nonrelativistic theory

$$\boldsymbol{j}(\boldsymbol{r}, t) \propto \sum_{\sigma} \left(\frac{\partial}{\partial \boldsymbol{r}'} - \frac{\partial}{\partial \boldsymbol{r}} \right) g^{<}(xt, x't) \Big|_{x=x'} , \tag{20.1}$$

where we neglect the diamagnetic current and use $x = (\boldsymbol{r}, \sigma)$ ($\sigma = \pm 1$) and take $\boldsymbol{r}' = \boldsymbol{r}$ after the differential operation. Our task is thus to calculate the Green's function $g^{<}$ including all interactions. For that purpose we notice that the full Keldysh Green's function $G(1, 2)$ ($1 = (x_1, t_1)$) satisfies the Dyson equation

$$\left[\mathrm{i}\frac{\partial}{\partial t_1} - h(1) - A(1) - B(1) - V_H(1) \right] G(1, 2) - \int_c \mathrm{d}3\; \Sigma(1, 3) G(3, 2) = \delta_c(1, 2) \; . \tag{20.2}$$

In the above equation $\int_c$ means the time integration, and δ_c is the delta function on the closed path. One electron operator A (B) describes the interaction between the probe XFEL pulse (strong pump laser pulse) and an electron in the system.

As demonstrated by (20.2), the total Green's function G includes the influence from A. Let G_B be the Green's function only affected by the pump pulse B, which satisfies the Dyson equation

$$\left[\mathrm{i}\frac{\partial}{\partial t_1} - h(1) - B(1) - V_H(1)\right] G_B(1,2) - \int_c \mathrm{d}3\ \Sigma_B(1,3) G_B(3,2) = \delta_c(1,2)\ , \tag{20.3}$$

where Σ_B is the electron self-energy with no influence from the XFEL x-rays.

We measure stationary photoelectron current which should be caused by 2nd order processes in regard to probe pulse A. By use of the asymptotic formulas of $f_{\boldsymbol{k}}^-$, the photoemission intensity $I_{\boldsymbol{k}}$ measuring the momentum $\boldsymbol{k}$ is given

$$I_{\boldsymbol{k}} \propto \Im\left[\int \mathrm{d}1\ \mathrm{d}2\ f_{\boldsymbol{k}}^{-*}(1) X^{<}(1,2) f_{\boldsymbol{k}}^{-}(2)\right]\ , \tag{20.4}$$

where $f_{\boldsymbol{k}}^-$ is the damping photoelectron wave function under the influence of the optical potential. The lowest order term of $X^<$ is given by

$$X^{<}(1,2) = A(1) g_B^{<}(1,2) A(2).$$

The Green's function $g_B^<(1,2)$ has information on nonequilibrium dynamics after the pump pulse irradiation, which is also written in terms of the time-dependent Dyson orbitals g_m for the core-hole state $|m^*\rangle$

$$g_m(1) = \langle m^*|\psi(1)|0\rangle = \langle m^*|\hat{U}(t_1,t_0)^\dagger \hat{\psi}(1) \hat{U}(t_1,t_0)|0\rangle, \tag{20.5}$$
$$\hat{U}(t_1,t_0) = T\exp\left[-\mathrm{i}\int_{t_0}^{t_1} \mathrm{d}2\ B(2)\hat{\rho}(2)\right]\ .$$

For the core excitation from ϕ_c, the Dyson orbital g_m is approximated

$$g_m(1) \approx \phi_c(x_1) S_m(t_1),\ S_m(t_1) = \langle m^*|\hat{U}(t_1,t_0)^\dagger \hat{b}(1) \hat{U}(t_1,t_0)|0\rangle\ , \tag{20.6}$$

where b is the annihilation operator for the core level ϕ_c. The time-dependent intrinsic amplitude $S_m(t_1)$ has information about the dynamics after the pump excitation. The main photoemission band is thus given

$$I_{\boldsymbol{k}} \propto \left|\int \mathrm{d}1\ f_{\boldsymbol{k}}^{-*}(1) A(1) \phi_c(x_1) S_0(t_1)\right|^2\ . \tag{20.7}$$

The time integration over t_1 can be carried out for some pulse function $a(t_1)$ in $A(1) = a(t_1)\Delta(x_1)$.

In the high-energy (EXAFS) region, photoelectrons are mainly scattered from core electrons, which are insensitive to the details of the electronic structures. We thus assume that in the EXAFS region the photoelectrons are scattered by each atom in a decaying time-dependent atomic configuration, and we can apply multiple scattering expansion for the calculation of the amplitude $\langle f_{\boldsymbol{k}}^{-}|\Delta|\phi_c\rangle$ in (20.7) which depends on t_1 through the temporal atomic arrangement. This time dependence is much more slower than those in the intrinsic amplitudes $S_m(t_1)$.

We next take the screened Coulomb interaction W into account, which yields extrinsic losses: Both $W^{>}$ and $W^{<}$ contribute to the extrinsic losses. The loss functions $W^{>}$ and $W^{<}$ are written in terms of the fluctuation potential δv_l related to $0 \to l$ bosonic excitation,

$$\mathrm{i}W^{>}(1,2) = \sum_{l} \delta v_l^{*}(1)\, \delta v_l(2), \tag{20.8}$$

$$\delta v_l(1) = \int \mathrm{d}3\, \frac{\langle l|\delta\rho(3)|0\rangle}{|\boldsymbol{r}_1 - \boldsymbol{r}_3|}\,. \tag{20.9}$$

Summing up important extrinsic loss terms, we obtain a similar formula to the stationary XPS loss bands including the interference between the intrinsic and extrinsic losses

$$I_{\boldsymbol{k}}^{m} \propto \left|\langle f_{\boldsymbol{k}}^{-}|\Delta|\phi_c\rangle S_m(t) + \langle f_{\boldsymbol{k}}^{-}|\delta v_m g_B^c \Delta|\phi_c\rangle S_0(t)\right|^2. \tag{20.10}$$

In the second amplitude in $|\ldots|^2$ the causal Green's function g_B^c describes the photoelectron propagation inside solids, and the fluctuation potential δv_m is responsible for the extrinsic excitation $|0\rangle \to |m\rangle$.

20.3 Quasi-boson Approximation

Further approximation must be introduced to calculate the pump-probe XPS spectra given by (20.7) and (20.10). Within quasi-boson approximation, the pump pulse excitation operator $B(t)$ is simply written in terms of boson operator a_q and $a_q^{\dagger}$

$$B(t) = b(t) \sum_{q} \left(B_q a_q + B_q^{*} a_q^{\dagger}\right).$$

The expansion coefficient B_q is calculated by use of solutions of RPA equation and matrix elements of electron-photon interaction operator Δ. This approximation allows us to calculate analytically the time evolution operator $\hat{U}$ without further approximation. With aid of the above approximation, the photoemission intensity I_k^0 is thus given for the very short pump- and probe-pulses.

$$I_k^0 \propto \left| M(\boldsymbol{k}; Q(t_A)) L_0(t_A) \right|^2, \quad M(\boldsymbol{k}; Q(t_A)) = \langle f_k^- | \Delta | \phi_c \rangle_{t_A} . \tag{20.11}$$

where instantaneous atomic configuration is represented as $Q(t_A)$ at time t_A when the probe pulse is applied. The multiple scattering matrix M can describe the photoelectron diffraction reflecting atomic configuration $Q(t_A)$. Rapid change with t_A should be observed for $L_0(t_A)$, which is due to electronic dynamics after the pump pulse irradiation.

20.4 Discussion

A first principle many-body time-dependent theory is given here to describe pump-probe ultrafast photoemission spectra based on a Keldysh Green's function approach. Rabi-type rapid oscillation is caused by electron excitation after the pump pulse excitation, and slow change in photoelectron diffraction pattern reflects the instantaneous structural change. It is important to note that they cannot be separated out as indicated by (20.11). Poor time resolution of XFEL excitation smears out the rapid oscillation: Only the information on local structural change could be obtained.

References

1. J.K. Freericks, H.R. Krishnamurthy, Th Pruschke, Phys. Rev. Lett. **102**, 13640 (2009)
2. B. Moritz, T.P. Devereaux, J.K. Freericks, Phys. Rev. B **81**, 165112 (2010)
3. J. Braun, R. Rausch, M. Potthoff, J. Minar, H. Ebert, Phys. Rev. B **91**, 035119 (2015)
4. T.D. Lee, Phys. Rev. Lett. **111**, 027401 (2013)
5. D. Rolle, R. Boll, M. Adolph et al., J. Phys. B **47**, 124085 (2014)
6. K. Nakajima, T. Teramoto, H. Akagi, T. Fujikawa, T. Majima, S. Minemoto, K. Ogawa, H. Sakai, T. Togashi, K. Tono, S. Tsuru, K. Wada, M. Yabashi, A. Yagishita, Sci. Rep. **15**, 14065 (2015)
7. M. Kazama, J. Adachi, H. Shinotsuka, M. Yamazaki, Y. Ohori, A. Yagishita, T. Fujikawa, Chem. Phys. **373**, 261 (2010)
8. M. Kazama, T. Fujikawa, N. Kishimoto, T. Mizuno, J. Adachi, A. Yagishita, Phys. Rev. A **87**, 063417 (2013)
9. A.I. Kuleff, L. Cederbaum, J. Phys. B **47**, 124002 (2014)
10. T. Fujikawa, K. Niki, J. Electron Spectrosc. Relat. Phenom. **206**, 74 (2016)

Chapter 21
Treatment of Thermal Effects by Means of the Alloy Analogy Model

Hubert Ebert, Jürgen Braun, Ján Minár and Sergiy Mankovsky

Abstract A scheme is presented that allows to account for finite temperature effects when calculating response functions or spectroscopic properties of solids. The approach is based on the alloy analogy model and treats thermally induced disorder as chemical disorder in substitutional alloys by means of the Coherent Potential Approximation (CPA). The reliability of the approach is demonstrated by calculations of the temperature dependent resistivity of ferromagnetic Fe. Calculations of angle-resolved photoemission show for increasing temperature the expected smearing of band-like features in the spectra, i.e. a transition from the so-called UPS to the XPS regime.

21.1 Introduction

Thermally induced lattice vibrations are often accounted for in the theory of spectroscopy by means of the Debye–Waller factor $W(T)$. This applies in particular for x-ray absorption (XAS) [1] and photoemission (PES) [2, 3]. For magnetic solids one has in addition to account for spin fluctuations [4] that contribute to the thermally induced disorder in a system and influence all spectroscopic properties this way.

H. Ebert (✉) · J. Braun · S. Mankovsky
Department Chemie, Ludwig-Maximilians-Universität München,
Butenandtstr. 5-13, 81377 Munich, Bavaria, Germany
e-mail: Hubert.Ebert@cup.uni-muenchen.de

J. Braun
e-mail: Juergen.Braun@cup.uni-muenchen.de

S. Mankovsky
e-mail: Sergiy.Mankovskyy@cup.uni-muenchen.de

J. Minár
New Technologies-Research Center, University of West Bohemia,
Univerzitní 8, 306 14 Plzeň, Czech Republic
e-mail: jminar@ntz.czu.cz

D. Sébilleau et al. (eds.), *Multiple Scattering Theory for Spectroscopies*,
Springer Proceedings in Physics 204,
https://doi.org/10.1007/978-3-319-73811-6_21

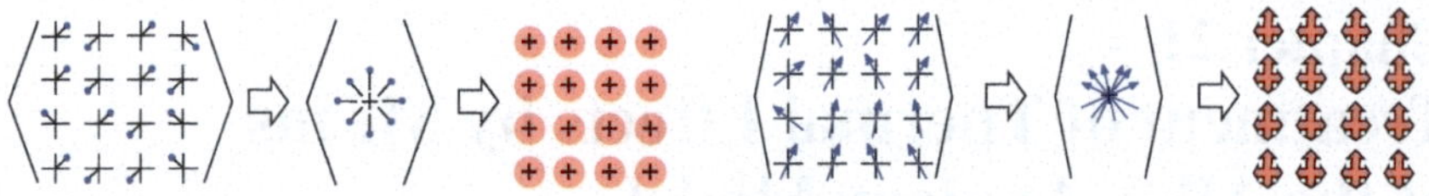

Fig. 21.1 Accounting for thermal disorder by means of an effective CPA medium. *Left*: the continuous distribution $P(\Delta \boldsymbol{R}_n(T))$ of atomic displacement vectors is replaced by a discrete set of vectors $\Delta \boldsymbol{R}_v(T)$ with probabilities x_v. *Right*: the continuous distribution $P(\hat{\boldsymbol{e}}_n)$ of orientations of the magnetic moments is replaced by a discrete set of orientation vectors $\hat{\boldsymbol{e}}_f$ with probabilities x_f [7]

The Coherent Potential Approximation (CPA) [5] in combination with multiple scattering theory or the KKR (Korringa–Kohn–Rostoker) formalism, respectively, is a well established scheme to deal with chemical disorder in substitutional alloys. Recently, this approach to get a configurational average was applied when dealing with thermally induced disorder [6]. Here the application of the corresponding alloy analogy model is demonstrated for the calculation of transport properties [7] and photoemission spectra [8].

21.2 The Alloy Analogy Model

The central idea of the corresponding alloy analogy model to deal with finite temperature effects is sketched in Fig. 21.1.

Thermally induced lattice vibrations are assumed to be much slower than the electronic propagation and for that reason they can be accounted for in a quasi static manner. For that purpose they are represented by a discrete set of displacement vectors $\Delta \boldsymbol{R}_v(T)$ with the probability x_v [7]. In a similar way spin fluctuations are represented by a discrete set of orientation vectors $\hat{\boldsymbol{e}}_f$ with probabilities x_f, that may be set up e.g. on the basis of Monte Carlo simulations [7].

In terms of the KKR formalism the single-site t-matrix $\underline{t}$ of a displaced atom can be related to that of the unshifted atom by means of the U-transformation [9] as given in (21.1), where the underline indicates matrices with respect to the combined index $\Lambda = (\kappa, \mu)$ with the relativistic spin-orbit and magnetic quantum numbers κ and μ, respectively [10].

$$\underline{t}_v = \underline{U}(\Delta \boldsymbol{R}_v)\, \underline{t}\, \underline{U}(\Delta \boldsymbol{R}_v)^{-1} \tag{21.1}$$

$$\underline{t}_f = \underline{R}(\hat{\boldsymbol{e}}_f)\, \underline{t}\, \underline{R}(\hat{\boldsymbol{e}}_f)^{-1} \tag{21.2}$$

$$\underline{t}_{vf} = \underline{U}(\Delta \boldsymbol{R}_v)\, \underline{R}(\hat{\boldsymbol{e}}_f)\, \underline{t}\, \underline{R}(\hat{\boldsymbol{e}}_f)^{-1} \underline{U}(\Delta \boldsymbol{R}_v)^{-1}\,. \tag{21.3}$$

Equation (21.2) gives in a similar way the impact of the rotation of a local moment in terms of rotation matrices $\underline{R}(\hat{\boldsymbol{e}})$, while (21.3) combines the impact of rotation and displacement. With this, the components of a quasi alloy are defined and the standard CPA equations

$$\underline{\tau}_{\mathrm{CPA}} = \sum_{vf} x_{vf}\, \underline{\tau}_{vf} \tag{21.4}$$

$$\underline{\tau}_{vf} = \left[(\underline{t}_{vf})^{-1} - (\underline{t}_{\mathrm{CPA}})^{-1} + (\underline{\tau}_{\mathrm{CPA}})^{-1}\right]^{-1} . \tag{21.5}$$

can be used to determine the single-site t-matrix $\underline{t}_{\mathrm{CPA}}$ representing the CPA medium. The corresponding CPA scattering path operator $\underline{\tau}_{\mathrm{CPA}}$ is obtained solving the associated multiple scattering problem.

21.3 Transport Properties

Representing the electronic structure of a solid in terms of the retarded Green's function $G^{+}(E)$ the symmetric part of the conductivity tensor for $T = 0\,\mathrm{K}$ can be expressed by means of the Kubo–Greenwood equation:

$$\sigma_{\mu\nu} = \frac{\hbar}{\pi N \Omega} \operatorname{Tr} \left\langle \hat{j}_{\mu}\, \Im G^{+}(E_{\mathrm{F}})\, \hat{j}_{\nu}\, \Im G^{+}(E_{\mathrm{F}}) \right\rangle_{\mathrm{c}} , \tag{21.6}$$

with $\hat{j}$ the current density operator [10], Ω the volume of the unit cell and E_{F} the Fermi energy. For a disordered system the bracket $\langle ... \rangle_{\mathrm{c}}$ in (21.6) implies a configurational average. The corresponding KKR-CPA based scheme for alloys [11] was adopted recently to deal with thermally induced disorder making use of the alloy analogy model [7].

Figure 21.2 (left) shows results of calculations of the temperature (T) dependent resistivity $\rho(T)$ of ferromagnetic bcc-Fe.

Including only thermal vibrations ($\rho_v(\mathrm{vib})$) one gets a nearly linear increase with T. Accounting for spin fluctuations on the basis of Monte Carlo (MC) simulations the corresponding resistivity ($\rho_{\mathrm{MC}}(\mathrm{fluct})$) increases up to the critical temperature T_{C} and then saturates. The dashed curve gives the sum of $\rho_v(\mathrm{vib})$ and $\rho_{\mathrm{MC}}(\mathrm{fluct})$. In particular at higher temperatures this sum deviates from the resistivity that is obtained if both sources of thermal disorder are accounted for simultaneously.

The right panel of Fig. 21.2 shows the resistivity $\rho(T)$ obtained on the basis of MC simulations that are found in reasonable good agreement with the experimental data. Modeling the spin fluctuations on the basis of the experimental magnetization curve $M(T)$ leads finally to a very good agreement with experiment demonstrating that the alloy analogy model allows to account for thermal effects in a quantitative way.

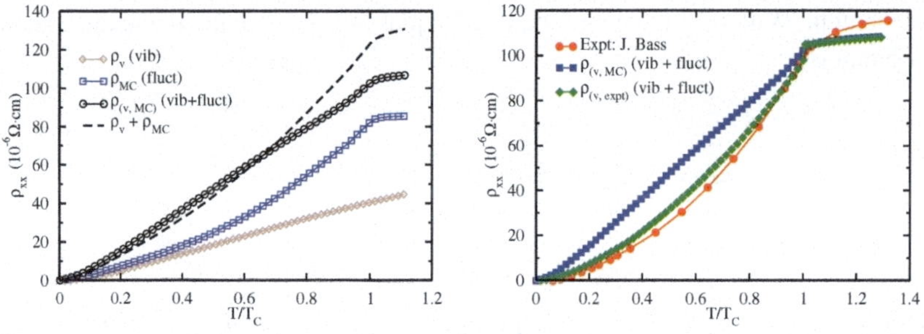

Fig. 21.2 Resistivity $\rho(T)$ of bcc-Fe. *Left*: accounting for vibrations (*diamonds*), spin fluctuations from MC simulations (squares) and both (*circles*). The *dashed line* gives the sum of the first two curves. *Right*: resistivity obtained by accounting for thermal vibrations and spin fluctuations using MC data (*squares*) and the experimental magnetic moment $M_{\text{exp}}(T)$ (*diamonds*) compared with experiment (*circles*) [7]

21.4 Photoemission

The one-step model of angle-resolved photoemission (AR-PES) was originally implemented for ordered solids using the KKR formalism [12]. This platform allowed to generalize the approach to deal with disordered alloys making use of the CPA [13, 14]. Recently, the one-step model could be extended to deal with thermal effects exploiting the alloy analogy model [8]. This way the emission intensity $\langle I(\epsilon_f, \boldsymbol{k}, T)\rangle$ depending on the final state energy ϵ_f and wave vector $\boldsymbol{k}$ as well as temperature T can be written as:

$$\langle I(\epsilon_f, \boldsymbol{k}, T)\rangle = \langle I^{\text{a}}(\epsilon_f, \boldsymbol{k}, T)\rangle + \langle I^{\text{m}}(\epsilon_f, \boldsymbol{k}, T)\rangle + \langle I^{\text{inc}}(\epsilon_f, \boldsymbol{k}, T)\rangle \ . \quad (21.7)$$

The first term is atomic-like, i.e. it has a local character. The other ones are determined by multiple scattering events and this way reflect the electronic structure of the investigated solid. The second coherent term allows to map the dispersion relation while the third incoherent term reflects the degree of disorder in the system. In fact, inspection of the explicit expressions for the photo current shows that the coherent term dominates for low T and photon energies, while the incoherent term dominates for high T or photon energies giving rise to a DOS (density of states) like spectrum. This expected behavior is indeed confirmed by the calculated AR-PES spectra for Au(111) and Pt(111) in Fig. 21.3.

As one can see with increasing temperature a smooth transition from the band structure like UPS to the DOS like XPS regime is observed. This occurs the quicker the higher the photon energy and the lower the Debye temperature Θ_{D} (Au: $\Theta_{\text{D}} = 165\,\text{K}$, Pt: $\Theta_{\text{D}} = 230\,\text{K}$).

The experimental energy distribution curves (EDCs) of W(110) ($\Theta_{\text{D}} = 400\,\text{K}$) for the photon energy 870 eV in Fig. 21.4 clearly show the expected transition to

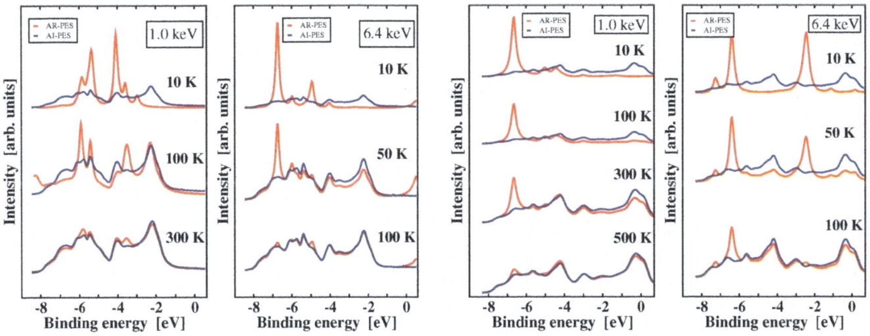

Fig. 21.3 Calculated normal emission angle-resolved (AR-PES, *red curves*) and angle-integrated spectra (AI-PES, *blue curves*) for Au(111) (*left*) and Pt(111) (*right*) at two different photon energies ($h\nu = 1.0$ and 6.4 eV) for the several temperatures [8]

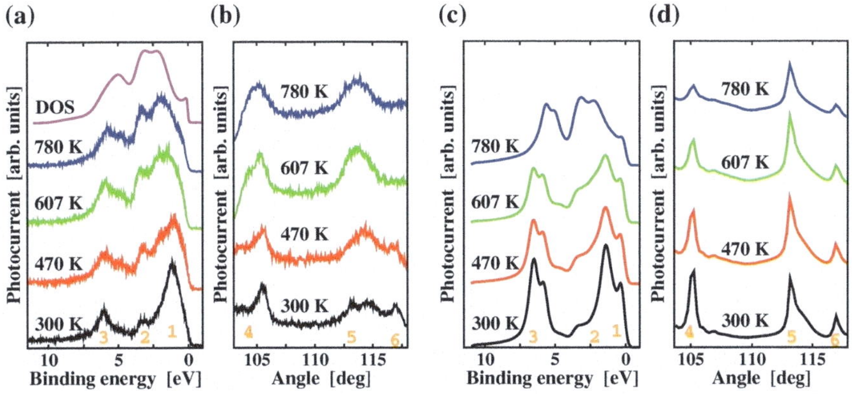

Fig. 21.4 **a** Experimental energy distribution curves (EDCs) for W(110) with a comparison to broadened DOS (*topmost curve*). **b** Experimental temperature-dependent momentum distribution curves (MDCs). **c**, **d** Corresponding theoretical results [8]

the XPS regime with increasing temperature [13]; i.e. the band like features of the spectra observed at low temperatures are continuously washed out due to increasing thermal disorder. This behavior is obviously well reproduced by the calculations based on the alloy analogy model. This also applies for the momentum distribution curves (MDCs).

For the experimental spectra finite temperature leads to a smearing out of the spectral features but does not diminish the total intensity. This applies obviously also for the theoretical AR-PES spectra shown in Figs. 21.3 and 21.4. This is because the alloy analogy model leads to an effective medium that does not violate the optical theorem. This fundamental requirement is not satisfied by simpler approaches that account for finite temperatures by multiplying the single-site t-matrices with the Debye–Waller factor $W(T)$ [3].

References

1. J.J. Rehr, Radiat. Phys. Chem. **75**, 1547 (2006)
2. N.J. Shevchik, Phys. Rev. B **16**, 3428 (1977)
3. C.G. Larsson, J.B. Pendry, J. Phys. C Solid State Phys. **14**, 3089 (1981)
4. I. Delgadillo, H. Gollisch, R. Feder, Phys. Rev. B **50**, 15808 (1994)
5. P. Soven, Phys. Rev. **156**, 809 (1967)
6. H. Ebert, S. Mankovsky, D. Ködderitzsch, P.J. Kelly, Phys. Rev. Lett. **107**, 066603 (2011)
7. H. Ebert et al., Phys. Rev. B **91**, 165132 (2015)
8. J. Braun et al., Phys. Rev. B **88**, 205409 (2013)
9. A. Lodder, J. Phys. F Metal Phys. **6**, 1885 (1976)
10. M.E. Rose, *Relativistic Electron Theory* (Wiley, New York, 1961)
11. W.H. Butler, Phys. Rev. B **31**, 3260 (1985)
12. J.F.L. Hopkinson, J.B. Pendry, D.J. Titterington, Comput. Phys. Commun. **19**, 69 (1980)
13. J. Braun, J. Minár, F. Matthes, C.M. Schneider, H. Ebert, Phys. Rev. B **82**, 024411 (2010)
14. P.J. Durham, J. Phys. F Metal Phys. **11**, 2475 (1981)

Chapter 22
Local Geometry by XANES and RXS

Joaquín García Ruiz

Abstract A brief overview of the capability of x-ray absorption near edge structure (XANES) and resonant x-ray scattering (RXS) in the hard x-ray region to determine the local geometrical structure around the photoabsorbing atom is presented.

22.1 Introduction. Geometry Beyond the Radial Distribution Function

X-ray absorption spectroscopy (XAS) is the appropriate technique to investigate the neighbourhood of a photoabsorber atom embedded in a medium [1]. The x-ray absorption process is a transition between two quantum states: from an initial state with an x-ray and a core electron to a final state with no x-ray, a core-hole and a photoelectron. The inner shell electron is well described by atomic wave functions and the photoelectron final states wave function for an isolated atom is well described by a free spherical wave outgoing from the atom. The final states outgoing wave becomes from the superposition of the free outgoing wave and the scattered waves by the atoms surrounding the absorber one. This interference between the outgoing wave and the scattered ones produces minima and maxima in space. Such a simplified picture describes the multiple scattering approximation to the XAS spectrum [2–5], where the cross section can be factorized as: $\sigma = \sigma_0(1 + \sum \chi_n)$. Here σ_0 represents the cross section for the isolated atom (the free outgoing wave function) and $\sum \chi_n$ comes from the processes in which the photoelectron is dispersed $n - 1$ times by the neighbour atoms before interfering with the outgoing wave at the origin. These multiple scattering events carry information on the topological order around the absorber, i.e. interatomic distances and angles between bonds. Such multiple scattering signals are always present in the XAS spectrum but their contribution rapidly decreases with increasing the energy (or wave vector $\mathbf{q}$).

The information on the local geometry contained in XAS comes from multiple scattering processes of the outgoing wave. Joined to this, direct information on the

J. García Ruiz (✉)
Instituto de Ciencia de Materiales de Aragòn CSIC-Univ, Zaragoza, Spain
e-mail: jgr@unizar.es

D. Sébilleau et al. (eds.), *Multiple Scattering Theory for Spectroscopies*, Springer Proceedings in Physics 204,
https://doi.org/10.1007/978-3-319-73811-6_22

geometry can be obtained from the tensor character of the absorption cross section [6]. Therefore in anisotropic materials the spectrum is dependent on the relative orientation of the electric field polarization of the x-ray beam to the sample orientation. In the EXAFS region interatomic distances on the direction of the electric field vector can be obtained [1]. On the other hand the energy of the absorption edge highly depends on the polarized direction. This strong difference at the edge allows to detect by RXS the scattered intensity for photon energies at the absorption edge in forbidden reflections [7]. Within this contribution, we will shortly review some experimental probes of these facts. i.e. local geometry determined by (i) multiple s-cattering description (ii) polarized absorption spectra and (iii) ordering of distortions by resonant x-ray scattering.

22.2 Local Geometry in XANES Spectrum. Multiple Scattering

The interpretation of the XANES spectra was matter of controversy in the eighties. Some studies suggested that the multiple scattering range is very small questioning the capability of XANES to probe high-order correlation functions. One of the difficulties to resolve this problem was that photoabsorbing atoms in a solid are surrounded by atoms in successive coordination shells in such a way that the contribution of distant shells interfere with that coming from multiple scattering of the nearest neighbours atoms. The simplest way to resolve this problem was by comparing the XAS spectra of transition metal ions in water solution where only the first coordination shell is relevant. This controversy was resolved by comparing the Mn K-edge absorption spectra of a water solution of potassium permanganate where MnO_4^- ion is tetrahedrally coordinated and a water solution of Mn_2^+ where this ion is octahedrally coordinated [5]. These two spectra after rescaling the energy and renormalizing the amplitude shows the same sinusoidal behaviour in the energy region sensitive only to the pair correlation function (EXAFS region). A strong difference is observed in the XANES region due to the different geometrical arrangement of the ligand atoms around the metal ion in the two complexes.

As illustration, Fig. 22.1 shows the expansion in the different MS contributions of the $[Mn(H_2O)_6]^{2+}$ octahedral cluster compared with the experimental spectrum. As we can see the main line A for octahedral clusters is determined by a full multiple scattering resonance where all the multiple scattering contributions are in phase [5, 8]. Multiple scattering theory is nowadays the usual method to analyze the X-ANES spectra. There is a lot of codes to calculate the XANES spectra as: MCMS, GNXAS, MXAN, FDMNES, FEFF, etc. The reader is addressed to other chapters of this book to know the recent improvements of the different codes. Theoretical fitting of the XANES spectra can be performed [9] and it is generalized the implementation of the MS contributions in the analysis of the extended part of the XAS spectrum (EXAFS) [10].

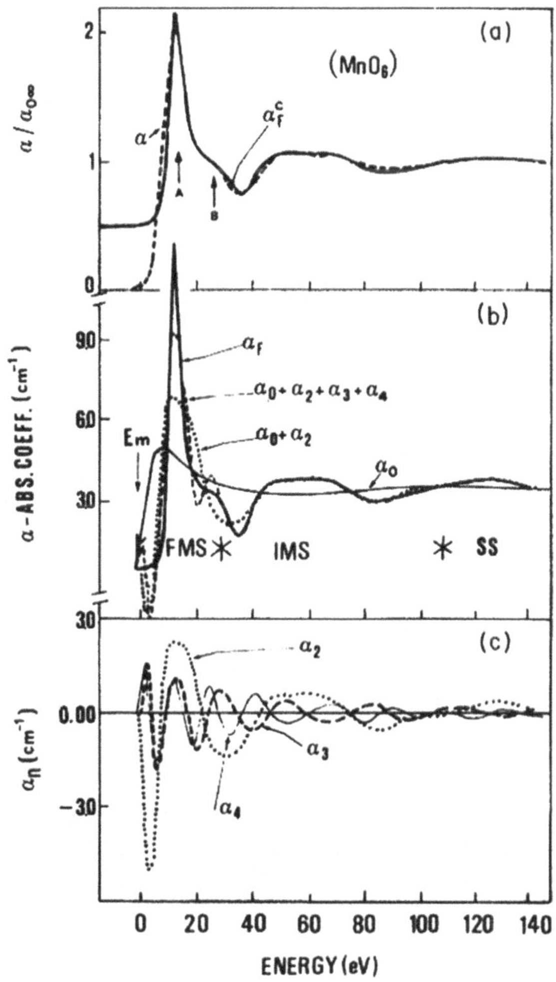

Fig. 22.1 Comparison between calculated and experimental spectra of $[Mn(H_2O)_6]^{2+}$ ion and expansion in its multiple scattering contributions. Source [5]. Reproduced with permission from American Physical Society

22.3 Anisotropy in the Local Geometry from Polarized XANES Spectra

As it is shown, the local geometry can be retrieved from XANES spectra through the theoretical simulation using MS framework. But independently of this fact, the absorption coefficient is a symmetric tensor of rank 2 in the dipole approximation related to the electric field vector of the incident x-ray beam ($\mathbf{e}_q$). In principle, six different components are necessary to completely describe it. Depending on the local symmetry the number of components is reduced in such a way that for octahedral or tetrahedral symmetries the tensor is diagonal and the three components are equal, i.e. it behaves as a scalar. For tetragonal symmetry and taken the reference frame as the tetragonal axis and the two directions perpendicular to it, the tensor is also diagonal with two different components, along the tetragonal axis and perpendicular to it [6]. It is obvious that XAS spectra on oriented samples will provide direct information on the local symmetry. In the case of the EXAFS region where the pair distribution function is obtained, the observed path contribution is factorized by the scalar product of $\mathbf{e_q}$ and the bond direction [1]. For example, the EXAFS contribution to the surrounding atoms located perpendicular to $\mathbf{e_q}$ is exactly zero. In the XANES spectra differences come mainly from the energy position of the absorption edge, an anisotropic shift is observed. As a matter of illustration, $La_{1-x}Sr_{1+x}MnO_4$ compounds present a tetragonal distorted local structure around the Mn atom. The experiments were performed with the polarization of the incident beam parallel and perpendicular to the *c*- axis. XANES spectra for the three single crystals show an anisotropic splitting being larger the for $x = 0$ sample and the smaller for the $x = 0.5$ one [11]. The magnitude of this splitting correlates with the larger Mn-O interatomic distance along the *c*-axis for the $x = 0$ composition. Therefore, the anisotropic splitting measures the degree of tetragonal distortion of the MnO_6 octahedron. Needless to say that from the polarized EXAFS spectra, the in-plane Mn-O interatomic distances are determined when $\mathbf{e_q}$ is perpendicular to the *c*-axis whereas the out of plane Mn-O distances are obtained for $\mathbf{e_q}$ parallel to *c*.

22.4 Ordering of the Local Distortions by Resonant X-Ray Scattering (RXS)

RXS combines absorption and diffraction as they have in common the x-ray atomic scattering factor (ASF), f, which is usually written as: $f = f_0 + f' + if''$ [1]. It contains an energy independent part, f_0, corresponding to the classical Thomson scattering and two energy-dependent terms, f' and f'', also known as the atomic anomalous scattering factor. RXS occurs when the x-ray energy is tuned near the absorption edge of an atom in the crystal. We recall here the intimate relationship between the atomic anomalous scattering factor (ASF) and the absorption coefficient, i.e. the imaginary part of the ASF is proportional to the absorption coefficient, whereas the real part is related to the imaginary part through the Kramers–Kronig transformation. Therefore, the ASF has the same tensorial character as the absorption coefficient. The polarization dependence of ASF is in the origin of the observation of RXS intensity in forbidden reflections. These reflections that are forbidden by symmetry due to the scalar character of the Thomson scattering. They are forbidden by symmetry elements with translation components (screw axes and glide planes) and appear due to the presence of local anisotropy (sometimes assigned to orbital ordering (OO)). As we have seen in the previous section, the anisotropy is mainly reflected in the anisotropic shift of the polarized XANES spectra (and the ASF) and therefore, the enhancement of the scattered intensity just appear close to the absorption threshold.

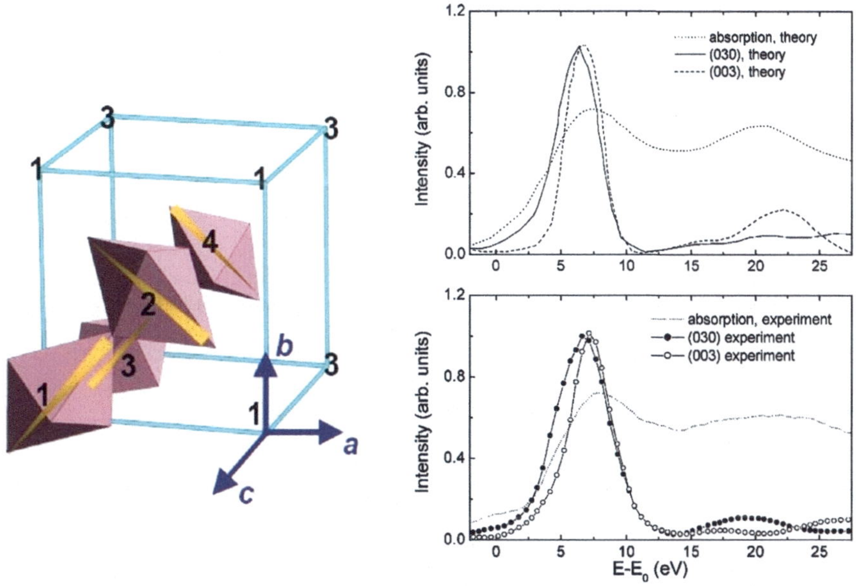

Fig. 22.2 Left: Schematic picture of the four Mn atoms in the *Pbnm* crystal cell indicating the order of the long Mn-O bond. Right: Calculated RXS of the (030) and (003) reflections. Upper panel: Calculations with MXAN. Lower panel: Experimental RXS spectra. Source [12]. Reproduced with permission from American Physical Society

An illustrative example is the RXS at the forbidden reflections in the $LaMnO_3$ [12]. The crystallographic symmetry of this compound is *Pbnm* and (odd, 0, 0), (0, odd, 0) and (0, 0, odd) reflections are forbidden. This is due to the fact that their structure factor comes from the difference between equivalent atoms in the unit cell. The difference in the ASF among equivalent Mn atoms is originated by the different orientation of the local tetragonal distortion, as it is depicted in Fig. 22.2. Therefore, scattered reflected intensity will be observed at energies close to the absorption edge due to the anisotropic shift between the two polarizations parallel and perpendicular to the local tetragonal axis. As it is shown in Fig. 22.2 maximum of the scattered intensity occurs just at the white line. The energy dependence of the scattered intensity of the (030) and (003) forbidden reflections together with and the calculated ones using the MXAN code [13] are shown in Fig. 22.2.

22.5 Summary

Within this short paper we pretend to briefly illustrate the capability of the hard x-ray spectroscopies to determine the local geometry around a specific atom. There is a lot of points that are not discussed, such as DAFS, DANES, or high energy resolution

XAS spectroscopies. The reader can be addressed to the large bibliography on these subjects, for instance the book by J.A. van Bokhoven and C. Lamberti [14]. Regarding the multiple scattering theory and its recent developments they can refer to the other papers contained in this book.

References

1. D.C. Koningsberger, R. Prins (eds.), *X-ray Absorption; Principles, Applications, Techniques of EXAFS, SEXAFS and XANES* (Wiley, New York, 1988)
2. C.R. Natoli, M. Benfatto, S. Doniach, Phys. Rev. A **34**, 4682 (1986)
3. J.J. Rehr, R.C. Albers, Rev. Mod. Phys. **72**, 621 (2000)
4. A. Filipponi, A. Di Cicco, C.R. Natoli, Phys. Rev. B **52**, 15122 (1995)
5. M. Benfatto et al., Phys. Rev. B **34**, 5774 (1986)
6. C. Brouder, J. Phys. Condens. Matter **2**, 701 (1990)
7. M. Blume, *Resonant Anomalous X-ray Scattering, Theory and Applications*, ed. by G. Materlik, C.J. Spark, K. Fisher (Elsevier Science B. V., North-Holland, Amsterdam, 1994)
8. J. García, A. Bianconi, M. Benfatto, C.R. Natoli, J. de Physique C **8**, 49 (1986)
9. M. Benfatto, S. Della Longa, J. Synchrotron Radiat. **8**, 1087 (2001)
10. T.E. Westre et al., JACS **117**, 1566 (1995)
11. J. Herrero Martin et al., Phys. Rev. B **72**, 085106 (2005)
12. G. Subías et al., Phys. Rev. B **75**, 235101 (2007)
13. M. Benfatto, S. Della Longa, C.R. Natoli, J. Synchrotron Radiat. **10**, 51 (2003)
14. J.A. van Bokhoven, C. Lamberti (eds.), *X-ray Absorption and X-ray Emission Spectroscopy. Theory and Applications* (Wiley, New York, 2016)

Chapter 23
Does an Electron Pair Diffract as a Whole?

Giovanni Stefani

Abstract The correlated behavior of the electrons in atoms, molecules and solids has been, and still is, a crucial issue for the quantum mechanics. Lately, experiments in which either a photon or an electron ionizes a system generating a pair of unbound electrons have shown to be particularly sensitive to correlations. In this paper a brief selection of such experiments is presented aiming at highlighting conditions under which the two final electrons behave as a single entity rather than two independent ones. It is concluded that whenever the pair generation occurs on an attosecond timescale the pair is strongly correlated, i.e. it is entangled, and the electrons continuously share between them the energy in excess to the pair creation threshold.

Quantum mechanics is the sound background on which most of nowadays technology rests. However, in spite of this undeniable success, there are still questions that after a century wait for an answer. Among them, one nut resulted hard to crack for both theory and experiment: The correlated motion of electrons.

From the experimental point of view correlations are elusive as they influence marginally the spectral response of the conventional spectroscopies. Hence the effort to develop new classes of spectroscopies more sensitive to electron correlations. These spectroscopies have as a common denominator the presence of two unbound electrons in the final state (2 electrons out). It is exactly the interaction of these electrons that shapes the cross-section making them very sensitive to correlations. Archetypal *2 electrons out* processes are: (1.) the electron impact ionization and (2.) the resonant double photoionization. In the latter processes, the photon energy is sufficient to create an intermediate core hole that eventually auto-ionizes generating an Auger electron. We shall see that the lifetime of the intermediate state plays a crucial role in generating coherent or incoherent photoelectron-Auger electron pairs. In one of such 2 electrons out processes, usually termed as Angle-Resolved Auger-Photoelectron Coincidence Spectroscopy (AR-APECS) [1, 2] (and references

G. Stefani (✉)
Dipartimento di Scienze and Unitá CNISM Universitá Roma Tre,
Via della Vasca Navale 84, 00146 Roma, Italy
e-mail: stefani@fis.uniroma3.it

D. Sébilleau et al. (eds.), *Multiple Scattering Theory for Spectroscopies*,
Springer Proceedings in Physics 204,
https://doi.org/10.1007/978-3-319-73811-6_23

therein quoted) a monochromatic light beam impinges on the target, the unbound electrons are separately discriminated in energy and momentum and detected, if and only if, they are coincident in time, i.e. originate from the same ionization event and not from uncorrelated ones. In the electron impact case a monochromatic electron beam initiates an ionization process where the two final unbound electrons are again detected in coincidence, thus generating a so called (e,2e) experiment [3, 4] (and references therein quoted), and continuously share energy and momentum in an entangled way. Surprisingly enough, even such conceptually simple experiments pose key questions that we still struggle to answer and that can be summarized as it follows.

1. Are the two unbound electrons entangled? We know from the very definition of entanglement that if two particles interact they cannot anymore be described by independent wave functions and that by measuring one of them the measure of the other one will be partly or totally determined.
2. In modeling 2-electron out processes, is it possible to use a single particle approach, which is more suitable for solids, instead of the density matrix that is usually employed for central potentials?
3. When the unbound electron pair diffracts from a lattice, is the entanglement preserved?
4. When the electron pair diffracts, do the two electrons behave as independent particles or as a single quasi-particle?

In order to tackle these problems for solid targets within the single particle approach, the first concern is to establish validity of this model in describing the photoelectron-Auger electron pair in a simple atomic target. Lets take a closed-shell atom like Argon and focus on the Auger $L_3M_{23}M_{23}$ angular distribution as measured in coincidence with a well specific direction of the parent L_3 photoelectron. In short, an AR-APECS experiment is performed on Ar initiated by a monochromatic linearly polarized light beam. If we were to measure the Auger angular distribution alone we would find one quantization axis only, i.e. the light polarization axis. When detecting in coincidence the electron pair the photoelectron establishes a second quantization axis and the Auger angular distribution is accordingly modified. In other words, selecting the photoelectron direction implies selecting a subset of the core ionic states from which the following Auger originates with a non-statistical distribution. Natoli and coworkers [5] have used a single-particle scattering approach for describing the angular correlation between a photoelectron and the subsequent Auger electron from atomic targets. This method was proposed as an alternative approach with respect to the usual density matrix formalism, since it is more convenient for the extension to the solid-state cases. They have derived a tensor expression for the cross section and an equivalent expression in terms of convenient angular functions has been treated for the case of linearly polarized light. This method was used to calculate the aforementioned coincident Auger decay in an Ar atom described in the single configuration Dirac-Fock scheme. Results were found to compare favorably with experimental data [6] for different final angular momentum states of the doubly charged ion and for different kinematical conditions.

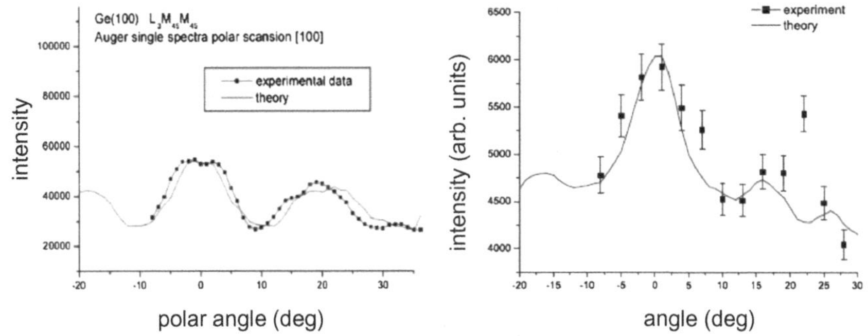

Fig. 23.1 *Left panel.* Theoretical curve for Ge(100) $L_3M_{45}M_{45}$ Auger electron angular distribution and comparison with experimental data. *Right panel.* Theoretical curve for Ge(100) $L_3M_{45}M_{45}$ AR-APECS and comparison with experimental data. From [6]

Natoli and coworkers extended the successful single-particle scattering approach to solids, Ge in particular, by conjugating it with multiple scattering diffraction of the Auger-photoelectron pair from the reconstructed Ge(100) surface 2×1 and within the 2-step approximation, i.e. photoemission and Auger processes are assumed to be incoherent. In this way, they have developed a generalization of the multiple-scattering formalism to deal with AR-APECS in the solid state. Which consists in merging the exact atomic treatment of the angular correlations between the two electrons and the single-particle approach, on which the multiple scattering description of condensed matter relies. This allows the recovering, even in extended systems, of the entangled form of the electron-pair wave function characterizing the coincidence angular diffraction pattern. In Fig. 23.1 the numerical calculations performed for Ge(100) $L_3M_{45}M_{45}$ Auger Electron angular Distribution (AED) and for Ge(100) $L_3M_{45}M_{45}$ AR-APECS are compared with the experimental data from [7].

The experiment allowed to select specific photoelectron directions and to correlate them to specific Auger directions. For a mild discrimination of the photoelectron direction is evident that the Auger angular distribution as measured in coincidence with the photoelectron (Fig. 23.1 right panel) differs markedly from the Diffraction Pattern shown by single AED (Fig. 23.1 left panel). What is relevant is that both experiments, AED and AR-APECS, are well accounted by the theory that uses the atomic model to describe entanglement of the electron pair while letting the pair diffract incoherently from the crystal lattice. Out of these results it can be concluded that difference between AED and AR-APECS is evidence for entanglement, i.e. evidence for the electron pair wave-function not being separable and that, at high energies characteristic of the AR-APECS experiment on Ge $L_3M_{45}M_{45}$ (Auger = 1139 eV Photoelectron = 250 eV), the electrons correlated in the pair diffract independently from the crystal. A somewhat different panorama is painted when dealing with electron impact ionization that in the final state features a pair of correlated electrons as well. In this case, a monochromatic electron beam impinges on the target and the two final electrons are detected coincident in time and discriminated in energy

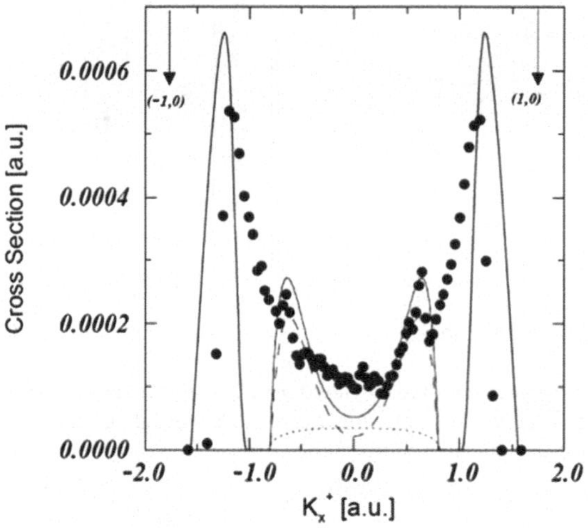

Fig. 23.2 The (*e*,2*e*) experimental cross-section on Cu(001) (*full dots*) at fixed incident energy of 85 eV and as a function of the relative center of mass momentum (K^+). Depicted are the separate contributions of amplitudes for direct pair emission (*dotted curve*) and the amplitude for the pair scattering from the lattice potential (*dashed curve*). Calculations using the coherent sum (*solid curve*) are also shown. The calculations are performed for infinite energy and angular resolution of the detectors. *Arrows* indicate the positions of the diffracted beams. From [8]

and momentum, i.e. an (*e*,2*e*) experiment is performed. Lets consider the experiment performed by Kirschner and coworkers [8] where the electron pairs were generated by a low energy (below 100 eV) electron beam, scattered off a Cu(001) and Fe(110) clean surfaces, and detected by time-of-flight electron spectrometers. Experimental results are compared with a model that accounts for diffraction of the pair from the crystal lattice. In this case, instead of the individual free electron momenta, momentum of the center of mass of the pair is considered. In other words, the probability of detecting correlated electron pairs is calculated and measured as a function of the relative center of mass momentum (K^+). The electron-pair energy-sharing spectra shown in Fig. 23.2 display peculiar features associated with the diffraction from the lattice of the pair as a whole. Indeed, the (*e*,2*e*) cross-section peaks when the K^+-component parallel to the sample surface matches surface reciprocal lattice vectors. In other words, the electron pair diffracts as a whole rather than as two independent particles. At this stage, we are brought to think that the Ge AR-APECS experiment is well accounted by a two-step model, photoionization incoherently followed by Auger decay, and that at high energy the two electrons of the pair, though entangled, diffract independently out of the crystal lattice. On the contrary, the (*e*,2*e*)-process performed at moderate energy, is eminently a one step process, i.e. a single interaction determines energies and momenta of the electron pair, we must use a single quasi-particle that diffracts out of the lattice in order to describe the electron pair behavior. Hence, we are left with a dilemma: is it the electrons kinetic energy or the

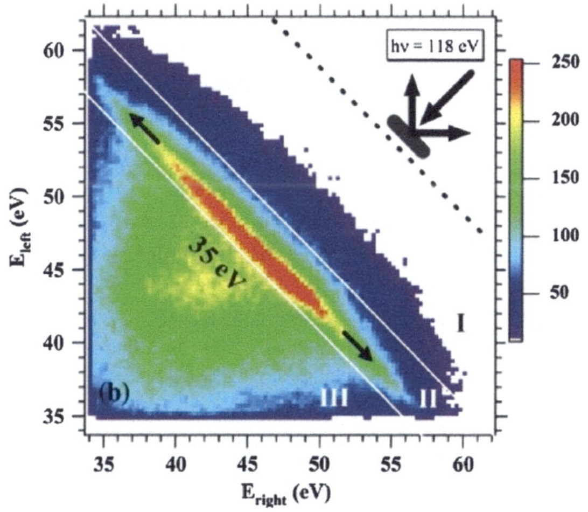

Fig. 23.3 The coincidence 2D-energy distributions of the $N_{23}VV$ Auger-photoelectron pair on the Ag(100) surface. The photon energy is 118 eV. The dashed lines indicate the position of the maximum sum energy ($E_{max} = E_{left} + E_{right}$). Most of the coincidence intensity is confined within the pair of solid lines. The prominent intensity band has a width of 35 eV, as marked by the *arrows*, and the otherwise sharp energy separation of the Auger from the photoelectron is lost. From [9]

coherence of the pair creation the critical parameter that determines the quasi-particle rather than the independent particles behavior?

A clear answer to this question comes from an APECS experiment performed on the Ag(100) surface [9] where the $N_{23}VV$ and $M_{45}VV$ Auger line shapes have been investigated by measuring the 2D-energy-distribution of the time correlated Auger-photoelectron pairs generated in the same ionization event. The two processes under consideration end up in the same 2-valence holes manifold of final states but they differ in the core hole intermediate state, $4p$ and $3d$ respectively. In Fig. 23.3, the coincidence 2D energy spectrum for the $N_{23}VV$ transition is reported. It is well known that the Ag $3d$ photoemission peak is sharp in energy and the relative core hole is to be considered as a quasi-stationary one. Hence, in the coincidence 2D-energy-distributions Auger and photoelectrons will pile up in sharp and well defined energy regions [9] and the whole process is well described by a two-step model. On the contrary, the $4p$ photoemission line deviates from the quasi-particle picture and has a linewidth of 13 eV. The 2D-energy-distribution that follows the $4p$ excitation is reported in Fig. 23.3. The photon energy is 118 eV and the dashed lines indicate the position of the maximum sum energy ($E_{max} = E_{left} + E_{right}$). Most of the coincidence intensity is confined within the pair of solid lines. The prominent intensity band has a width of 35 eV, as marked by the arrows, and the otherwise sharp energy separation of the Auger from the photoelectron is lost. Obviously, it does not make sense to speak of an Auger or photoelectron in this decay as energy and momentum are continuously shared between the two final electrons. The fast dynamics of the core ionization process is responsible for them to behave as a strongly correlated pair (one-step process) rather than two independent particles (two-step process). Hence, it is the lifetime of the core intermediate state that dictates for the electron pair to behave either as a single entity or as two independent ones.

A single answer can then be given to the four questions listed at the beginning. Whenever the electron pair generation, be that (*e*,2*e*) or APECS, is well accounted by a one-step model, such as a (*e*,2*e*), the two final electrons will behave as a whole, whereas they behave as separate entities when generated through a two-step process, such as an APECS event involving a Core Valence Valence Auger transition.

Acknowledgements Partial financial support through SIMDALEE2 Sources, Interaction with Matter, Detection and Analysis of Low Energy Electrons 2 Marie Skłodowska Curie FP7-PEOPLE-2023-INT Grant #606988, is greatly acknowledged.

References

1. G. Stefani, R. Gotter, A. Ruocco, F. Offi, F. Da Pieve, S. Iacobucci, A. Morgante, A. Verdini, A. Liscio, H. Yao, R.A. Bartynski, J. Electron Spectrosc. Relat. Phenom. **141**, 149 (2004)
2. R. Gotter, A. Ruocco, A. Morgante, D. Cvetko, L. Floreano, F. Tommasini, G. Stefani, *Nuclear Instruments and Methods in Physics Research Section A: Accelerators, Spectrometers, Detectors and Associated Equipment* (Elsevier, Amsterdam, 2001), p. 1468
3. S. Iacobucci, L. Marassi, R. Camilloni, S. Nannarone, G. Stefani, Phys. Rev. B **51**, 10252(R) (1995)
4. S.A. Keifets, S. Iacobucci, A. Ruocco, R. Camilloni, G. Stefani, Phys. Rev. B **57**, 7360 (1998)
5. F. Da Pieve, S. Di Matteo, D. Sébilleau, R. Gunnella, G. Stefani, C.R. Natoli, Phys. Rev. A **75**, 52704 (2007). https://doi.org/10.1103/PHYS-REVA.75.052704
6. F. Da Pieve, D. Sébilleau, S. Di Matteo, R. Gunnella, R. Gotter, A. Ruocco, G. Stefani, C.R. Natoli, Phys Rev. B **78**, 035122 (2008). https://doi.org/10.1103/PHYS-REVB.78.035122
7. R. Gotter, A. Ruocco, M.T. Butterfield, S. Iacobucci, G. Stefani, R.A. Bartynski, Phys. Rev. B **67**, 033303 (2003)
8. J. Berakdar, S.N. Samarin, R. Hermann, J. Kirschner, Phys. Rev. Lett. **81**, 3535 (1998)
9. Z. Wei, F.O. Schumann, C.H. Li, L. Behnke, G. Di Filippo, G. Stefani, J. Kirschner, Phys. Rev. Lett. **113**, 267603 (2014)

Chapter 24
A Multiple Scattering Approach to the EELS Cross-Section

Didier Sébilleau, Junqing Xu, Rakesh Choubisa and Calogero R. Natoli

Abstract We derive a general spherical wave multiple scattering expression of the EELS cross-section. Contrarily to most of the previous theoretical frameworks, this approach treats all the electrons involved on an equal footing with respect to multiple scattering. The main point in our results is that the cross-section depends not only on Coulomb integrals on the absorbing atom, but also on contributions from all the other atoms. We show that these external contributions should be restricted to neighbouring atoms.

24.1 Introduction

Electron-Energy-Loss Spectroscopy (EELS) is a versatile tool that has already a long history in condensed matter. In this spectroscopy, a beam of electrons is shone on a material where it can loose energy through the Coulomb excitation of electrons. It is usually performed in transmission mode (using for instance a Transmission Electron Microscope – TEM) where it can excite different physical phenomena such as phonons, plasmons or intraband and interband electronic transitions. Therefore,

D. Sébilleau (✉)
Institut de Physique de Rennes (IPR), Univ Rennes, CNRS, UMR 6251, F-35000 Rennes, France
e-mail: didier.sebilleau@univ-rennes1.fr

J. Xu (✉)
Department für Geo- und Umweltwissenschaften, Ludwig Maximilians Universität München, Theresienstrasse 41, 80333 München, Deutschland
e-mail: Junqing.Xu@lrz.uni-muenchen.de

R. Choubisa (✉)
Department of Physics, Birla Institute of Technology and Science, Pilani 333 031, Rajasthan, India
e-mail: rchoubisa@pilani.bits-pilani.ac.in

C. R. Natoli (✉)
Theory Group, INFN-Laboratori Nazionali di Frascati, c. p. 13, 00044, Frascati, Italy
e-mail: calogero.natoli@lnf.infn.it

D. Sébilleau et al. (eds.), *Multiple Scattering Theory for Spectroscopies*, Springer Proceedings in Physics 204,
https://doi.org/10.1007/978-3-319-73811-6_24

EELS can give access to the electronic structure of the material. A second way to perform an EELS experiment is in the reflection mode. In this case, provided the kinetic energy of the electrons involved is rather low, EELS can become extremely sensitive to the local geometry near the surface. This was demonstrated experimentally some years ago [1–3]. However, this approach, although very promising, was not pursued due to the lack of a suitable theoretical framework. Indeed, EELS is generally interpreted within the simple and appealing model developed by Kincaid, Meixner and Platzman [4] where the incoming and the outgoing electrons are treated as plane waves so that multiple scattering is only considered for the ejected electron. This model has enjoyed a widespread popularity because it reduces to an expression of the cross-section which is basically that of X-ray Absorption Spectroscopy (XAS), allowing to compute it with standard XAS codes. However, in the energy range where EELS could become a useful surface probe (energies of the order of 20–200 eV, such as in the available experimental works [1–3]), the neglect of multiple scattering effects in the description of the incoming and outgoing electrons is clearly too strong an approximation. Further models have been developed in order to account for the multiple scattering in the incoming and outgoing beams but they are either approximate (multiple scattering effect treated as a crystal field) [5], or relying on the periodicity of the system [2, 6]. This is why we propose here a new theoretical framework where all electrons are treated on an equal footing and that can be applied to any sort of materials.

24.2 The Generic EELS Cross-Section

The starting point of our approach is the distorted wave Born approximation of the double differential EELS cross-section which can be written as [7]

$$\frac{\mathrm{d}^2\sigma}{\mathrm{d}E\,\mathrm{d}\Omega_{k_{\mathrm{sc}}}} = 4\pi^4 \frac{k_{\mathrm{sc}}}{k_{\mathrm{in}}} \sum_{0,m_c,n} \left| \langle \phi_n \,|\, \langle \varphi^-_{\mathrm{sc}} \,|\, \mathcal{K} \,|\, \phi^0_{L_c} \rangle \,|\, \varphi^+_{\mathrm{in}} \rangle \right|^2 \delta\left(E_0 + E_{\mathrm{in}} - E_n - E_{\mathrm{sc}}\right) \,, \tag{24.1}$$

where $|\,\varphi^+_{\mathrm{in}}\rangle$, $|\,\phi^0_{L_c}\rangle$ and $|\,\varphi^-_{\mathrm{sc}}\rangle$ are respectively the incoming electron state, the core state that is to be excited and the scattered electron state. $|\,\varphi^+_{\mathrm{in}}\rangle$ and $|\,\varphi^-_{\mathrm{sc}}\rangle$ take into account the interaction of the incoming/outgoing electron with the potential V_S of the sample. As the excited electron is not detected, we must sum over all the excited electron states $|\,\phi_n\rangle$ that satisfy the energy conservation $E_n = E_{\mathrm{in}} + E_0 - E_{\mathrm{sc}}$. Here, E in $\mathrm{d}E$ is the energy loss $E_{\mathrm{in}} - E_{\mathrm{sc}}$. Note that here we have supposed all the other electrons to be *spectators* so that we can work solely in a two-electron space.

$\mathcal{K}$ is the potential responsible for the energy loss, i.e. the Coulomb potential with a suitable antisymmetrizer so that the direct and exchange terms in (24.1) are correctly accounted for.

We introduce now the two-electron operator

$$-\frac{1}{\pi}\,|\,\varphi^-_{\mathrm{sc}}\rangle\langle\varphi^-_{\mathrm{sc}}\,| \otimes \Im\left[G^+(E_n)\right] = -\frac{1}{\pi}\,\mathcal{W}\,, \tag{24.2}$$

in order to simplify the notation.

We expand the square modulus in (24.1) to obtain

$$\frac{\mathrm{d}^2\sigma}{\mathrm{d}E\mathrm{d}\Omega_{\boldsymbol{k}_{\mathrm{sc}}}} = -4\pi^3 \frac{k_{\mathrm{sc}}}{k_{\mathrm{in}}} \sum_{0,m_c} \sum_{\gamma,\delta} \sum_{\gamma',\delta'} \langle \varphi_{\mathrm{in}}^{\gamma'+} \mid \langle \phi_{L_c}^0 \mid \left(\mathcal{K}^\dagger \mathcal{W}_{\delta\delta'} \mathcal{K}\right) \mid \phi_{L_c}^0 \rangle \mid \varphi_{\mathrm{in}}^{\gamma+} \rangle \, . \tag{24.3}$$

The incoming state is now labelled $\mid \varphi_{\mathrm{in}}^{\gamma+} \rangle$ to account for the fact that, due to the long-range nature of the Coulomb potential, the incoming electron can be in the vicinity of atom γ when it interacts with the core state $\mid \phi_{L_c}^0 \rangle$ located on atom 0. Likewise, we have indexed the scattering state by δ. The double sum over (δ, γ) and (δ', γ') comes from the fact that the sum over (δ, γ) is *inside* the square modulus in (24.1).

Expression (24.3) differs from the equation previously derived by one of us [8] where the long-range character of the Coulomb interaction was implicitly neglected.

We note the similarity with the XAS cross-section which can be expressed as

$$\sigma_a = -\frac{2m}{\hbar^2} 8\pi\, \alpha\, \hbar\omega_{\boldsymbol{q}} \sum_{0,m_c} \langle \phi_{L_c}^0 \mid \mathcal{K}^\dagger \Im\left[G^+(k_{\mathrm{ex}})\right] \mathcal{K} \mid \phi_{L_c}^0 \rangle \, . \tag{24.4}$$

24.3 The Generalized Cross-Section in a Spherical Wave Representation

We can now use the closure relation of the spherical wave basis as defined in Chap. 1 as

$$\langle \boldsymbol{r} | kL \rangle = k\sqrt{\frac{2}{\pi}} \mathrm{i}^\ell j_\ell(kr) Y_L(\hat{\boldsymbol{r}}) \, , \tag{24.5}$$

and insert it into (24.3) to obtain finally

$$\frac{\mathrm{d}^2\sigma}{\mathrm{d}E\mathrm{d}\Omega_{\boldsymbol{k}_{\mathrm{sc}}}} = \frac{2^7}{\pi^2} \frac{k_{\mathrm{sc}} k_{\mathrm{ex}}}{k_{\mathrm{in}}} \sum_{0,m_c} \sum_{\substack{\beta,L_\beta \\ \alpha,L_\alpha}} \Im\left[A_{m_c L_\alpha}^{0\alpha\,*}(\boldsymbol{k}_{\mathrm{in}}, \boldsymbol{k}_{\mathrm{sc}}) \tau_{L_\alpha L_\beta}^{\alpha\beta}(k_{\mathrm{ex}}) A_{m_c L_\beta}^{0\beta}(\boldsymbol{k}_{\mathrm{in}}, \boldsymbol{k}_{\mathrm{sc}}) \right] , \tag{24.6}$$

where

$$A_{m_c L_\alpha}^{0\alpha}(\boldsymbol{k}_{\mathrm{in}}, \boldsymbol{k}_{\mathrm{sc}}) = \sum_{\gamma,L_\gamma} \sum_{\delta,L_\delta} \left[B_{L_\gamma}^{\gamma\,(+)}(\boldsymbol{k}_{\mathrm{in}}) \left(B_{L_\delta}^{\delta\,(-)}(\boldsymbol{k}_{\mathrm{sc}}^*) \right)^* M_{L_\delta L_\gamma}^{L_\alpha L_c} \right] , \tag{24.7}$$

with the multiple scattering amplitudes on atom α given by

$$B_{L_\alpha}^{\alpha\,(\pm)}(\boldsymbol{k}) = \sum_{i,L_i} \tau_{L_\alpha L_i}^{\alpha i\,\pm}(k)\, Y_{L_i}^*(\hat{\boldsymbol{k}})\, \mathrm{e}^{\mathrm{i}\boldsymbol{k}\cdot\boldsymbol{R}_i} \, . \tag{24.8}$$

τ is the scattering path operator defined in Chap. 1 and $M_{L_\delta L_\gamma}^{L_\alpha L_c}$ is the *regular* Coulomb interaction matrix element.

Here, we have supposed the sample potential V_S to be *real* so that we have no extra term involving *irregular* Coulomb matrix elements.

24.4 The Regular Coulomb Matrix Elements

The first step in the calculation of the Coulomb matrix elements is to replace the Coulomb potential V_C by a *screened* Coulomb potential

$$V_C^s(\boldsymbol{r}_1, \boldsymbol{r}_2) = \frac{e^2}{4\pi\varepsilon_0} \frac{\mathrm{e}^{-\frac{|\boldsymbol{r}_1 - \boldsymbol{r}_2|}{\lambda}}}{|\boldsymbol{r}_1 - \boldsymbol{r}_2|} . \tag{24.9}$$

We recover the Coulomb potential when $\lambda \longrightarrow +\infty$. Otherwise, we can consider the screening constant λ as an adjustable parameter, or set it to a fixed value such as the Thomas–Fermi value $\lambda_{\mathrm{TF}} = \sqrt[3]{\pi/12}\,\sqrt{r_s}$, with r_s in units of the Bohr radius a_0. The reason for such a replacement is that we recognize now in the screened Coulomb potential a quantity proportional to a Green's function

$$V_C^s(\boldsymbol{r}_1, \boldsymbol{r}_2) = -\frac{e^2}{\varepsilon_0}\, G_0^+\left(\boldsymbol{r}_1, \boldsymbol{r}_2; \frac{\mathrm{i}}{\lambda}\right) . \tag{24.10}$$

We can now expand it using the well-known one-site or two-site expansions of Green's functions [9]. This gives finally for the *direct* integrals on the absorbing atom 0 and between atom i and atom 0

$$\left\{\begin{array}{l} D_{L_i \bar{L}_i}^{L_0 L_c}(0) = -\dfrac{N_c e^2}{\lambda \varepsilon_0} \left(\dfrac{\pi}{2}\right)^{\frac{3}{2}} \displaystyle\sum_L \tilde{\rho}(\ell_i, \ell_0 | \ell | \ell_c, \bar{\ell}_i)\; G(L\bar{L}_i | L_i)\; G(L_0 L | L_c) \\ \qquad \text{with the pattern} \qquad \downarrow\ \downarrow\ \downarrow\ \downarrow\ \downarrow \\ \qquad\qquad\qquad\qquad\quad \text{sc}\ \ \text{ex}\ \ \tfrac{1}{\lambda}\ \ \text{co}\ \ \text{in} \\ D_{L_i \bar{L}_i}^{L_0 L_c}(i) = \dfrac{N_c e^2}{\lambda \varepsilon_0} \left(\dfrac{\pi}{2}\right)^{\frac{3}{2}} \displaystyle\sum_{L_i', L_0'} (-1)^{\ell_i'}\, \tilde{G}_{L_i' L_0'}^{i0}\; \rho(\ell_i | \ell_i' | \bar{\ell}_i)\; \rho(\ell_0 | \ell_0' | \ell_c) \times \\ \qquad \text{with the pattern} \qquad \downarrow\ \downarrow\ \downarrow \quad \downarrow\ \downarrow\ \downarrow \\ \qquad\qquad\qquad\qquad\quad \text{sc}\ \ \tfrac{1}{\lambda}\ \ \text{in} \quad \text{ex}\ \ \tfrac{1}{\lambda}\ \ \text{co} \\ \qquad\qquad G(L_i' \bar{L}_i | L_i)\; G(L_0' L_0 | L_c) , \end{array}\right. \tag{24.11}$$

where $\tilde{\rho}(\ell_i, \ell_0|\ell|\ell_c, \bar{\ell}_i)$ and $\rho(\ell_0|\ell_0'|\ell_c)$ are radial integrals. The $\tilde{G}_{L_i' L_0'}^{i0}$ are the matrix elements of the propagator of the Coulomb field (24.10), and $G(L\bar{L}_i|L_i)$ is a Gaunt coefficient (with angular momenta on the right-hand side of | complex-conjugated).

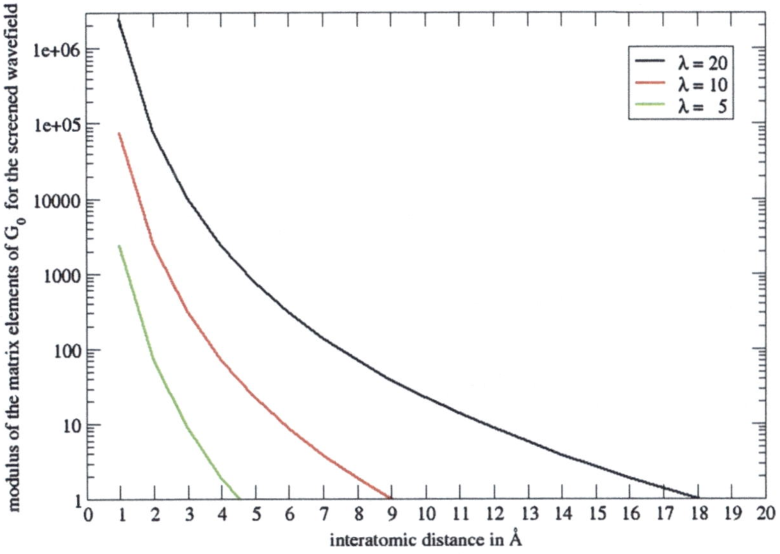

Fig. 24.1 Magnitude of the matrix elements of the Coulomb field for $\ell = 2$ and $m = -2$, and for different values of the screening parameter λ as a function of the interatomic distance.

When we take into account the fact that the core state wave function $\phi^0_{L_c}(\boldsymbol{r})$ in these radial integrals is *localized* on the absorbing atoms, we find the *atomic selection rules*

$$\begin{cases} \beta = 0 \\ i = \gamma = \delta , \end{cases} \tag{24.12}$$

which simplify considerably the expression of the cross-section (24.6). Corresponding results for the exchange matrix elements are derived in a similar way.

The point we want to emphasize here is that a priori *all the atoms* in the cluster contribute to the Coulomb matrix elements in the expression (24.6) of the cross-section. This is a consequence of the long-range nature of the Coulomb potential. However, as we know that in materials Coulomb potentials are screened, let us figure out whether all atoms contribute significantly or just a few shells of neighbours around the absorbing atom 0.

From (24.11), we see that the main difference in magnitude between the diagonal term $D^{L_0 L_c}_{L_i \bar{L}_i}(0)$ and the off-diagonal contributions $D^{L_0 L_c}_{L_i \bar{L}_i}(i)$ lies in the matrix elements of the screened Coulomb field. Figure 24.1 gives an example of the evolution of such a matrix elements as a function of the distance atoms 0 and i for the case $L = (2, -2)$. The y-axis has a logarithmic scale. We see clearly that for different values of the screening parameter λ, the magnitude of the matrix elements drops very quickly to zero. This is a good indication that probably only the very first few shells of neighbours need to be taken into account in the calculation of the EELS cross-section.

References

1. M. De Crescenzi, F. Antonangeli, C. Bellini, R. Rosei, Phys. Rev. Lett. **50**, 1949 (1983)
2. M. De Crescenzi, L. Lozzi, P. Picozzi, S. Santucci, M. Benfatto, C.R. Natoli, Phys. Rev. B **39**, 8409 (1989)
3. L.S. Caputi, O. Comite, A. Amoddeo, G. Chiarello, S. Scalese, E. Colavita, L. Papagno, Phys. Rev. Lett. **77**, 1059 (1996)
4. B.M. Kincaid, A.E. Meixner, P.M. Platzman, Phys. Rev. Lett. **40**, 1296 (1978)
5. J.J.M. Michiels, J.E. Inglesfield, C.J. Noble, V.M. Burke, P.G. Burke, Phys. Rev. Lett. **78**, 2851 (1997)
6. D.K. Saldin, Phys. Rev. Lett. **60**, 1197 (1988)
7. D. Sébilleau, R. Gunnella, Z.-Y. Wu, S. Di Matteo, C.R. Natoli, J. Phys. Condens. Matter Top. Rev. **18**, R175–R230 (2006)
8. D. Sébilleau, Surf. Sci. **482–485**, 729–734 (2001)
9. D. Sébilleau, Phys. Rev. B **61**, 14167 (2000)

Chapter 25
About Spin-Orbit in the Multiple Scattering Theory

Yves Joly

Abstract This paper presents the treatment of the spin-orbit interaction to calculate x-ray spectroscopies. This term is here more specifically developed inside the Multiple Scattering Theory. We give the general paths resulting in a Schrödinger-like couple of equations, whose solutions are given in terms of ℓ, m, σ (spin) and s (solution) quantum numbers. The connection with the multiple scattering amplitudes is then given as well as the more general expression when dealing with complex self-energies. Finally the response functions for x-ray absorption near edge structure is given as it is implemented in the FDMNES code.

25.1 Introduction

When studying material with heavy atoms, it is most often necessary to perform a relativistic calculation. Fully relativistic calculations starting from the Dirac equation contains by itself the spin-orbit interaction term. Because spin-orbit is one of the strongest term in the relativistic extension, and even more importantly because the x-ray magnetic circular dichroism (XMCD) signal is non zero at K edge only because of this term, it is mandatory to include it in the simulation codes.

The purpose of this paper is to present the general paths allowing the ab initio simulation of x-ray spectroscopies taking into account the spin-orbit component of the Hamiltonian for the photoelectron. This theoretical framework is the one used in the FDMNES [1] code which allows such calculation involving absorption or scattering processes at energies near characteristic absorption edges.

In the next section, we show how the relativistic approach is used to treat the spin-orbit term to give first the amplitude of the radial Schrödinger-like equations and second the multiple scattering amplitudes. Section 25.3 treats the specific cases of complex self-energies and Sect. 25.4 gives an example of a response function as it is implemented in the FDMNES code.

Y. Joly (✉)
Inst NEEL, CNRS & Univ. Grenoble Alpes, 38042, Grenoble, France
e-mail: yves.joly@neel.cnrs.fr

D. Sébilleau et al. (eds.), *Multiple Scattering Theory for Spectroscopies*,
Springer Proceedings in Physics 204,
https://doi.org/10.1007/978-3-319-73811-6_25

25.2 Relativistic Approach with Spin-Orbit of MST

To give the response function of core spectroscopies, one needs first to give a convenient description of the electronic states inside the absorbing atoms, then the connection with multiple scattering amplitude can be performed. The important case of complex self-energy is also given.

25.2.1 Solution in the Atom

Following Wood and Boring [2] we explicitly solve only the large components instead of the four determined by the Dirac-Slater calculations. This reformulation gives a couple of Schrödinger like equations, including the spin-orbit effect, closely akin but improving the Pauli equation:

$$(-\Delta + V - E + \Omega)\,\varphi = 0\,, \tag{25.1}$$

with:

$$\Omega = -\frac{\alpha^2}{4}\left((V-E)^2 - B\left[\nabla V \cdot \nabla + \mathrm{i}\boldsymbol{S}\cdot(\nabla V \times \nabla)\right]\right) = 0\,, \tag{25.2}$$

where the potential, $V = V(\boldsymbol{r},\sigma)$, and the wave function, $\varphi = \varphi(\boldsymbol{r},\sigma)$, are spin, σ, dependent. $\boldsymbol{S}$ is a vector whose components are the Pauli matrices, and $B = B(\boldsymbol{r},\sigma) = 1/\left(1 - \alpha^2/4\,(V-E)^2\right)$, α being the fine structure constant and E the photoelectron energy. The last term depending on $\boldsymbol{S}$ and proportional to the potential gradient represents the spin-orbit effect. By its S_x and S_y matrices, it connects the spin up and spin down components of the wave function.

For its radial expression, this equation leads to the coupled system:

$$\begin{aligned} \left(-\frac{\partial^2}{\partial r^2} + G_{O\ell,\frac{1}{2}} + mG_{\mathrm{so},\frac{1}{2}}\right) u_{\ell,m,s,\frac{1}{2}} + s_r G_{\mathrm{so},\frac{1}{2}} u_{\ell,m+1,s,-\frac{1}{2}} &= 0 \\ s_r G_{\mathrm{so},-\frac{1}{2}} u_{\ell,m,s,\frac{1}{2}} + \left(-\frac{\partial^2}{\partial r^2} + G_{O\ell,-\frac{1}{2}} + (m+1)\,G_{\mathrm{so},-\frac{1}{2}}\right) u_{\ell,m+1,s,-\frac{1}{2}} &= 0 \end{aligned} \tag{25.3}$$

where ℓ and m are the usual orbital quantum numbers and we have introduced $s_r = \sqrt{(\ell - m)(\ell + m + 1)}$. When $m = \ell$ and $\sigma = \frac{1}{2}$ or $m + 1 = -\ell$ and $\sigma = -\frac{1}{2}$, the two equations are decoupled and give respectively pure up or pure down spin states. In the other cases the two solutions, indexed by $s = \pm\frac{1}{2}$, contain both, spin down and spin up components, respectively, $u_{\ell,m+1,s,-\frac{1}{2}}$, and, $u_{\ell,m,s,\frac{1}{2}}$. One uses the simple radial dependence for the spin components, $V_\sigma = V_\sigma(r)$ and $B_\sigma = B_\sigma(r)$, valid in the spherically symmetric area inside the atoms. So, we define:

$$G_{O\ell\sigma} = V_\sigma - E + \frac{\ell(\ell+1)}{r^2} - \frac{\alpha^2}{4}(V_\sigma - E)^2 - \frac{\alpha^2}{4}B_\sigma \frac{\partial V_\sigma}{\partial r}\left(\frac{\partial}{\partial r} - \frac{1}{r}\right)$$

$$G_{\mathrm{so}\sigma} = \frac{\alpha^2}{4}B_\sigma \frac{1}{r}\frac{\partial V_\sigma}{\partial r}, \tag{25.4}$$

At the origin, the 2 solutions tend to:

$$u_{\ell,m+\frac{1}{2}-\sigma,s,\sigma}(r) = A_{|s-\sigma|} r^{\sqrt{\left(\ell+\frac{1}{2}+s\right)^2-(\alpha Z)^2}}, \tag{25.5}$$

with $A_0 = \sqrt{\frac{\ell-m}{2\ell+1}}$ and $A_1 = \sqrt{\frac{\ell+m+1}{2\ell+1}}$.

25.2.2 *Normalization of the Radial Wave Functions*

As shown by D. Sébilleau in the introductory chapter, to get the atomic scattering amplitude and the corresponding amplitude of the radial wave function, one just has to use the continuity of the wave function at the muffin radius, R, between the solutions given above inside the atom and the known solution outside the atoms where the potential is constant. When spin-orbit interaction is there, it is slightly more complex, because during the scattering process, spin-flip is possible. We define $\bar{f}_\ell(k_\sigma r) = \sqrt{k_\sigma/\pi}\, f_\ell(k_\sigma r)$, for $f_\ell = j_\ell$ or h_ℓ^+, (the Bessel and outgoing Hankel functions) where the photoelectron wave vector, k_σ, is set spin dependent for more generality. $k_\sigma/\pi = 2\pi\delta_V$ where $\delta_V/2$ is the density of states in vacuum in atomic units per spin, and the remaining factor 4π comes with the angular integral. In the introductory chapter, this normalization is called the *normalization to one state per Rydberg*. One must consider a couple of equations corresponding to the scattering of generalized Bessel functions, $\bar{j}_\ell(k_{s'}r)\, Y_\ell^m \chi_{s'}$ fully spin polarized $s' = \pm\frac{1}{2}$, and giving 2 outgoing generalized Hankel functions $\bar{h}_\ell^+(k_\sigma r)\, Y_\ell^{m+\frac{1}{2}-\sigma}\chi_\sigma$, including thus a spin-flip term. The continuity gives:

$$\sum_{\sigma\sigma'} a_{\ell,m+\frac{1}{2}-s'}\, b_{\ell,m+\frac{1}{2}-\sigma,\sigma',\sigma}(R)\,\chi_\sigma = \bar{j}_\ell(k_{s'}R)\,\chi_{s'} - \mathrm{i}\times$$
$$\sum_s t_{\ell,m+\frac{1}{2}-s,s}^{\ell,m+\frac{1}{2}-s',s'}\, \bar{h}_\ell^+(k_s R)\,\chi_s, \tag{25.6}$$

where $t_{\ell,m+\frac{1}{2}-s,s}^{\ell,m+\frac{1}{2}-s',s'}(E)$ are the atomic scattering amplitudes, χ_s is the spin projector, and we used $b(E,r) = u(E,r)/r$. The normalization corresponds to a change in the solution basis to get the radial function amplitudes equal to the atomic scattering amplitudes, as in the case without spin-orbit, but for both spin components. Its interest in MST is that it remains true when the atom is embedded in a cluster, and thus for the multiple scattering amplitude. *Mutatis mutandis* using:

$$\Delta_{f_\ell} = W\left(\bar{f}_\ell, b_{m,\frac{1}{2},\frac{1}{2}}\right) W\left(\bar{f}_\ell, b_{m+1,-\frac{1}{2},-\frac{1}{2}}\right) - W\left(\bar{f}_\ell, b_{m,-\frac{1}{2},\frac{1}{2}}\right) W\left(\bar{f}_\ell, b_{m+1,\frac{1}{2},-\frac{1}{2}}\right) , \tag{25.7}$$

where W is the Wronskian, the atomic scattering amplitudes are given by:

$$t^{\ell,m+\frac{1}{2}-s',s'}_{\ell,m+\frac{1}{2}-s,s} = \frac{4s'i}{\Delta_{h^+_\ell}} \sum_\sigma \sigma W\left(\bar{j}_\ell, b_{\ell,m+\frac{1}{2}-s,\sigma,s}\right) W\left(\bar{h}^+_\ell, b_{\ell,m-\frac{1}{2}+s',-\sigma,-s'}\right) , \tag{25.8}$$

whereas the normalized radial wave functions are given by:

$$\bar{b}_{\ell,m+\frac{1}{2}-\sigma,s,\sigma} = -\frac{4s}{\pi R^2 \Delta_{j_\ell}} \sum_{\sigma'} \sigma' W\left(\bar{j}_\ell, b_{\ell,m+\frac{1}{2}-s,\sigma',s}\right) b_{\ell,m-\frac{1}{2}+\sigma,-\sigma',-\sigma} . \tag{25.9}$$

25.2.3 Relation with Green's Function and Multiple Scattering Amplitudes

When the atom is embedded inside a cluster, the projection of the electronic states inside the atom up to a radius where we supposed the potential spherically symmetric can be described by an expansion in spherical harmonics. Using the (ℓ, m, s, σ) quantum numbers, plus a change in index, $m \rightarrow m - \frac{1}{2} + \sigma$, the wave functions for any state n can be written as:

$$\varphi_n(\boldsymbol{r}, E) = \sum_L a^n_L(E) R_L(\boldsymbol{r}, E) , \tag{25.10}$$

with $L = (\ell, m + \sigma - s, s)$ or (ℓ, m, s, σ), $a^n_L = a^n_{\ell,m+\sigma-s,s}$ and $R_L = b_{\ell m s \sigma} Y^m_\ell \chi_\sigma$. When spin-orbit can be neglected, solution and spin indexes merge, that is $s = \sigma$.

Here we generalize the MST formulation giving the relation between the amplitudes a^n_L, and the Green's function. When self-energy and potential are real, the latter is given by:

$$G^+\left(\boldsymbol{r}, \boldsymbol{r}'\right) = \sum_{LL'} G^+_{LL'}\left(\boldsymbol{r}, \boldsymbol{r}'\right) = -\pi \sum_{LL'} R_L \tau_{LL'} R^\dagger_{L'} , \tag{25.11}$$

where symbol † means complex conjugate on the spherical harmonics but not on the radial wave function. With the normalization done above, and thanks to the optical theorem, this relation can be written:

$$\sum_n \left|a_L^n R_L\right\rangle\left\langle a_{L'}^n R_{L'}\right| \delta\left(E-E_n\right) = \frac{\mathrm{i}}{2\pi}\left(G_{LL'}^{+} - G_{L'L}^{+\dagger}\right)$$
$$= -\frac{\mathrm{i}}{2} R_L \left(\tau_{LL'} - \tau_{L'L}^{*}\right) R_{L'}^{\dagger}\,, \qquad (25.12)$$

where $\delta\left(E-E_n\right)$ is the Kronecker symbol (and not the density of states) to stress that the summation is over the states with energy E. Note that when neglecting spin-orbit (and thus $s=\sigma$), τ is symmetric and diagonal in spin, and one recovers:

$$\sum_n a_{\ell,m,\sigma}^n a_{\ell',m',\sigma}^{n*} \delta\left(E-E_n\right) = -\Im\left[\tau_{\ell,m,\sigma}^{\ell',m'}\right]. \qquad (25.13)$$

25.3 Complex Energy

It is often necessary to use complex energy, to simulate the finite lifetime of both the core-hole and photoelectron states. Another and equivalent way is to make a first part of the calculation with a real energy and make afterward the broadening with a convolution by a Lorentzian. Nevertheless, when calculating at negative energy or when the electronic states and very localized in energy, such way would need extremely small energy step. It is then better to use a formal width, typically 0.1 eV obtained by an imaginary part of the self-energy of the same value.

We know, that in this case, the second solution, of the radial Schrödinger equation, the so-called irregular one, does not diverge anymore [3] at the atomic core, becoming then physical and must be taken into account, as it can be seen in the expression of the Green's function:

$$\frac{1}{\pi}G_{LL'}^{+}\left(\boldsymbol{r},\boldsymbol{r}'\right) = -R_L\tau_{LL'}R_L'^{\dagger} + \delta_{LL'}R_L\left(\boldsymbol{r}_<\right)S_{L'}^{\dagger}\left(\boldsymbol{r}_>\right) \qquad (25.14)$$
$$= -R_L\left(\tau_{LL'} - \delta_{LL'}t_{LL'}\right)R_{L'}^{\dagger} - \delta_{LL'}R_L\left(\boldsymbol{r}_<\right)t_{LL'}U_{L'}^{\dagger}\left(\boldsymbol{r}_>\right)\,,$$

where $\boldsymbol{r}_{<(>)}$ means the lesser (greater) between $\boldsymbol{r}$ and $\boldsymbol{r}'$. S_L and U_L are the irregular solutions obtained by inward integration from the atom sphere radius R matching respectively $\bar{j}_\ell\,(kr)\,Y_\ell^m$ and $-\mathrm{i}\bar{h}_\ell^{+}\,(kr)\,Y_\ell^m$ for both spin. The integration is m and s dependent and couples $\left(m, \frac{1}{2}\right)$ with $\left(m+1, -\frac{1}{2}\right)$ orbitals as for the regular solutions.

25.4 Implementation in the FDMNES Code

25.4.1 The FDMNES *Code*

The FDMNES code can calculate spectra involving one photon processes such as X-ray Absorption Near Edge Structure (XANES) and X-ray Emission Spectroscopy

(XES) as well as polarization dependent ones such as XMCD, X-ray Magnetic Linear Dichroism (XMLD), X-ray Natural linear or circular dichroism (XNLD or XNCD). It can also calculate spectra involving two photon processes like Resonant Elastic X-ray Scattering (REXS also called RXD or DANES) as well as Non-resonant Inelastic X-ray Scattering (NRIXS or Raman X-ray Scattering).

To solve the states probed by the photoelectron, FDMNES uses two different DFT techniques: the multiple scattering theory (MST) explained in the present book and the Finite Difference Method (FDM) [4]. Both are here real space techniques applied to a cluster surrounding the absorbing atom. In FDMNES, FDM is more precise that MST because the latter is used within the muffin-tin approximation in which the potential is taken as spherically symmetric around the atoms and constant between them, the full-potential MST approach developed by Hatada and Natoli [5] (seen in Chap. 3) being not included. The availability of both techniques within the same code offers common grounds for the comparison of their respective performances, speediness for MST and precision for FDM.

25.4.2 Formula for Spectroscopy

FDMNES uses the formalism given above to get the response functions for the different spectroscopies. For example the cross section in XANES is given by:

$$\begin{aligned} \sigma &= 4\pi^2\,\alpha\,\hbar\omega_{\boldsymbol{q}} \sum_{ng} |\langle n\,|o|\,g\rangle|^2\,\delta\,(E-E_n) \\ &= 4\pi^2\,\alpha\,\hbar\omega_{\boldsymbol{q}} \sum_{LL'g} \left\langle g\,|o^*|\,\frac{\mathrm{i}}{2\pi}\left(G^{+}_{LL'} - G^{+\dagger}_{L'L}\right)|o|\,g\right\rangle , \end{aligned} \tag{25.15}$$

where $\hbar\omega_{\boldsymbol{q}}$ is the photon energy. $o = \boldsymbol{e}_{\boldsymbol{q}} \cdot \boldsymbol{r}\,(1 + \mathrm{i}/2\,\boldsymbol{q}\cdot\boldsymbol{r})$, at second order, is the transition operator, $\boldsymbol{e}_{\boldsymbol{q}}$ and $\boldsymbol{q}$ being here the photon polarization and wave vector.

Acknowledgements Calogero R. Natoli is especially thanked for its teaching on MST, used in the FDMNES code. Particularly, he discovered independently from Wood and Boring the presented approach of spin-orbit and my knowledge on this subject directly comes from him.

References

1. O. Bunau, Y. Joly, J. Phys. Condens. Matter. **21**, 345501 (2009). http://www.neel.cnrs.fr/fdmnes
2. J.H. Wood, A.M. Boring, Phys. Rev. B **18**, 2701 (1978)
3. T.A. Tyson, K.O. Hodgson, C.R. Natoli, M. Benfatto, Phys. Rev. B **46**, 5997 (1992)
4. Y. Joly, Phys. Rev. B **63**, 125120 (2001)
5. K. Hatada, K. Hayakawa, M. Benfatto, C.R. Natoli, J. Phys. Condens. Matter **22**, 185501 (2010)

Chapter 26
Implementation of Exact Diagonalization in KKR+DMFT

Ján Minár, Igor Di Marco and Jindřich Kolorenč

Abstract We describe an implementation of the LDA+DMFT method in the SPR-KKR code. The auxiliary impurity model that is at the heart of the dynamical mean-field theory is solved by iterative diagonalization (the Lanczos method). We illustrate that the implemented scheme accurately models the electronic structure of Mott insulators, exemplified here by NiO.

26.1 Introduction

Complex oxides are among the most interesting systems in condensed matter physics due to the intricate interplay of strong Coulomb repulsion, hybridisation with ligand states, and crystal field [1]. In particular, the electronic structure of transition-metal monoxides has been the subject of intense experimental and theoretical investigation for a long time [2]. The surge of interest arose when it stood clear that one-electron band-structure calculations predict these oxides to be metallic, whereas experiments show that they are insulators. During the last decades, much progress has been made in developing advanced methods for electronic-structure calculations that offer an adequate description of the transition-metal monoxides. Self-interaction correction [3], GW [4], finite clusters [5], and dynamical mean-field theory (DMFT) [1] are among the most successful approaches. Especially DMFT has been applied to several

J. Minár (✉)
New Technologies-Research Center, University of West Bohemia, Univerzitní 8, 306 14 Plzeň, Czech Republic
e-mail: jminar@ntc.zcu.cz

I. Di Marco
Division of Materials Theory, Department of Physics and Astronomy, Uppsala University, Box 516, 75120 Uppsala, Sweden
e-mail: igor.dimarco@physics.uu.se

J. Kolorenč
Institute of Physics, Czech Academy of Sciences, Na Slovance 2, 182 21, Praha 8, Czech Republic
e-mail: kolorenc@fzu.cz

D. Sébilleau et al. (eds.), *Multiple Scattering Theory for Spectroscopies*, Springer Proceedings in Physics 204,
https://doi.org/10.1007/978-3-319-73811-6_26

transition-metal monoxides by different groups and with great success (see [6] and references within).

In Chap. 4 of this book, a detailed description of the Korringa–Kohn–Rostoker (KKR) method and its combination with DMFT is presented. The KKR method for the electronic-structure calculations is based on the multiple scattering theory, and addresses the Kohn–Sham problem directly in terms of the one-electron Green's function $G(\boldsymbol{r}, \boldsymbol{r}', E)$. The KKR implementation of the LDA+DMFT scheme is built on the solution of a Dyson-like equation which includes a self-energy $\Sigma(\boldsymbol{r}, \boldsymbol{r}', E)$ [7]. This is done by solving the single-site scattering Schrödinger (or Dirac) problem. In this way, we account for the general non-local, site-diagonal, complex and energy-dependent self-energy $\Sigma(\boldsymbol{r}, \boldsymbol{r}', E)$ already when calculating the basis functions. It allows us to directly exploit all the advantageous features of the KKR Green's function method when performing LDA+DMFT calculations. Consequently, one can investigate electronic correlation effects in a wide range of systems [8].

So far, our implementation was applicable only to moderately correlated systems, for instance to transition metal alloys, since the auxiliary impurity model arising in DMFT was solved by methods based on the perturbation theory, such as SPTF [9]. The reason for the limitation to the perturbative methods is the fact that other impurity solvers usually provide the self-energy either at real-energy points or at (complex) Matsubara frequencies. In the same time, the energy integration of $G(\boldsymbol{r}, \boldsymbol{r}', E)$, which is used for obtaining the charge density, is done along a particular contour in the complex-energy plane. Therefore, the self-energy has to be analytically continued from the points provided by the solver to this integration contour, which is possible only if the Green's function has good analytic properties. Unfortunately, this is not always true. For instance, quantum Monte Carlo techniques generally lead to numerically noisy self-energies and hence their combination with KKR method has not been possible.

In the following, we present a recent implementation of the exact diagonalization impurity solver into the KKR+DMFT method. This approach is suitable for description of strongly correlated systems. As an illustration, we apply the new implementation to the electronic structure of NiO.

26.2 Formalism

Even though the KKR method is based on Green's functions, it is more didactic to outline the LDA+DMFT scheme in the standard Hamiltonian formalism. The LDA+DMFT method is based on the assumption that the LDA band structure, which we denote as H_{LDA}, represents a non-interacting lattice model. Then, the mean-field terms corresponding to the Coulomb interaction among the $3d$ electrons are replaced with explicit two-body interaction vertices. Here we use the spherically symmetric Coulomb vertex U_{sph} and the removed mean-field term, usually labelled as double-counting correction U_{DC}, is approximated in the so-called fully localised limit [10, 11]. This construction leads to an effective Hubbard model $H_{\mathrm{Hub}} = H_{\mathrm{LDA}} + \sum_n \left(U^n_{\mathrm{sph}} - U^n_{\mathrm{DC}}\right)$, where n runs over the transition-metal atoms.

The Hubbard model is solved using DMFT, that is, the many-body effects are taken into account only locally, separately for each of the 3d shells, by means of a momentum-independent self-energy Σ. This self-energy is evaluated in an auxiliary impurity model that consists of one fully interacting 3d shell (the impurity) embedded in a self-consistent non-interacting medium $H_{\mathrm{MF}} = H_{\mathrm{LDA}} + \sum_n \left(\Sigma^n - U^n_{\mathrm{DC}}\right)$ [12].

In our implementation, the auxiliary model is discretized and solved by the Lanczos method [13]. The discrete model can be written as $H_{\mathrm{imp}} = H'_{\mathrm{imp}} + U_{\mathrm{sph}}$ where the non-interacting part H'_{imp} reads as

$$H'_{\mathrm{imp}} = \sum_{ij} H^{ij}_{\mathrm{loc}}\, d^\dagger_i d_j + \sum_{IJ} H^{IJ}_{\mathrm{bath}}\, b^\dagger_I b_J + \sum_{iJ} V^{iJ}\, d^\dagger_i b_J + \mathrm{h.c.} \tag{26.1}$$

The lower-case indices label the d orbitals, the upper-case indices label orbitals of the effective medium that is usually referred to as bath. We limit the bath to contain only 10 orbitals, and therefore the local hamiltonian H_{loc}, the bath hamiltonian H_{bath} as well as the hybridisation V are all 10×10 matrices. The discretized model roughly corresponds to a finite cluster that includes one 3d shell and those linear combinations of the nearest-neighbour ligand orbitals that hybridise with the 3d shell [14]. Despite being relatively small, such finite models were shown to accurately represent the local electronic structure in insulating oxides [13, 14].

The parameter matrices H_{loc}, H_{bath} and V entering (26.1) are determined by a procedure outlined in [13]. The main idea is to match the asymptotic expansion of the d-electron Green's function G_{loc}, evaluated in the impurity model H'_{imp}, to the asymptotic expansion of the d-electron Green's function $\mathcal{G}$, corresponding to the self-consistent medium H_{MF} [12, 13, 15]. The expansion of G_{loc} starts as $G_{\mathrm{loc}}(z) = 1/z + H_{\mathrm{loc}}/z^2 + \cdots$ where z is the complex energy. To find the matching expansion of $\mathcal{G}$, one needs to calculate the first few moments of $\mathcal{G}(z)$ along the real axis, namely $M_m = \int E^m \mathcal{G}(E + \mathrm{i}0)\,\mathrm{d}E$ for $0 \le m \le 3$. In practice, these integrals are evaluated along a semi-circular contour depicted in Fig. 26.1.

The bath Green's function $\mathcal{G}$ is calculated as a projection of the KKR Green's function $G(\boldsymbol{r}, \boldsymbol{r}', E)$ onto a local basis $\phi_L(\boldsymbol{r}, E_{\mathrm{ref}})$. In the case of NiO, we used d-like regular single-site solutions of the Kohn–Sham–Dirac equations normalised to one. This basis function is calculated at a reference energy E_{ref} for the magnetic field set to zero in the relativistic case. In the full-potential calculations, the non-spherical parts of the potential have to be suppressed in the definition of the local orbitals $\phi_L(\boldsymbol{r}, E_{\mathrm{ref}})$. The choice of the reference energy E_{ref} turned out to be crucial for stability of the exact-diagonalization scheme. In the previous applications of KKR+DMFT, E_{ref} was calculated from the centre of mass of the d band. In order to get the correct number of localised d electrons, we alternatively propose to use an energy for which the single-site phase shift shows a resonance. This choice leads to a reduction of the wave-function tail extending into neighbouring atoms.

During the self-consistency cycle, $\mathcal{G}$ is calculated on two semi-circular contours shown in Fig. 26.1. When using the exact-diagonalization scheme, the analytical

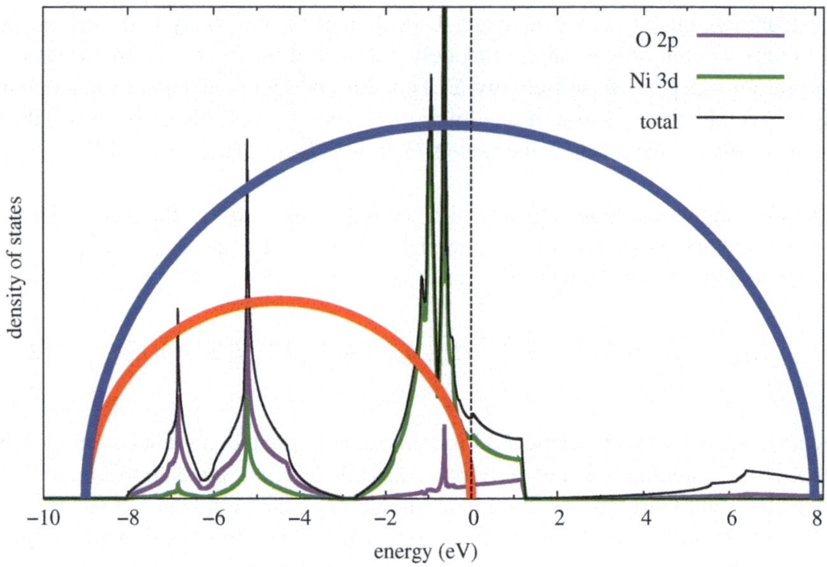

Fig. 26.1 The *contour* in the complex-energy plane that is used for integration of the moments M_m (*blue line*), and its position relative to the single-particle spectrum (represented by the total and partial LDA densities of states of NiO). This contour has to encircle *all* 3*d* states and hence it differs from the integration path that is used to evaluate the number of occupied states (*red line*)

continuation, like the Padé method [16], is avoided. In addition, the determination of the matrices H_{loc}, H_{bath} and V directly from the moments of $\mathcal{G}$ can be efficiently done also if the spin-orbital coupling is considered. In that case, the matrices become non-diagonal which causes the usual fitting schemes to fail.

26.3 Application to NiO

The electronic structure of NiO is a typical example of the failure of LDA to produce an insulating paramagnetic ground state. The inclusion of the antiferromagnetic order leads to a small band gap, but it is still far from the experimental value. In Fig. 26.2 we present the total density of states of NiO calculated by means of the full-potential KKR method [17]. The parameters of the Coulomb interaction in the 3*d* shells are $U = 8\,\text{eV}$ and $J = 1.0\,\text{eV}$. The static, energy-independent self-energy that one obtains in LSDA+U is sufficient for the band gap to match the experimental data. Nevertheless, it is clear from Fig. 26.2 that several features of the photoemission and inverse photoemission experiments can not be properly described in this approach. Inclusion of the local dynamical correlations leads to a much improved description. In particular, the satellite structures seen at about $-10\,\text{eV}$ are properly reproduced by the LDA+DMFT method. Additionally, no artificial magnetic order

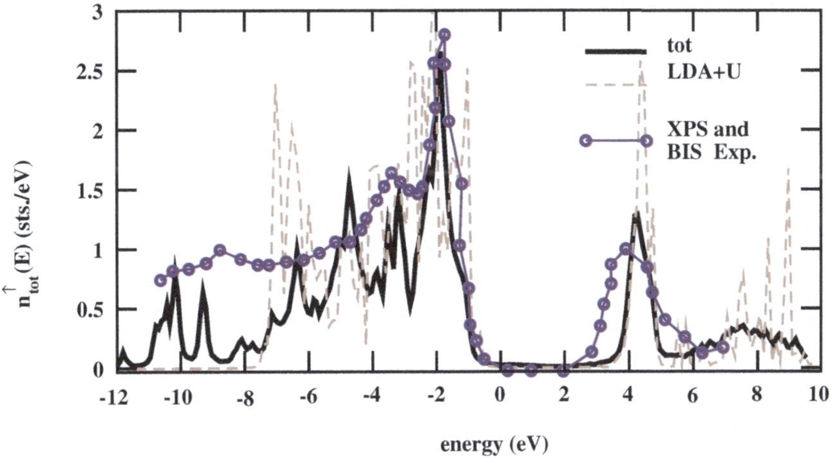

Fig. 26.2 The total density of states of NiO compared with the experimental angle-integrated photoemission and inverse photoemission [21]

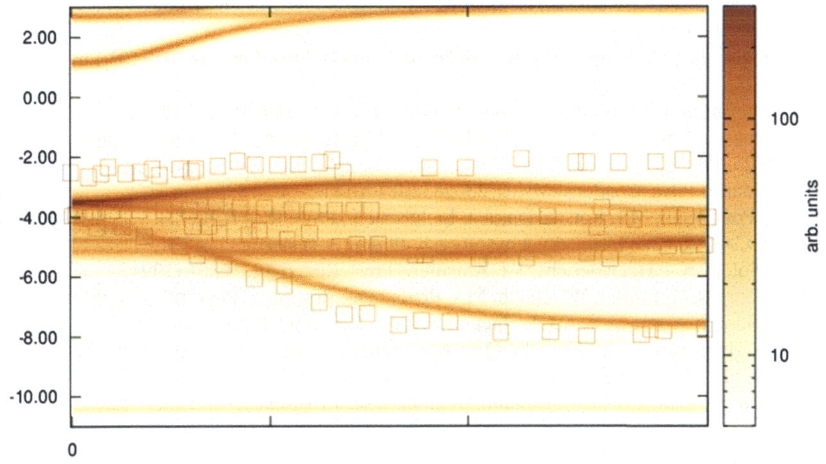

Fig. 26.3 Bloch spectral function of NiO along Γ–X direction calculated by self-consistent KKR+DMFT method. The experimental data measured by the angle-resolved photoemission are represented by *squares* [22]

is required. In Fig. 26.3, the Bloch spectral function of NiO is shown and compared to the corresponding data from angle-resolved photoemission experiments. We find that the experimental points along Γ–X direction are located at energies that are well reproduced by our LDA+DMFT scheme. Our results are in a very good agreement with previous LDA+DMFT calculations that employed the exact-diagonalization solver [6] and the quantum Monte Carlo solver [18]. All these results show that the theoretical framework outlined above can be used to successfully reproduce all features of the valence band of NiO, including the satellites at high binding energy. Thus, the present calculations provide a basis for a future study of the angle-resolved photoemission spectra of NiO by means of the one-step model [19, 20]. Such an investigation, which would include matrix elements, final-state effects and surface effects is still missing in modern literature.

Acknowledgements J.K. acknowledges financial support from the Czech Science Foundation under the grant number 15-05872J. We thank the DFG for financial support via FOR1346. J.M. was supported by the CEDAMNF project (CZ.02.1.01/0.0/0.0/15_003/0000358), co-funded by the ERDF as part of the OP RDE program of the Ministry of Education, Youth and Sports (Czech Republic).

References

1. K. Held, Adv. Phys. **56**, 829 (2007)
2. B. Brandow, Adv. Phys. **26**, 651 (1977)
3. Z. Szotek, W.M. Temmerman, H. Winter, Phys. Rev. B **47**(7), 4029 (1993)
4. F. Aryasetiawan, O. Gunnarsson, Phys. Rev. Lett. **74**(16), 3221 (1995). https://doi.org/10.1103/PhysRevLett.74.3221
5. A. Fujimori, F. Minami, Phys. Rev. B **30**, 957 (1984). https://doi.org/10.1103/PhysRevB.30.957
6. P. Thunström, I. Di Marco, O. Eriksson, Phys. Rev. Lett. **109**, 186401 (2012)
7. J. Minár, L. Chioncel, A. Perlov, H. Ebert, M.I. Katsnelson, A.I. Lichtenstein, Phys. Rev. B **72**, 045125 (2005)
8. J. Minár, J. Phys. Condens. Matter **23**, 253201 (2011)
9. L.V. Pourovskii, M.I. Katsnelson, A.I. Lichtenstein, Phys. Rev. B **72**(11), 115106 (2005)
10. M.T. Czyżyk, G.A. Sawatzky, Phys. Rev. B **49**, 14211 (1994)
11. I.V. Solovyev, P.H. Dederichs, V.I. Anisimov, Phys. Rev. B **50**, 16861 (1994)
12. A. Georges, G. Kotliar, W. Krauth, M.J. Rozenberg, Rev. Mod. Phys. **68**, 13 (1996)
13. J. Kolorenč, A.B. Shick, A.I. Lichtenstein, Phys. Rev. B **92**, 085125 (2015)
14. M.W. Haverkort, M. Zwierzycki, O.K. Andersen, Phys. Rev. B **85**, 165113 (2012)
15. Q. Si, M.J. Rozenberg, G. Kotliar, A.E. Ruckenstein, Phys. Rev. Lett. **72**, 2761 (1994)
16. J. Schött, I.L.M. Locht, E. Lundin, O. Grånäs, O. Eriksson, I. Di Marco, Phys. Rev. B **93**, 075104 (2016)
17. H. Ebert, D. Ködderitzsch, J. Minár, Rep. Prog. Phys. **74**(9), 096501 (2011)
18. J. Kuneš, V.I. Anisimov, S.L. Skornyakov, A.V. Lukoyanov, D. Vollhardt, Phys. Rev. Lett. **99**, 156404 (2007)
19. J. Braun, Rep. Prog. Phys. **59**, 1267 (1996)
20. J. Minár, J. Braun, H. Ebert, J. Electron. Spectrosc. Relat. Phenom. **189**, 129 (2013)
21. G.A. Sawatzky, J.W. Allen, Phys. Rev. Lett. **53**(24), 2339 (1984)
22. Z.X. Shen, R.S. List, D.S. Dessau, B.O. Wells, O. Jepsen, A.J. Arko, R. Barttlet, C.K. Shih, F. Parmigiani, J.C. Huang, P.A.P. Lindberg, Phys. Rev. B **44**, 3604 (1991)

Chapter 27
Cumulant Approach for Inelastic Losses in X-ray Spectra

John J. Rehr and Joshua J. Kas

27.1 Introduction

The theory of inelastic losses in x-ray spectra has long been of interest. These losses include plasmons, particle-hole pairs, Debye–Waller factors, and other many-body effects that damp and shift the spectra. The losses give rise to additional features known as satellites that are not present in independent electron or quasi-particle (QP) approaches. These features lead to a path- and energy-dependent many-body amplitude reduction factor $S_0^2(k, R)$ in the extended x-ray absorption fine structure (EXAFS) [1],

$$\chi(k) = \sum_R \frac{|f_{\rm eff}(k)|}{kR^2} \sin(2kR + \Phi_k)|S_0(k, R)|^2 e^{-2R/\lambda_k} e^{-2\sigma^2 k^2} \,. \tag{27.1}$$

Typically $|S_0^2| \approx 0.8 - 0.9$. Formally such effects can be treated with many-body techniques such as CI, DMFT, or the multi-channel multiple-scattering formalism of Natoli et al. [2], but these methods are all computationally intensive. As a result, these effects are usually neglected on the belief that the error is small or only contributes a smooth background, and hence conventional theories of x-ray spectra are typically only semi-quantitative.

Two classes of inelastic losses have been identified: (i) *intrinsic* losses (the static part of S_0^2) from excitations due to the sudden creation of a core hole, such as shake-up, shake-off excitations; and (ii) *extrinsic* losses (including the mean-free path term λ_k), which arise from similar excitations during the propagation of the photo-electron.

J. J. Rehr (✉) · J. J. Kas
University of Washington, Seattle, WA 98195-1560, USA
e-mail: jjr@uw.edu

J. J. Kas
e-mail: hebhop@uw.edu

D. Sébilleau et al. (eds.), *Multiple Scattering Theory for Spectroscopies*,
Springer Proceedings in Physics 204,
https://doi.org/10.1007/978-3-319-73811-6_27

Interference between these losses is also important. The extrinsic losses are often approximated in terms of an inelastic mean free path λ_k, which is related to the imaginary part of the photoelectron self-energy [3, 4].

Recently a new approach has been developed to treat these effects based on a particle-hole cumulant expansion and the quasi-boson approximation. Remarkably the method can also account for effects of vibrations and other neutral (bosonic) excitations such as charge-transfer satellites. We refer to the full paper for more details [5]. Here we briefly summarize the main results and explain how this approach can be used to calculate the many-body amplitude reduction factor S_0^2.

Briefly, the cumulant expansion for the one particle Green's function $g(t)$ is an alternative to the Dyson equation, and is based on an exponential representation in the time-domain $g(t) = g^0(t)\mathrm{e}^{C(t)}$, where $g^0(t)$ is the non-interacting Green's function and $C(t)$ is the cumulant. This expansion was treated extensively by Hedin and collaborators [6, 7], and a new derivation based on a functional differential equation has recently been developed [8, 9]. However, the single-particle Green's function alone is inadequate to describe x-ray spectra, which involves the simultaneous creation of both a particle and a hole. Consequently a generalization to an analogous exponential representation for the *particle-hole Green's function* is needed, $G_K(t) = G_K^0(t)\mathrm{e}^{\tilde{C}_K(t)}$, where the particle-hole cumulant $\tilde{C}_K(t)$ is calculated to second order in the couplings to the excitations in the system. The structure of G_K is related to the effective Green's function for x-ray spectra introduced by Campbell et al. [10], transformed to the time-domain. Here $K = (c, k)$ labels the transition from a given core-level $|c\rangle$ to a photoelectron state $|k\rangle$. A related cumulant model for the 2-particle Green's function has recently been introduced by Zhou et al. [9].

The real-time representation of $G_K(t)$ considerably simplifies the theory. The formalism leads directly to an expression for the many-body XAS $\mu(\omega)$ at photon energy $\hbar\omega$ as a convolution of the spectrum calculated in the presence of a static core hole with an effective particle-hole spectral function $A_K(\omega)$

$$\mu(\omega) = \int \mathrm{d}\omega'\, A_{K'}(\omega')\mu^0(\omega - \omega')\,. \tag{27.2}$$

Here $\mu^0(\omega)$ is the independent-particle XAS calculated in the presence of a core-hole, $K' = (c, k(\omega - \omega'))$, and $A_{K'}(\omega) = -(1/\pi)\Im\,[G_{K'}(\omega)]$. We now discuss how a similar convolution [(49) in [10]] over the XAS fine structure $\chi_K(\omega)$ yields $S_0^2(R)$. Effects of thermal vibrations and disorder can be included implicitly by averaging over the structural variations. Convolutions similar to that in (27.2) have also been used to incorporate inelastic losses in the XPS photocurrent $J_k(\omega)$ [6, 11, 12]. Inelastic losses beyond the independent-particle approximation are embedded in the cumulant $\tilde{C}_K(t)$. Partitioning the cumulant into intrinsic, extrinsic, and interference terms then facilitates practical calculations. This factorization of the particle-hole Green's function G_K is analogous to that in the classic treatment of the x-ray edge singularities by Nozières and de Dominicis [13]. Similarly, our generalized treatment also accounts for edge-singularities from low-energy particle-hole excitations in metals [5].

27.2 Particle-Hole Spectral Function

The generalized cumulant $\tilde{C}_K(t)$ is obtained by transforming equation (32) of Campbell et al. to the time-domain, and matching the leading terms in powers of the quasi-boson coupling constants,

$$\tilde{C}_K(t) = \int d\omega\, \gamma_K(\omega)(\mathrm{e}^{\mathrm{i}\omega t} - \mathrm{i}\omega t - 1)\ . \tag{27.3}$$

Practical calculations of the kernel γ_K are based on a partition of $C_K(t)$ into intrinsic (c), extrinsic (k), and interference terms (kc), respectively, i.e.,

$$\tilde{\gamma}_K(\omega) = \gamma_c(\omega) + \gamma_k(\omega) + \gamma_{ck}(\omega), \tag{27.4}$$

$$\tilde{C}_K(t) = C_c(t) + C_k(t) + C_{ck}\ . \tag{27.5}$$

The Landau representation [14] of (27.3) ensures that the particle-hole spectral function

$$\tilde{A}_K(\omega) = -\frac{1}{\pi}\Im\left[\int \mathrm{d}t\, \mathrm{e}^{\mathrm{i}\omega t}\, \tilde{G}^0_K(t)\mathrm{e}^{\tilde{C}_K(t)}\right], \tag{27.6}$$

remains normalized with an invariant centroid. Thus the effect of the bosonic excitations is a transfer of spectral weight from the main peak to the satellites while the overall strength is conserved. Note that lifetime broadening due to the photoelectron interactions is included naturally, while the core-hole lifetime is included by adding a damping term, $-\Gamma_c|t|$, to the cumulant. This representation is similar to that in the treatment of inelastic losses in XPS [6, 11, 15].

27.3 EXAFS Reduction Factor S_0^2

In the usual MS theory, [16] the XAFS spectrum $\chi^{(0)}(\omega)$ is a rapidly varying energy dependent factor in the one-particle expression for the x-ray absorption,

$$\mu^{(0)}(\omega) = \mu_0^{(0)}(\omega)[1 + \chi^{(0)}(\omega)]\ , \tag{27.7}$$

where $\mu_0^{(0)}$ is the generally smooth absorption from the central embedded atom alone, in the absence of MS. The many body XAFS function $\chi(\omega) = (\mu - \mu_0)/\mu_0$ then becomes

$$\chi(\omega) \approx \int \mathrm{d}\omega' \tilde{A}(\omega, \omega')\chi^{(0)}(\omega - \omega')\ , \tag{27.8}$$

where $\tilde{A}(\omega, \omega') = A_K(\omega')$ with $\omega = (1/2)k^2 - E_c$. If interference is neglected, the particle-hole Green's function would simply be a product of the core-hole Green's function $g_c(t) = g_c^0(t)e^{C_c(t)}$, and the damped final state Green's function in the presence of a core hole $\tilde{g}_k(t) = \tilde{g}_k^0(t)e^{C_k(t)}$. This approximation would imply that the intrinsic and extrinsic losses are independent and additive, but that yields XAS satellite strengths that are generally too large. Consequently the interference terms are essential. They provide an energy dependence which tends to cancel the extrinsic and intrinsic losses near threshold, due to the opposite signs of the hole and photoelectron charges, while at very high energies only the intrinsic losses remain. This difference is characterized as an adiabatic to sudden transition, and can be used to justify the adiabatic approximation and the usual neglect of inelastic losses near threshold, i.e., well below the characteristic excitation energy ω_p of order 10–15 eV.

The full many body XAS $\mu(\omega)$ can then be expressed as a convolution of an independent particle XAS with a spectral function as in (27.2), where $\mu^0(\omega)$ is the independent particle XAS calculated in the presence of a core hole. The net effect of the convolution over a particle-hole spectral amplitude $\tilde{A}(\omega, \omega')$ in (27.8) is clearly a decreased XAFS amplitude and a phase shifted oscillatory signal compared to the one-particle XAFS $\chi^{(0)}$. In the single scattering approximation the oscillatory energy dependence of $\chi_{\rm qp}(\omega)$ enters primarily through the complex exponential $\Im\{\exp[i2k(\omega)R]\}$, where R is an interatomic distance and $k(\omega) = \sqrt{2\omega}$ is the photoelectron wave vector. Neglecting the smoothly varying vectors, the result of the convolution can be written in terms of a complex-valued amplitude factor $S^2(\omega, R) = |S(\omega, R)|^2 \exp(i\psi(\omega, R))$, which is given by an energy dependent *phasor sum* over the effective normalized spectral function,

$$S_0^2(R) = \int_0^{\omega} d\omega' \tilde{A}(\omega, \omega') e^{i2[k(\omega-\omega')-k(\omega)]R} \ . \tag{27.9}$$

The *many-body phase factor* $\psi_k(\omega)$ is usually small but can be important. An additional factor from the core-core overlap factor and from edge-singularity enhancement may also be needed in some cases, but this factor is usually near unity and neglected in this summary. The qualitative behavior of $S^2(\omega, R)$ can be understood as follows: At low energies compared with the excitation energy ω_p, the satellite terms strongly cancel so $A(\omega, \omega') \approx \delta(\omega - \omega')$ and hence, $S_0^2(R) \to 1$. At high energies the sudden approximation prevails, and $A \approx A_{\rm qp} + A_{\rm intr}$, which has a strong satellite structure. However, the phase difference $2[k(\omega - \omega') - k(\omega)]$ between the primary channel and satellite becomes small at high energies ($\omega \gg \omega_p$) so that also $S^2(\omega, R) \to 1$, with a correction of order $(\omega_p R/\sqrt{\omega})$ similar to an additional mean-free path term. At intermediate energies comparable to ω_p, the value of $S^2(\omega, R)$ has a minimum.

As an example, we show the experimental XAS for fcc Al metal compared to the calculated results using the FEFF9 code including the cumulant convolution, and those of the single particle calculation (Fig. 27.1 top). Both calculations agree fairly well with experiment, although the single particle spectrum does not contain enough broadening at about 1590 eV, where the dip is slightly too large. The figure (bottom panel) also shows the many-body amplitude reduction factor including extrinsic

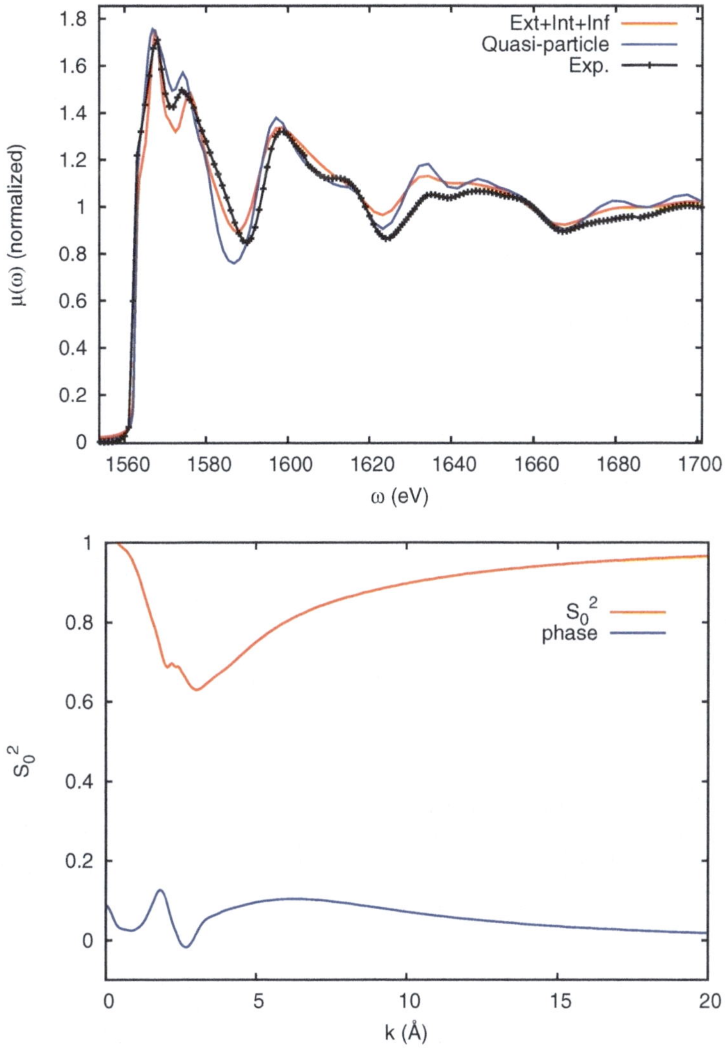

Fig. 27.1 (*top*) Theoretical Al K-edge XAS spectrum compared to the quasi-particle theory in this work and experimental data [17]; and (*bottom*) the many-body amplitude factor for the first shell R including extrinsic losses $|\tilde{S}_0^2| = |S_0^2(R)|e^{-2R/\lambda_k}$ as well as the associated phase ψ_k in radians. The top figure was adapted from that in [5]

losses, $|\tilde{S}_0^2|e^{-2R/\lambda_k}$, and the associated phase shift as a function of EXAFS wavenumber k. Note the limiting behavior $S_0^2 \to 1$ at both low and high k as explained above. In addition, we see appreciable reduction in the amplitude at the minimum $k \approx 3(\text{Å}^{-1})$ where $S_0^2 \approx 0.65$.

27.4 Summary

The particle-hole cumulant expansion with a partition of the cumulant into extrinsic, intrinsic and interference contributions. provides practical approach for calculating inelastic losses due to intrinsic, extrinsic and interference effects in x-ray spectra. These losses are included in the spectra in terms of a convolution with a particle-hole spectral function that accounts for their energy dependence. The cumulant approach simplifies the formalism and facilitates practical calculations. The theory elucidates both their behavior and the differences between the spectral functions for XAS and XPS which may be important to their interpretation. The cumulant approach can also account for edge singularities in the spectrum. Physically, the treatment of inelastic losses here is analogous to an *excitonic polaron*, i.e., the interaction of the particle-hole created in photoexcitation with the density fluctuations produced by the particle-hole system. This is in contrast to the *electronic polaron* described by the GW approximation [6], where the single-particle excitations arise from the much stronger density fluctuations due to a core-hole.

Acknowledgements We thank G. Bertsch, C. Draxl, C.R. Natoli, L. Reining, E. Shirley, and T. Devereaux for comments and suggestions. This work was supported by DOE Grant DE-FG03-97ER45623 (JJR and JJK).

References

1. J.J. Rehr, E.A. Stern, R.L. Martin, E.R. Davidson, Phys. Rev. B **17**, 560 (1978)
2. C.R. Natoli, M. Benfatto, C. Brouder, M.F.R. López, D.L. Foulis, Phys. Rev. B **42**, 1944 (1990)
3. L. Hedin, Phys. Rev. **139**(3A), A796 (1965)
4. J.J. Kas, A.P. Sorini, M.P. Prange, L.W. Cambell, J.A. Soininen, J.J. Rehr, Phys. Rev. B **76**, 195116 (2007), https://doi.org/10.1103/PhysRevB.76.195116
5. J.J. Kas, J.J. Rehr, J.B. Curtis, Phys. Rev. B **94**, 035156 (2016)
6. L. Hedin, J. Phys. Condens. Matter **11**, R489 (1999)
7. F. Aryasetiawan, L. Hedin, K. Karlsson, Phys. Rev. Lett. **77**, 2268 (1996)
8. M. Guzzo, G. Lani, F. Sottile, P. Romaniello, M. Gatti, J.J. Kas, J.J. Rehr, M.G. Silly, F. Sirotti, L. Reining, Phys. Rev. Lett. **107**, 166401 (2011)
9. J. Zhou, J. Kas, L. Sponza, I. Reshetnyak, M. Guzzo, C. Giorgetti, M. Gatti, F. Sottile, J. Rehr, L. Reining, J. Chem. Phys. **143**, 184109 (2015)
10. L. Campbell, L. Hedin, J.J. Rehr, W. Bardyszewski, Phys. Rev. B **65**, 064107 (2002)
11. L. Hedin, J. Michiels, J. Inglesfield, Phys. Rev. B **58**, 15565 (1998)
12. M. Calandra, J.P. Rueff, C. Gougoussis, D. Céolin, M. Gorgoi, S. Benedetti, P. Torelli, A. Shukla, D. Chandesris, C. Brouder, Phys. Rev. B **86**, 165102 (2012)
13. P. Nozières, C.T. de Dominicis, Phys. Rev. **178**, 1097 (1969)
14. L. Landau, J. Phys. USSR **8**, 201 (1944)
15. W. Bardyszewski, L. Hedin, Phys. Scr. **32**, 439 (1985)
16. J.J. Rehr, J.J. Kas, M.P. Prange, A.P. Sorini, Y. Takimoto, F. Vila, Comput. Rendus Phys. **10**, 548 (2009)
17. P. Lagarde. Private communication (1999)

Chapter 28
Screened KKR

Krisztián Palotás and László Szunyogh

Abstract The concept of screening in the Korringa–Kohn–Rostoker (KKR) multiple scattering electronic structure method for solids is briefly presented. The main advantages of the screened KKR (SKKR) method and recent applications are highlighted.

28.1 Introduction

Since the publications of the seminal works of Korringa [1] and Kohn and Rostoker [2] with the aim at describing the electronic structure of periodic solids, the Korringa–Kohn–Rostoker (KKR) Green-function-based multiple scattering method has been considerably developed over the years [3–5]. This methodological progress has been fueled by the steadily growing computational (hardware) facilities and constantly developing efficiency in the software implementations of the KKR method. An important step has been the introduction of the so-called screening transformation [6, 7] that for systems with at best two-dimensional translational symmetry provides a fast and, in principle, exact numerical solution through the casting of big (infinite) matrices into a block-tridiagonal form. The Screened Korringa–Kohn–Rostoker (SKKR) method has been invented in the 1990s in the groups of Peter Weinberger at the Technical University of Vienna and László Szunyogh at the Technical University of Budapest, where it is still hosted and being continuously developed.

K. Palotás
Department of Theoretical Physics, Budapest University of Technology and Economics, Budafoki út 8., Budapest 1111, Hungary
e-mail: palotas@phy.bme.hu

K. Palotás
Slovak Academy of Sciences, Institute of Physics, Dúbravská cesta 9., 84511 Bratislava, Slovakia

L. Szunyogh (✉)
Department of Theoretical Physics and MTA-BME Condensed Matter Research Group, Budapest University of Technology and Economics, Budafoki út 8., Budapest 1111, Hungary
e-mail: szunyogh@phy.bme.hu

D. Sébilleau et al. (eds.), *Multiple Scattering Theory for Spectroscopies*, Springer Proceedings in Physics 204,
https://doi.org/10.1007/978-3-319-73811-6_28

While a very detailed description of the SKKR method can be found in [8], in this chapter we briefly present the underlying concept and highlight the main advantages of the SKKR method together with selected recent applications.

28.2 The KKR Method

Unlike other popular electronic structure calculation methods using electron wave functions, the KKR formalism relies on the Green's function that has a conceptual advantage of straightforward evaluation of physical properties. The basic ingredients of the KKR method formulated within multiple scattering theory [4] are the single-site scattering t–matrices and the structure constants. The single-site t-matrix is an angular momentum representation of the single-site T–operator formally obtained in the following way: Assuming that a single-particle Hamiltonian, composed in the spirit of the Kohn–Sham formulation [9] of density functional theory, is given as a sum of an unperturbed (H_0) and a Hermitian perturbation (V) part, $H = H_0 + V$, the corresponding resolvents (with z a complex energy argument, and I the identity operator) are $G_0(z) = (zI - H_0)^{-1}$ and $G(z) = (zI - H)^{-1}$. They can be related to each other as $G(z) = G_0(z) + G_0(z)T(z)G_0(z)$ by defining the T–operator $T(z) = V(I - G_0(z)V)^{-1} = (I - VG_0(z))^{-1}V$, which implies $G_0(z)T(z) = G(z)V$ and $T(z)G_0(z) = VG(z)$. If only a single scattering potential centered at the lattice position $\boldsymbol{R}_n$ is present in the system then $\underline{t}^n(z)$ represents the single-site scattering t-matrix with respect to the product of the spherical Bessel functions and spherical harmonics, $j_L(z, \boldsymbol{r}) = j_\ell(\sqrt{z}r)Y_L(\hat{r})$ with the abbreviation $L = (\ell, m)$.

Generalizing the system by introducing non-overlapping potentials at various positions in space, e.g., at atomic sites, that individually act as single-site scatterers, their common effect is described by multiple scattering theory. The multi-site T–operator describes all possible scattering events: $T = \sum_n t^n + \sum_{nm} t^n G_0(1 - \delta_{nm})t^m + \sum_{nmo} t^n G_0(1 - \delta_{nm})t^m G_0(1 - \delta_{mo})t^o + \ldots$, which can be recast to $T = \sum_{nm} \tau^{nm}$ by introducing τ the so-called scattering path operator [3]. This implies that the multiple scattering can be taken into account as simple as $G = G_0 + \sum_{nm} G_0 \tau^{nm} G_0$. The real-space structure constants, $G^{nm}_{0,LL'}(z)$, result from the two-center expansion of $G_0(z)$, i.e. for $n \neq m$ and $\boldsymbol{r} = \boldsymbol{r}_n + \boldsymbol{R}_n$ and $\boldsymbol{r}' = \boldsymbol{r}'_m + \boldsymbol{R}_m$: $G_0(z; \boldsymbol{r}_n + \boldsymbol{R}_n, \boldsymbol{r}'_m + \boldsymbol{R}_m) = \sum_{LL'} j_L(z, \boldsymbol{r}_n) G^{nm}_{0,LL'}(z) j_{L'}(z, \boldsymbol{r}'_m)^{\times}$, where the superscript $^{\times}$ denotes that only the spherical harmonics are conjugated. Using a site-angular momentum supermatrix representation, the matrix of the structure constants is denoted by $\underline{\underline{G}}_0(z) = \underline{G}^{nm}_0(z)(1 - \delta_{nm})$, while the t-matrices can also be structured in a site-angular momentum supermatrix, $\underline{\underline{t}}(z) = \underline{t}^n(z)\delta_{nm}$, i.e. block-diagonal in the site index.

The fundamental KKR equation defines the site-angular momentum supermatrix representation of the scattering path operator, $\underline{\underline{\tau}}(z)$, in terms of the single-site t-matrix and the structure constant matrix as

$$\underline{\underline{\tau}}(z) = \left(\underline{\underline{t}}(z)^{-1} - \underline{\underline{G}}_0(z) \right)^{-1} . \tag{28.1}$$

Finally, the single-particle Green's function is obtained from the scattering path operator as

$$\begin{aligned} G(z; \boldsymbol{r}_n + \boldsymbol{R}_n, \boldsymbol{r}'_m + \boldsymbol{R}_m) = \sum_{LL'} Z^n_L(z, \boldsymbol{r}_n) \tau^{nm}_{LL'}(z) Z^m_{L'}(z, \boldsymbol{r}'_m)^{\times} - \\ \delta_{nm} \sum_{L} Z^n_L(z; \boldsymbol{r}_{n,<}) J^n_L(z; \boldsymbol{r}_{n,>})^{\times} , \end{aligned} \tag{28.2}$$

where Z^n_L and J^n_L are properly normalized regular and irregular scattering solutions, respectively, and $r_< = \min(r, r')$ and $r_> = \max(r, r')$. From the above Green's function the physical quantities can straightforwardly be calculated.

28.3 Screening Transformations in the KKR Method

Solving the KKR equation (28.1) for systems in reduced dimension requires the inversion of a big matrix, and its computational time scales with N^3 (N being the size of the system in real space). This is highly unfavorable at a large number of scatterers (atoms) and one possible way to overcome the problem is to use a *screening transformation*. Here, a reference potential, V_r, is added to the Hamiltonian of the unperturbed system, $H'_0 = H_0 + V_r$, such that $H = H_0 + V = (H_0 + V_r) + (V - V_r) = H'_0 + V'$ with $V' = V - V_r$. The resolvent of H can then be expressed as $G(z) = G_0(z)(I + VG(z)) = G'_0(z)(I + V'G(z))$, where $G'_0(z) = (zI - H'_0)^{-1} = (zI - H_0 - V_r)^{-1}$. Once the potential V_r is repulsive, $G'_0(z)$ gets localized in real space for $\Re(z) < V_r$, which makes the calculation of $G_0(z)$ feasible.

Turning to multiple scattering, such a reference potential is written as a superposition of non-overlapping potentials $V_r = \sum_n V^n_r(\boldsymbol{r}_n)$. If the corresponding single-site t-matrices are $\underline{t}^n_r(z)$ then the Green's function matrix of the reference system, termed as *screened structure constant*, is obtained as

$$\underline{\underline{G}}_r(z) = \underline{\underline{G}}_0(z) \left(\underline{\underline{I}} - \underline{\underline{t}}_r(z) \underline{\underline{G}}_0(z) \right)^{-1} . \tag{28.3}$$

By defining the *screened t-matrix* as the difference, $\underline{\underline{t}}_\Delta(z) = \underline{\underline{t}}(z) - \underline{\underline{t}}_r(z)$, a formally equivalent equation to (28.1) can be obtained,

$$\underline{\underline{\tau}}_\Delta(z) = \left(\underline{\underline{t}}_\Delta(z)^{-1} - \underline{\underline{G}}_r(z) \right)^{-1} , \tag{28.4}$$

related to τ as

$$\underline{\underline{\tau}}(z) = \underline{\underline{t}}(z) \left(\underline{\underline{t}}_\Delta(z)^{-1} \underline{\underline{\tau}}_\Delta(z) \underline{\underline{t}}_\Delta(z)^{-1} + \left(\underline{\underline{t}}(z)^{-1} - \underline{\underline{t}}_\Delta(z)^{-1} \right) \right) \underline{\underline{t}}(z) . \tag{28.5}$$

By using repulsive V_r^n *screening potentials*, (28.3) can be solved so that $\underline{G}_r^{nm}(z) \simeq \underline{0}$ for all $|\underline{R}_n - \underline{R}_m| > d$, where d is a distance of some atomic spacings, i.e., the structure constants in (28.4) are indeed *screened* and their solution reduces to the inversion of a sparse matrix [6, 7]. In case of layered systems, the corresponding matrix gets block-tridiagonal and the required computational time scales with N [10]. Using (28.5), the τ-matrix in *screened representation* can be transformed to the physical representation and the Green's function is obtained from (28.2).

28.4 Suitability of the SKKR Method

The above described strategy of the SKKR method can uniformly be used for a non-relativistic or fully relativistic angular momentum expansion, for spin-polarized systems and for scattering potentials of spherically symmetrical or arbitrary (full-potential) shape. The SKKR method has traditionally been used for layered systems, i.e., materials with two-dimensional translational symmetry, like thin films, multilayers, surfaces and interfaces [8]. This combined with the Coherent Potential Approximation (CPA) [3] enables the investigation of substitutionally disordered random alloys described by an effective medium. The Embedded Cluster Method (ECM) [11] enables the investigation of real-space nanostructures, like impurities, surface islands, atomic contacts. The SKKR within a fully relativistic spin-polarized description is extremely suitable to study diverse magnetic properties, like local spin/orbital moments, magnetic anisotropy (MAE), domain walls [12, 13], interlayer exchange coupling (IEC), tensorial exchange interactions [14], spin wave (magnon) spectroscopy [14, 15]. Combined with the linear response Kubo-Greenwood theory, electrical (e.g., conductance, magnetoresistance) and magneto-optical transport (e.g., Kerr spectroscopy) properties can be studied at a fully relativistic first principles level. For a more detailed and structured overview the reader is referred to [8].

It is important to note that the massively parallelized KKRnano program package [16], developed mainly in the Research Center Jülich in Germany, takes advantage of the SKKR concept to provide an order-N electronic structure code suitable for studying solid-state systems consisting of tens of thousands of atoms. For more details, the reader is referred to Chap. 17 of this book.

28.5 Using SKKR to Explore Complex Magnetism of Nanostructures

Recent advancements in experimental techniques of imaging complex magnetic structures in real space above surfaces using spin-polarized scanning tunneling microscopy [17] require theoretical efforts for the understanding of the formation of complex magnetic patterns in thin films. Based on the SKKR method, the parameters of a classical spin Hamiltonian, including tensorial exchange interactions, can be extracted using either the relativistic torque method (RTM) [14] or the spin cluster expansion (SCE) technique [18]. A suitable spin Hamiltonian is

$$H = -\frac{1}{2}\sum_{i\neq j}\frac{1}{m_i m_j}\boldsymbol{m}_i \underline{\underline{J}}_{ij}\boldsymbol{m}_j + \sum_i \frac{1}{m_i^2}\boldsymbol{m}_i \underline{\underline{K}}_i \boldsymbol{m}_i - \sum_i \boldsymbol{m}_i \cdot \boldsymbol{b}_{\rm ext} , \tag{28.6}$$

with $\boldsymbol{m}_i$ the classical spin moment of atom i, $\underline{\underline{J}}_{ij}$ the exchange tensor, $\underline{\underline{K}}_i$ the on-site anisotropy matrix, and $\boldsymbol{b}_{\rm ext}$ the external magnetic field. Based on this, the magnetic ground state can be estimated in the following way: The energy of a spin spiral with propagation vector $\boldsymbol{q}$ corresponds to the maximal eigenvalue of the Fourier transform of the exchange tensor. When calculating spin spiral energies by sweeping $\boldsymbol{q}$ in the Brillouin zone, the maximal obtained value corresponds to the estimated magnetic ground state. Another method to find the ground state magnetic structure is based on the zero temperature (deterministic) Landau–Lifshitz–Gilbert (LLG) equation of atomistic spin dynamics,

$$\frac{\partial \boldsymbol{m}_i}{\partial t} = -\frac{\gamma}{1+\alpha^2}\boldsymbol{m}_i \times \boldsymbol{b}_i - \frac{\alpha\gamma}{(1+\alpha^2)m_i}\boldsymbol{m}_i \times (\boldsymbol{m}_i \times \boldsymbol{b}_i) , \tag{28.7}$$

where γ is the gyromagnetic ratio, α the Gilbert damping, and the effective field is $\boldsymbol{b}_i = -\partial H/\partial \boldsymbol{m}_i$.

Following this multiscale approach based on the spin Hamiltonian parameters obtained within the SKKR method, the complex ground states of a variety of magnetic thin films have been obtained ranging from spin spirals to skyrmions, for example: 1–4 monolayers (ML) of Fe on Ir(001) surface [19]; 1 ML Fe on different substrates composed of 5*d* elements [20]; PdFe double layer on Ir(111) [21]; 1 ML Fe on Rh(001) surface [22]; 1 ML Fe on W(110) and Ta(110) substrates [23]. Theoretical analysis provides information on the relative importance and competition of isotropic exchange, Dzyaloshinskii–Moriya, and in certain cases of biquadratic and higher order spin interactions, partly considering the effect of layer relaxations as well. Extending the LLG equation (28.7) to include thermal effects on the effective field [24, 25], temperature dependence of the magnetic states can also be studied [23, 26, 27].

The above examples illustrate the importance of the SKKR method to contribute to the theoretical understanding of complex magnetism at the atomic scale.

Acknowledgements Financial support of the SASPRO Fellowship of the Slovak Academy of Sciences (project no. 1239/02/01), the Hungarian State Eötvös Fellowship of the Tempus Public Foundation (contract no. 2016-11) and the National Research, Development and Innovation Office of Hungary under Project No. K115575 is gratefully acknowledged.

References

1. J. Korringa, Physica **XIII**, 392 (1947)
2. W. Kohn, N. Rostoker, Phys. Rev. **94**, 1111 (1954)
3. B.L. Gyorffy, Phys. Rev. B **5**, 2382 (1972)
4. J.S. Faulkner, G.M. Stocks, Phys. Rev. B **21**, 3222 (1980)
5. H. Ebert, D. Ködderitzsch, J. Minár, Rep. Prog. Phys. **74**, 096501 (2011)
6. L. Szunyogh, B. Újfalussy, P. Weinberger, J. Kollár, Phys. Rev. B **49**, 2721 (1994)
7. R. Zeller, P.H. Dederichs, B. Újfalussy, L. Szunyogh, P. Weinberger, Phys. Rev. B **52**, 8807 (1995)
8. J. Zabloudil, R. Hammerling, L. Szunyogh, P. Weinberger, *Electron Scattering in Solid Matter*, 1st edn. (Springer, Berlin, 2005)
9. W. Kohn, L.J. Sham, Phys. Rev. **140**, A1133 (1965)
10. E.M. Godfrin, J. Phys. Condens. Matter **3**, 7843 (1991)
11. B. Lazarovits, L. Szunyogh, P. Weinberger, Phys. Rev. B **65**, 104441 (2002)
12. S. Gallego, P. Weinberger, L. Szunyogh, P.M. Levy, C. Sommers, Phys. Rev. B **68**, 054406 (2003)
13. L. Balogh, K. Palotás, L. Udvardi, L. Szunyogh, U. Nowak, Phys. Rev. B **86**, 024406 (2012)
14. L. Udvardi, L. Szunyogh, K. Palotás, P. Weinberger, Phys. Rev. B **68**, 104436 (2003)
15. L. Udvardi, L. Szunyogh, Phys. Rev. Lett. **102**, 207204 (2009)
16. KKRnano project (2016), http://www.fz-juelich.de/ias/jsc/EN/Expertise/High-Q-Club/KKRnano/_node.html. Cited 30 Sep 2016
17. R. Wiesendanger, Rev. Mod. Phys. **81**, 1495 (2009)
18. L. Szunyogh, L. Udvardi, J. Jackson, U. Nowak, R. Chantrell, Phys. Rev. B **83**, 024401 (2011)
19. A. Deák, L. Szunyogh, B. Ujfalussy, Phys. Rev. B **84**, 224413 (2011)
20. E. Simon, K. Palotás, B. Ujfalussy, A. Deák, G.M. Stocks, L. Szunyogh, J. Phys. Condens. Matter **26**, 186001 (2014)
21. E. Simon, K. Palotás, L. Rózsa, L. Udvardi, L. Szunyogh, Phys. Rev. B **90**, 094410 (2014)
22. A. Deák, K. Palotás, L. Szunyogh, I.A. Szabó, J. Phys. Condens. Matter **27**, 146003 (2015)
23. L. Rózsa, L. Udvardi, L. Szunyogh, I.A. Szabó, Phys. Rev. B **91**, 144424 (2015)
24. L. Rózsa, L. Udvardi, L. Szunyogh, J. Phys. Condens. Matter **25**, 506002 (2013)
25. L. Rózsa, L. Udvardi, L. Szunyogh, J. Phys. Condens. Matter **26**, 216003 (2014)
26. G. Hasselberg, R. Yanes, D. Hinzke, P. Sessi, M. Bode, L. Szunyogh, U. Nowak, Phys. Rev. B **91**, 064402 (2015)
27. L. Rózsa, E. Simon, K. Palotás, L. Udvardi, L. Szunyogh, Phys. Rev. B **93**, 024417 (2016)

Chapter 29
Magnetic Dipole Term T_z and its Importance for Analysing XMCD Spectra

Ondřej Šipr

Abstract Magnetism of nanostructures is often studied by means of x-ray magnetic circular dichroism (XMCD). The XMCD sum rules are a very powerful tool but they allow for the spin magnetic moment μ_{spin} to be determined only in combination with the magnetic dipole term T_z. This chapter presents few examples demonstrating that neglecting T_z could in some cases lead to completely wrong conclusions about the trends of μ_{spin} with the size of the system or with the magnetisation direction. Further, we inspect conditions that have to be met so that eliminating of T_z from the XMCD sum rules is possible.

29.1 Introduction

Magnetism of diluted and low-dimensional systems is one of strongly pursued research areas. Magnetic properties of large assemblies of nanostructures are macroscopic and there are thus no principal problems with investigating them experimentally. However, the effort to understand trends in nanomagnetism invokes the need to study systems such as small clusters of well-defined sizes. The magnetisation of individual clusters of just few atoms cannot be measured by macroscopic methods. It can be, however, probed indirectly by relying on spectroscopy.

One of the most powerful methods in this respect is x-ray magnetic circular dichroism (XMCD). It consists in measuring the difference of the absorption of left- and right-circularly polarised x-rays in a magnetised sample while the energy of the incident x-rays is varied. XMCD spectra are often analysed by means of the sum rules, which link integrals of XMCD and x-ray absorption spectral peaks to local spin and orbital magnetic moments. Recent progress in magnetism of atomic-sized systems is mostly associated with the application of the XMCD sum rules [1–3].

O. Šipr (✉)
Institute of Physics, Czech Academy of Sciences, Cukrovarnická 10,
162 53 Praha, Czech Republic
e-mail: sipr@fzu.cz

D. Sébilleau et al. (eds.), *Multiple Scattering Theory for Spectroscopies*,
Springer Proceedings in Physics 204,
https://doi.org/10.1007/978-3-319-73811-6_29

The strength of the sum rules is that they provide separate information about the orbital magnetic moment μ_{orb} and the spin magnetic moment μ_{spin} of the photoabsorbing atom [4, 5]. However, extracting μ_{orb} and, especially, μ_{spin} from XMCD spectra is not straightforward. Considering the most common case of the $L_{2,3}$ edge spectra and a sample magnetised along the α direction ($\alpha = x, y, z$), the spin magnetic moment sum rule can be written as [5]

$$\frac{3}{I} \int \left(\Delta\mu_{L_3} - 2\Delta\mu_{L_2}\right) \mathrm{d}E = \frac{\mu_{\mathrm{spin}} + 7T_\alpha}{n_h}, \tag{29.1}$$

where $\Delta\mu_{L_{2,3}}$ are the differences $\Delta\mu = \mu^{(+)} - \mu^{(-)}$ between absorption coefficients for the left and right circularly polarised light propagating along the α direction, I is the integral over the isotropic absorption spectrum, μ_{spin} is the local spin magnetic moment (its d component, to be precise), and n_h is the number of holes in the d band. The term T_α is the expectation value of the intra-atomic spin dipole operator for the valence d electrons. It is often called the magnetic dipole term in the literature. As the magnetisation is typically in the $\alpha = z$ direction, one often speaks simply about the T_z term.

The magnetic dipole T_α term can be written as [6, 7]

$$\begin{aligned} T_\alpha &= -\frac{\mu_B}{\hbar} \langle \hat{T}_\alpha \rangle \\ &= -\frac{\mu_B}{\hbar} \left\langle \sum_\beta Q_{\alpha\beta} S_\beta \right\rangle, \end{aligned} \tag{29.2}$$

with

$$Q_{\alpha\beta} = \delta_{\alpha\beta} - 3 r^0_\alpha r^0_\beta \,, \tag{29.3}$$

being the quadrupole moment operator and S_α being the spin operator. The T_α term cannot be easily determined by experiment and its occurrence in (29.1) thus poses a problem. This chapter is devoted to illustrating what kind of complications the T_α term may cause and under which conditions dealing with the T_α term can be by-passed.

29.2 Computational Framework

The results shown in this chapter were obtained within the ab initio spin density functional theory, relying on the local spin density approximation (LSDA) with the Vosko, Wilk and Nusair parameterisation for the exchange and correlation potential [8]. The electronic structure is described, including all relativistic effects, by the Dirac equation, which is solved using the spin polarised relativistic multiple-scattering or Korringa–Kohn–Rostoker (KKR) Green function formalism [9] as implemented in the SPR-TB-KKR code [10]. The potentials were treated within the atomic sphere approximation (ASA). More details about the Green function KKR method can be found in Chap. 4.

We present here several data obtained for monolayers, clusters and adatoms deposited on surfaces. Technically, the calculations for supported monolayers were accomplished by means of the tight-binding or screened KKR method [11]. The substrate is modelled by a finite slab typically of 15 atomic layers, the vacuum is represented by about 5 layers of empty sites. Calculations for clusters and adatoms were accomplished by means of embedded impurity formalism: first one has to calculate the electronic structure of the host system (clean surface) and then a Dyson equation for an embedded impurity cluster is solved [12]. The embedded impurity cluster defines the zone in which the electrons are allowed to react to the presence of deposited atom or cluster; there is an unperturbed host beyond this zone. Typically the embedded cluster contains about 150 sites.

29.3 Impact of the T_z Term on the Apparent Dependence of μ_{spin} on the Cluster Size and Shape

For bulk systems, the T_α term can be often neglected (provided that the spin-orbit coupling is not very strong [13]). However, for low-dimensional systems it can be significant [14–16]. To make matters worse, the T_α term cannot be considered just as an additive correction that for similar systems simply shifts the values of μ_{spin} by approximately the same amount. We will demonstrate below that neglecting T_α for a sequence of supported magnetic clusters can lead to erroneous conclusions regarding the dependence of the average μ_{spin} on the cluster size and shape [17].

As the orientation of the magnetic field is not varied for the examples presented in this part, we assume $\alpha = z$ and speak simply about the T_z term in Sect. 29.3.

29.3.1 *Effect of the Cluster Size*

To demonstrate our point, let us consider a series of supported Co clusters on Au(111). The Co_N clusters are planar, containing $N = 1$–7 atoms. Their shapes are compact. Our focus is on comparing how μ_{spin} per atom and $[\mu_{\text{spin}} + 7T_z]/n_h$ per atom depend on the cluster size. This is shown in Fig. 29.1, where these values are shown averaged over all sites of the respective cluster [17]. One can see that the trends for μ_{spin} and for $[\mu_{\text{spin}} + 7T_z]/n_h$ are exactly opposite.

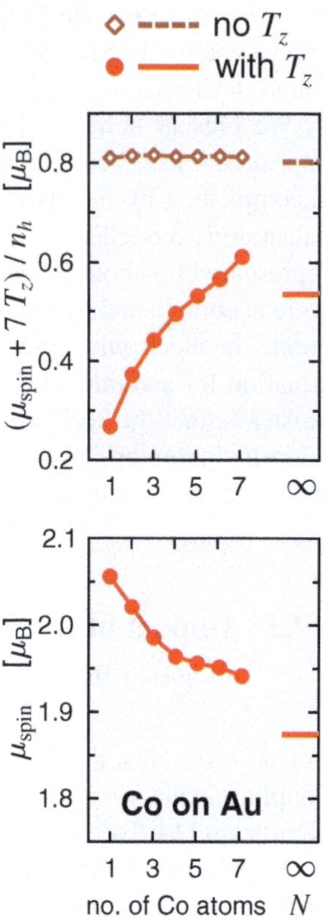

Fig. 29.1 Dependence of μ_{spin} (*lower panel*) and $(\mu_{\text{spin}} + 7T_z)/n_h$ (*upper panel*) per atom on the number of atoms N in compact Co_N clusters on the Au(111) surface. In the upper panel, data which take T_z into account are shown via the *full line* and *filled circles*, while data obtained if T_z is neglected are shown via the *dashed line* and *open diamonds*. Results for a complete monolayer $(N = \infty)$ are shown via *horizontal lines* at the right border of each panel. Figure reproduced from [17]

To prove that the increase of $[\mu_{\text{spin}} + 7T_z]/n_h$ is caused by the cluster-size dependence of the magnetic dipole T_z term and not of the number of holes n_h, we inspect also how the quantity μ_{spin}/n_h per atom depends on N. The results are shown in upper panel of Fig. 29.1 (broken line with open diamonds). It is evident that indeed it is the T_z term that makes $[\mu_{\text{spin}} + 7T_z]/n_h$ per atom to increase with increasing size of the clusters.

The quantity $[\mu_{\text{spin}} + 7T_z]/n_h$ appears in the XMCD spin sum rule (29.1). The results shown above demonstrate that for some combinations of the cluster and substrate material, ignoring the variations in T_z would lead to a *completely false estimate* of how μ_{spin} per atom depends on the cluster size.

29.3.2 Effect of the Cluster Shape

We restricted ourselves to compact clusters in Sect. 29.3.1, so that results for clusters of different sizes are comparable. If clusters of different shapes were involved,

Fig. 29.2 Dependence of μ_{spin} (*lower panel*) and $(\mu_{\text{spin}} + 7T_z)/n_h$ (*upper panel*) per atom on the shapes of Co_4 clusters on Au(111). The cluster shapes are depicted below the *horizontal axis*. Figure reproduced from [17]

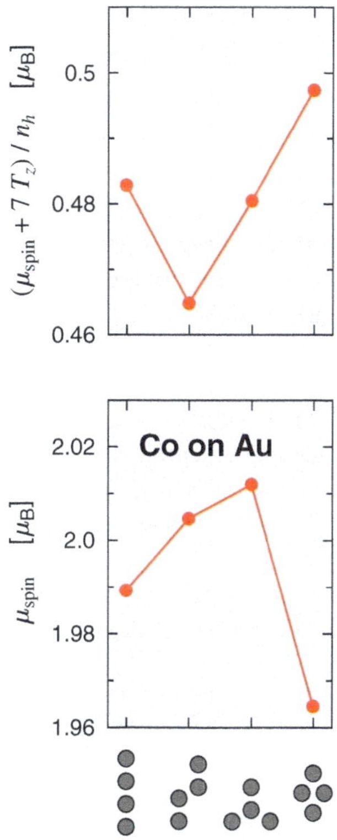

the values would be different: one can anticipate this by noticing that changing the cluster shape changes the coordination numbers of atoms and that, in turn, should result in different μ_{spin} and, presumably, also T_z [18–20]. To demonstrate to what extent the variance in shape may influence the results, we calculated average μ_{spin} and average $[\mu_{\text{spin}} + 7T_z]/n_h$ for Co_4 clusters of different shapes. Again, these clusters are supported by the Au(111) surface. The results are shown in Fig. 29.2 [17]. Similarly as when inspecting the dependence on the cluster size, we observe that the trends for μ_{spin} and for $[\mu_{\text{spin}} + 7T_z]/n_h$ differ.

We have thus demonstrated that knowledge of the variations of T_z with cluster size and shape is essential for application of the XMCD spin sum rule. For transition metal clusters, the magnetic dipole term T_z is not just a minor factor that only mildly affects the analysis. On the contrary, $T_z$may change the overall picture completely. It may cause that experiment will provide results which are seemingly counterintuitive. Whether the T_z term will actually cause that the size-dependences of μ_{spin} and $[\mu_{\text{spin}} + 7T_z]/n_h$ will be opposite or not depends both on the clusters and on the substrate and is difficult to be guessed beforehand [17].

29.4 Impact of the T_α Term on the Apparent Dependence of μ_{spin} on the Direction of the Magnetisation

Neglecting T_α and its angular dependence could introduce spurious anisotropy of μ_{spin} for low-dimensional systems [6, 21–23]. We illustrate this point on the case of Co adatoms and monolayers on Pd(100), Pd(111) and Pd(110) surfaces. The corresponding structure diagrams are shown in Fig. 29.3 (for adatoms, only one Co atom is kept obviously). Two hollow adatom positions are possible for the (111) surface, differing by the position of the adatom with respect to the sub-surface layer; we consider the fcc position here.

The spin sum rule (29.1) provides a means to extract the quantity $\mu_{spin} + 7T_\alpha$ from XMCD spectra. The moment μ_{spin} practically does not depend on the direction of the magnetisation. The T_α term, on the other hand, depends on it considerably. If the value of T_α is large so that it cannot be neglected, the sum-rule-related combination $\mu_{spin} + 7T_\alpha$ depends on the magnetisation direction. This dependence can be quite striking, as it is illustrated by the results summarised in Table 29.1 [23]. One can see that depending on the choice of the surface, the difference between $\mu_{spin} + 7T_\alpha$ evaluated for in-plane and out-of-plane orientations of the magnetisation may reach

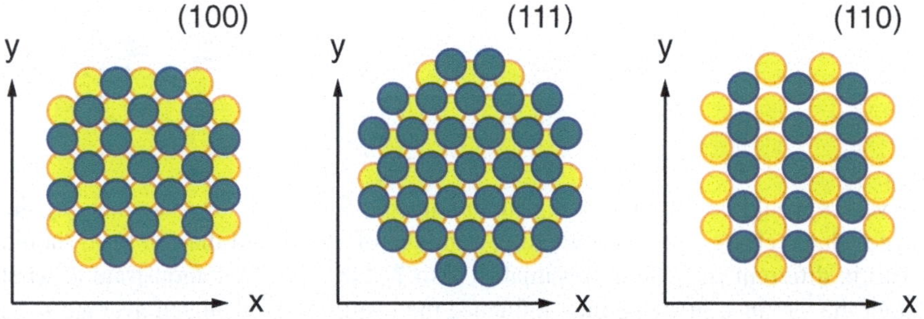

Fig. 29.3 Structure diagrams for a Co monolayer on Pd(100), Pd(111) and Pd(110). The orientation of the x and y coordinates used throughout Sect. 29.4 is also shown. Figure reproduced from [23]

Table 29.1 Dependence of the sum $\mu_{spin} + 7T_\alpha$ on the direction of the magnetisation α ($\alpha = x, y, z$) for Co adatoms and monolayers deposited on three crystallographically different Pd surfaces [23]. The bare moment μ_{spin} is also shown for comparison

System	μ_{spin}	$\mu_{spin} + 7T_x$	$\mu_{spin} + 7T_y$	$\mu_{spin} + 7T_z$
Adatom on Pd(100)	2.29	2.40	2.40	2.12
Adatom on Pd(111)	2.35	2.91	2.91	1.33
Adatom on Pd(110)	2.20	2.06	3.19	1.42
Monolayer on Pd(100)	2.09	2.13	2.13	1.94
Monolayer on Pd(111)	2.02	2.27	2.27	1.56
Monolayer on Pd(110)	2.15	2.07	2.60	1.73

50–100%. For the (110) surface, there is also a surprisingly large in-plane anisotropy of $\mu_{\rm spin} + 7T_\alpha$, especially for the case of the adatom.

We summarise this section by noting that the magnetic dipole term T_α depends substantially on the direction of the magnetisation (and also on the crystallographic orientation of the substrate). As a results, a strong apparent anisotropy of $\mu_{\rm spin}$ may be falsely deduced from the XMCD sum rules if the T_α term is neglected.

29.5 Influence of the Spin-Orbit Coupling: When the T_α Term Can Be Eliminated

In principle, $\mu_{\rm spin}$ can be obtained if the T_α term is calculated and inserted into (29.1). However, one would really have to do the calculation for each system which is studied, because the T_α term is quite sensitive to details of the electronic structure [14, 15, 17] and taking it from calculations done for only similar systems might not be reliable. At the same time, performing the calculations for the system in question might be difficult or impractical.

The good news is that there appears to be a way to eliminate the T_α term from XMCD analysis, relying solely on experiment: either by performing a series of angle-dependent XMCD measurements or by performing just a single XMCD measurement at the right geometry. The key lies in decoupling the quadrupole moment operator $\hat{Q}$ and the spin operator $\hat{S}$ in (29.2). Namely, if the influence of the spin-orbit coupling (SOC) on T_α can be neglected and some not very severe restrictions on the symmetry are satisfied, the T_α term can be expressed as [24]

$$T_\alpha = \sum_m \frac{1}{2} \langle Y_{2m} | \hat{Q}_{\alpha\alpha} | Y_{2m} \rangle \, \mu_{\rm spin}^{(m)} . \tag{29.4}$$

Here $\mu_{\rm spin}^{(m)}$ is the spin magnetic moment resolved into components according to the magnetic quantum number m. The matrix elements $\langle Y_{2m} | \hat{Q}_{\alpha\alpha} | Y_{2m} \rangle$ depend on the direction α along which the sample is magnetically saturated. The values of $\langle Y_{2m} | \hat{Q}_{\alpha\alpha} | Y_{2m} \rangle$ together with further discussion of (29.4) can be found elsewhere [7, 23, 24].

If (29.4) is valid, further relations for the T_α term can be derived which can be used to eliminate T_α from the sum rule (29.1). One way to achieve this is by performing three XMCD measurements and making use of the relation [24]

$$T_x + T_y + T_z = 0 . \tag{29.5}$$

Furthermore, if the system has higher than twofold symmetry around the z axis, the magnetic dipole term depends on the polar angle θ as [24, 25]

$$T_\theta \approx 3\cos^2\theta - 1 , \tag{29.6}$$

Table 29.2 The ratio $7(T_x + T_y + T_z)/\mu_{spin}$ for Co monolayers and adatoms on noble metals (111) surfaces [29]

Substrate	$7(T_x + T_y + T_z)/\mu_{spin}$	
	Monolayer	Adatom
Cu	0.01	0.21
Pd	0.02	0.07
Ag	0.02	0.37
Pt	0.01	0.10
Au	0.01	0.28

(the magnetic dipole term is denoted T_θ for a general sample magnetisation direction). Equation (29.6) can be employed for another way of eliminating T_θ from the XMCD sum rules: One has to make the XMCD measurement with circularly polarised x-rays coming in the direction of the magic angle 54.7°. The point to be emphasised is that eliminating T_α from the sum rule analysis is possible only if the effect of SOC on T_α can be neglected.

This raises a question whether the influence of SOC on T_α can be neglected in common circumstances. There are theoretical indications that this influence can be sometimes important. For example, it was found that (29.5) is strongly violated for free-standing Co wires [26]. Considering more realistic materials, violation of (29.6) was predicted for a Pt monolayer magnetised by the underlying Fe substrate [27]. There have been also experimental indications that the SOC may be important for the T_α term: violation of (29.5) was observed for magnetite nanoparticles [28].

To provide a comprehensive picture to what extent the SOC is important for T_α in thin films, adatoms or clusters, we show below theoretical results for Co monolayers and adatoms on Cu, Pd, Ag, Pt, and Au (111) surfaces. This selection guarantees that one accounts for effects connected with the change of dimensionality as well as for effects connected with the change of chemical environment.

29.5.1 Sum over Magnetic Dipole Term Components T_α

The first test of the influence of SOC on the T_α term is checking the validity of (29.5). The motivation for this test comes from the spin moment sum rule (29.1), in which μ_{spin} appears only in combination with $7T_\alpha$, as $\mu_{spin} + 7T_\alpha$. The key indicator is thus the ratio $7T_\alpha/\mu_{spin}$. Table 29.2 shows this ratio summed over all three coordinates, $\sum_{\alpha=x,y,z} 7T_\alpha/\mu_{spin}$. If the influence of SOC on T_α was completely neglected, this quantity would be zero. One can see from Table 29.2 that for Co monolayers the condition (29.5) is fulfilled with a high accuracy. Equation (29.5) is thus valid and the T_α term for monolayers can be eliminated from the XMCD spin sum rule.

The situation changes for Co adatoms. It is obvious from Table 29.2 that the ratio $\sum_\alpha 7T_\alpha/\mu_{\text{spin}}$ is significantly larger for the adatoms than for the corresponding monolayers. For Pd and Pt substrates the breakdown of (29.5) is modest. For Cu, Ag, and Au substrates this breakdown is substantial. Elimination of the T_α term from the XMCD spin moment sum rule in not possible in these situations.

29.5.2 Angular Dependence of Magnetic Dipole Term

Another test whether the influence of SOC on the magnetic dipole term T_θ is significant can be performed by inspecting the full angular dependence of T_θ. If the influence of SOC can be neglected, this dependence should satisfy (29.6).

Figure 29.4 shows the T_θ term for Co monolayers and adatoms on Cu, Pd, Ag, Pt, and Au (111) surfaces calculated while varying the angle θ between the magnetisation direction and the surface normal [29]. To test the validity of (29.6), we attempted to fit the ab initio data to

$$A\,(3\cos^2\theta - 1)\,, \tag{29.7}$$

(dashed lines in Fig. 29.4). The fit is quite accurate except for Co adatoms on Cu, Ag, and Au. In these cases the T_θ dependence can be fitted with the function

$$A\,(3\cos^2\theta - 1 + B)\,, \tag{29.8}$$

(full lines in Fig. 29.4). The fact that the T_θ dependence can be fitted by (29.6) only if a rigid shift (represented by the constant B) is introduced presents another evidence that the magnetic dipole term sum rule (29.5) is not universally valid for supported $3d$ systems. Likewise, for systems where the parameter B in (29.8) is important, the term T_θ does not vanish at the *universal magic angle* 54.7°. Rather, the magnetisation tilt angle for which T_θ vanishes differs from substrate to substrate: it is 45° for a Co adatom on Cu, 13° for an adatom on Ag, and 42° for an adatom on Au. For other systems explored here, it is close to 54.7°.

The results presented in Sect. 29.5 demonstrate that even for atoms with low SOC such as Co, the influence of SOC on T_α in certain environments can be so large that (29.5)–(29.6) cannot be used. While for the monolayers the influence of SOC on T_α can be neglected for any substrate, the situation is different for the adatom. Here for some substrates (29.5)–(29.6) are valid but for other substrates they are not. The crucial factor turns out to be the ratio between the SOC strength and the bandwidth, ξ/W_{3d} [24, 29, 30]. The width of the adatom $3d$ band is apparently just about what is required to make the ratio ξ/W_{3d} large enough so that SOC becomes important. Consequently for some substrates (such as Pd, Pt) (29.5)–(29.6) still can be used while for others (such as Cu, Ag, Au) they cannot. Generally, one can expect that for systems with considerable overlap between adatom and substrate density of states around the Fermi level the influence of SOC on T_α can be neglected not only for monolayers but also for adatoms. Otherwise (29.5)–(29.6) should rather not be used [29].

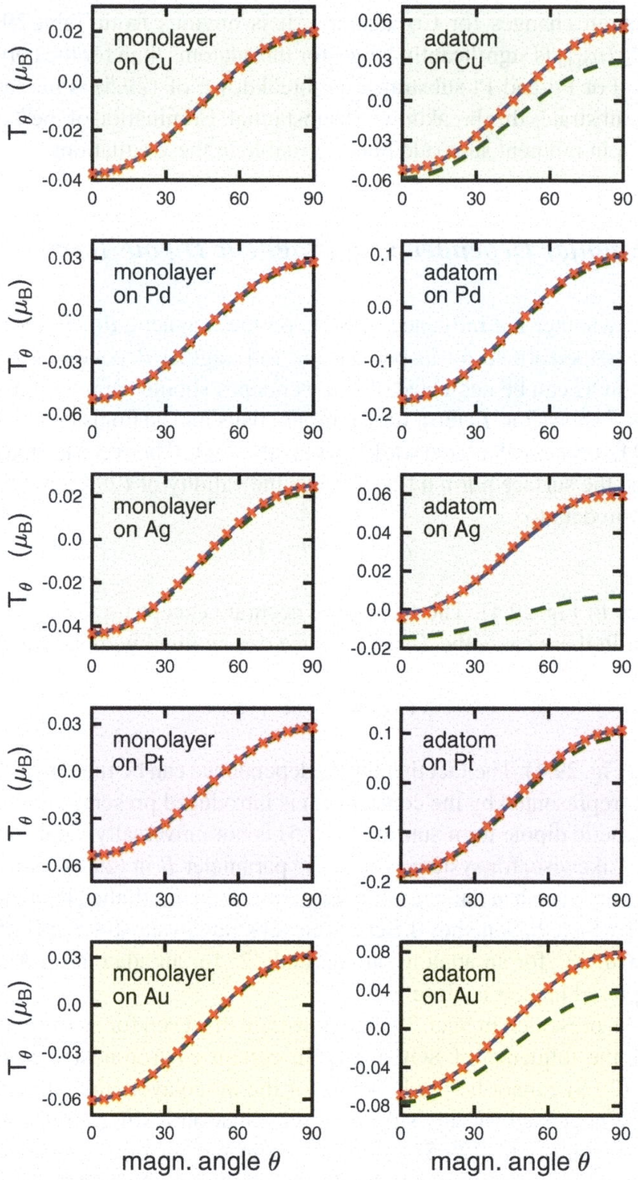

Fig. 29.4 Dependence of the magnetic dipole term T_θ on the magnetisation angle θ for Co monolayers (*left panels*) and Co adatoms (*right panels*) on different substrates. Ab initio results are shown by *crosses*, fits to $A(3\cos^2\theta - 1 + B)$ are shown by *full lines*, fits to $A(3\cos^2\theta - 1)$ are shown by *dashed lines*. Both fits are practically undistinguishable except for the cases of adatoms on Cu, Ag, or Au. Figure reproduced from [29]

29.6 Conclusions

The magnetic dipole term T_α is not just a minor factor that affects the sum rules analysis of XMCD spectra. On the contrary, it may change the overall picture completely. Neglecting the T_α term could lead, e.g., to wrong conclusions concerning the dependence of the magnetic moment on the cluster size or concerning the apparent dependence of the spin magnetic moment on the direction of the magnetisation. Whether this will actually happen or not depends both on the clusters and on the substrate and is difficult to be guessed beforehand.

In case that the influence of spin-orbit coupling on T_α can be neglected, the T_α term can be eliminated from the XMCD sum rules either by relying on the $T_x + T_y + T_z = 0$ relation or by making use of the magic angle $\theta = 54.7°$. For $3d$ transition metal systems, the influence of SOC on T_α can be neglected only if these systems are sufficiently large (wires, monolayers, large clusters). If the system contains just a few $3d$ atoms (as is the case of adatoms or small supported clusters), the influence of SOC on T_α may be significant. This further depends on the hybridisation between states of the $3d$ atoms and of the substrate, especially around the Fermi level: if the hybridisation is only weak, the role of the SOC is enhanced while if the hybridisation is strong, the role of the SOC is suppressed.

Acknowledgements The author would like to thank to Hubert Ebert, Jan Minár and Jiří Vackář for numerous stimulating discussions and collaboration on this line of research. Support by the Ministry of Education, Youth and Sport (Czech Republic) within the project LD15097 is gratefully acknowledged.

References

1. P. Gambardella, S. Rusponi, M. Veronese, S.S. Dhesi, C. Grazioli, A. Dallmeyer, I. Cabria, R. Zeller, P.H. Dederichs, K. Kern, C. Carbone, H. Brune, Science **300**, 1130 (2003)
2. F. Donati, A. Singha, S. Stepanow, C. Wäckerlin, J. Dreiser, P. Gambardella, S. Rusponi, H. Brune, Phys. Rev. Lett. **113**, 237201 (2014)
3. F. Donati, S. Rusponi, S. Stepanow, C. Wäckerlin, A. Singha, L. Persichetti, R. Baltic, K. Diller, F. Patthey, E. Fernandes, J. Dreiser, Ž. Šljivančanin, K. Kummer, C. Nistor, P. Gambardella, H. Brune, Science **352**, 318 (2016)
4. B.T. Thole, P. Carra, F. Sette, G. van der Laan, Phys. Rev. Lett. **68**, 1943 (1992)
5. P. Carra, B.T. Thole, M. Altarelli, X. Wang, Phys. Rev. Lett. **70**, 694 (1993)
6. J. Stöhr, J. Electron. Spectrosc. Relat. Phenom. **75**, 253 (1995)
7. J. Stöhr, J. Magn. Magn. Mater. **200**, 470 (1999)
8. S.H. Vosko, L. Wilk, M. Nusair, Can. J. Phys. **58**, 1200 (1980)
9. H. Ebert, D. Ködderitzsch, J. Minár, Rep. Prog. Phys. **74**, 096501 (2011)
10. H. Ebert, R. Zeller, The SPR-TB-KKR package, version Feb05 (2005), http://olymp.cup.uni-muenchen.de/ak/ebert/SPR-TB-KKR
11. R. Zeller, P.H. Dederichs, B. Újfalussy, L. Szunyogh, P. Weinberger, Phys. Rev. B **52**, 8807 (1995)
12. S. Bornemann, J. Minár, S. Polesya, S. Mankovsky, H. Ebert, O. Šipr, Ph. Transit. **78**, 701 (2005)
13. S.P. Collins, D. Laundy, C.C. Tang, G. van der Laan, J. Phys. Condens. Matter **7**, 9325 (1995)

14. R.Q. Wu, A.J. Freeman, Phys. Rev. Lett. **73**, 1994 (1994)
15. M. Komelj, C. Ederer, J.W. Davenport, M. Fähnle, Phys. Rev. B **66**, 140407(R) (2002)
16. S. Stepanow, A. Mugarza, G. Ceballos, P. Moras, J.C. Cezar, C. Carbone, P. Gambardella, Phys. Rev. B **82**, 014405 (2010)
17. O. Šipr, J. Minár, H. Ebert, Europhys. Lett. **87**, 67007 (2009)
18. O. Šipr, M. Košuth, H. Ebert, Phys. Rev. B **70**, 174423 (2004)
19. P. Mavropoulos, S. Lounis, R. Zeller, P.H. Dederichs, S. Blügel, Appl. Phys. A **82**, 103 (2006)
20. O. Šipr, S. Bornemann, J. Minár, S. Polesya, V. Popescu, A. Šimůnek, H. Ebert, J. Phys. Condens. Matter **19**, 096203 (2007)
21. D. Weller, J. Stöhr, R. Nakajima, A. Carl, M.G. Samant, C. Chappert, R. Megy, P. Beauvillain, P. Veillet, G.A. Held, Phys. Rev. Lett. **75**, 3752 (1995)
22. J. Bartolomé, F. Bartolomé, L.M. García, G. Filoti, T. Gredig, C.N. Colesniuc, I.K. Schuller, J.C. Cezar, Phys. Rev. B **81**, 195405 (2010)
23. O. Šipr, S. Bornemann, H. Ebert, S. Mankovsky, J. Vackář, J. Minár, Phys. Rev. B **88**, 064411 (2013)
24. J. Stöhr, H. König, Phys. Rev. Lett. **75**, 3748 (1995)
25. G. van der Laan, Phys. Rev. B **57**, 5250 (1998)
26. C. Ederer, M. Komelj, J.W. Davenport, M. Fähnle, J. Electron. Spectrosc. Relat. Phenom. **130**, 97 (2003)
27. T. Oguchi, T. Shishidou, Phys. Rev. B **70**, 024412 (2004)
28. D. Schmitz, C. Schmitz-Antoniak, A. Warland, M. Darbandi, S. Haldar, S. Bhandary, O. Eriksson, B. Sanyal, H. Wende, Sci. Rep. **4**, 5760 (2014)
29. O. Šipr, J. Minár, H. Ebert, Phys. Rev. B **94**, 144406 (2016)
30. O. Šipr, S. Mankovsky, S. Polesya, S. Bornemann, J. Minár, H. Ebert, Phys. Rev. B **93**, 174409 (2016)

Erratum to: Multiple Scattering Theory for Spectroscopies

Didier Sébilleau, Keisuke Hatada and Hubert Ebert

Erratum to:
D. Sébilleau et al. (eds.), *Multiple Scattering Theory for Spectroscopies*, Springer Proceedings in Physics 204, https://doi.org/10.1007/978-3-319-73811-6

In the original version of the book, belated correction from author to update the frontmatter with revised acknowledgement content has to be incorporated. The erratum book has been updated with the change.

The update online version of this book can be found at
https://doi.org/10.1007/978-3-319-73811-6

D. Sbilleau et al. (eds.), *Multiple Scattering Theory for Spectroscopies*, Springer Proceedings in Physics 204, https://doi.org/10.1007/978-3-319-73811-6_30

Index

D. Sébilleau et al. (eds.), *Multiple Scattering Theory for Spectroscopies*,
Springer Proceedings in Physics 204,
https://doi.org/10.1007/978-3-319-73811-6

T

W

X

The manufacturer's authorised representative in the EU is Springer Nature Customer Service Centre GmbH, Europaplatz 3, 69115 Heidelberg, Germany. If you have any concerns regarding our products, please contact ProductSafety@springernature.com

Printed and bound by CPI Group (UK) Ltd, Croydon, CR0 4YY
05/10/2026
02228138-0016